Fundamentals of Structural Dynamics

Keith D. Hjelmstad

Fundamentals of Structural Dynamics

Theory and Computation

 Springer

Keith D. Hjelmstad
Arizona State University
Tempe, AZ, USA

ISBN 978-3-030-89946-2   ISBN 978-3-030-89944-8   (eBook)
https://doi.org/10.1007/978-3-030-89944-8

This Springer imprint is published by the registered company Springer Nature Switzerland AG
The registered company address is: Gewerbestrasse 11, 6330 Cham, Switzerland

To Kara

Preface

Structural dynamics is a foundational topic for structural engineers. As commercial software dominates the practice of engineering, it is more important than ever for engineers to understand how the tools they use are made. This book is my suggestion that the *learning* of structural dynamics should embrace computation as a way to prepare for modern practice. This book is aimed at seniors and first-year graduate students in civil, mechanical, and aerospace engineering specializing in structures, but is accessible to anyone with a solid background in calculus, differential equations, and solid mechanics to the level represented in most undergraduate engineering curricula.

As computers have become more capable, higher fidelity in simulation of structural response has become more feasible. In many dynamic environments (e.g., earthquakes), structures are *expected* to respond nonlinearly. Modern design approaches are increasingly oriented toward performance-based design. This book incorporates nonlinearity as a natural part of modeling structural response. Most of the codes in the book are equipped to deal with nonlinearity in the equations of motion. If you want linear response, you can always limit the algorithm to one iteration, but you can never get a linear code to show you nonlinear response.

The book is built around the powerful synergy between traditional mathematics and computing. It is one thing to be able to solve problems manually using techniques of algebra and calculus, and yet another to express the solution strategy as a computer code. Almost everyone who has engaged computational mechanics seriously knows that mathematical formulation and computation can be strong allies. These two acts inform each other. Mathematical derivation provides the basis for creating a code; writing code gives a window to see what the theory implies. The two together are great pedagogical partners.

Structural dynamics is full of surprises that are hard to appreciate by looking at equations. The reader of this book should not be satisfied simply to learn the theories that govern the dynamic behavior of structures. The subject is best learned by implementing the ideas in a framework where you can do something with them. The punchline of every theory should be a computer code that brings that theory to life. If, at the end of your quest to understand a concept in dynamics you have a code that faithfully executes it, then you have created an opportunity to try things out to see what happens.

This book is more than an introduction to the theory of structural dynamics. It is also a comprehensive introduction to how to organize the complex computations associated with the subject. As such, the book is full of code. I selected MATLAB as the computation environment because it is available at most engineering schools and provides easy entry into the matrix computations that are at the core of structural dynamics theory. It also provides an environment where expressing results graphically is easy to do.

Computer code is like poetry. Every poet has a distinctive style, but all are seekers of direct and succinct expression. I have tried to write codes that bring out the essence of the computational algorithms and to place them in close proximity to where the theory is presented, using notation that makes the correlation as obvious as possible. The codes are intended to amplify the theory, not to confuse it.

The codes in the book are organized so that the input and output are subdued as much as possible. In fact, most of the functions that do the graphical output are not included in the book. Most commercial codes have many more lines of code devoted to the graphical user interface than to the core analytical engine. The purpose of this book is to see the engine—to take a look under the hood—so that when one moves on to the world of commercial codes, you have some idea of what those codes are doing with the theory. That said, the complete set of codes (including graphics) are available at SpringerLink or by contacting the author at Keith.Hjelmstad@asu.edu.

The book follows a developmental path, starting with a brief review of Newtonian dynamics. The first chapter refreshes those ideas and introduces the notation used throughout the book. The approach to formulating rigid-body dynamics shows up in our study of trusses, beams, and frames. Even though each of these contexts is about flexible members, rigidity lurks inside in the form of modeling assumptions and kinematic hypotheses.

In Chap. 2, we take what might seem like a detour. It is very clear that numerical integration of equations of motion is inevitable in structural dynamics because we are interested in complex load forms like earthquakes and in nonlinear response of structures. So, why not introduce numerical methods right at the start? This chapter gives the reader an opportunity to understand how numerical integration of differential equations works and what the limitations are. Chapter 2 gives a very detailed introduction to the implementation and performance of three common methods, one of which (Newmark's method) will be used extensively throughout the remainder of the book.

Chapter 3 starts the study of structural dynamics in earnest through the lens of the single-degree-of-freedom (SDOF) system. The beauty of this simple system is that it provides a framework for understanding dynamic phenomena with a modest mathematical investment. Important concepts like natural vibration, damping, and resonance all manifest in the context of the SDOF system. It is also a good starting point for introducing the concepts of earthquake ground motions and nonlinear constitutive response. In many ways, Chap. 3 is a primer for structural dynamics.

Chapters 4, 5, 6, and 7 extend the ideas of the SDOF system in the simplest possible context—the shear building (which we refer to as the NDOF system because it has N degrees of freedom). The study of the shear building introduces concepts associated with systems with multiple degrees of freedom (like multiple natural frequencies and mode shapes) without the mathematical overhead of the complex geometries of real structures. In Chap. 4, we start with a fairly intuitive extension of the formulation done for the SDOF system. In Chaps. 5 and 6, we introduce an approach to formulating equations of motion based upon the *principle of virtual work* and the direct assembly of equations for nonlinear problems with earthquake excitation. In Chap. 7, we take a brief look at some special analysis techniques for dealing with large, complex systems.

Chapter 8 introduces truss structures. Now, structural form starts to play a more central role in dynamic response. Seeing how geometry and topology relate to the formulation of the equations of motion is a key step in understanding how real structural systems work. This step takes us from structures that are simple both in geometry and mechanical behavior to structures that are complicated in geometry, but with elements that still exhibit simple mechanical behavior. As trusses are discrete systems, all of the techniques for solving problems learned in the earlier chapters directly apply.

Chapter 9 launches into the world of continuous (*not* discrete) systems. Up to this point, all the elasticity of a structure was considered to be massless and all the mass was considered to be rigid. In reality, all structural members have distributed mass and elasticity. One of the key phenomena that these systems exhibit which was not explicitly examined in the discrete systems is *wave propagation*. The study of the axial bar problem is the simplest context in which this phenomenon manifests. In this chapter we formulate *partial* differential equations of motion and methods of solution for those equations. We develop both the classical separation of variables method as well as a numerical spatial discretization via the Ritz method. The codes that emerge from the axial bar problem provide a template for the more complex systems of subsequent chapters.

Chapters 10 through 13 take us down a path similar to the one we traveled for the axial bar, now with the full mechanical complexity of beam theory. In Chap. 10, we derive the dynamic theory of planar beams. Chapter 11 examines the important problem of *wave propagation* and presents classical approaches to solving the partial differential equations of linearized beam theories. The focus in this chapter is on three particular approximations—the Bernoulli-Euler beam, the Rayleigh beam, and the Timoshenko beam. Chapter 12 continues to investigate the linear beam theories, but now through the lens of the Ritz approximation to deal with the spatial part of the problem. Here the problem is reformulated in the context of the principle of virtual work and the finite element method, allowing the solution to a much broader class of excitations. In Chap. 13, we embrace the full nonlinear theory of beams and formulate a code that can execute the computations based upon the principle of virtual work and the Ritz method.

Chapter 14 extends the beam formulation to frames—structures that can have multiple elements framing into the same node and the elements can have any orientation. This chapter is the culmination of a long crescendo of the previous chapters. The code presented in this chapter is a complete nonlinear planar frame code capable of solving a very broad array of problems. Implementation of this code provides the reader with a tool to explore an unlimited range of questions about the dynamic response of structures, including instability.

There are several concepts that have a longitudinal span across the book—concepts like numerical quadrature, interpolation, the solution of eigenvalue problems, and the solution of nonlinear systems of algebraic equations. These concepts show up everywhere, so they are covered in a series of appendices. These stand-alone appendices should be consulted early and often as a support for developments throughout the book.

Appendix A introduces *Newton's method* for solving nonlinear algebraic equations. Almost *all* of our theories result in nonlinear equations of motion. Time discretization converts these nonlinear differential equations into nonlinear algebraic equations that are amenable to Newton's method. Throughout the book, the nonlinear formulations are reduced to *residual* and *tangent* form as presented in this appendix. Creating a standard framework to deal with nonlinearity helps to make it a natural feature of structural modeling.

Newton's method requires the computation of the tangent matrix, which is generally best done using the *directional derivative*. Appendix B covers the directional derivative and provides some examples of how to do these derivatives. Directional derivatives are also used extensively in the later chapters of the book (again in support of Newton's method).

The *eigenvalue problem* is a core concept in structural dynamics. For linear problems, it provides both physical understanding of the system through natural frequencies (eigenvalues) and mode shapes (eigenvectors) and a route to computation (modal analysis). The eigenvalue problem is also central to our modeling of damping in systems with multiple degrees of freedom. Appendix C gives a brief introduction to the eigenvalue problem. While some features of the problem, like orthogonality of eigenvectors, are also covered in the main text, the basic question of how these problems are actually solved is reserved for the appendix. In the book's codes, of course, we rely on the amazing and highly optimized `eig` function in MATLAB to carry out the computations.

The Ritz method for continuous systems relies on the concept of *interpolation* of functions. Appendix D gives a fairly detailed introduction to interpolation and how it is implemented in the finite element approach to the Ritz discretization of the spatial part of the partial differential equations of motion. This appendix provides a standard approach to interpolation with *bubble functions* that is used in all of the codes in Chaps. 9 through 14.

The success of structural analysis codes hinges on our ability to characterize the structure in terms that can be manipulated in the computational environment—that is, the *data* structures that define the *physical* structure. Appendix E establishes the framework for the data structures used throughout the codes in the book. These data structures are the key to localization and the direct assembly of the equations of motion.

The Ritz approach to spatial discretization in conjunction with the principle of virtual work requires integration of certain quantities over the spatial domain of the body. In the codes presented in the book, all such integrals are done numerically. Numerical quadrature provides a conceptually simple way to think about the many definite integrals that appear in the virtual work formulations. Appendix F gives an introduction to *numerical quadrature*, culminating in a function that is used in all codes associated with Chaps. 9 through 14.

The fundamental role of this book is in *learning* about structural dynamics—from understanding the incumbent phenomena to developing strategies for computing results. So many aspects of structural dynamics appear in the context of particular problems. As such, we do not attempt to provide a guide to how structural dynamics is currently practiced, sticking as close as possible to durable and universal concepts. The book does contain practical advice on executing structural dynamics computations, common visualization approaches for dynamics, and topics like response spectrum methods for earthquake engineering. Those concepts are scattered throughout the book. The aim is to provide a solid theoretical background for understanding structural dynamics practice, particularly as it evolves over time.

The computational framework presented in this book is, to some extent, forward-looking. In the practice of structural dynamics, nonlinear analysis is not yet as common as we make it appear in this book. It is, however, likely to be in the future. This book aims to prepare future generations of structural engineers for a profession where performance-based design is more prevalent. We have abandoned the coverage of topics that don't support that broader goal—like dwelling extensively on linearity and special techniques for hand computations. We believe that if there is something that needs to be computed, we will ask the computer to do it. In that context, linearity and hand computations move to the important role of code verification. The learner still needs to do derivations and computations with pen and paper to gain understanding, but the evolution of the subject of structural dynamics has shifted the role of those tasks and techniques.

I am fond of saying that the person who learns the most from a book is the one who wrote it. The writing of the book was a delightful exercise in the interplay between deriving theory and writing code to implement it. As one would expect, the derivations told me how to write the codes. But frequently it was the code that told me there was an error in my derivation. I cannot even begin to catalog the things that I learned about structural dynamics that I thought I already knew. Returning to fundamentals and doing the derivations always cleaned up my sloppy thinking.

If I have one wish for the reader, it is to have the same sort of experience I have had writing it. Don't just read the book; work it out. Implement every algorithm. Get into the habit of debugging your thinking by debugging a computer program. Don't think of a derivation as the end point of learning, but rather the starting point for computing. In the same vein, don't think about the completion of a code as the end of the solution process, but rather the beginning of the verification process. You will learn more through trying to verify your codes than you will from writing them in the first place. Finally, don't view code verification as the end of the computing process. Use the codes to promote your understanding of structural dynamics. The development of engineering judgment is the process of trying stuff, getting results, and then convincing yourself why the world works that way. This book is the starting point for building an environment in which you can learn structural dynamics. Enjoy!

Tempe, AZ, USA   Keith D. Hjelmstad
June 2021

Contents

Chapter 1
Foundations of Dynamics

To establish a starting point for our exploration of structural dynamics, we first review the fundamental ideas of the dynamics of particles and rigid bodies. The theories in this book are built on Newton's laws of motion for particles and Euler's extension of those laws to rigid bodies. In each derivation of the equations of motion for a structural element or system, we will identify a typical particle and track its motion as part of the development of the theory. In most cases, we will identify a rigid body (e.g., the cross section in a beam) that plays an important role in describing the motion. Hence, a review of the foundations of dynamics is in order. This brief tour through the basic concepts will also allow us to set up some of the notation used throughout the book.

A *particle* is a body with mass assumed to be concentrated at a single point. Newton's laws were built for particles. The *mass* of the particle is a property that measures its resistance to translation when subjected to force. A *rigid body* is a collection of particles distributed over a spatial domain of finite size and constrained in a way that the distance between any two particles remains constant even as the system moves. While rotation is immaterial to particles, it is important to the motion of rigid bodies. Each particle in a rigid body experiences a different acceleration when rotating due to its position in the body. Euler's laws of motion account for the spatially distributed mass and can be derived from Newton's laws.

All real materials deform, so Euler's laws cannot adequately characterize the dynamic response of structures. Structural mechanics models represent a partial relaxation of the rigid body assumption. In beam theory, for example, we assume that cross sections act as rigid bodies while allowing adjacent cross sections to experience relative deformation. It is important to keep in mind that the main purpose of introducing any constraint is to reduce the *mathematical* complexity of the resulting equations of motion. In each case, the assumption must be physically justified.

© The Author(s), under exclusive license to Springer Nature Switzerland AG 2022   1
K. D. Hjelmstad, *Fundamentals of Structural Dynamics*,
https://doi.org/10.1007/978-3-030-89944-8_1

Notation Most of the notation used in this book will be defined at the point of first use, but here is a starter set. Scalars will be denoted in *italic* font, vectors and matrices in **bold** font, and computer code variables and functions in `typewriter` font. When needed, we will set up a coordinate system to define the components of a vector. Vector addition is done through the *head-to-tail rule*, and vectors can be multiplied using either the dot ($\cdot$) or cross ($\times$) product. The unit orthogonal base vectors $\{e_1, e_2, e_3\}$ define the coordinate directions. A vector can then be described, e.g., as $x = x_1 e_1 + x_2 e_2 + x_3 e_3$, where x_1, x_2, and x_3 are the *components* of the vector x. Time derivatives will be indicated both with the Leibniz notation $d(\bullet)/dt$ and with Newton's notation where a time derivative is indicated by a dot above the variable, e.g., $\dot{u}(t)$. Derivatives will also be denoted with Lagrange's notation wherein a prime indicates differentiation with respect to the argument of the function, e.g., $f'(x)$.

1.1 Kinematics of Particles

Kinematics is the study of motion apart from its cause. A particle can be represented as a point in space whose location can be described by a position vector $x(t)$, as shown in Fig. 1.1. The position is measured relative to a fixed (but arbitrary) point $\mathcal{O}$ in the space called the *origin* in an inertial frame, which is required for Newton's laws to be valid. The time t is measured relative to some arbitrary beginning, which we will usually take as $t = 0$.

The *velocity* $v(t)$ is the rate at which position $x(t)$ changes with respect to time. The velocity vector is tangent to the *path of motion*. To prove this assertion, consider the difference in position of a particle at times t and $t + \Delta t$, as shown in Fig. 1.1b. Dividing this difference by Δt and taking the limit as Δt approaches zero gives the velocity

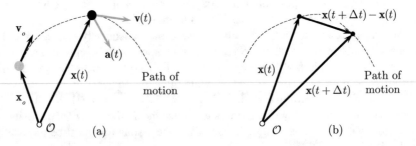

Fig. 1.1 *Kinematics* is the study of motion. (**a**) The state of the motion of a particle is characterized by the position $x(t)$, velocity $v(t)$, and acceleration $a(t)$. The particle has initial position x_o and initial velocity v_o. Note that the velocity vector is tangent to the *path of motion*. (**b**) Velocity is defined as the rate of change of the position vector, which is the limit of secant vectors

$$\mathbf{v}(t) = \lim_{\Delta t \to 0} \frac{\mathbf{x}(t + \Delta t) - \mathbf{x}(t)}{\Delta t} = \frac{d\mathbf{x}}{dt} \tag{1.1}$$

As the time increment gets smaller, the secant vector $\mathbf{x}(t + \Delta t) - \mathbf{x}(t)$ gets closer to the tangent to the path and becomes tangent in the limit. The scalar magnitude of the velocity vector is often called the *speed*, which we represent as

$$v(t) = \|\mathbf{v}(t)\| = \sqrt{\mathbf{v}(t) \cdot \mathbf{v}(t)}$$

The *acceleration* $\mathbf{a}(t)$ is the rate of change of velocity

$$\mathbf{a}(t) = \lim_{\Delta t \to 0} \frac{\mathbf{v}(t + \Delta t) - \mathbf{v}(t)}{\Delta t} = \frac{d\mathbf{v}}{dt} \tag{1.2}$$

Acceleration is also a vector, and its magnitude and direction influence where the particle will go next. In general, the position, velocity, and acceleration vectors do not point in the same direction, but in special cases such alignment is possible (e.g., when the motion is constrained to be along a straight line). The position, velocity, and acceleration together comprise the *kinematic state* of the particle.

It is possible to continue to differentiate the position vector, ad infinitum. Derivatives beyond the second (acceleration) do not play a role in Newtonian mechanics but might have interest for other reasons not related to the evolution of the motion. The third derivative of position is often referred to as *jerk* and the fourth derivative as *jounce*. These quantities are useful in characterizing human perception of motion.

For the path of motion to be uniquely specified we must know both the initial position $\mathbf{x}(0) = \mathbf{x}_o$ and the initial velocity $\mathbf{v}(0) = \mathbf{v}_o$ (i.e., the position and velocity at time $t = 0$) or some equivalent information (e.g., the position and velocity at other times in the motion, not necessarily coincident).

Differential Relationships   Velocity is the rate of change of position. Acceleration is the rate of change of velocity and is, therefore, the second derivative of position:

$$\mathbf{v}(t) = \frac{d\mathbf{x}}{dt}, \qquad \mathbf{a}(t) = \frac{d\mathbf{v}}{dt} = \frac{d^2\mathbf{x}}{dt^2}$$

Observe that these are *differential relationships*. That means that they relate instantaneous quantities (i.e., quantities that change with time will have certain values at certain times, but they hold that value only for that instant). This sort of relationship exists in stark contrast to quantities that are related by algebraic equations, wherein the relationships (even if nonlinear) hold forever. Differential relationships *evolve*. Newton and Leibniz invented calculus precisely to deal with these kinds of relationships.

It is important to remember that position, velocity, and acceleration are not independent. It is also important to remember that the solution of differential

Fig. 1.2 *Graphical interpretation.* The integrals that relate position, velocity, and acceleration of the particle can be imagined as "areas under the curve." The change in velocity is the area under the acceleration curve; the change in displacement is the area under the velocity curve

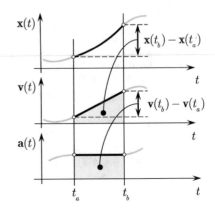

equations is quite different from solving algebraic equations (although we will often do a lot of algebra in the pursuit of such solutions).

Integral Relationships The position, velocity, and acceleration of a particle are also related to each other through integral relationships. Roughly speaking, position is the integral (more precisely, antiderivative) of velocity and velocity is the integral of acceleration. Formally, we can write

$$\mathbf{v}(t) = \mathbf{v}_o + \int_0^t \mathbf{a}(\tau)\,d\tau$$

$$\mathbf{x}(t) = \mathbf{x}_o + \int_0^t \mathbf{v}(\tau)\,d\tau$$

(1.3)

where τ is a dummy variable of integration representing time. The initial position $\mathbf{x}_o$ and initial velocity $\mathbf{v}_o$ appear as the *constants of integration*. Evaluating these expressions at time $t = 0$ confirms this choice (the integral of any function from zero to zero is zero). These relationships are illustrated graphically in Fig. 1.2.

The expression for velocity can be substituted into the equation for position to provide an integral relationship directly between position and acceleration. To wit,

$$\mathbf{x}(t) = \mathbf{x}_o + \int_0^t \left(\mathbf{v}_o + \int_0^\tau \mathbf{a}(\eta)\,d\eta \right) d\tau$$

where η is also a dummy variable representing time. The double integral can be integrated by parts to give the equivalent form

$$\mathbf{x}(t) = \mathbf{x}_o + \mathbf{v}_o\,t + \int_0^t (t - \tau)\,\mathbf{a}(\tau)\,d\tau$$

(1.4)

Note that the variable of integration is τ and the t that appears in the integrand is held constant during integration. Observe that in the absence of acceleration, the position changes linearly in time in accord with the initial velocity.

Example Consider a particle that starts from rest and experiences an acceleration in the form $\mathbf{a}(t) = 12\,t^2\mathbf{e}_1$, where $\mathbf{e}_1$ is the unit base vector. The position can be computed from Eq. 1.4, noting that $\mathbf{x}_o = \mathbf{0}$ and $\mathbf{v}_o = \mathbf{0}$, as

$$\mathbf{x}(t) = \int_0^t (t-\tau)\left(12\,\tau^2\right)\mathbf{e}_1\,d\tau = 12\left(\tfrac{1}{3}t^4 - \tfrac{1}{4}t^4\right)\mathbf{e}_1 = t^4\mathbf{e}_1$$

The final result can be easily verified by differentiating twice to get the acceleration from position. Note the importance of distinguishing between the real time t and the dummy variable of integration τ in the process.

1.2   Kinetics of Particles

While kinematics is the study of motion apart from the cause of the motion, *kinetics* is the study of the cause. Newton posited that *force* causes motion, and it does so by affecting the acceleration of the particle. Hence, Newton's second law is

$$\mathbf{F}(t) = m\,\mathbf{a}(t) \tag{1.5}$$

where $\mathbf{F}(t)$ is the net force acting on the particle, m is the mass of the particle, and $\mathbf{a}(t)$ is the acceleration of the particle. The net force on a particle can be the resultant of several forces acting individually and can be caused by various phenomena, including force at a distance (e.g., gravity) and surface traction (e.g., contact with a solid object or fluid).

If the net force is a known function of time, then we can simply substitute $\mathbf{a}(t) = \mathbf{F}(t)/m$ in Eqs. 1.3 or 1.4 and integrate to find the velocity and position. An example of when this is possible is *projectile motion* when the net force acting on the particle is only the force of gravity. In problems with constraints (e.g., a bead moving along a wire), the net force is not known a priori because the reaction associated with the constraint is unknown. In such a case, the equations of kinetics and kinematics must be used together to solve the problem.

Impulse–Momentum Force is an instantaneous quantity that can change with time. *Impulse* is the aggregate effect of a force acting over time. If the mass m is constant, we can write Newton's second law in the form

$$\mathbf{F}(t) = \frac{d}{dt}\left(m\mathbf{v}(t)\right) \tag{1.6}$$

Fig. 1.3 Impulse is the area under the force versus time curve. It represents the action of a force over time

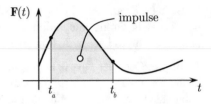

The product of mass and velocity, $m\mathbf{v}(t)$, is called the *momentum* of the particle. Hence, another way of describing Newton's second law is that force is equal to the rate of change of momentum. For this reason, we often refer to the equations of motion implied by Newton's second law as *balance of momentum*.

To find the total *impulse* between times t_a and t_b, multiply Eq. 1.6 through by dt and integrate from time t_a to time t_b to get

$$\int_{t_a}^{t_b} \mathbf{F}(t)\, dt = \int_{t_a}^{t_b} d(m\mathbf{v}) = m\mathbf{v}(t_b) - m\mathbf{v}(t_a) \tag{1.7}$$

As illustrated in Fig. 1.3, the integral of the force between the times t_a and t_b is the impulse. It is, roughly speaking, the area under the force versus time curve— roughly, because force is a vector and the geometric notion of area under the curve applies only to scalar functions of scalar variables. We refer to the difference $m\mathbf{v}(t_b) - m\mathbf{v}(t_a)$ as the *change in momentum*. Hence, impulse is equal to change in momentum. This result a direct result of Newton's second law. Therefore, it is true for any particle acted upon by any force.

The impulse–momentum relationship is particularly useful in *impact* problems, where the contact force acts over a very short period of time. During that short time, the position changes very little. For a rigid body, impact occurs over an infinitesimally short amount of time, causing an instantaneous change in velocity with no change in position.

1.3 Power, Work, and Energy

Newton provided a framework for modeling force and motion through the lens of balance of momentum. We can get another view of mechanics by considering the concepts of *power*, *work*, and *energy*. In this section we will define those terms and show how they relate to the Newtonian balance of momentum concept. In particular, we show that conservation of energy is the first integral of the equations of motion $\mathbf{F} = m\mathbf{a}$.

Kinetic Energy of a Particle Kinetic energy is the energy of motion. As such, it depends only on the velocity $\mathbf{v}$ of the particle, not on its position $\mathbf{x}$ or acceleration $\mathbf{a}$. We use the symbol T for kinetic energy and define it as one half the mass times the velocity dotted with itself

$$T = \tfrac{1}{2}m(\mathbf{v} \cdot \mathbf{v}) = \tfrac{1}{2}mv^2 \tag{1.8}$$

where v is the magnitude of the velocity or *speed*. Kinetic energy is a scalar quantity with no associated direction. At a given time, a particle simply possesses a kinetic energy. It is a measure of the kinematic state of the particle. Observe that by virtue of the definition, the smallest kinetic energy a particle can possess is zero, which is the case when the particle is at rest.

Power and Work of a Force   Define the *power* of a force to be the dot product of the force and velocity. To wit,

$$P(t) = \mathbf{F}(t) \cdot \mathbf{v}(t) \tag{1.9}$$

One interesting and important feature of this definition is that a force has no power if it acts perpendicular to the velocity. This situation could occur if the force is one of constraint (e.g., a normal force associated with a roller support or a body sliding without friction on a surface). A force also has no power if the velocity is zero (which is generally the case for a reaction force).

   Work is the integral of the power over some period of time. From that perspective, then, power is the rate at which a force does work, and work is the accumulation over time of the power of a force. The definition of work of a force (between times t_a and t_b) is

$$W_a^b = \int_{t_a}^{t_b} P(t)\, dt = \int_{t_a}^{t_b} \mathbf{F}(t) \cdot \mathbf{v}(t)\, dt \tag{1.10}$$

A force does no work if it acts perpendicular to the path of motion. Power and work are simply definitions. We still need to show how they are useful in mechanics.

Power and Kinetic Energy   The relationship between work and energy can be derived from Newton's second law. Write *power* in the form

$$\mathbf{F} \cdot \mathbf{v} = \left(m\frac{d\mathbf{v}}{dt} \right) \cdot \mathbf{v} \tag{1.11}$$

Let us compute the rate of change of kinetic energy. By the product rule for differentiation, noting that the dot product is commutative, we can write the time rate of change of kinetic energy T as

$$\frac{dT}{dt} = \frac{d}{dt}\left(\tfrac{1}{2}m\,\mathbf{v}\cdot\mathbf{v}\right) = \tfrac{1}{2}m\left(\mathbf{v}\cdot\frac{d\mathbf{v}}{dt} + \frac{d\mathbf{v}}{dt}\cdot\mathbf{v}\right) = \left(m\frac{d\mathbf{v}}{dt}\right)\cdot\mathbf{v}$$

The last expression is exactly the same as the right side of Eq. 1.11. Thus,

$$\mathbf{F} \cdot \mathbf{v} = \frac{dT(\mathbf{v})}{dt} \tag{1.12}$$

In other words, power is the rate of change of kinetic energy. One might view this result as the reason to define kinetic energy in the first place. It is also the route to making the definitions of power and work useful in mechanics.

Work and Kinetic Energy To compute the work of a force, multiply Eq. 1.12 through by dt and integrate from t_a to t_b. Equation 1.10 then gives

$$W_a^b = \int_{t_a}^{t_b} (\mathbf{F} \cdot \mathbf{v}) \, dt = \int_{t_a}^{t_b} dT = T(t_b) - T(t_a). \tag{1.13}$$

Therefore, the work done by the force $\mathbf{F}$ in moving between state a and state b is equal to the difference in kinetic energy between those two states. Work is what causes kinetic energy to change. If work is positive, then the kinetic energy increases. If work is negative, then the kinetic energy decreases.

In some circumstances, the work of a force can be computed without direct reference to time. The force can sometimes be expressed as $\mathbf{F}(\mathbf{x}(t)) = \mathbf{F}(\mathbf{x})$. In other words, we can express the force as purely a function of position. In such a case we can use a change of variable and let $\mathbf{v} \, dt = d\mathbf{x}$. Then,

$$W_a^b = \int_{\mathbf{x}_a}^{\mathbf{x}_b} \mathbf{F}(\mathbf{x}) \cdot d\mathbf{x} \tag{1.14}$$

If the force is constant, it can be moved out of the integral. Now, $d\mathbf{x}$ is an exact differential and the integral can be evaluated as

$$\int_{\mathbf{x}_a}^{\mathbf{x}_b} \mathbf{F} \cdot d\mathbf{x} = \mathbf{F} \cdot \int_{\mathbf{x}_a}^{\mathbf{x}_b} d\mathbf{x} = \mathbf{F} \cdot \left(\mathbf{x}_b - \mathbf{x}_a \right)$$

That is, the work is the force dotted with the change in position $\mathbf{x}_b - \mathbf{x}_a$, which is the distance that the force moves. Thus, work is force times distance if the force is constant throughout the motion. The form of Eq. 1.14 does not exactly illuminate *how* one might do that computation in the presence of constraints, but it is generally clear what do in a particular problem, as the following example shows.

Example Consider the situation shown in Fig. 1.4 where a block moves up an inclined plane under the action of a force $f(t)\mathbf{p}$, resisted by gravity and friction. The coefficient of friction is μ. The constant unit vectors $\mathbf{p}$ and $\mathbf{q}$ are oriented in the direction of the plane and perpendicular to it. The motion up the plane is described as $\mathbf{x}(t) = z(t)\mathbf{p}$, where $z(t)$ is a scalar function of time. By differentiation, the velocity is $\mathbf{v}(t) = \dot{z}(t)\mathbf{p}$. From balance of linear momentum in the $\mathbf{q}$ direction we can determine that the normal force $N = mg \, \mathbf{e}_2 \cdot \mathbf{q}$ is constant.

Consider the computation of the work done by the friction force:

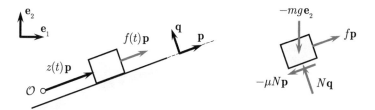

Fig. 1.4 *Example.* A block of mass m slides on a plane with coefficient of friction μ under a force $f(t)$ directed up the plane. This example shows how the computation of work can be expressed in terms of distance, rather than time

$$W_a^b = \int_{t_a}^{t_b} (-\mu N \mathbf{p}) \cdot (\dot{z}(t)\,\mathbf{p})\,dt = -\mu N \int_{z_a}^{z_b} dz$$

The change of variable in this case is $dz = \dot{z}\,dt$, which implies $\mathbf{v}\,dt = \dot{z}\,\mathbf{p}\,dt$. Thus, $d\mathbf{x} = \mathbf{p}\,dz$. Of course, the dot product of $\mathbf{p}$ with itself can be done independently of the integration process and reduces to the integral of a scalar function.

Potential Energy Certain forces have the special property that the work done by the force does not depend upon the path of the motion and can be characterized by a *potential energy* function $U(\mathbf{x}(t))$ such that

$$\mathbf{F} \cdot \mathbf{v} = -\frac{dU}{dt}$$

We can think of the potential energy U as being associated with the force $\mathbf{F}$. Now, compute the work done by the force between state a and state b as

$$W_a^b = \int_{t_a}^{t_b} (\mathbf{F} \cdot \mathbf{v})\,dt = -\int_{\mathbf{x}_a}^{\mathbf{x}_b} dU = U(\mathbf{x}_a) - U(\mathbf{x}_b)$$

Observe that if the work done by the force is positive, then the potential energy decreases. This is simply a result of how we have defined the potential energy, but it puts us in a position to have balance of momentum be consistent with conservation of energy, wherein loss of potential energy coincides with a gain of kinetic energy, and vice versa.

A potential energy function $U(\mathbf{x})$ exists only for certain types of forces, which we refer to as *conservative forces*. The conditions for the existence of $U(\mathbf{x})$ are evident from the chain rule of differentiation. To wit,

$$\mathbf{F} \cdot \mathbf{v} = -\frac{dU}{dt} = -\frac{\partial U}{\partial \mathbf{x}} \cdot \frac{d\mathbf{x}}{dt} = -\frac{\partial U}{\partial \mathbf{x}} \cdot \mathbf{v}$$

By association, the negative of the force is equal to the gradient of the potential energy function. The gradient of U can be computed, in components relative to base vectors $\{e_1, e_2, e_3\}$ as

$$\frac{\partial U}{\partial \mathbf{x}} = \frac{\partial U}{\partial x_1} \mathbf{e}_1 + \frac{\partial U}{\partial x_2} \mathbf{e}_2 + \frac{\partial U}{\partial x_3} \mathbf{e}_3$$

where we have used the notation $U(\mathbf{x}) = U(x_1, x_2, x_3)$ to represent the scalar function of the three coordinates that locate the position of the point in space. The gradient is, essentially, the slope of the energy function. The steeper the slope, the higher the force.

The Gravity Potential (Near the Surface of the Earth) The force due to gravity is associated with a potential energy function and, thus, is a conservative force. The potential function takes the particular form

$$U(\mathbf{x}) = mg\, \mathbf{x} \cdot \mathbf{e}_g + U_o \tag{1.15}$$

where g is the acceleration of gravity and $\mathbf{e}_g$ is a unit vector pointing in the positive direction associated with gravity, i.e., the direction associated with increasing potential energy. Notice that the constant U_o is arbitrary, which means that you can select the location of zero potential energy, a place we refer to as *datum*, anywhere. Now,

$$\mathbf{F} = -\frac{\partial U}{\partial \mathbf{x}} = -mg\, \mathbf{e}_g \tag{1.16}$$

which is a constant force acting in the negative direction of gravity (downward in Fig. 1.5). Note that mg is the *weight* associated with the mass in the gravitational field.

The Newtonian Gravity Potential The previous example of the gravitational potential is a special case of Newton's law of gravitation. The potential energy function for the gravitational attraction of two masses m_1 and m_2 is

Fig. 1.5 *Gravitational potential.* If we define the vertical distance as $h(\mathbf{x}) = \mathbf{x} \cdot \mathbf{e}_g$ as the height above datum, we can express the potential energy of the gravitational force in the customary form $U = mgh$

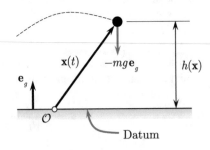

Fig. 1.6 *Newtonian gravity.*
Newton hypothesized that the
force of gravity acting
between two masses was
proportional to the product of
their masses divided by the
square of the distance
between them

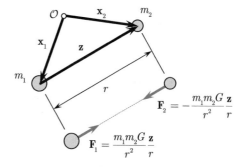

$$U(\mathbf{z}) = -\frac{Gm_1m_2}{r(\mathbf{z})} \qquad (1.17)$$

where G is the universal gravitational constant, $\mathbf{z} = \mathbf{x}_2 - \mathbf{x}_1$ is the vector from the first mass to the second, and $r(\mathbf{z}) = \sqrt{\mathbf{z} \cdot \mathbf{z}}$ is the scalar distance between the masses, as shown in Fig. 1.6.

The gradient of the potential energy gives the force of gravity acting on each mass

$$\mathbf{F}_1 = -\frac{\partial U}{\partial \mathbf{x}_1} = -\frac{\partial U}{\partial r}\frac{\partial r}{\partial \mathbf{x}_1} = \frac{Gm_1m_2}{r^2}\left(\frac{\mathbf{z}}{r}\right)$$
$$\mathbf{F}_2 = -\frac{\partial U}{\partial \mathbf{x}_2} = -\frac{\partial U}{\partial r}\frac{\partial r}{\partial \mathbf{x}_2} = -\frac{Gm_1m_2}{r^2}\left(\frac{\mathbf{z}}{r}\right) \qquad (1.18)$$

where the derivatives of r with respect to $\mathbf{x}_1$ and $\mathbf{x}_2$ are

$$\frac{\partial r}{\partial \mathbf{x}_1} = -\frac{\mathbf{z}}{r}, \qquad \frac{\partial r}{\partial \mathbf{x}_2} = \frac{\mathbf{z}}{r}$$

A good way to execute this computation is to apply the *directional derivative* to the equation defining r, as shown in Appendix B. Equations 1.18 tells us that the magnitude of the force of gravity acting between two masses is inversely proportional to the square of the distance between them and is directed along the line that joins them.

It is instructive to compute the work done by the two forces. To wit,

$$W_a^b = \int_{t_a}^{t_b}\left(\mathbf{F}_1 \cdot \dot{\mathbf{x}}_1 + \mathbf{F}_2 \cdot \dot{\mathbf{x}}_2\right)dt = -Gm_1m_2\int_{t_a}^{t_b}\frac{\mathbf{z} \cdot \dot{\mathbf{z}}}{r^3}dt = \frac{Gm_1m_2}{r}\bigg|_{t_a}^{t_b}$$

where we have used the expressions for $\mathbf{F}_1$ and $\mathbf{F}_2$ from Eq. 1.18, noted that $\dot{\mathbf{z}} = \dot{\mathbf{x}}_2 - \dot{\mathbf{x}}_1$, and recognized that

$$\frac{d}{dt}\left(\frac{1}{r}\right) = -\frac{\mathbf{z} \cdot \dot{\mathbf{z}}}{r^3}$$

The work done by the gravitational forces is $W_a^b = U(t_a) - U(t_b)$, as promised. The potential energy difference between states a and b accounts for all of the work done by the forces associated with gravity.

If the first mass is the Earth, then we can see that $g = Gm_1/r^2$, where m_1 is the mass of the Earth and r is the radius of the Earth. The mass of the Earth is approximately 5.972×10^{24} kg and the radius of the Earth is approximately 6.371×10^6 m. The gravitational constant is 6.674×10^{-11} m³ kg⁻¹ s⁻². Carrying out the calculation gives the familiar $g = 9.81$ m/s².

The Linear Spring One of the common idealizations used in mechanics is the *linear spring*. This model asserts that the magnitude of the force in the spring is equal to a spring modulus k times the change in length of the spring. We can show that the spring force can be derived from a potential energy function

$$U = \tfrac{1}{2}k\Delta^2 \qquad (1.19)$$

where $\Delta = L - L_o$ is the change in the length of the spring, with L_o being the length of the spring when it is not stretched. Let $\mathbf{z} = \mathbf{x}_2 - \mathbf{x}_1$ be the vector from one end of the spring to the other, as shown in Fig. 1.7. The current length of the spring, therefore, is $L = \sqrt{\mathbf{z} \cdot \mathbf{z}}$. The forces acting on the masses, caused by stretch of the spring, are

$$\mathbf{F}_1 = -\frac{\partial U}{\partial \mathbf{x}_1} = k\Delta\frac{\mathbf{z}}{L}, \qquad \mathbf{F}_2 = -\frac{\partial U}{\partial \mathbf{x}_2} = -k\Delta\frac{\mathbf{z}}{L} \qquad (1.20)$$

as shown in Fig. 1.7. The derivation is very similar to the gravitational potential. The magnitude of the spring force is $k\Delta$ and the force acts along the direction of the axis of the spring. Notice that the force depends only on the change in length and not on the original length. In some circumstances it is not necessary to know the original length of the spring to solve a problem involving a spring (e.g., when the direction of the axis of the spring does not change due to the motion).

Fig. 1.7 *Linear spring.* A linear spring develops force in proportion to the change in length, with the force directed along the axis of the spring. The constant of proportionality is the spring constant k

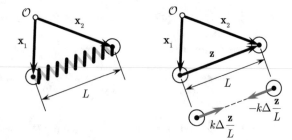

1.4 Conservation of Energy

The connection between power and kinetic energy was a direct outcome of Newton's second law. The concept of potential energy of a force was a construction that allowed the work done by a conservative force to be evaluated as the difference in potential energy between the start and end of the process. We can eliminate the work integral between the two definitions to arrive at what we call an *energy principle*. To wit,

$$\int_{t_a}^{t_b} \left(\mathbf{F} \cdot \mathbf{v} \right) dt = U(t_a) - U(t_b) = T(t_b) - T(t_a)$$

Moving the two terms at state a to the left side and the two terms at state b to the right side we have

$$T(t_a) + U(t_a) = T(t_b) + U(t_b)$$

The sum of kinetic and potential energy at a given state is often called the *total energy*. The above equation simply states that in any motion the total energy is conserved (as long as the force is conservative).

The concept of conservation of energy can be used to derive the equations of motion of a system. Let the total energy be the sum of the kinetic and potential energy, and let energy be conserved. Then,

$$E = T + U = \text{constant} \tag{1.21}$$

Taking the time derivative of Eq. 1.21, we get

$$\frac{dE}{dt} = \frac{dT}{dt} + \frac{dU}{dt} = 0$$

since the derivative of a constant is zero. Let us work this through using what we know about kinetic and potential energy. In particular,

$$\frac{dU}{dt} + \frac{dT}{dt} = \frac{\partial U}{\partial \mathbf{x}} \cdot \frac{d\mathbf{x}}{dt} + \frac{d}{dt} \left(\tfrac{1}{2} m \, \mathbf{v} \cdot \mathbf{v} \right)$$

$$= -\mathbf{F} \cdot \mathbf{v} + m \frac{d\mathbf{v}}{dt} \cdot \mathbf{v}$$

$$= \left(-\mathbf{F} + m\mathbf{a} \right) \cdot \mathbf{v} = 0$$

The velocity $\mathbf{v}$ is not always zero. If it were, that would imply no motion. Therefore, in order for energy to be conserved at all times, we must have

$$\mathbf{F} - m\mathbf{a} = \mathbf{0}$$

Hence, conservation of energy implies Newton's second law. The derivation above can be run in the opposite direction, too. Start with $-\mathbf{F} + m\mathbf{a} = \mathbf{0}$. Multiply this equation by the integrating factor $\mathbf{v}$ and integrate with respect to time

$$\int_0^t (-\mathbf{F} + m\mathbf{a}) \cdot \mathbf{v} \, dt = \int_0^t \frac{d}{dt}\left(U + T\right) dt = E + \text{constant}$$

Consequently, $E = T + U$ is called the *first integral of the equation of motion* (as long as the forces can be derived from a potential energy function, of course).

1.5   Dynamics of Rigid Bodies

Euler extended Newton's ideas to rigid bodies of finite size. The equations of dynamics for finite-sized bodies is important to practical engineering applications because all bodies of interest are finite in size. The equations of motion for a rigid body can be built by recognizing that the body is an assembly of particles that individually obey Newton's laws. We will employ a *typical particle* construction to derive equations of motion for practical rigid body problems.

Balance of Linear Momentum Consider a body $\mathcal{B}$ of mass density ρ with boundary $\partial\mathcal{B}$. The body is subjected to body force $\mathbf{b}$ in $\mathcal{B}$ and traction $\mathbf{t}$ applied on $\partial\mathcal{B}$, as shown in Fig. 1.8. Let us isolate a small region $\mathcal{B}_e$ of the body with boundary $\partial\mathcal{B}_e$. Balance of linear momentum for this *typical particle* is, according to Newton's second law,

$$\int_{\partial\mathcal{B}_e} \mathbf{t}_e \, dA + \int_{\mathcal{B}_e} \mathbf{b}_e \, dV = \int_{\mathcal{B}_e} \rho_e \ddot{\mathbf{x}}_e \, dV \qquad (1.22)$$

where $\mathbf{b}_e$, $\mathbf{t}_e$, and ρ_e are the body force, surface traction, and density respectively, associated with particle e. The integrals are over the surface and domain of the particle as indicated.

Fig. 1.8 A *rigid body* with a typical particle $\mathcal{B}_e$ with traction $\mathbf{t}_e$ and body force $\mathbf{b}_e$. The position of the particle is given by the position vector $\mathbf{x}$ relative to the origin $\mathcal{O}$. For balance of angular momentum, we sum moments about the point $\mathcal{C}$. The moment arm is the vector $\mathbf{r}$

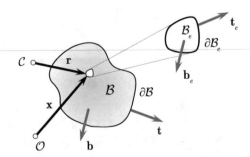

Each particle in the body responds according to its own version of Eq. 1.22. We can add up these equations for all particles that comprise the body. First, note that

$$\sum_e \int_{\partial B_e} \mathbf{t}_e \, dA = \int_{\partial B} \mathbf{t} \, dA$$

In other words, the sum of the integrals of traction over the surfaces of the particles that make up the body is equal to the integral of the traction acting on the exterior surface of the body. The proof goes as follows. Each point on the surface of a particle has a traction. At locations where two particles meet, those particles have equal and opposite tractions that act at the same point in space, by Newton's third law. When added, these tractions cancel out pairwise. The only surfaces that do not have a canceling partner are those surfaces on the exterior boundary of the body.

The volume integrals over the particles simply add up to volume integrals over the entire body. To wit,

$$\sum_e \int_{B_e} \mathbf{b}_e \, dV = \int_B \mathbf{b} \, dV, \qquad \sum_e \int_{B_e} \rho_e \, \ddot{\mathbf{x}}_e \, dV = \int_B \rho \, \ddot{\mathbf{x}} \, dV$$

With these observations, the sum of Eqs. 1.22 over all particles in the body yields

$$\int_{\partial B} \mathbf{t} \, dA + \int_B \mathbf{b} \, dV = \int_B \rho \, \ddot{\mathbf{x}} \, dV \qquad (1.23)$$

This equation is known as *balance of linear momentum* for the rigid body.

Balance of Angular Momentum For a particle, balance of angular momentum does not give new information beyond balance of linear momentum. For finite-sized bodies it does. Let us first compute the balance of angular momentum of the particle B_e about point C, as shown in Fig. 1.8. To wit,

$$\int_{\partial B_e} (\mathbf{r}_e \times \mathbf{t}_e) \, dA + \int_{B_e} (\mathbf{r}_e \times \mathbf{b}_e) \, dV = \int_{B_e} (\mathbf{r}_e \times \rho_e \, \ddot{\mathbf{x}}_e) \, dV \qquad (1.24)$$

where $\mathbf{r}_e$ is the moment arm for particle e. Observe that in the limit, as the volume of B_e goes to zero, Eq. 1.24 reduces to

$$\mathbf{r}_e \times \left(\int_{\partial B_e} \mathbf{t}_e \, dA + \int_{B_e} \mathbf{b}_e \, dV - \int_{B_e} \rho_e \, \ddot{\mathbf{x}}_e \, dV \right) = \mathbf{0}$$

because all the moment arm vectors $\mathbf{r}_e$ reduce to the same vector. The quantity in parentheses is zero due to balance of linear momentum.

For a rigid body, we can sum Eqs. 1.24 over all particles in the body and note that the integrals over the internal boundaries of the particles cancel out for the same reason as previously noted. The integrals over the volume of the particles add up, and the resulting equation is

$$\int_{\partial B} (\mathbf{r} \times \mathbf{t}) \, dA + \int_{B} (\mathbf{r} \times \mathbf{b}) \, dV = \int_{B} (\mathbf{r} \times \rho \ddot{\mathbf{x}}) dV \tag{1.25}$$

This equation is known as *balance of angular momentum*.

Balance of linear and angular momentum, Eqs. 1.23 and 1.25, respectively, constitute a complete set of equations of motion for the rigid body. In two dimensions, it takes exactly three time-dependent parameters to characterize the motion of a rigid body. In three dimensions it takes six. We have exactly the right number of equations of balance of momentum to determine the evolution of those parameters. The reason we consider the entire body is because of the assumption of rigidity. While we can find internal resultants by taking a free body diagram of a portion of the body, we cannot determine the state of internal stress because of the rigidity constraint. Hence, elucidating the overall motion of the body is the primary goal in these types of problems.

In practice, we will set up these integrals for each problem, noting that the traction **t** and body force **b** are known. We will derive the acceleration of a *typical particle* from a description of the position vector of that particle, in terms of the generalized motion variables appropriate to the problem. Integration over the surface and domain of the body will yield the *equations of motion*.

Energy of a Rigid Body The kinetic and potential energies of a rigid body are simply the sum of the kinetic and potential energies of the particles that comprise the body. Hence, the *kinetic energy* is

$$T = \int_{B} \tfrac{1}{2} \rho \, (\mathbf{v} \cdot \mathbf{v}) \, dV \tag{1.26}$$

and the *potential energy* due to gravity (i.e., the body force)

$$U = \int_{B} \rho g \, (\mathbf{x} \cdot \mathbf{e}_g) \, dV \tag{1.27}$$

Conservation of energy works the same for a rigid body as it does for a particle. In general, the kinetic and potential energy of a rigid body depend upon the time-dependent variables that describe the motion (i.e., the variables from which we construct the position vector of the typical particle).

1.6 Example

Perhaps the best way to see how the process of formulating equations of motion goes is through a particular example. In this section we go through the details of the derivation of the equations of motion using the typical particle construction.

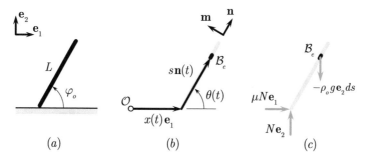

Fig. 1.9 *Sliding rod example.* The rod of length L starts from rest at an angle φ_o and falls under its own weight on a surface with a coefficient of friction μ. Find the equations of motion. Note that we will assume that the mass is distributed along a line with mass per unit length of ρ_o (**a**) initial position of rod, (**b**) description of motion of a typical particle, and (**c**) free body diagram showing the force acting on the typical particle

Through this example you will see how the kinematic constraints are represented and how the domain integrals are executed.

Consider the sliding rod problem shown in Fig. 1.9a. The rod has length L, mass per unit length ρ_o (density times area), and starts from rest at an initial angle of φ_o measured from the horizontal. The coefficient of friction between the rod and the surface is μ. Find the state at time zero.

Figure 1.9b shows the geometry of the bar at time t. The origin is located on the surface at $\mathcal{O}$ and the distance to the lower end of the bar is characterized by the time-dependent variable $x(t)$. The angle that the rod makes with the horizontal is characterized by the time-dependent variable $\theta(t)$. The initial conditions for this system are

$$x(0) = 0, \quad \dot{x}(0) = 0, \quad \theta(0) = \varphi_o, \quad \dot{\theta}(0) = 0 \qquad (1.28)$$

The choice $x(0) = 0$ is arbitrary, as we can select any point to be the origin. This specific choice means that the origin is the point directly under the lower end of the bar at time $t = 0$.

The *typical particle* $\mathcal{B}_e$ is located a distance s from the lower end of the rod. The unit vectors $\mathbf{n}$ and $\mathbf{m}$ can be expressed in terms of the angle θ as

$$\mathbf{n}(t) = \cos\theta(t)\mathbf{e}_1 + \sin\theta(t)\mathbf{e}_2$$
$$\mathbf{m}(t) = -\sin\theta(t)\mathbf{e}_1 + \cos\theta(t)\mathbf{e}_2$$

Taking the time derivative of these expressions gives $\dot{\mathbf{n}} = \dot{\theta}\mathbf{m}$ and $\dot{\mathbf{m}} = -\dot{\theta}\mathbf{n}$. With these definitions we can describe the position of the typical particle as

$$\mathbf{x}(s, t) = x(t)\mathbf{e}_1 + s\,\mathbf{n}(t) \qquad (1.29)$$

The velocity and acceleration of the typical particle can be determined by differentiation of the position vector as

$$\dot{\mathbf{x}}(s,t) = \dot{x}(t)\mathbf{e}_1 + s\,\dot{\theta}(t)\mathbf{m}(t)$$

$$\ddot{\mathbf{x}}(s,t) = \ddot{x}(t)\mathbf{e}_1 + s\big(\ddot{\theta}(t)\mathbf{m}(t) - \dot{\theta}^2(t)\mathbf{n}(t)\big) \tag{1.30}$$

Figure 1.9c shows the free body diagram of the system with the reaction forces at the left end and the weight shown only for the typical particle (every particle has a weight but we will attend to that through our summation process by integrating from 0 to L). The friction force is assumed to be acting to the right because we anticipate that the bar will slide to the left. We will need to verify this assumption once the solution is obtained. If it is not correct, then we can change the assumption and solve the problem again. Balance of linear momentum gives

$$\mu N\mathbf{e}_1 + N\mathbf{e}_2 - \int_0^L \rho_o g\,\mathbf{e}_2\,ds = \int_0^L \rho_o\,\ddot{\mathbf{x}}(s,t)\,ds \tag{1.31}$$

We can substitute the expression for the acceleration of the typical particle from Eq. 1.30 and carry out the integration with respect to s to get

$$\mu N\mathbf{e}_1 + N\mathbf{e}_2 - \rho_o g L\mathbf{e}_2 = \rho_o L\ddot{x}\,\mathbf{e}_1 + \tfrac{1}{2}\rho_o L^2\big(\ddot{\theta}\mathbf{m} - \dot{\theta}^2\mathbf{n}\big) \tag{1.32}$$

This equation is balance of linear momentum. Notice that it is a vector equation, which means that it functions algebraically as two scalar equations (one for each component of the vectors). We do not write our equations in the (x, y) components, as is often done in elementary texts, because it is not always evident what algebraic steps are needed to make headway with solving the system. It is always possible to dot the equation in any direction, which is equivalent to projecting the equations along that direction.

We can balance angular momentum about the lower end of the rod. In this case the two reaction forces have no moment arm, and the moment arm to the typical particle is $\mathbf{r} = s\mathbf{n}$. Hence,

$$-\int_0^L \big(s\mathbf{n} \times \rho_o g\,\mathbf{e}_2\big)\,ds = \int_0^L \big(s\mathbf{n} \times \rho_o\,\ddot{\mathbf{x}}\big)\,ds \tag{1.33}$$

Noting that $\mathbf{n} \times \mathbf{e}_2 = \cos\theta\,\mathbf{e}_3$ and that $\mathbf{n} \times \ddot{\mathbf{x}} = \big(s\,\ddot{\theta} - \ddot{x}\sin\theta\,\big)\mathbf{e}_3$, we can evaluate the integrals in Eq. 1.33 to get

$$-\tfrac{1}{2}\rho_o g L^2 \cos\theta\,\mathbf{e}_3 = -\tfrac{1}{2}\rho_o L^2\ddot{x}\sin\theta\,\mathbf{e}_3 + \tfrac{1}{3}\rho_o L^3\ddot{\theta}\,\mathbf{e}_3 \tag{1.34}$$

Now, we can dot Eq. 1.32 with $\mathbf{e}_1$ and $\mathbf{e}_2$ and Eq. 1.34 with $\mathbf{e}_3$ to get our final system of equations

$$\mu N - \rho_o L \ddot{x} + \tfrac{1}{2}\rho_o L^2 \left(\ddot{\theta}\sin\theta + \dot{\theta}^2\cos\theta\right) = 0$$
$$N - \tfrac{1}{2}\rho_o L^2 \left(\ddot{\theta}\cos\theta - \dot{\theta}^2\sin\theta\right) = \rho_o g L \qquad (1.35)$$
$$-\tfrac{1}{2}\rho_o L^2 \ddot{x}\sin\theta + \tfrac{1}{3}\rho_o L^3 \ddot{\theta} = -\tfrac{1}{2}\rho_o g L^2 \cos\theta$$

Since the bar starts at rest, $\dot{\theta}(0) = 0$ and $\theta(0) = \varphi_o$. We can solve for the normal force, linear acceleration, and angular acceleration at $t = 0$. Let

$$N_o = N(0), \qquad \ddot{x}_o = \ddot{x}(0), \qquad \ddot{\theta}_o = \ddot{\theta}(0)$$

The equations of motion reduce to

$$\mu N_o - \rho_o L \ddot{x}_o + \tfrac{1}{2}\rho_o L^2 \sin\varphi_o\, \ddot{\theta}_o = 0$$
$$N_o - \tfrac{1}{2}\rho_o L^2 \cos\varphi_o\, \ddot{\theta}_o = \rho_o g L$$
$$-\tfrac{1}{2}\rho_o L^2 \sin\varphi_o\, \ddot{x}_o + \tfrac{1}{3}\rho_o L^3 \ddot{\theta}_o = -\tfrac{1}{2}\rho_o g L^2 \cos\varphi_o$$

which is a system of three equations in three unknowns. These equations can be solved algebraically to find the initial accelerations and normal force. It is also worth noting that Eqs. 1.35 are nonlinear differential equations, which is the most common outcome for rigid-body dynamics. There are no known classical methods to solve most of these equations, but they are easily solved numerically.

1.7   The Euler–Lagrange Equations

We introduced energy methods earlier in this chapter with a simple consideration of *conservation of energy* and showed that it yielded the equations of motion for the system. Unfortunately, for nonlinear systems with more than one degree of freedom, the approach we took using conservation of energy does not always provide an unambiguous route to determining the equations of motion. This limitation can be resolved through the Euler–Lagrange equations.

Let us assume that the time-dependent unknowns of a problem can be collected in the array $\mathbf{q}(t)$. This set of unknowns could include translations and rotations (they are not all of the same character, but they all depend exclusively on time). With this notation, we can write the kinetic and gravitational potential energy as

$$T(\mathbf{q}, \dot{\mathbf{q}}) = \int_B \tfrac{1}{2}\rho\, \dot{\mathbf{x}}(\mathbf{q}, \dot{\mathbf{q}}) \cdot \dot{\mathbf{x}}(\mathbf{q}, \dot{\mathbf{q}})\, dV$$

$$U(\mathbf{q}) = \int_B \rho g\, \mathbf{x}(\mathbf{q}) \cdot \mathbf{e}_g\, dV$$

The key point is that the kinetic energy can be thought of as a function of both $\mathbf{q}$ and $\dot{\mathbf{q}}$, while the potential energy is only a function of $\mathbf{q}$ (and that is true of any potential energy function, not just the gravitational potential).

Define the *Lagrangian* functional[1] to be the difference between kinetic and potential energy. To wit,

$$\mathcal{L}(\mathbf{q}, \dot{\mathbf{q}}) = T(\mathbf{q}, \dot{\mathbf{q}}) - U(\mathbf{q}) \tag{1.36}$$

Contrast this quantity with the total energy E, which is the sum of kinetic and potential energy. One way to think about dynamical systems is that they constantly convert kinetic energy to potential energy and vice versa. The law of conservation of energy says that the system does this in a way that makes the total energy (i.e., $T + U$) constant. For some systems, there are an infinite number of ways that energy could be exchanged while still conserving it. Hamilton's principle adds an additional consideration to determine which one of those possibilities is what actually happens.

Let us define the *Hamiltonian* functional to be the integral of the Lagrangian between two times t_a and t_b

$$\mathcal{H}(\mathbf{q}, \dot{\mathbf{q}}) = \int_{t_a}^{t_b} \mathcal{L}(\mathbf{q}, \dot{\mathbf{q}}) \, dt \tag{1.37}$$

This quantity is often called the *action integral*. Hamilton suggested that among all possible motions $\mathbf{q}(t)$, the actual motion is the one that minimizes the Hamiltonian. This principle, usually referred to as the *principle of least action*, makes intuitive sense if you think of dynamical systems as passing kinetic and potential energy back and forth. Among the many possible ways that can happen, the one that accords with nature is the one that minimizes the difference between the two. This is an efficiency argument, of sorts, claiming that nature finds a way to convert energy with the smallest fluctuations possible.

We can find the minimum using the directional derivative (see Appendix B for a refresher on the directional derivative). Let $\mathbf{p}(t)$ be an arbitrary variation of the function $\mathbf{q}(t)$. Then the directional derivative is

$$\mathcal{D}\mathcal{H}(\mathbf{q}, \dot{\mathbf{q}}) \cdot \mathbf{p} = \frac{d}{d\varepsilon} \Big[\mathcal{H}(\mathbf{q} + \varepsilon \mathbf{p}, \dot{\mathbf{q}} + \varepsilon \dot{\mathbf{p}}) \Big]_{\varepsilon = 0} \tag{1.38}$$

Substitute the kinetic and potential energies into the directional derivative. Using the chain rule of differentiation and evaluating the result at $\varepsilon = 0$ gives

$$\mathcal{D}\mathcal{H}(\mathbf{q}, \dot{\mathbf{q}}) \cdot \mathbf{p} = \int_{t_a}^{t_b} \left(\frac{\partial T}{\partial \dot{\mathbf{q}}} \cdot \dot{\mathbf{p}} + \frac{\partial T}{\partial \mathbf{q}} \cdot \mathbf{p} - \frac{\partial U}{\partial \mathbf{q}} \cdot \mathbf{p} \right) dt$$

---

[1] A *functional* is a function of functions. In this case the functional takes as arguments the functions $\mathbf{q}(t)$ and $\dot{\mathbf{q}}(t)$. The output of a functional is a number. The study of functionals is called the *calculus of variations*.

Integrating the first term by parts yields

$$\mathcal{DH}(\mathbf{q}, \dot{\mathbf{q}}) \cdot \mathbf{p} = \int_{t_a}^{t_b} \left[-\frac{d}{dt}\left(\frac{\partial T}{\partial \dot{\mathbf{q}}}\right) + \frac{\partial T}{\partial \mathbf{q}} - \frac{\partial U}{\partial \mathbf{q}} \right] \cdot \mathbf{p}\, dt + \left[\frac{\partial T}{\partial \dot{\mathbf{q}}} \cdot \mathbf{p} \right]_{t_a}^{t_b}$$

To find the minimum, the directional derivative must be zero for all choices of the arbitrary function $\mathbf{p}$. To eliminate the last term, we restrict the virtual motions $\mathbf{p}(t)$ to be zero at times t_a and t_b. Using the *fundamental theorem of the calculus of variations* we get the Euler–Lagrange equations

$$\frac{d}{dt}\left(\frac{\partial T}{\partial \dot{\mathbf{q}}}\right) - \frac{\partial T}{\partial \mathbf{q}} + \frac{\partial U}{\partial \mathbf{q}} = \mathbf{0} \qquad (1.39)$$

The proof is simple. If $\mathbf{p}$ can be anything, then we could set it equal to the term in square brackets. That results in the dot product of $\mathbf{p}$ with itself, which must be positive. The only way for a positive quantity to integrate to zero is if that quantity is, itself, equal to zero. The Euler–Lagrange equations yield the proper equations of motion for any conservative system (i.e., a system where the forces possess a potential energy function). The equations are exactly equivalent to those derived from balance of momentum.

1.8 Summary

This chapter has introduced the concepts of dynamics starting from Newton's laws for a particle and extending them to rigid bodies through the *typical particle* construction. Newton's laws are adequate to establish the equations of motion of any system in solid mechanics.

In the early chapters we will consider the single-degree-of-freedom model in which rotation does not play a part. In that context, we will only need balance of linear momentum. In our study of discrete systems with multiple degrees of freedom, we will investigate systems comprising masses that do not rotate (i.e., the *shear building*) and systems with point masses connected by massless elastic elements (i.e., *truss structures*). These systems will only require balance of linear momentum.

When we get to the analysis of continuous systems, we will invoke the typical particle strategy as the means of deriving the equations of motion. The axial bar problem will not involve rotation, but the beam problem will. For the beam problem, we will need balance of angular momentum to complete the derivation.

Chapter 2
Numerical Solution of Ordinary Differential Equations

The equations of motion that emanate from Newton's laws for particles and rigid bodies are second order ordinary differential equations, and are often nonlinear. In our study of the dynamic response of structural systems we will encounter problems with complex forcing functions (e.g., earthquakes) and nonlinearity in the response (e.g., material yielding). While there are methods available to solve linear ordinary differential equations in a classical sense, many practical problems are not amenable to those techniques. We will rejoice when a classical solution is available (and squeeze as much insight from it as we can), but we will also prepare for the more common situation where numerical analysis of the equations of motion is required. While it may seem like a bit of an interruption to turn our attention to the numerical solution of ordinary differential equations, we do so to lay the groundwork for developing the tools that we will need to make progress with many of the problems of structural dynamics. In this chapter we consider a few of the more common numerical integrators and examine their behavior in some detail.

We will motivate the analysis of numerical integrators with the simplest problem we can think of—a linear first order differential equation. That will set up the discussion of second order differential equations, which are more important for dynamics. Among the many methods available, we will focus on the *generalized trapezoidal rule* for the first order system. We will use this same method and add the *central difference method* and *Newmark's method* to the mix when we move on to the second order system.

As with all numerical integrators for differential equations, we will be concerned with *accuracy* of the approximation (how well the integrator approximates the actual solution for finite values of the time step h and how it converges to the exact solution as h approaches zero) and *stability* (the absence of numerical artifacts that cause

Electronic Supplementary Material The online version of this chapter (https://doi.org/10. 1007/978-3-030-89944-8_2) contains supplementary material, which is available to authorized users.

the solution to perform erratically or for errors to grow exponentially with time). In structural dynamics we will observe two important phenomena associated with the performance of the numerical integrator: (1) artificial *amplification* or *decay* of the response and (2) *period elongation* where the apparent period of the computed response is different from the exact period predicted by the classical solution. We will also see that numerical integrators can exhibit *numerical instability* where the computed numerical solution diverges due to the mathematical properties of the discrete equations. These phenomena will be explored by applying the numerical integrators to linear homogeneous equations, which are problems where the exact solution is known.

2.1 Why Numerical Methods?

Finding classical solutions to differential equations is difficult in all but the simplest circumstances (e.g., linear equations). That is why we resort to numerical methods. In numerical integration we discretize the problem to find the response at certain points in time (and then connect the dots to visualize the continuous response history). If we are trying to approximate a function $u(t)$, then we will find a sequence of values

$$\{u_0, u_1, u_2, \ldots, u_n, \ldots, u_N\}$$

where u_n represents the value of the exact function $u(t_n)$ at the specific time t_n. In general, u_n will only be an approximation of the actual response function evaluated at that time. Figure 2.1 shows a discrete representation of a function $u(t)$. While it is possible to develop methods with variable step sizes, in this book we will always take equally spaced time points, so that $h = t_{n+1} - t_n$ is constant for all n. The final time $t_N = t_f$ will establish the end of the analysis.

The derivatives of the primary response variable (e.g., velocity and acceleration) will appear in our formulations. In a similar fashion, we will designate these as

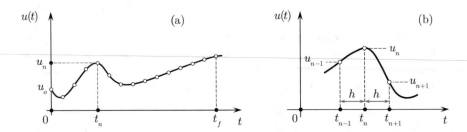

Fig. 2.1 *Discrete time series.* A time series is a sequence of points in time that represent an underlying function. (**a**) The complete time series, showing the initial value u_o, the value u_n at time t_n, and the final time t_f. (**b**) Close-up of three adjacent points, showing their respective values and the size of the time step, h

$$v_n \leftarrow \dot{u}(t_n), \quad a_n \leftarrow \ddot{u}(t_n)$$

Again, the discrete values only represent an approximation to the exact values. We call the set of values $\{u_n, v_n, a_n\}$ the *state* at time t_n. In the continuous setting, the displacement, velocity, and acceleration are related to each other through differential relationships. In the discrete setting we view u_n, v_n, and a_n as independent variables and we need to add additional relationships among them (beyond the equation of motion) that cause them to emulate the differential relationships. Those relationships distinguish one numerical integrator from another.

In the discrete world, derivatives are approximated by differences and integrals are approximated by simple mensuration formulas for area under the curve (e.g., trapezoids). The key constraint is that only the values at the discrete points in the approximation can be used.

Numerical solutions introduce approximation error, and we need to understand how those errors manifest in order to be competent users of numerical methods. On the simple end of the spectrum is the idea that as we take a finer and finer time discretization (smaller time steps), we should get closer and closer to the exact solution. On the more complicated end we have the problem of potential numerical instability of the integration scheme. Each numerical integrator has its own performance features.

The Methods The three numerical integrators that we will focus on for structural dynamics are summarized in the box below:

Numerical Integrators for Structural Dynamics

- *Generalized Trapezoidal Rule*

$$v_{n+1} = v_n + h\big[\alpha\, a_n + (1-\alpha)\, a_{n+1}\big]$$
$$u_{n+1} = u_n + h\big[\alpha\, v_n + (1-\alpha)\, v_{n+1}\big]$$

- *Newmark's Method*

$$v_{n+1} = v_n + h\big[\gamma\, a_n + (1-\gamma)\, a_{n+1}\big]$$
$$u_{n+1} = u_n + h\, v_n + h^2\big[\beta\, a_n + \big(\tfrac{1}{2} - \beta\big)a_{n+1}\big]$$

- *Central Difference Method*

$$u_{n+1} - u_{n-1} = 2h\, v_n$$
$$u_{n+1} - 2u_n + u_{n-1} = h^2 a_n$$

The quantities α, β, and γ are parameters, set by the user, associated with the numerical method. Changing the values of these parameters changes the behavior of the method.

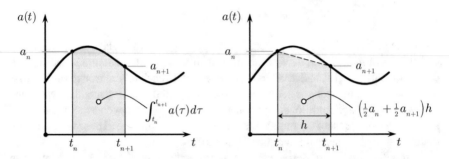

Fig. 2.2 *The trapezoidal rule.* The idea behind the trapezoidal rule is to approximate the integral using the area of the trapezoid defined by connecting two adjacent points in the time series

It is reasonable to ask where these numerical methods come from. Can they be derived? The answer to that question is more nuanced than one might first imagine, and that is the main reason for the analysis that we undertake in this chapter. However, the methods *are* very simply motivated and we can gain some insight from considering the derivation of the methods.

From Eq. 1.3 we can see that the velocity at time t_{n+1} can be computed from the velocity at time t_n and the integral of the acceleration as

$$v(t_{n+1}) = v(t_n) + \int_{t_n}^{t_{n+1}} a(\tau)\,d\tau \qquad (2.1)$$

That relationship is exact. The problem is that we do not know how the acceleration varies during that time period. The situation is depicted in Fig. 2.2. The integral we seek is the area under the curve. We know a_n and we seek a_{n+1}. An integral of a scalar function represents the area under the curve. Let us approximate the variation of the acceleration between two time points as a linear function. To simplify the computations, let

$$\xi = \frac{\tau - t_n}{h}$$

Now, the acceleration between the time points can be expressed as

$$a(\xi) = (1-\xi)\,a_n + \xi\,a_{n+1} \qquad (2.2)$$

Even though we still do not know the value of a_{n+1}, we can execute the integral as

$$\int_{t_n}^{t_{n+1}} a(\tau)\,d\tau = \int_0^1 \left((1-\xi)a_n + \xi\,a_{n+1}\right)h\,d\xi = \left(\tfrac{1}{2}a_n + \tfrac{1}{2}a_{n+1}\right)h$$

Of course, the result is simply the area of a trapezoid defined by the linear interpolation of the two discrete accelerations. This method is called the *trapezoidal rule* of integration. We can generalize this formula by asserting that

$$\int_{t_n}^{t_{n+1}} a(\tau)\,d\tau \approx [\alpha\,a_n + (1-\alpha)\,a_{n+1}]\,h$$

where $\alpha \in [0, 1]$ is a parameter. This parameterization creates a family of integrators that we call the *generalized trapezoidal rule*. One member of that family, $\alpha = 0.5$, is the trapezoidal rule. If we take $\alpha = 1$, then we are approximating the integral with the left Riemann sum and if we take $\alpha = 0$, then we are approximating the integral with the right Riemann sum, which you might recall from the introduction to integration in calculus. The main reason for generalizing the rule is to open up additional possibilities for the method. This line of reasoning gives rise to the first equation of the generalized trapezoidal rule relating velocity to acceleration

$$v_{n+1} = v_n + h\Big[\alpha\,a_n + (1-\alpha)\,a_{n+1}\Big]$$

Equation 1.3 also shows that the change in displacement is equal to the area under the velocity curve. Thus,

$$u(t_{n+1}) = u(t_n) + \int_{t_n}^{t_{n+1}} v(\tau)\,d\tau \tag{2.3}$$

We can approximate the velocity as the linear interpolation of v_n and v_{n+1} and integrate to get the second equation of the generalized trapezoidal rule relating displacement and velocity

$$u_{n+1} = u_n + h\Big[\alpha\,v_n + (1-\alpha)\,v_{n+1}\Big]$$

The value of α would not have to be the same for the first and second equations of the generalized trapezoidal rule, but we will take them to be the same.

From Fig. 2.2, it is evident that the area of the trapezoid is not the same as the actual integral unless the variation of the real function is linear. But if the two points are taken close together, then the error is small. Hence, the size of the time step h is crucial to the accuracy of the approximation.

To integrate a given function between two limits, we would apply the generalized trapezoidal rule successively to segments of length h going from the lower limit of the integral to the upper limit. Within the context of integration of the differential equation, we step forward from known initial conditions $\{u_o, v_o\}$, computing the new acceleration from the old state and then moving on to the next time step.

To get some insight on the origins of the Newmark relationships, recall that Eq. 1.4 allows us to write an expression for the displacement in terms of an integral of the acceleration. Integrating from t_n to t_{n+1} gives

$$u(t_{n+1}) = u(t_n) + v(t_n)h + \int_{t_n}^{t_{n+1}} (t_{n+1} - \tau) a(\tau) d\tau \qquad (2.4)$$

which is exact. Using the same change of variable as we did before, we can write

$$t_{n+1} - \tau = t_{n+1} - (t_n + \xi h) = (1 - \xi)h$$

Assuming the linear variation of the acceleration of Eq. 2.2, we can evaluate the integral of the acceleration to be

$$\int_{t_n}^{t_{n+1}} (t_{n+1} - \tau) a(\tau) d\tau \approx \int_0^1 \left((1-\xi)^2 a_n + \xi(1-\xi) a_{n+1} \right) h^2 d\xi$$

$$= \left(\tfrac{1}{3} a_n + \tfrac{1}{6} a_{n+1} \right) h^2$$

We can generalize this formula by asserting that

$$\int_{t_n}^{t_{n+1}} (t_{n+1} - \tau) a(\tau) d\tau \approx \left[\beta \, a_n + \left(\tfrac{1}{2} - \beta \right) a_{n+1} \right] h^2$$

where $\beta \in [\, 0, 0.5 \,]$ is a parameter. This parameterization, taken in conjunction with the generalized trapezoidal rule for velocity, creates a family of integrators that we call the *Newmark method*

$$v_{n+1} = v_n + h \left[\gamma \, a_n + (1-\gamma) \, a_{n+1} \right]$$

$$u_{n+1} = u_n + h \, v_n + h^2 \left[\beta \, a_n + \left(\tfrac{1}{2} - \beta \right) a_{n+1} \right]$$

If the acceleration is taken as linear between the two time points, as we did in the derivation, then the coefficients that result are $\beta = 1/6$ and $\gamma = 1/2$. This method is often referred to as the *linear acceleration method*. If the acceleration is taken as constant and equal to the average of the two accelerations at the ends of the time step, then the coefficients that result are $\beta = 1/4$ and $\gamma = 1/2$. This method is often referred to as the *constant average acceleration method*. Intuitively, one might imagine that the linear acceleration is better than the constant acceleration method, but our analysis of the methods in the sequel will show that the behavior of the numerical methods require some additional considerations that make this conclusion a little more complex.

The central difference method comes from approximating the derivatives as differences. In the case of velocity, the centered difference $(u_{n+1} - u_{n-1})/2h$ is used to approximate the first derivative of $u(t)$ at t_n. In essence, we are taking the secant between the two points adjacent to t_n to approximate the tangent to the curve at t_n. In the case of acceleration, the difference of the right difference $(u_{n+1} - u_n)/h$

and the left difference $(u_n - u_{n-1})/h$ approximates the second derivative of $u(t)$ at t_n.

What is common about all three methods is that they each produce two equations relating the three state variables. These two equations will be supplemented with the equation of motion to complete an *algebraic* system of equations. Each method has numerical integration parameters that must be specified by the user (i.e., α, β, and γ). The behavior of the method depends on the values of the parameters (as we shall see) and the time step size h.

Generalized trapezoidal rule and Newmark are both *single step methods*. That means that the new state can be completely determined from the previous state and the equation of motion. Central difference is a two-step method. That means that you need information from the previous *two* states to compute the new state. Single step methods are called *self starting* because the initial conditions are sufficient to execute the first step of the algorithm. Multistep methods are often not self starting, but some manipulation of the equations can be formulated to create starting conditions that are consistent with the initial conditions.

The Complete Set of Equations The generalized trapezoidal rule and the Newmark method both give two equations relating the previous state $\{u_n, v_n, a_n\}$ to the new state $\{u_{n+1}, v_{n+1}, a_{n+1}\}$. But that gives only two equations for three unknowns. The third equation needed to solve the system of equations is the equation of motion at step $n + 1$, which we write as

$$g\left(u_{n+1}, v_{n+1}, a_{n+1}\right) = 0$$

where the function g represents the discrete equation of motion, which depends upon the original governing differential equation. For example, the linear single-degree-of-freedom oscillator comes from the differential equation

$$m\ddot{u}(t) + c\dot{u}(t) + ku(t) = 0$$

where m, c, and k are physical parameters. For the discrete equation, we can define the residual function

$$g\left(u_n, v_n, a_n\right) = ma_n + cv_n + ku_n$$

Now, the equation $g(u_{n+1}, v_{n+1}, a_{n+1}) = 0$ is an assertion that the state variables satisfy the discrete equation of motion at step $n + 1$. This is an algebraic equation because we are only trying to satisfy the equation of motion at a single point in time.

For many dynamic problems, the equations of motion are nonlinear. That is the motivation for writing the discrete equations in *residual form* $g(x) = 0$. To solve the nonlinear algebraic equations we will appeal to Newton's method, as outlined in Appendix A. Now, with three equations and three unknowns, we can solve the discrete set of equations. The time integration equations (e.g., generalized trapezoidal rule) create an algebraic equation consistent with the original differential equation. Newton's method is simply a way to deal with the algebraic equations that emanate from this discretization process. Observe that the two equations representing integration in time are always linear.

Initial Acceleration The equations of dynamics are *initial value problems*, which means that we know the initial displacement u_o and initial velocity v_o. Initial acceleration *cannot* be specified as an initial condition. It must be determined from the equations of motion (otherwise it might conflict with the equations of motion). At time $t = 0$ we can write

$$g(u_o, v_o, a_o) = 0$$

which we can solve for a_o, knowing u_o and v_o. Even if the equations of motion are nonlinear, this equation is usually very simple to solve because the inertial term is always linear. The equation might look like

$$m a_n + r(u_n, v_n) = 0$$

Therefore, the initial acceleration can be computed as $a_o = -r(u_o, v_o)/m$.

Before we embark on our analysis of the performance of the numerical methods, it is instructive to see how the methods are implemented in a simple dynamic analysis code. We will look at a specific example to illustrate the concepts.

2.2   Practical Implementation

Consider the simple uniform pendulum of length L shown in Fig. 2.3. The time-dependent variable for this system is the angle $\theta(t)$ of the pendulum measured from vertical. The differential equation governing the motion is

$$\ddot{\theta}(t) + C \sin \theta(t) = 0$$

where $C = 3g/2L$ is a constant with g being the acceleration of gravity.

This nonlinear differential equation can be solved using Newmark's method for the time integrator along with Newton's method to solve the resulting nonlinear algebraic equations. To write the discrete equations of motion, let us define the discrete approximations of angle, angular velocity, and angular acceleration as follows

$$u_n \leftarrow \theta(t_n), \qquad v_n \leftarrow \dot{\theta}(t_n), \qquad a_n \leftarrow \ddot{\theta}(t_n)$$

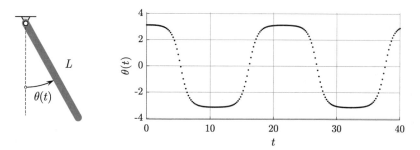

Fig. 2.3 *Simple pendulum.* The response of a simple pendulum computed using Newmark's method with $\beta = 0.25$, $\gamma = 0.5$, and time step $h = 0.2$. The physical parameters are $L = 10$ and $g = 9.81$. The pendulum starts from rest at an initial angle of $\theta_o = 3.13$ (nearly vertical). The convergence tolerance for Newton's method is 10^{-8}

The discrete equation of motion and the Newmark equations are

$$g(u_{n+1}, v_{n+1}, a_{n+1}) = a_{n+1} + C \sin(u_{n+1}) = 0$$

$$u_{n+1} = b_n + \zeta a_{n+1}$$

$$v_{n+1} = c_n + \eta a_{n+1}$$

where $b_n = u_n + hv_n + \beta h^2 a_n$ and $c_n = v_n + \gamma h a_n$ are the parts of the new angle and angular velocity, respectively, in the Newmark equations that involve information from step n and $\zeta = (0.5 - \beta)h^2$ and $\eta = (1 - \gamma)h$ are integration constants. If we knew a_{n+1}, then we could compute both u_{n+1} and v_{n+1} from the Newmark relationships. However, both u_{n+1} and a_{n+1} show up in the equation of motion. We can solve this problem iteratively by first assuming a value for a_{n+1}, using that value to compute u_{n+1} by Newmark, testing to see if the equation of motion is satisfied, and improving the estimate of a_{n+1} using Newton's update if it is not.

Let $x = a_{n+1}$ be the angular acceleration at step $n + 1$ that we seek. The residual $g(x)$ and tangent $A = dg/dx$ for Newton's method for this problem are given by

$$g(x) = x + C \sin(b_n + \zeta x)$$

$$A(x) = 1 + \zeta C \cos(b_n + \zeta x)$$

The Newton update is (see Appendix A)

$$x^{i+1} = x^i - g(x^i)/A(x^i)$$

We start the iteration with $x^0 = a_n$ and continue until the norm of the residual is smaller than a predefined tolerance. The value a_{n+1} is taken as the converged value of x^N, where N is the number of Newton iterations required to obtain convergence for that time step.

Figure 2.3 shows the response of the pendulum for the values of the parameters given in the figure caption. The solution is the discrete time series shown by the points on the plot. It is important to remember that we are computing values only at discrete points in time. Of course, in plotting we could connect the dots to give a curve that looks like a continuous function. The nonlinearity is evident in the results. Since the initial angle puts the pendulum nearly upright, the nonlinearity of the $\sin\theta$ term has a large influence. The natural period of the linearized system (i.e., with $\sin\theta \approx \theta$) is $T = 5.132$ s, which is quite different from the observed period of over 20 s for the nonlinear pendulum. The code that solves this problem is given as follows:

Code 2.1 MATLAB code for numerically integrating the equation governing the simple pendulum using Newmark's method

```
clear; clc;

%. Physical parameters
   gravity = 9.81; L = 10.0; C = 3*gravity/(2*L);

%. Numerical analysis parameters
   h = 0.2; beta = 0.25; gamma = 0.5; tol = 1.e-8;
   eta = (1-gamma)*h; zeta = (0.5-beta)*h^2;

%. Set final time and compute number of steps
   tf = 40;  nSteps = ceil(tf/h)+1;

%. Initialize time, angle, and angular velocity
   t = 0; uo = 3.13; vo = 0.0;

%. Set old values, compute initial acceleration
   uold = uo;  vold = vo;
   aold = -C*sin(uo);

%. Initialize storage for plotting results
   Store = zeros(nSteps,3);

%. Loop over time steps
   for i=1:nSteps

%.... Store values for later plotting
      Store(i,:) = [t,uold,vold];

%.... Set up values for new time step
      t = t + h;
      bn = uold + vold*h + beta*h^2*aold;
      cn = vold + gamma*h*aold;
      anew = aold; err = 1;

%.... Compute new acceleration by Newton iteration
      while err>tol
         unew = bn + zeta*anew;
         g = anew + C*sin(unew);
         A = 1 + zeta*C*cos(unew);
         anew = anew - g/A;
         err = norm(g);
      end

%.... Compute final angle and angluar velocity
      unew = bn + zeta*anew;
      vnew = cn + eta*anew;

%.... Update state to prepare for next time step
```

```
      aold = anew;   vold = vnew;   uold = unew;

    end % loop over time steps

    PlotPendulum(Store);
```

This example shows the simplicity and power of numerical integration for solving dynamics problems. You can see that we carry out the computation keeping track of only the current state (step n, which we call old) and the next state (step $n + 1$, which we call new). This same code organization will be used throughout the book. We name the variables u, v, and a, with either an old or new designation to represent the time-dependent variable, its first derivative, and its second derivative, respectively.

The graphical output is contained in the function PlotPendulum, which is not included here to avoid distracting from the simplicity of the analysis code structure. Suffice it to say that the function simply produces the graph in Fig. 2.3. Graphical output is essential but tangential to the discussion of numerical methods.

There are several things to notice about this practical implementation. First, we name the current state (uold, vold, aold) and the next state (unew, vnew, anew). The initial values of uold and vold are determined by the initial conditions, but the starting value for aold must be computed from the equation of motion. The solution is carried out as a loop over time steps. For each step, the state must satisfy the equation of motion (at the new time) as well as the Newmark relationships. The solution of the nonlinear algebraic equation that represents the equation of motion is done as a Newton iteration. Once the Newton iteration has converged, we update the state by putting the new values into the old storage locations.

We store the values we need for graphical output in the array Store. While we store all values in this code segment, it is often not necessary to store all of the computed values. For example, accuracy of the solution might require a very small h, but the graphical output might not require that many points to produce a clear visual result. In such a circumstance, you can set up a strategy for storing the computed values every now and then rather than at every time step. It is also possible to plot values on the fly as you compute them, which reduces the need for storage, but in most computing environments that strategy is not very inefficient.

2.3 Analysis of a First Order Equation

We will study the generalized trapezoidal rule in the context of a simple first order differential equation we can solve exactly. Along the way we will encounter a recursion equation for the numerical method, which we can also solve exactly for the test problem. We will analyze the recursion equation to see what it tells us about the method. This warm-up exercise will help to set the stage for the analysis of the more complex second order problems covered in the following section.

Let us start our investigation with a very simple differential equation—a first order linear ordinary differential equation with constant coefficients (the one-dimensional heat equation):

$$\dot{u}(t) + c\,u(t) = 0 \qquad u(0) = u_o \tag{2.5}$$

where c is a physical constant and u_o is the known initial value of the time-dependent function $u(t)$.

Classical Solution The standard approach to solving this differential equation classically is to assume a solution of the form $u(t) = Ae^{\lambda t}$. Substituting this expression into the differential equation shows that the only possible value of λ is $\lambda = -c$. Using the initial condition, we find that $A = u_o$. Therefore, the classical solution to this differential equation is

$$u(t) = u_o e^{-ct} \tag{2.6}$$

Clearly, for positive c, the nature of this response is exponential decay. The physical constant c determines how fast the response decays from the initial value. For discrete time points at spacing h, the time is $t_n = nh$ and the classical solution can be written as

$$u_n = u(nh) = u_o e^{-cnh} = u_o \left(e^{-ch}\right)^n \tag{2.7}$$

We will use this exact solution as a basis to judge our numerical solution to the problem.

Numerical Solution The discrete version of the first order differential equation can be expressed as

$$v_n + c\,u_n = 0 \tag{2.8}$$

where $u_n = u(t_n)$ and $v_n = \dot{u}(t_n)$ represent the values of the function $u(t)$ and its first derivative at time t_n. The generalized trapezoidal rule approximates the relationship between the function and its first derivative as

$$u_{n+1} = u_n + h\big[\alpha\,v_n + (1-\alpha)\,v_{n+1}\big] \tag{2.9}$$

where h is the size of the time step and α is a user-specified parameter. By substitution of Eq. 2.8, we can eliminate v_n and v_{n+1} from Eq. 2.9 to get

$$u_{n+1} - u_n - h\big[\alpha\,(-cu_n) + (1-\alpha)\,(-cu_{n+1})\big] = 0$$

Rearranging terms, we get

$$\left(1 + (1-\alpha)ch\right)u_{n+1} - \left(1 - \alpha ch\right)u_n = 0 \tag{2.10}$$

What we have managed to do is convert the discrete equations into a single *recursion equation*—i.e., an equation involving the value of the unknown state variable at adjacent time steps. In this process, we have eliminated v_n, but it can always be computed from Eq. 2.8 if needed.

Exact Solution of Recursion Equations It turns out that we can solve linear recursion equations with constant coefficients exactly. Consider the simple equation

$$C_1 u_{n+1} - C_2 u_n = 0 \tag{2.11}$$

where C_1 and C_2 are constants. To solve the recursion equation, let us investigate a solution of the form

$$u_n = A z^n \tag{2.12}$$

where A and z are constants yet to be determined. Plug the candidate solution into the recursion equation and factor out common terms to get

$$C_1 \left(A z^{n+1}\right) - C_2 \left(A z^n\right) = A z^n \left[C_1 z - C_2\right] = 0 \tag{2.13}$$

If $z=0$ or $A=0$, then all $u_n = 0$, and the solution is simply zero for all time. Furthermore, we will not be able to satisfy our initial condition. The only other way Eq. 2.13 can be satisfied is if

$$C_1 z - C_2 = 0 \tag{2.14}$$

We call this the *characteristic equation* of the discrete system. We solve the characteristic equation to find the value of z that yields a solution to the original recursion equation. In this case,

$$z = \frac{C_2}{C_1} \tag{2.15}$$

This is the *only* value of z that is consistent with the recursion equation. Since we have an initial value problem, we know the value u_o as given data. We can use the initial conditions to determine A. In fact, it is pretty straightforward to see that since $u_o = A z^0 = A$. Hence,

$$u_n = u_o \left(\frac{C_2}{C_1}\right)^n \tag{2.16}$$

What is significant about this result is that we have found the *exact* solution to the recursion equation. These values are what the numerical method will compute, even though we would not actually implement the method in this manner (as noted in Code 2.1). It also gives us a means of seeing what the numerical solution looks like *between* the discrete time points we are computing, which can be helpful for understanding the method for larger time steps.

We can recognize that our first order test problem has $C_2 = 1 - \alpha ch$ and $C_1 = 1 + (1-\alpha)ch$. Hence,

$$z = \frac{C_2}{C_1} = \frac{1 - \alpha ch}{1 + (1-\alpha)ch} \tag{2.17}$$

Therefore, the exact numerical solution to this problem is

$$u_n = u_o \left[\frac{1 - \alpha ch}{1 + (1-\alpha)ch} \right]^n \tag{2.18}$$

The exact classical solution is given by Eq. 2.6. So, the numerical solution is obviously different from the exact solution. We can make a few observations. First, the classical solution depends only on the physical parameter c while the numerical solution depends upon the physical parameter c and the numerical analysis parameters h and α. Second, we can see that for time steps where

$$0 < h < \frac{1}{c\alpha} \tag{2.19}$$

the solution always decays for positive values of c. That is good because the classical solution always decays. On the other hand, for

$$h > \frac{1}{c\alpha} \tag{2.20}$$

the numerical solution changes sign every time step. That is bad because the numerical solution does not behave like the classical solution. We refer to such a phenomenon as a *numerical instability*. The issue of numerical stability is different from the issue of accuracy of the approximation.

One way to investigate the *accuracy* of a numerical integrator is to examine the difference between the Taylor series expansions of the exact classical solution and the numerical solution. As a warm-up, note the following series:

$$e^{-x} = 1 - x + \tfrac{1}{2}x^2 - \tfrac{1}{6}x^3 + \tfrac{1}{24}x^4 - \cdots$$

$$\frac{1 - \tfrac{1}{2}x}{1 + \tfrac{1}{2}x} = 1 - x + \tfrac{1}{2}x^2 - \tfrac{1}{4}x^3 + \tfrac{1}{8}x^4 - \cdots$$

By making the association $x = ch$, we can see that the difference between the Taylor series for the exact and approximate expressions, which we call *error*, is (for the specific case $\alpha = 0.5$)

$$error = u_o \left[e^{-ch} - \frac{1 - \frac{1}{2}ch}{1 + \frac{1}{2}ch} \right] = \frac{1}{12} u_o (ch)^3 + O(h^4)$$

Where the notation $O(h^n)$ or "order of h^n" means that the additional terms in the series are all of polynomial order higher than h raised to the nth power, which is important for small values of h as higher and higher powers produce smaller and smaller numbers. Note that h always appears multiplied by c. Therefore, the time step that gives adequate accuracy depends upon the physical parameter of the system. The size of the error also depends upon the initial value u_o.

The trapezoidal rule ($\alpha = 0.5$) matches the first three terms of the Taylor series of the exact classical solution. We call that a *second order method* because the error at each step is proportional to h^3 and over n steps the overall error is proportional to h^2 because $t_n = nh$. The implication of the error analysis is that if we cut the time step in half, the error will reduce by a factor of 4. For other values of α we have

$$error = u_o \left[e^{-ch} - \frac{1 - \alpha ch}{1 + (1-\alpha)ch} \right] = u_o \left(\alpha - \frac{1}{2} \right) (ch)^2 + O\left(h^3 \right)$$

Therefore, the generalized trapezoidal rule is only second order accurate for $\alpha = 0.5$.

The analysis of accuracy also suggests that the main route to improving a numerical approximation is to reduce the time step h. As h gets smaller, the error gets smaller. A numerical integrator is *consistent* with the classical equations if the numerical solution converges to the exact solution in the limit as h approaches zero. In general, we do not know in advance what value of h will give a sufficiently accurate solution. So, one strategy is to pick h arbitrarily, compute the solution, cut h in half, and compute the solution again. If the two solutions are the same, then h is sufficiently small. If not, then repeat the process.

2.4 Analysis of Second Order Differential Equations

The second order differential equations of dynamics behave differently from the first order equations. In this section we look at how the analysis of our three numerical integration methods goes for second order systems, and explore the performance of these methods.

Consider the second order initial value problem (the undamped oscillator in free vibration)

$$\ddot{u}(t) + \omega^2 u(t) = 0, \qquad u(0) = u_o, \qquad \dot{u}(0) = v_o \tag{2.21}$$

as the test problem, where ω is a physical constant and u_o and v_o are the known values of the initial conditions. Using the approach outlined previously, we can find the classical solution (we will go through the derivation in detail in the next chapter)

$$u(t) = u_o \cos \omega t + \frac{v_o}{\omega} \sin \omega t \tag{2.22}$$

This classical solution will guide us in understanding the behavior of our numerical methods. Let us successively consider three different time integration schemes: (1) the central difference method, (2) the generalized trapezoidal rule, and (3) Newmark's method (which is the most commonly used method in structural dynamics). In the discrete equations we will use the associations

$$u_n \leftarrow u(t_n), \qquad v_n \leftarrow \dot{u}(t_n), \qquad a_n \leftarrow \ddot{u}(t_n)$$

to represent the discrete values of displacement, velocity, and acceleration.

2.4.1   The Central Difference Method

The central difference method is based upon approximating the velocity and acceleration with the difference operators

$$v_n = \frac{u_{n+1} - u_{n-1}}{2h}, \qquad a_n = \frac{u_{n+1} - 2u_n + u_{n-1}}{h^2} \tag{2.23}$$

These equations can be used in conjunction with the discrete version of the equation of motion

$$a_n + \omega^2 u_n = 0 \tag{2.24}$$

Substituting Eq. 2.23(b) into 2.24 and multiplying through by h^2, we transform the differential equation into a discrete equation involving only the displacement at three adjacent time points

$$\left(u_{n+1} - 2u_n + u_{n-1}\right) + \omega^2 h^2 u_n = 0$$

Rearranging terms, we can put this difference equation into the form

$$u_{n+1} - \left(2 - \omega^2 h^2\right) u_n + u_{n-1} = 0 \tag{2.25}$$

which is a three term recursion relationship. Contrast that with the first order system that gave rise to a two term recursion. Once again, we assume that the solution of the recursion equation is in the form $u_n = Az^n$. Substituting this expression into the recursion equation and factoring out common terms, we get

$$\left[z^2 - \left(2 - \omega^2 h^2 \right) z + 1 \right] A z^{n-1} = 0$$

If $z = 0$ or $A = 0$, then $u_n = 0$ for all n, which will not satisfy the initial conditions. Therefore, we can identify the *characteristic equation* for the central difference integrator for the test problem as

$$z^2 - \left(2 - \omega^2 h^2 \right) z + 1 = 0 \qquad (2.26)$$

The solutions to this equation are the only values of z that are consistent with the recursion equation. Solving the quadratic equation gives roots

$$z = 1 - \tfrac{1}{2}\omega^2 h^2 \pm i\omega h \sqrt{1 - \tfrac{1}{4}\omega^2 h^2} \qquad (2.27)$$

where $i = \sqrt{-1}$ is the imaginary unit. It is evident that there are two roots to the characteristic equation and for small values of ωh they are complex conjugates of each other. Because the roots of the characteristic equation come out as complex conjugates, it will prove convenient to put them into Euler's form for complex numbers. We can write the value of z in the following equivalent forms

$$z = a \pm ib = r(\cos\varphi \pm i \sin\varphi) \qquad (2.28)$$

where $a = r\cos\varphi$ and $b = r\sin\varphi$ are the real and imaginary parts of the complex number, respectively. The *modulus r* and the *phase angle φ* are

$$r = \sqrt{a^2 + b^2}, \qquad \tan\varphi = \frac{b}{a}$$

In the present case, the constants a and b are

$$a = 1 - \tfrac{1}{2}\omega^2 h^2 \qquad b = \omega h \sqrt{1 - \tfrac{1}{4}\omega^2 h^2} \qquad (2.29)$$

and $r = 1$ (verify this!). The angle φ is defined through the relationship[1]

$$\tan\varphi = \frac{b}{a} = \frac{\omega h \sqrt{1 - \tfrac{1}{4}\omega^2 h^2}}{1 - \tfrac{1}{2}\omega^2 h^2} \qquad (2.30)$$

---

[1] It is worth noting that φ can be computed from any of the three trigonometric relationships $\tan\varphi = b/a$, $\cos\varphi = a/r$, and $\sin\varphi = b/r$. In fact, the cosine version will often prove to be the most reliable because of how the inverse trigonometric functions are typically implemented in computer languages.

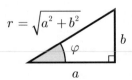

Fig. 2.4 *Complex numbers.* The relationships among the complex number parameters a, b, r, and φ come from this triangle construction. A complex number can be represented as $a + ib$ or $r(\cos \varphi + i \sin \varphi)$. Note that the following relationships are true: $\tan \varphi = b/a$, $\sin \varphi = b/r$, and $\cos \varphi = a/r$

The implication of the importance of the values of r and φ will become evident in the sequel. Figure 2.4 shows the relationship of these quantities.

Because the characteristic equation has two roots, the general solution to the difference equation must be the sum of two terms, $A_1 z_1^n$ and $A_2 z_2^n$, with $z_1 = a + ib$ and $z_2 = a - ib$. Substituting the complex form of the two roots of the characteristic equation, the exact solution to the difference equation is

$$u_n = A_1(a + ib)^n + A_2(a - ib)^n \tag{2.31}$$

This equation can be equivalently expressed in polar form as

$$u_n = r^n \left[A_1(\cos \varphi + i \sin \varphi)^n + A_2(\cos \varphi - i \sin \varphi)^n \right] \tag{2.32}$$

One of the reasons that the polar form is convenient is because of the identity

$$(\cos \varphi \pm i \sin \varphi)^n = \cos n\varphi \pm i \sin n\varphi \tag{2.33}$$

The proof of this identity relies on the two *angle sum* trigonometric identities

$$\cos(x + y) = \cos x \cos y - \sin x \sin y$$

$$\sin(x + y) = \cos x \sin y + \sin x \cos y$$

Letting $x = n\varphi$ and $y = \varphi$, these identities take the form

$$\cos\big((n+1)\varphi\big) = \cos\big(n\varphi + \varphi\big) = \cos n\varphi \cos \varphi - \sin n\varphi \sin \varphi$$

$$\sin\big((n+1)\varphi\big) = \sin\big(n\varphi + \varphi\big) = \cos n\varphi \sin \varphi + \sin n\varphi \cos \varphi$$

Assuming that the identity 2.33 holds for n, let us compute it for $n + 1$. First expand the product and then apply the trigonometric identity, as follows:

$$(\cos \varphi \pm i \sin \varphi)^{n+1} = (\cos n\varphi \pm i \sin n\varphi)(\cos \varphi \pm i \sin \varphi)$$

$$= \cos n\varphi \cos \varphi - \sin n\varphi \sin \varphi$$

$$\pm i \left(\cos n\varphi \sin \varphi + \sin n\varphi \cos \varphi\right)$$

$$= \cos \left((n+1)\varphi\right) \pm i \sin \left((n+1)\varphi\right)$$

Equation 2.33 is obviously true for $n = 1$. Thus, by induction, the identity is proved. Using Eq. 2.33, we can rewrite Eq. 2.32 in the form

$$u_n = r^n \left[A_1(\cos n\varphi + i \sin n\varphi) + A_2(\cos n\varphi - i \sin n\varphi)\right]$$

Collecting terms, we arrive at the final (polar) form

$$u_n = r^n \left(B_1 \cos n\varphi + B_2 \sin n\varphi\right) \tag{2.34}$$

where $B_1 = A_1 + A_2$ and $B_2 = i(A_1 - A_2)$ are constants that we can determine using the initial conditions $u(0) = u_o$ and $\dot{u}(0) = v_o$. We will find that B_1 and B_2 are both real numbers, which implies that A_1 and A_2 are complex numbers whose sum is purely real and whose difference is purely imaginary.

The recursion equation for u_n does not involve the velocity v_n, so we will need to do some work to get the initial velocity into our equations. From the discrete equation for velocity, Eq. 2.23(a), and the discrete equation of motion, Eq. 2.25, we have, respectively, for $n = 0$,

$$u_1 - u_{-1} = 2hv_o$$
$$u_1 + u_{-1} = \left(2 - \omega^2 h^2\right)u_o \tag{2.35}$$

One strange aspect of these equations is that they introduce the displacement at time step $n = -1$ (i.e., u_{-1}). That is clearly a fictitious quantity, but if we let it into our equations, we will be able to work it out. In particular, if we add the two equations together and solve for u_1, we get

$$u_1 = hv_o + \left(1 - \tfrac{1}{2}\omega^2 h^2\right)u_o = hv_o + au_o \tag{2.36}$$

where a is identified from Eq. 2.29. This equation gives the displacement at step $n = 1$ in terms of the initial displacement and velocity.

From Eq. 2.34 with $n = 0$ we find that $B_1 = u_o$. Similarly, using that same equation with $n = 1$ we can write

$$u_1 = r\left(B_1 \cos \varphi + B_2 \sin \varphi\right) = au_o + B_2 b \tag{2.37}$$

where we have noted that $a = r\cos\varphi$, $b = r\sin\varphi$, and we have substituted the previous result that $B_1 = u_o$. The values of u_1 from Eqs. 2.36 and 2.37 must be equal. Thus, $au_o + hv_o = au_o + B_2 b$. Therefore,

$$B_2 = \frac{hv_o}{b} = \frac{hv_o}{\omega h \sqrt{1 - \frac{1}{4}\omega^2 h^2}} \tag{2.38}$$

recalling the expression for b from Eq. 2.29. Substitute these values back into the polar form of the discrete solution (i.e., Eq. 2.34) to get the final expression for the exact solution to the recursion equation

$$u_n = u_o \cos n\varphi + \frac{v_o}{\omega \sqrt{1 - \frac{1}{4}\omega^2 h^2}} \sin n\varphi \tag{2.39}$$

There are some striking similarities to the exact solution of the differential equation, which is

$$u(t_n) = u_o \cos \omega t_n + \frac{v_o}{\omega} \sin \omega t_n$$

First, observe that both are oscillatory. Noting that $t_n = hn$ we can write the numerical solution as

$$u_n = u_o \cos \tilde{\omega} t_n + \frac{v_o}{\omega \sqrt{1 - \frac{1}{4}\omega^2 h^2}} \sin \tilde{\omega} t_n \tag{2.40}$$

where the *apparent frequency* associated with the numerical method is

$$\tilde{\omega} = \frac{\varphi}{h} \tag{2.41}$$

and φ is defined in Eq. 2.30. We can observe that the solution is well behaved (i.e., oscillatory) as long as

$$1 - \frac{1}{4}\omega^2 h^2 > 0 \quad \text{or} \quad h < \frac{2}{\omega} \tag{2.42}$$

This is the stability limit of the central difference method and it defines the largest time step that gives a response consistent with the classical solution.

Figure 2.5 gives results for the central difference method for two specific cases. The black dots are the results of numerical integration, the heavy black line is the exact classical solution, and the thin dotted black line is a plot of the function

$$u_h(t) = u_o \cos \tilde{\omega} t + \frac{v_o}{\omega \sqrt{1 - \frac{1}{4}\omega^2 h^2}} \sin \tilde{\omega} t$$

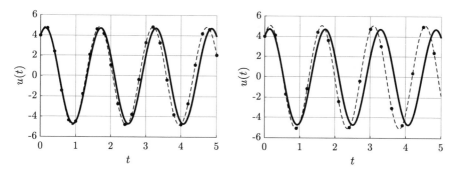

Fig. 2.5 *Performance of the central difference method.* The response of the system for two different values of α. The parameters are $\omega = 4$, $u_o = 4$, and $v_o = 10$. On the left the time step is $h = 0.2$. On the right the time step is $h = 0.3$

which is a continuous version of the exact analysis of the numerical method. The black dots fall exactly on the dotted curve, as they should. The value of having the continuous version of the discrete solution is that you can see where the solution would fall between the discrete points. That can be helpful for interpreting results with large time steps or for time steps that lead to unstable solutions.

The case on the left in Fig. 2.5 has $h = 0.2$. It is evident that the computed solution is fairly close to the exact classical solution. The numerical solution has a shorter period than the exact. It is also evident that the amplitude of the solution is greater than the exact solution but remains constant over time. This result is consistent with the observation that $r = 1$ but the amplitude is a function of ωh (see Eq. 2.40). The offending term is the one associated with the initial velocity. One can verify this observation by setting $v_o = 0$ and computing the results. The case on the right has $h = 0.3$. The amplitude magnification is greater than it was for $h = 0.2$ and the apparent period is even shorter. This example shows the effect of the size of the time step.

In Fig. 2.6 we examine the response around the stability limit of $h_{crit} = 0.5$. The case on the left has $h = 0.499$ and the case on the right has $h = 0.501$, the former just below and the latter just above the critical time step. While the time step below h_{crit} gives a very inaccurate result, it is stable (i.e., the amplitude does not grow with time). The numerical solution in this case is the victim of the amplitude amplification (the v_o term in Eq. 2.40). When the time step is above h_{crit} the amplitude of the solution grows with time.

The performance of the integrator just under the stability limit shows the value of the continuous version of the numerical solution. If only the discrete points for the numerical solution had been shown in Fig. 2.6 on the left, one might be tempted to conclude that the solution was growing, and hence unstable. In fact, the amplitude of the solution is constant and only a quirk of sampling suggests growth. The response on the right actually *is* unstable, as evidenced by the dotted curve.

When the time step is greater than the critical time step, the roots of the characteristic equation (Eq. 2.26) are both real, so the above analysis that assumes

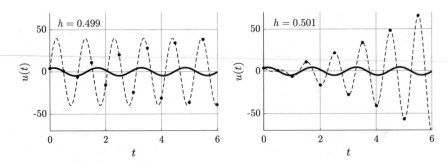

Fig. 2.6 *Performance of the central difference method.* The response of the system for two different time steps. The parameters are $\omega = 4$, $u_o = 4$, and $v_o = 10$. On the left the time step is $h = 0.499$. On the right the time step is $h = 0.501$. The critical time step is $h_{crit} = 0.500$

the roots are complex must be revised to reflect it. In this case, we have

$$z = 1 - \tfrac{1}{2}\omega^2 h^2 \pm \omega h \sqrt{\tfrac{1}{4}\omega^2 h^2 - 1} \tag{2.43}$$

One can prove that both roots are negative (in addition to being real), and at least one of them has absolute magnitude greater than one. Hence, the solution grows exponentially with time, which is the definition of instability.

It should be clear that for the parameters given, the time step required for *accuracy* is much smaller than the time step needed for *stability*. In general, if you generate enough points to get a good plot of an oscillating function, the time step should be well below the critical time step. However, when we examine systems with many degrees of freedom, we will find that some of the frequencies are very large and will therefore limit the time step, even if those frequencies do not contribute important response for the given loading.

2.4.2 The Generalized Trapezoidal Rule

The second numerical integrator that we will analyze is the *generalized trapezoidal rule*. The generalized trapezoidal rule for the second order system has the form

$$v_{n+1} = v_n + h\left[\alpha\, a_n + (1-\alpha)\, a_{n+1}\right] \tag{2.44}$$

$$u_{n+1} = u_n + h\left[\alpha\, v_n + (1-\alpha)\, v_{n+1}\right] \tag{2.45}$$

Substitute the first into the second to get

$$u_{n+1} = u_n + h v_n + h^2\alpha(1-\alpha)a_n + h^2(1-\alpha)^2 a_{n+1} \tag{2.46}$$

Next, substitute $a_n = -\omega^2 u_n$ from the equation of motion at the two time steps to get

$$D_1 u_{n+1} - D_2 u_n - h v_n = 0 \tag{2.47}$$

where $D_1 = 1 + (1-\alpha)^2 \omega^2 h^2$ and $D_2 = 1 - \alpha(1-\alpha)\omega^2 h^2$. At two adjacent time steps we can write

$$D_1 u_{n+1} - D_2 u_n - h v_n = 0$$
$$D_1 u_n - D_2 u_{n-1} - h v_{n-1} = 0$$

The difference of these equations gives

$$D_1 u_{n+1} - (D_1 + D_2) u_n + D_2 u_{n-1} - h (v_n - v_{n-1}) = 0$$

Finally, substituting Eqs. 2.44 and Eq. 2.24, we can eliminate the velocity terms to get, after some simplification,

$$D_1 u_{n+1} - 2 D_2 u_n + D_3 u_{n-1} = 0 \tag{2.48}$$

where

$$D_1 = 1 + (1-\alpha)^2 \omega^2 h^2$$
$$D_2 = 1 - \alpha(1-\alpha)\omega^2 h^2 \tag{2.49}$$
$$D_3 = 1 + \alpha^2 \omega^2 h^2$$

Assuming a solution in the form $u_n = A z^n$, we get the characteristic equation that determines z as

$$D_1 z^2 - 2 D_2 z + D_3 = 0 \tag{2.50}$$

with solution

$$z = \frac{D_2 \pm \sqrt{D_2^2 - D_1 D_3}}{D_1} \tag{2.51}$$

Noting that $D_2^2 - D_1 D_3 = -\omega^2 h^2$, we can put this result in standard complex number form $z = a \pm ib$ if we define a and b as

$$a = \frac{1 - \omega^2 h^2 \alpha(1-\alpha)}{1 + \omega^2 h^2 (1-\alpha)^2}, \qquad b = \frac{\omega h}{1 + \omega^2 h^2 (1-\alpha)^2} \tag{2.52}$$

Since $b > 0$, there is no possibility of a time step that does not yield an oscillatory solution. Thus, this scheme is *unconditionally stable*. In this case, the amplitude, or spectral radius, is

$$r = \sqrt{a^2 + b^2} = \sqrt{\frac{D_3}{D_1}} = \sqrt{\frac{1 + \alpha^2 \omega^2 h^2}{1 + (1-\alpha)^2 \omega^2 h^2}} \tag{2.53}$$

For $\alpha = 0.5$, $r = 1$ and the discrete solution has a constant amplitude. For $\alpha > 0.5$ the response grows exponentially and for $\alpha < 0.5$ the response decays exponentially. We can identify the angle φ for the polar representation of the roots of the characteristic equation to be

$$\tan \varphi = \frac{\omega h}{1 - \alpha(1-\alpha)\omega^2 h^2} \tag{2.54}$$

As it was in the case of the central difference method, the general solution to the recursion equation for the generalized trapezoidal rule is

$$u_n = r^n \left(B_1 \cos n\varphi + B_2 \sin n\varphi \right) \tag{2.55}$$

where, as usual, $a = r \cos \varphi$ and $b = r \sin \varphi$. We can determine B_1 and B_2 by using the initial values of displacement u_o and velocity v_o. For the values $n = 0$ and $n = 1$, Eq. 2.55 gives

$$u_o = B_1, \qquad u_1 = B_1 a + B_2 b \tag{2.56}$$

In order to use the second equation, we need to evaluate u_1. From the equation of motion at $n = 0$ and $n = 1$, we know the accelerations to be $a_o = -\omega^2 u_o$ and $a_1 = -\omega^2 u_1$. Using Eq. 2.46 with $n = 0$ gives

$$u_1 = u_o + h v_o - \omega^2 h^2 \alpha (1-\alpha) u_o - \omega^2 h^2 (1-\alpha)^2 u_1$$

Solving this equation for u_1 gives

$$u_1 = \frac{\left(1 - \alpha(1-\alpha)\omega^2 h^2\right) u_o + h v_o}{1 + (1-\alpha)^2 \omega^2 h^2} = a u_o + \frac{v_o}{\omega} b$$

Equating this expression for u_1 with that obtained in Eq. 2.56, we get

$$B_2 = \frac{v_o}{\omega}$$

Substituting back into the spectral form of the solution to the recursion equation, we get

$$u_n = r^n \left[u_o \cos n\varphi + \frac{v_o}{\omega} \sin n\varphi \right] \tag{2.57}$$

Noting that the time $t_n = nh$ we can write this equation as

$$u_n = r^{(t_n/h)} \left[u_o \cos \tilde{\omega} t_n + \frac{v_o}{\omega} \sin \tilde{\omega} t_n \right]$$

which gives rise to the continuous version of the exact solution to the discrete equations

$$u_h(t) = r^{(t/h)} \left[u_o \cos \tilde{\omega} t + \frac{v_o}{\omega} \sin \tilde{\omega} t \right] \tag{2.58}$$

the spectral radius r can be computed from Eq. 2.53. As before, we have defined the *apparent frequency* as

$$\tilde{\omega} = \frac{\varphi}{h}$$

where φ can be computed from Eq. 2.54.

Figure 2.7 shows results for the generalized trapezoidal rule using the same system we explored for the central difference method. As before, the black dots are the results of numerical integration, the heavy black line is the exact classical solution, and the thin dotted black line is the continuous version of the analysis of the numerical method (Eq. 2.58). The black dots fall exactly on the thin black dotted curve, as expected.

The plot on the left has time step $h = 0.2$ and the plot on the right has $h = 0.3$ (which are the same values used for the central difference method) with $\alpha = 0.5$

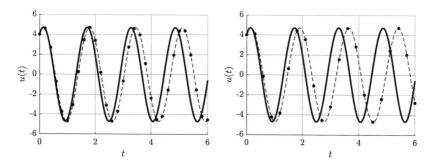

Fig. 2.7 *Influence of time step size on the generalized trapezoidal rule.* The response of the system for two different time steps. The parameters are $\omega = 4$, $u_o = 4$, and $v_o = 10$. On the left the time step is $h = 0.2$. On the right it is $h = 0.3$. For both cases the integration parameter is $\alpha = 0.5$

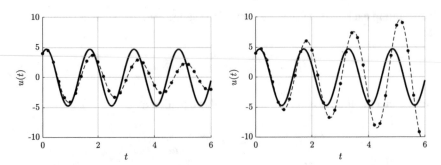

Fig. 2.8 *Influence of α on the generalized trapezoidal rule.* The response of the system for two different values of α. The parameters are $\omega = 4$, $u_o = 4$, and $v_o = 10$. On the left the time step is $\alpha = 0.45$. On the right it is $\alpha = 0.55$. For both cases the time step is $h = 0.2$

for both. The amplitude of the numerical solution is exactly the same as the exact classical solution. The reason this happens is not only because $r = 1$ for $\alpha = 0.5$, but also because the amplitude of the term that multiplies r^n in Eq. 2.57 does not depend upon the time step and has coefficients equal to those of the classical solution. This feature is quite different from the situation in the central difference method, which had amplitude magnification due to the v_o term.

The effect of varying the value of the integration parameter α is shown in Fig. 2.8. The plot on the left has $\alpha = 0.45$. The plot on the right has $\alpha = 0.55$. The time step is $h = 0.2$ for both cases. Values of $\alpha \neq 0.5$ lead to either exponential growth of the numerical solution $(\alpha > 0.5)$, which we will consider unstable, or decay $(\alpha < 0.5)$. The phenomenon of decay is often called *numerical damping* because of its resemblance to physical damping, which also causes exponential decay of the solution (as we will see in the next chapter). The system in our test problem has no physical damping.

2.4.3   Newmark's Method

Newmark proposed the following integrator for second order systems:

$$v_{n+1} = v_n + h\left[\gamma\, a_n + (1-\gamma)\, a_{n+1}\right]$$
$$u_{n+1} = u_n + hv_n + h^2\left[\beta\, a_n + \left(\tfrac{1}{2}-\beta\right)a_{n+1}\right] \qquad (2.59)$$

where h is the size of the time step and the constants γ and β are the numerical integration parameters (the Newmark parameters). The nature of the integrator is determined by the values selected for these parameters (like α was for the generalized trapezoidal rule in our previous discussion).

We can follow the same strategy used previously for the analysis of Newmark's method. First, write the Newmark displacement relationship at time step $n+1$ and n and subtract them. Next, substitute the Newmark velocity relationship at time step $n+1$ to eliminate the velocities. Finally, substitute $a_n = -\omega^2 u_n$ (the equation of motion for the test problem) to eliminate the accelerations. The final result is the characteristic equation

$$D_1 z^2 - 2D_2 z + D_3 = 0$$

where, for the Newmark integrator,

$$D_1 = 1 + \omega^2 h^2 \left(\tfrac{1}{2} - \beta \right)$$

$$D_2 = 1 - \omega^2 h^2 \left(\tfrac{1}{4} + \beta - \tfrac{1}{2}\gamma \right) \qquad (2.60)$$

$$D_3 = 1 + \omega^2 h^2 \left(\gamma - \beta \right)$$

Using the quadratic formula, the roots $z_1 = a - ib$ and $z_2 = a + ib$ can be found, where

$$a = \frac{D_2}{D_1} \qquad b = \frac{\sqrt{D_1 D_3 - D_2^2}}{D_1}$$

The spectral radius can be computed from $r^2 = a^2 + b^2 = D_3/D_1$ to be

$$r = \sqrt{\frac{1 + (\gamma - \beta)\omega^2 h^2}{1 + (\tfrac{1}{2} - \beta)\omega^2 h^2}} \qquad (2.61)$$

A simple calculation shows that $r < 1$ if $\gamma < 0.5$. The stability does not depend upon the value of β. It is also clear from Eq. 2.61 that for $\gamma = 0.5$ the spectral radius is $r = 1$.

The apparent frequency $\tilde{\omega} = \varphi/h$ of Newmark's method comes from the phase angle

$$\tan \varphi = \frac{b}{a} = \frac{\omega h \sqrt{1 - \omega^2 h^2 \left(\tfrac{1}{4} \left(\gamma - \tfrac{1}{2} \right)^2 + \beta - \tfrac{1}{2}\gamma \right)}}{1 - \omega^2 h^2 \left(\tfrac{1}{4} + \beta - \tfrac{1}{2}\gamma \right)} \qquad (2.62)$$

In the next section, we will examine graphically how the apparent frequency depends upon γ and β.

Following a process similar to that done for the generalized trapezoidal rule, we can compute the exact numerical solution for the Newmark integrator

$$u_n = r^n \left[u_o \cos n\varphi + \frac{V_o}{\omega} \sin n\varphi \right]$$

where

$$V_o = \frac{v_o h + \frac{1}{2}\left(\gamma - \frac{1}{2}\right)\omega^2 h^2 u_o}{h\sqrt{1 + \frac{1}{2}\omega^2 h^2 \left[\gamma - 2\beta - \frac{1}{2}\left(\gamma - \frac{1}{2}\right)^2\right]}}$$

Observe that in the limit as $h \to 0$, $V_o \to v_o$. The continuous exact solution for the Newmark integrator is

$$u_h(t) = r^{(t/h)} \left[u_o \cos \tilde{\omega} t + \frac{V_o}{\omega} \sin \tilde{\omega} t \right]$$

This integrator, like the central difference integrator, exhibits an amplitude magnification due to the V_o term, even if $r = 1$. If we take $\gamma = 0.5$, then the expressions for V_o and $\tan \varphi$ reduce to

$$V_o = \frac{v_o}{\sqrt{1 + \omega^2 h^2 \left(\frac{1}{4} - \beta\right)}}, \qquad \tan \varphi = \frac{\omega h \sqrt{1 + \omega^2 h^2 \left(\frac{1}{4} - \beta\right)}}{1 - \beta \omega^2 h^2}$$

These results tell us that if $\beta < 0.25$ the amplitude of the numerical solution will be less than the classical solution and the apparent frequency will be greater than the actual natural frequency. On the other hand, if $\beta > 0.25$ the reverse would be true.

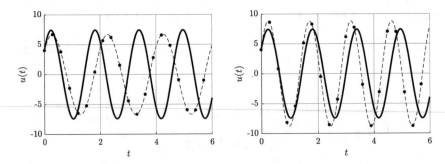

Fig. 2.9 *Amplitude magnification in Newmark's method.* The response of the system for two different values of β with $\gamma = 0.5$. The parameters are $\omega = 4$, $u_o = 4$, and $v_o = 25$. On the left the time step is $\beta = 0$. On the right is $\beta = 0.5$. For both cases the time step is $h = 0.3$

Figure 2.9 confirms these observations. The plot on the left has $\beta = 0$ and the one on the right has $\beta = 0.5$. The amplitude is constant for the numerical solution in both cases because $r = 1$, but in the first case the amplitude is diminished due to V_o and in the second case it is augmented. The effect is primarily activated by the initial velocity, but there is a small effect caused by the initial displacement, too. For $\beta = 0.25$ and $\gamma = 0.5$ we get $V_o = v_o$, but this case still exhibits period elongation.

Newmark vs. Generalized Trapezoidal Rule It should be evident that Newmark's method might be identical to the generalized trapezoidal rule for certain values of α, β, and γ. The two methods will be the same if the coefficients of the methods match exactly. That happens if $\alpha = \gamma$, $\alpha(1 - \alpha) = \beta$, and $(1 - \alpha)^2 = 0.5 - \beta$. Solving these equations gives the unique values

$$\alpha = \tfrac{1}{2}, \quad \beta = \tfrac{1}{4}, \quad \gamma = \tfrac{1}{2}$$

The two methods are identical when these parameters are used.

2.5 Performance of the Methods

We have observed that the three numerical methods (central difference, generalized trapezoidal rule, and Newmark's method) give different approximate solutions to the differential equation $\ddot{u}(t) + \omega^2 u(t) = 0$. The main numerical analysis parameter shared by all methods is the time step h. Clearly, the smaller the time step, the closer the numerical approximation is to the exact solution. For generalized trapezoidal rule and Newmark's method, the additional parameters (α, β, and γ) also influence the accuracy and stability of the approximation.

The main features that these numerical methods exhibit come from the spectral radius r, which determines the stability and numerical damping of the method, and the apparent frequency $\tilde{\omega}$, which determines the period elongation or shortening. Each of these features is influenced by the value of ωh (in other words, it is not the time step itself, but its size relative to the natural frequency of the system).

Figure 2.10 shows the variation of the spectral radius for the three methods. The spectral radius of the central difference method is constant and equal to one, as we proved earlier, so the method does not have amplitude growth or decay for any value of time step. The generalized trapezoidal rule exhibits exponential growth (numerical instability) for any $\alpha > 0.5$. The spectral radius is equal to one for $\alpha = 0.5$. The method shows exponential decay (numerical damping) for values $\alpha < 0.5$. The smaller the value below 0.5, the more rapid is the rate of decay of the method. Figure 2.10c and d shows the spectral radius of Newmark's method for $\beta = 0$ and $\beta = 0.25$, respectively. Newmark's method exhibits exponential growth (numerical instability) for values of $\gamma > 0.5$ and exponential decay (numerical damping) for values of $\gamma < 0.5$. Newmark has a spectral radius of one for $\gamma = 0.5$, independent of β.

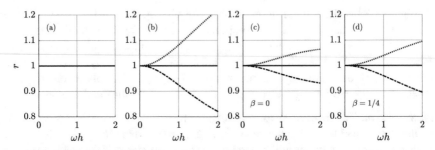

Fig. 2.10 *Spectral radius of the three numerical methods.* These plots show the spectral radius r for (**a**) central difference, (**b**) generalized trapezoidal rule for $\alpha = 0.4, 0.5, 0.6$ corresponding to dot-dash, solid, and dot, (**c**) Newmark $\beta = 0$, and (**d**) Newmark $\beta = 0.25$. For both (**c**) and (**d**), $\gamma = 0.4, 0.5, 0.6$ correspond to dot-dash, solid, and dot

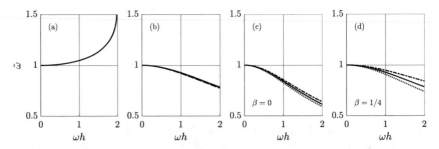

Fig. 2.11 *Apparent frequency of the three numerical methods.* These plots show the apparent frequency $\tilde{\omega}$ for (**a**) central difference, (**b**) generalized trapezoidal rule for $\alpha = 0.4, 0.5, 0.6$ corresponding to dot-dash, solid, and dot, (**c**) Newmark $\beta = 0$, and (**d**) Newmark $\beta = 0.25$. For both (**c**) and (**d**), $\gamma = 0.4, 0.5, 0.6$ correspond to dot-dash, solid, and dot

Figure 2.11 shows the variation of apparent frequency $\tilde{\omega}$ for the three methods. Observe what happens as the time step approaches the stability limit $\omega h = 2$ for central difference. This method always has period shortening (if the frequency increases, the period $T = 2\pi/\omega$ decreases). Near the stability limit there is an extreme departure from the actual natural frequency. The generalized trapezoidal rule exhibits period elongation, but the value of α does not influence it much. Newmark's method also exhibits period elongation. Apparently, β has more influence on period elongation than γ does.

Generalized trapezoidal rule and Newmark are identical when $\alpha = 0.5$, $\beta = 0.25$, and $\gamma = 0.5$. It appears that Newmark's method is more tolerant in the spectral radius but less tolerant with respect to period elongation than generalized trapezoidal rule. However, the period elongation can be adjusted with β in Newmark, and it really cannot be changed in the generalized trapezoidal rule.

2.6 Summary

We leave this chapter with one last important observation. Time-stepping methods are *not* implemented in structural dynamics codes by solving the recursion equation and computing $u_n = Az^n$. In fact, that approach is only feasible for the simple linear differential equations like the test problem. The main reason that we resort to numerical methods is because we cannot solve the equations by classical means for most problems (and any recursion equation we can solve exactly comes from a differential equation that we can easily solve classically).

The implementation of a numerical integrator is much simpler than the exact analysis of that same numerical integrator. In the practical implementation, we march forward in time using the previous state (or two states in the case of central difference) to compute the next state. Once the new state is found, we can update and repeat the calculation. For the single step methods, we need only store the current state to compute the new state. The framework for the computation will remain the same for all problems in structural dynamics.

There are many other numerical methods available to integrate the equations of structural dynamics. Research into these methods has been fruitful for decades. The three methods we have investigated are *implicit methods*, which means that we try to satisfy the equation of motion at time step $n + 1$ given the values at time step n (and before if needed). We will see that for systems with multiple degrees of freedom, implicit methods will require the solution of a system of equations at each time step.

As solving systems of equations is a computationally intensive operation (especially for large systems), researchers have also been interested in *explicit methods*. There are many explicit methods (including the Runge–Kutta family), each with different stability and accuracy characteristics. The main attraction of explicit methods is the avoidance of the solution of a system of equations. The main drawback is that they generally have much poorer stability characteristics compared with implicit methods.

Chapter 3
Single-Degree-of-Freedom Systems

Structural dynamics is the study of systems that are designed to oscillate under the influence of applied loads. The essential features of the response of an oscillating system can be observed in simplest form in a single-degree-of-freedom (SDOF) oscillator. Hence, we start our exploration of structural dynamics here, both to refresh our understanding of how to solve the differential equations that govern these types of systems and to see some of the basic behavioral features of oscillating systems. Among those features, we will find that a linear SDOF system possesses a natural frequency, which is determined by the physical properties of the system, and the system will resonate if excited by a sinusoidal force with a driving frequency equal to the natural frequency. We will examine the effects of viscous damping on the response of the system and consider various loading conditions, including earthquake excitation. Finally, we will extend the formulation to include nonlinear response of the restoring force elements and develop a framework for the numerical analysis of nonlinear systems that will be useful throughout the book.

This chapter will set the stage for the later chapters that deal with more complex systems. In the simple context of the SDOF system, we will run into many of the phenomena that come up in structural dynamics. In this setting we can gain an appreciation for the tools needed to analyze structural systems, as well as the behavior we expect to see when we solve them.

Electronic Supplementary Material   The online version of this chapter (https://doi.org/10.1007/978-3-030-89944-8_3) contains supplementary material, which is available to authorized users.

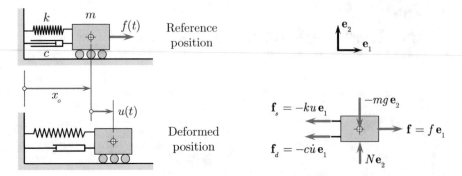

Fig. 3.1 *The SDOF system.* The SDOF system has a mass m, damping constant c, and spring stiffness k. The position of the spring is characterized by its displacement $u(t)$ relative to the unstretched position of the spring. The free body diagram is shown at right

3.1   The SDOF Oscillator

To understand vibrations we will consider the SDOF system shown in Fig. 3.1. The body has mass m, the viscous damper has modulus c, the spring has stiffness k, and the system is subjected to a time-dependent forcing function $f(t)$. The reference position is taken to be the place where the spring is unstretched. We will characterize the motion of the mass by the displacement $u(t)$, which measures the horizontal displacement of the center of the mass.[1] The position $\mathbf{x}(t)$ of the mass, relative to the inertial reference frame, is

$$\mathbf{x}(t) = x_o\mathbf{e}_1 + u(t)\mathbf{e}_1 \tag{3.1}$$

where x_o is the fixed distance from the origin to the center of the mass in the reference position. The selection of the reference position is arbitrary, but we will take it to be the place where the spring is unstretched. The motion is constrained by the rollers to be in the $\mathbf{e}_1$ direction. The velocity is $\dot{\mathbf{x}}(t) = \dot{u}(t)\mathbf{e}_1$ and the acceleration is $\ddot{\mathbf{x}}(t) = \ddot{u}(t)\mathbf{e}_1$, where each dot over a function denotes a derivative with respect to time t.

Newton's second law ($\mathbf{F} = m\mathbf{a}$) applied to this system gives the equation of motion. With reference to the free body diagram in Fig. 3.1, we get

$$-ku(t)\mathbf{e}_1 - c\dot{u}(t)\mathbf{e}_1 + f(t)\mathbf{e}_1 - mg\mathbf{e}_2 + N\mathbf{e}_2 = m\ddot{u}(t)\mathbf{e}_1 \tag{3.2}$$

Taking the dot product of the equation of motion with $\mathbf{e}_1$, we get the equation that governs the evolution of the function $u(t)$

---

[1] Actually, any point on the body can serve as the reference point because we assume that the body is rigid and does not rotate.

$$m\ddot{u}(t) + c\dot{u}(t) + ku(t) = f(t) \tag{3.3}$$

It is important to note that the function $N(t)$ is also unknown. Dotting the equation with $\mathbf{e}_2$ gives $N = mg$, which proves that $N(t)$ is actually constant in this case. We think of N as a *reaction force* associated with the constraint that the motion must be in the $\mathbf{e}_1$ direction. In general, we will have both motion and force unknowns in each dynamics problem.

To find a unique solution to the problem, we must also satisfy *initial conditions*, i.e., the values of displacement and velocity at $t = 0$. To wit,

$$u(0) = u_o, \qquad \dot{u}(0) = v_o \tag{3.4}$$

where u_o and v_o are the specified values of initial displacement and velocity, respectively. The solution to the problem must satisfy both the governing differential equation and the initial conditions.

3.2 Undamped Free Vibration

An *undamped* system has $c = 0$ and a system in *free vibration* has $f(t) = 0$. Thus, the equation governing the motion of the undamped SDOF system is

$$m\ddot{u}(t) + ku(t) = 0 \tag{3.5}$$

which is a linear second order ordinary differential equation with constant coefficients. Any differential equation in this class has a solution in the form of exponential functions. Hence, we can try a solution of the form $u(t) = Ae^{\lambda t}$, where A and λ are (as yet undetermined) constants.

To find out something about those constants and to understand why the solution should be in this form, note that the derivatives of $u(t)$ are simply

$$\dot{u}(t) = A\lambda e^{\lambda t}, \qquad \ddot{u}(t) = A\lambda^2 e^{\lambda t}$$

If we substitute the trial solution back into the equation of motion, we find that

$$\left(m\lambda^2 + k\right)Ae^{\lambda t} = 0$$

To satisfy this equation we have two possibilities. First, note that $e^{\lambda t}$ can never be zero. If $A = 0$, then $u(t) = 0$, which we usually call the *trivial solution* because it implies no motion, and is valid only if the initial displacement and velocity are zero. The only other possibility is

$$m\lambda^2 + k = 0 \tag{3.6}$$

This equation is called the *characteristic equation* and it gives us a means to determine λ. For notational simplicity let us define

$$\omega = \sqrt{\frac{k}{m}} \tag{3.7}$$

and note that we get two solutions to the characteristic equation:

$$\lambda_1 = i\omega \quad \lambda_2 = -i\omega \tag{3.8}$$

where $i = \sqrt{-1}$ is the imaginary unit. Because there are two possible values of the exponent, the general solution is the linear combination of both solutions. Hence, we can write the general solution in the form

$$u(t) = A_1 e^{\lambda_1 t} + A_2 e^{\lambda_2 t} = A_1 e^{i\omega t} + A_2 e^{-i\omega t} \tag{3.9}$$

Complex exponentials are oscillatory. We can put the general solution into a different form using Euler's identity $e^{i\varphi} = \cos\varphi + i\sin\varphi$. First, note that

$$e^{i\omega t} = \cos\omega t + i\sin\omega t, \qquad e^{-i\omega t} = \cos\omega t - i\sin\omega t$$

Now the solution to the differential equation can be recast as follows:

$$\begin{aligned} u(t) &= A_1 e^{i\omega t} + A_2 e^{-i\omega t} \\ &= A_1(\cos\omega t + i\sin\omega t) + A_2(\cos\omega t - i\sin\omega t) \\ &= (A_1 + A_2)\cos\omega t + i\,(A_1 - A_2)\sin\omega t \end{aligned}$$

Defining $B_1 = A_1 + A_2$ and $B_2 = i(A_1 - A_2)$, we can write

$$u(t) = B_1 \cos\omega t + B_2 \sin\omega t \tag{3.10}$$

Equation 3.10 is called the *general solution* to Eq. 3.5. This solution satisfies the equation of motion for any values of the constants B_1 and B_2.

Initial Conditions Let us see what the initial conditions imply. Differentiating the general solution, Eq. 3.10, the velocity is

$$\dot{u}(t) = -\omega B_1 \sin\omega t + \omega B_2 \cos\omega t$$

At $t = 0$ the displacement and velocity evaluate to

$$u(0) = B_1 \cos(0) + B_2 \sin(0) = B_1 = u_o$$
$$\dot{u}(0) = -\omega B_1 \sin(0) + \omega B_2 \cos(0) = \omega B_2 = v_o$$

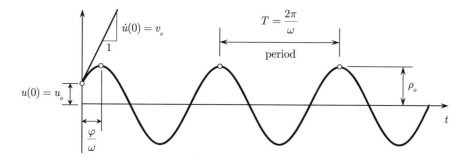

Fig. 3.2 *The response of the undamped SDOF system.* The response of the SDOF system is a sinusoidal oscillation with frequency ω

Hence, $B_1 = u_o$ and $B_2 = v_o/\omega$. Substituting these values back into Eq. 3.10 gives the unique solution

$$u(t) = u_o \cos \omega t + \frac{v_o}{\omega} \sin \omega t \tag{3.11}$$

This displacement function satisfies both the governing differential equations and the initial conditions.[2] The displacement as a function of time is shown in Fig. 3.2. The response starts at time zero with the initial displacement and has a slope equal to the initial velocity. The response from that point on is a pure sinusoidal function.

We can observe a few key things from this solution. First, the system oscillates at a frequency ω with no driving force at all. We call this the *natural frequency* of the system. It should be evident that the natural frequency came from the characteristic equation and it represents the *only* possibility for motion in free vibration. Another way to characterize the motion is with the *natural period of vibration*, defined as

$$T = \frac{2\pi}{\omega} \tag{3.12}$$

The period is the time between any two corresponding points on the curve, measuring the time it takes to return to the same point of response. Figure 3.2 shows T as the time between peaks of response.

It is also evident that the initial conditions determine the amplitude of oscillation and that the amplitude remains constant over time. To get more insight into the amplitude of vibration, let $u_o = \rho_o \cos \varphi$ and $v_o = \omega \rho_o \sin \varphi$. This is essentially a change of variables from rectangular to polar coordinates

---

[2] In keeping with common practice we note that $\cos \omega t = \cos(\omega t)$. When there is risk of ambiguity, we will include the parentheses. Otherwise, assume that the operation of sine and cosine is on the quantity ωt and not just ω.

where we replace (u_o, v_o) with (ρ_o, φ) as the motion parameters. If we substitute these expressions into Eq. 3.11, and make use of the trigonometric identity $\cos(x - y) = \cos x \cos y + \sin x \sin y$, we get

$$u(t) = \rho_o \cos(\omega t - \varphi) \tag{3.13}$$

where

$$\rho_o = \sqrt{u_o^2 + \left(\frac{v_o}{\omega}\right)^2} \qquad \varphi = \tan^{-1}\left(\frac{v_o}{\omega u_o}\right) \tag{3.14}$$

It should be evident from the form of this equation that ρ_o is the amplitude of the oscillation and that the angle φ is a *phase shift*. The phase shift is simply a measure of how far forward from $t = 0$ do you have to go to get to the first peak of the cosine function, i.e., the time where $\omega t - \varphi = 0$, as illustrated in Fig. 3.2.

3.3 Damped Free Vibration

Most real systems show a decay of motion with time in free vibration. The simplest model that does this is the spring/mass/damper system shown in Fig. 3.1. The damper accrues force in proportion to the velocity, with constant of proportionality c. The damped equation of motion (with no force applied) is

$$m\ddot{u}(t) + c\dot{u}(t) + ku(t) = 0 \tag{3.15}$$

We solve this equation in exactly the same way we did the undamped system. Assume a solution in the form $u(t) = Ae^{\lambda t}$. Differentiate $u(t)$ with respect to t to get the velocity $\dot{u}(t) = A\lambda e^{\lambda t}$ and acceleration $\ddot{u}(t) = A\lambda^2 e^{\lambda t}$. Substitute these into Eq. 3.15 to get

$$\left(m\lambda^2 + c\lambda + k\right)Ae^{\lambda t} = 0$$

Again, either $A = 0$ (implying no motion) or the exponent λ must satisfy the characteristic equation

$$m\lambda^2 + c\lambda + k = 0 \tag{3.16}$$

which is simply a quadratic equation, whose solution is

$$\lambda = \frac{-c \pm \sqrt{c^2 - 4mk}}{2m} \tag{3.17}$$

We can put this solution into a much more convenient form by making the following definitions:

$$\omega = \sqrt{\frac{k}{m}}, \qquad \xi = \frac{c}{2\sqrt{mk}}, \qquad \bar{\omega} = \omega\sqrt{1 - \xi^2} \qquad (3.18)$$

Notice that ω is defined in the same way as the undamped system. We call ξ the *damping ratio* and $\bar{\omega}$ the *damped natural frequency* for reasons that will soon become evident. A system with damping ratio less than one is called *underdamped*, the most common case for structural dynamics. With these definitions we can write the two solutions to the characteristic equation for the underdamped case ($\xi < 1$) as

$$\lambda_1 = -\xi\omega + i\bar{\omega}, \qquad \lambda_2 = -\xi\omega - i\bar{\omega}$$

With two roots to the characteristic equation, the general solution is

$$u(t) = A_1 e^{\lambda_1 t} + A_2 e^{\lambda_2 t} \qquad (3.19)$$

Unlike the undamped case, the complex roots have a real part and an imaginary part. To put the solution into its simplest form, we will use the law of exponents and apply Euler's relationship to convert the imaginary exponential to sinusoids. To wit,

$$e^{\lambda_1 t} = e^{(-\xi\omega + i\bar{\omega})t} = e^{-\xi\omega t} e^{i\bar{\omega}t} = e^{-\xi\omega t} (\cos\bar{\omega}t + i\sin\bar{\omega}t)$$

and

$$e^{\lambda_2 t} = e^{(-\xi\omega - i\bar{\omega})t} = e^{-\xi\omega t} e^{-i\bar{\omega}t} = e^{-\xi\omega t} (\cos\bar{\omega}t - i\sin\bar{\omega}t)$$

As before, let $B_1 = A_1 + A_2$ and $B_2 = i(A_1 - A_2)$. Now, the final form of the general solution to the damped equation of motion is

$$u(t) = e^{-\xi\omega t} (B_1 \cos\bar{\omega}t + B_2 \sin\bar{\omega}t) \qquad (3.20)$$

The damped natural frequency $\bar{\omega}$ appears in the arguments of the sine and cosine functions. Hence, the frequency of oscillation is $\bar{\omega}$ and not ω, as it was in the undamped case. Thus, damping affects the frequency of vibration. However, for small values of the damping ratio the damped natural frequency is nearly equal to the undamped natural frequency (e.g., for $\xi = 0.2$ the difference is roughly 2%).

We can use the initial conditions to find the unique solution to the problem. Again, using $u(0) = u_o$ and $\dot{u}(0) = v_o$ will allow us to determine the values of the constants B_1 and B_2. First, the initial displacement condition gives

$$u(0) = B_1 = u_o$$

thereby establishing the value of B_1. We can compute the velocity by differentiating the displacement with respect to time

$$\dot{u}(t) = -\xi\omega e^{-\xi\omega t}\,(B_1\cos\bar{\omega}t + B_2\sin\bar{\omega}t)$$
$$+ e^{-\xi\omega t}\,(-\bar{\omega}B_1\sin\bar{\omega}t + \bar{\omega}B_2\cos\bar{\omega}t)$$

Substituting the initial velocity we get

$$\dot{u}(0) = -\xi\omega B_1 + \bar{\omega}B_2 = v_o$$

Solving for B_2, using the value we already determined for B_1, gives

$$B_2 = \frac{v_o + \xi\omega u_o}{\bar{\omega}}$$

Substituting back into the original expression for the displacement gives

$$u(t) = e^{-\xi\omega t}\left(u_o\cos\bar{\omega}t + \frac{v_o + \xi\omega u_o}{\bar{\omega}}\sin\bar{\omega}t\right) \qquad (3.21)$$

This is the exact solution for damped free vibration of the SDOF oscillator. We can put this solution into polar form, as we did for the undamped case. To wit,

$$u(t) = \rho_o\,e^{-\xi\omega t}\cos(\bar{\omega}t - \varphi) \qquad (3.22)$$

where the amplitude and phase angle of the motion are defined as

$$\rho_o = \sqrt{u_o^2 + \left(\frac{v_o + \xi\omega u_o}{\bar{\omega}}\right)^2}, \qquad \varphi = \tan^{-1}\left(\frac{v_o + \xi\omega u_o}{\bar{\omega}u_o}\right) \qquad (3.23)$$

The similarities with the undamped case, Eq. 3.14, should be noted. In fact, Eq. 3.23 reduces to Eq. 3.14 when the damping ratio is zero.

Damping affects both the magnitude and phase angle of the oscillator. One way to interpret the response is to note that it is a sinusoidal oscillation that decays exponentially with time due to the $e^{-\xi\omega t}$ multiplier. The larger the damping ratio, the faster the decay. In most structural dynamics problems, the damping ratio will be fairly small (usually less than 0.2). A typical response of a damped system is shown in Fig. 3.3, which includes a plot of the bounding exponential decay function (positive and negative). Observe that the sinusoidal oscillation is contained within those bounds.

Overdamped Systems If $\xi \geq 1$, both roots of the characteristic equation (Eq. 3.17) are real, with values

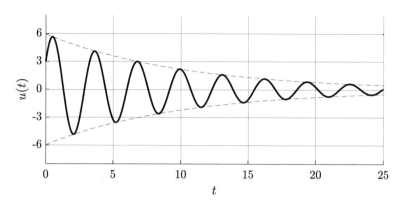

Fig. 3.3 *Damped motion of the SDOF system.* The response of the damped SDOF system is an exponentially decaying sinusoidal oscillation. This case has $\omega = 2$, $\xi = 0.05$, $u_o = 3$, and $v_o = 10$. The envelope is the function $\pm \rho_o \, e^{-\xi \omega t}$

$$\lambda_1 = -\omega\left(\xi + \sqrt{\xi^2 - 1}\right), \qquad \lambda_2 = -\omega\left(\xi - \sqrt{\xi^2 - 1}\right)$$

Both roots are negative because $\xi > \sqrt{\xi^2 - 1}$, which means that the solution will decay exponentially without oscillation. Implementing the boundary conditions gives

$$u(t) = \tfrac{1}{2}\left(u_o - \beta/\alpha\right)e^{-\omega(\xi+\alpha)t} + \tfrac{1}{2}\left(u_o + \beta/\alpha\right)e^{-\omega(\xi-\alpha)t}$$

where

$$\alpha = \sqrt{\xi^2 - 1}, \qquad \beta = \frac{v_o}{\omega} + \xi u_o$$

The case with $\alpha = 0$, which corresponds to $\xi = 1$, is called *critical damping*. The roots of the characteristic equation are equal in this case with

$$\lambda_1 = \lambda_2 = -\omega$$

Therefore, the solution takes the form

$$u(t) = \left(A_1 + A_2 t\right)e^{-\omega t}$$

Implementing the initial conditions gives

$$u(t) = \left[u_o + \left(\omega u_o + v_o\right)t\right]e^{-\omega t}$$

Typical responses for overdamped systems with various damping ratios are shown in Fig. 3.4. Observe that there is no oscillation for any of these damping values.

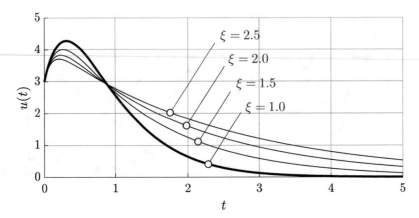

Fig. 3.4 *Overdamped motion of the SDOF system.* The response of the damped SDOF system is an exponentially decaying sinusoidal oscillation. This case has $\omega = 2$, $u_o = 3$, and $v_o = 10$ with various values of the damping ratio

The critically damped system (plotted with the heavier line) has the largest initial response, but damps out faster than the systems with higher damping. For very large values of the damping ratio, the solution tends to the constant value $u(t) = u_o$.

3.4 Forced Vibration

By accident of history, most students of structural dynamics have been steeped in statics where the loads are present with no indication of how they came to be there. In reality, loads come upon a system in a dynamic fashion and the nature of how that happens can influence the response. By studying the classical solution of the damped SDOF oscillator under a few different temporal variations, we can see many of the important dynamic phenomena in a very simple context, most of which manifest in more complex structures.

Recall the governing equation of the damped, forced system (Eq. 3.3):

$$m\ddot{u}(t) + c\dot{u}(t) + ku(t) = f(t)$$

In what follows, we will consider suddenly applied constant loading, sinusoidal loading, general periodic (but non-sinusoidal) loading, blast loading, and earthquake excitation. Along the way we will encounter important dynamic phenomena, like *resonance*, and we will get a dynamic perspective on static equilibrium.

Solution of Non-homogeneous Differential Equations The solution of a linear differential equation can be found as the sum of the *homogeneous solution* $u_h(t)$ and a *particular solution* $u_p(t)$:

$$u(t) = u_h(t) + u_p(t) \tag{3.24}$$

The homogeneous solution satisfies

$$m\ddot{u}_h(t) + c\dot{u}_h(t) + ku_h(t) = 0$$

The particular solution is any function $u_p(t)$ that satisfies

$$m\ddot{u}_p(t) + c\dot{u}_p(t) + ku_p(t) = f(t) \tag{3.25}$$

Initial conditions must be satisfied for $u(t)$, but not necessarily either of the other two individually.

The homogeneous solution is simply the solution to the equation when $f(t) = 0$. Of course, that is the solution that we have already found for the free vibration problem. Hence,

$$u_h(t) = e^{-\xi\omega t}\left(B_1 \cos \bar{\omega}t + B_2 \sin \bar{\omega}t\right) \tag{3.26}$$

The task that we face in solving the driven problem is to find a particular solution and to then satisfy the initial conditions for the general solution. Thus, we must have

$$u(0) = u_h(0) + u_p(0) = u_o$$
$$\dot{u}(0) = \dot{u}_h(0) + \dot{u}_p(0) = v_o$$

Since we already know the particular solution, we can use it to express the initial condition equations

$$u(0) = B_1 + u_p(0) = u_o$$
$$\dot{u}(0) = -\xi\omega B_1 + \bar{\omega}B_2 + \dot{u}_p(0) = v_o \tag{3.27}$$

These two equations are sufficient to determine the constants B_1 and B_2. It is evident that the nature of the particular solution influences the outcome of this calculation.

3.4.1 Suddenly Applied Constant Load

The first case to consider is the suddenly applied load that remains constant after it is applied. That is what happens when you step on the bathroom scales in the morning. The specific loading function for this case is

$$f(t) = p_o \tag{3.28}$$

where p_o is the (constant) magnitude of the load. The particular solution that we can try is $u_p(t) = C$ (i.e., a constant). Note that $\dot{u}_p(t) = 0$ and $\ddot{u}_p(t) = 0$. Substituting these back into Eq. 3.25 gives $kC = p_o$, which we can solve to give $u_p(t) = p_o/k$. Substituting this result back into Eqs. 3.27, we find the constants B_1 and B_2 to be

$$B_1 = u_o - \frac{p_o}{k}, \qquad B_2 = \frac{v_o}{\bar{\omega}} + \frac{\xi\omega}{\bar{\omega}}\left(u_o - \frac{p_o}{k}\right)$$

Therefore, the complete solution to the problem of the suddenly applied constant load is

$$u(t) = e^{-\xi\omega t}\left\{\left(u_o - \frac{p_o}{k}\right)\cos\bar{\omega}t + \left[\frac{v_o}{\bar{\omega}} + \frac{\xi\omega}{\bar{\omega}}\left(u_o - \frac{p_o}{k}\right)\right]\sin\bar{\omega}t\right\} + \frac{p_o}{k}$$

We can gain insight from this equation. For example, if we have initial conditions $u_o = p_o/k$ and $v_o = 0$, then the solution is $u(t) = p_o/k$, the *static* solution. What that implies is that if you displace the system in exactly the amount of the static deflection and then apply the load, there will be no oscillation. That is really what we assume in the statement of a typical statics problem.

Only the term in curly brackets multiplies $e^{-\xi\omega t}$. That term decays exponentially with time and the oscillations die out leaving only the static solution. This implies that the system oscillates about the static position and eventually settles into that position if the system is damped. Finally, note that if there is no damping and the system starts at rest ($u_o = 0$ and $v_o = 0$), then the displacement is

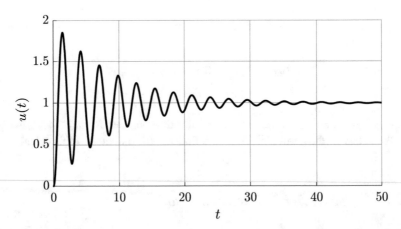

Fig. 3.5 *Damped SDOF system with suddenly applied constant load.* The damped SDOF system has mass $m = 1$, $\xi = 0.05$, $k = 5$, $p_o = 5$, $u_o = 0$, and $v_o = 0$. Note that the response damps out to the static response p_o/k

$$u(t) = \frac{p_o}{k}\left(1 - \cos \omega t\right)$$

Thus, the system oscillates between $u = 0$ and $u = 2p_o/k$. In other words, the effect of the suddenly applied load is that the maximum response is double the static response.

These phenomena are illustrated in Fig. 3.5, which shows the response for a particular set of input parameters given in the figure caption. Observe that the oscillations are bounded by the envelope of exponential decay defined by the function $e^{-\xi \omega t}$. Even for modest values of the damping ratio, the oscillations decay fairly quickly. Note, also, that damping takes a toll before the first positive peak so the maximum positive displacement does not reach two, as it would if undamped.

3.4.2 Sinusoidal Load

Next, consider the case where a steady sinusoidal load is applied, i.e,

$$f(t) = p_o \sin \Omega t \tag{3.29}$$

where p_o is the magnitude of the load and Ω is the driving frequency. For this loading case, we can assume a particular solution in the form

$$u_p(t) = A \sin \Omega t + B \cos \Omega t$$

The derivatives of this function are

$$\dot{u}_p(t) = \Omega\left(A \cos \Omega t - B \sin \Omega t\right), \qquad \ddot{u}_p(t) = -\Omega^2\left(A \sin \Omega t + B \cos \Omega t\right)$$

We seek values of A and B that will satisfy the equation of motion for all values of time t. Substituting the particular solution into the governing equation (Eq. 3.25), and rearranging terms, we get

$$\left(kA - m\Omega^2 A - c\,\Omega B - p_o\right)\sin \Omega t + \left(c\,\Omega A + kB - m\Omega^2 B\right)\cos \Omega t = 0$$

This equation can be satisfied only if the coefficients of $\sin \Omega t$ and $\cos \Omega t$ are zero independently. To wit, we must have

$$\begin{bmatrix} k - m\Omega^2 & -c\,\Omega \\ c\,\Omega & k - m\Omega^2 \end{bmatrix} \begin{Bmatrix} A \\ B \end{Bmatrix} = \begin{Bmatrix} p_o \\ 0 \end{Bmatrix}$$

Or, if we divide both equations through by k,

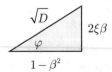

Fig. 3.6 *Amplitude and phase form of the particular solution.* The particular solution to the response to the sinusoidal loading can be put into a different form using the triangle construction shown here

$$\begin{bmatrix} 1-\beta^2 & -2\xi\beta \\ 2\xi\beta & 1-\beta^2 \end{bmatrix} \begin{Bmatrix} A \\ B \end{Bmatrix} = \frac{p_o}{k} \begin{Bmatrix} 1 \\ 0 \end{Bmatrix} \tag{3.30}$$

where $\beta = \Omega/\omega$ is the ratio of driving frequency Ω to the undamped natural frequency ω. We can solve this system of equations to find the values of the constants A and B to be

$$A = \frac{p_o}{k} \frac{1-\beta^2}{\left(1-\beta^2\right)^2 + (2\xi\beta)^2}, \qquad B = \frac{p_o}{k} \frac{-2\xi\beta}{\left(1-\beta^2\right)^2 + (2\xi\beta)^2}$$

Thus, the particular solution is

$$u_p(t) = \frac{p_o}{k} \frac{\left(1-\beta^2\right)\sin\Omega t - (2\xi\beta)\cos\Omega t}{\left(1-\beta^2\right)^2 + (2\xi\beta)^2} \tag{3.31}$$

The particular solution for the sinusoidal loading can be put into a slightly different form using the triangle construction shown in Fig. 3.6. Let

$$D = \left(1-\beta^2\right)^2 + \left(2\xi\beta\right)^2 \tag{3.32}$$

and define the angle φ through the relationships

$$\cos\varphi = \frac{1-\beta^2}{\sqrt{D}} \qquad \sin\varphi = \frac{2\xi\beta}{\sqrt{D}} \qquad \tan\varphi = \frac{2\xi\beta}{1-\beta^2}$$

Using these definitions, the particular solution can be rewritten as

$$u_p(t) = \frac{p_o}{k\sqrt{D}} \left(\sin\Omega t \cos\varphi - \cos\Omega t \sin\varphi \right)$$

Using the trigonometric identity $\sin(x-y) = \sin x \cos y - \cos x \sin y$, we can finally write the particular solution in the form

$$u_p(t) = \frac{p_o}{k\sqrt{D}} \sin(\Omega t - \varphi) \tag{3.33}$$

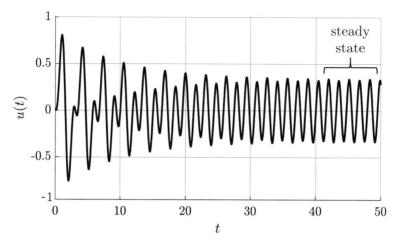

Fig. 3.7 *General response to sinusoidal forcing function.* The response to the sinusoidal driving force with initial conditions shows that the *transient* response damps out fairly quickly, leaving the *steady state* response. The system properties are $m = 1$, $k = 4$, $\xi = 0.05$, $\Omega = 4$, and $p_o = 4$

We can use the initial conditions to solve for the constants B_1 and B_2 for this case from Eq. 3.27. From Eq. 3.31 or Eq. 3.33, it is evident that

$$u_p(0) = -\frac{2\xi\beta}{D}\frac{p_o}{k} \qquad \dot{u}_p(0) = \frac{(1-\beta^2)\Omega}{D}\frac{p_o}{k}$$

Now, substitute these expressions into Eq. 3.27 to get

$$B_1 = u_o + \frac{2\xi\beta}{D}\frac{p_o}{k}, \qquad B_2 = \frac{v_o + \xi\omega B_1}{\bar{\omega}} - \frac{(1-\beta^2)\Omega}{\bar{\omega}D}\frac{p_o}{k}$$

Finally, we can plug these expressions back into the general equation to get the displacement function

$$u(t) = e^{-\xi\omega t}\left(B_1\cos\bar{\omega}t + B_2\sin\bar{\omega}t\right) + \frac{p_o}{k\sqrt{D}}\sin(\Omega t - \varphi) \tag{3.34}$$

It is evident that the exact expression for the displacement is pretty complicated. A typical result is shown in Fig. 3.7. One of the most important things to observe about the response is that the transient portion of the solution (i.e., the first term in Eq. 3.34) damps out to a steady state response (i.e., the second term in the equation).

The steady state response is what is left after the negative exponential term reduces to a very small value. What is left is the function

$$u_{ss}(t) = \frac{p_o}{k\sqrt{D}}\sin(\Omega t - \varphi)$$

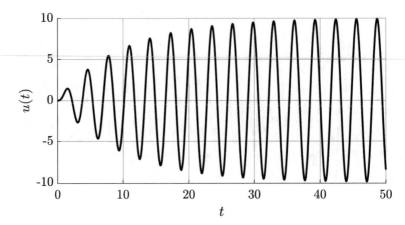

Fig. 3.8 *Resonance.* The response to the sinusoidal driving force at (or near) the natural frequency produces resonant response. The system properties are $m = 1$, $k = 4$, $\xi = 0.05$, $\Omega = 2$, and $p_o = 4$

which also happens to be the contribution of the particular solution. The steady state response is worth examining in a bit more detail because it holds the secret to *resonance*.

Dynamic Amplification For a system subjected to a sinusoidal driving function, we have the possibility that the driving frequency Ω can be at or near the natural frequency ω of the system. When the driving frequency is very near to the natural frequency, we get a phenomenon called *resonance* wherein the motion builds up to a large value even with a very small load magnitude. In the case shown in Fig. 3.8 the motion grows to an amplitude of about ten, even though the static displacement is one.

It is evident from the mathematical form of the solution what causes this phenomenon. The steady state solution includes a multiplier, the *amplification factor* $A(\xi, \beta)$, that is a function of the damping ratio ξ and the frequency ratio β. Let,

$$A(\xi, \beta) = \frac{1}{\sqrt{D}} = \frac{1}{\sqrt{\left(1 - \beta^2\right)^2 + (2\xi\beta)^2}} \tag{3.35}$$

The amplification function is shown in Fig. 3.9. The factor $A(\xi, \beta)$ defines the amplitude of the steady state motion. What happens in resonance is that the system response synchronizes with the loading function so that the power (force times velocity) is always positive. With only positive work, energy is continuously pumped into the system, causing the motions to grow. If the system starts at rest, the response grows through the transient phase and eventually settles into steady state with a pure sinusoidal motion with amplitude of the static response p_o/k times $A(\xi, \beta)$.

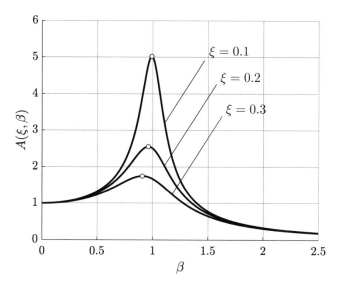

Fig. 3.9 *Response amplification under sinusoidal load.* Plot of $A(\xi, \beta)$ vs. β for damping ratios of $\xi = 0.1, 0.2, 0.3$ (top to bottom). The amplification factor multiplies the static response p_o/k to give the maximum response under the sinusoidal load. The frequency ratio is $\beta = \Omega/\omega$

Observe that the maximum magnification factor happens close to $\beta = 1$ (but not exactly). The location of the maximum can be determined as

$$\frac{dA}{d\beta} = 0 \quad \rightarrow \quad \beta_{max} = \sqrt{1-2\xi^2} \tag{3.36}$$

The value of the maximum is

$$A\left(\xi, \beta_{max}\right) = \frac{1}{2\xi\sqrt{1-\xi^2}} \tag{3.37}$$

which, for small values of ξ, can be very nearly approximated by $A(\xi, 1)$ as

$$A_{max} \approx \frac{1}{2\xi} \tag{3.38}$$

This is a very important result for damped resonant motion. The maximum amplitude is limited by the damping and is roughly equal to one over twice the damping ratio. If the damping ratio is $\xi = 0.05$, the amplification of the motion due to resonance would be limited to a factor of 10, as demonstrated in Fig. 3.8.

The resonant response of an undamped system differs from the damped system in important ways. The function $A(0, \beta)$ tends to infinity at $\beta = 1$, suggesting that there is no upper bound on the response. It is possible to prove that

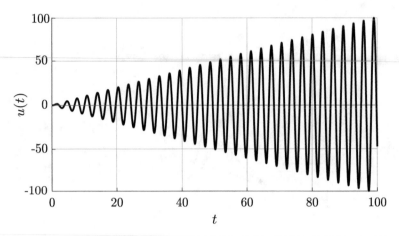

Fig. 3.10 *Resonant response without damping.* When there is no damping and $\Omega = \omega$, the response increases linearly without bound. The system properties are $m = 1$, $k = 4$, $\xi = 0$, $\Omega = 2$, and $p_o = 4$

the growth of the motion is linear with time, as shown in Fig. 3.10. This system has the same properties as the one in Fig. 3.8 except that the damping is zero.

Cosine Loading It is not difficult to verify that the particular solution for forced vibration for a cosine loading $f(t) = p_o \cos \Omega t$ is

$$u_p(t) = \frac{p_o}{k\sqrt{D}} \cos(\Omega t - \varphi)$$

This is also the steady state solution for the cosine loading. The main reason for including this loading case is to prepare for the general periodic loading, which we will solve using Fourier series.

3.4.3 General Periodic Loading

If we have a periodic forcing function that is not a pure sinusoidal function (e.g., the one shown in Fig. 3.11), we can build the solution using a Fourier series. Fourier's great idea was that any periodic function with period T could be represented as an infinite series of the form

$$p(t) = a_0 + \sum_{n=1}^{\infty} a_n \cos(\Omega_n t) + \sum_{n=1}^{\infty} b_n \sin(\Omega_n t) \tag{3.39}$$

Fig. 3.11 *Non-sinusoidal periodic load.* A periodic load repeats with period T. This example is a square wave

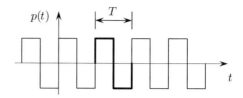

where the reference frequency is given as $\Omega_n = 2\pi n/T$. The coefficients $a_0, a_1, a_2, a_3, \ldots, b_1, b_2, b_3, \ldots$ then describe the unique properties of a particular function. The coefficients can be computed using the orthogonality properties of the sine and cosine functions. We will use the principle of superposition to get a solution to the steady-state motion of a damped oscillator under arbitrary periodic forcing.

The computation of the coefficients relies on the orthogonality of the sine and cosine function over the range of integration. The derivation can be found in almost any textbook that covers Fourier series. The coefficient a_0 can be found as

$$a_0 = \frac{1}{T} \int_0^T p(t)\, dt$$

It should be evident that the first coefficient is simply the average of the function over the period. The remaining coefficients can be computed as

$$a_n = \frac{2}{T} \int_0^T p(t) \cos(\Omega_n t)\, dt, \qquad b_n = \frac{2}{T} \int_0^T p(t) \sin(\Omega_n t)\, dt$$

The steady state response of the system to this general periodic loading is the superposition of the responses to each term in the loading. Therefore, the response is

$$u(t) = \frac{a_0}{k} + \sum_{n=1}^{\infty} \frac{1}{k\sqrt{D_n}} \left[a_n \cos\left(\Omega_n t - \varphi_n\right) + b_n \sin\left(\Omega_n t - \varphi_n\right) \right] \qquad (3.40)$$

where

$$D_n = (1 - n^2\beta^2)^2 + (2\xi n\beta)^2 \qquad \varphi_n = \tan^{-1}\left(\frac{2\xi n\beta}{1 - n^2\beta^2}\right)$$

and $\beta = 2\pi/T\omega$.

Example To see how the Fourier series approach works, consider the example of a square wave. The square wave loading function is defined for one cycle as

$$p(t) = \begin{cases} p_o & 0 < t < \frac{1}{2}T \\ -p_o & \frac{1}{2}T < t < T \end{cases}$$

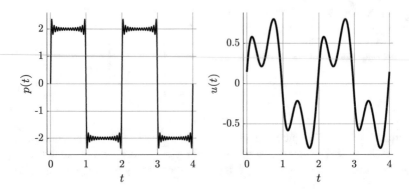

Fig. 3.12 *Fourier approximation of square wave.* The square wave loading function $p(t)$ is approximated with 15 terms at left. The steady state displacement is shown at right. The properties of the system are $T = 2$, $\omega = 8$, $\xi = 0.1$, $k = 5$, and $p_o = 2$

The coefficients can be evaluated as $a_0 = 0$, $a_n = 0$ (for all n), $b_n = 0$ (for $n = 2, 4, 6, \ldots$), and

$$b_n = \frac{4p_o}{n\pi} \qquad n = 1, 3, 5, \ldots$$

Thus, the approximate load function is

$$p(t) = \frac{4p_o}{\pi} \sum_{n=1,3,5} \frac{\sin(\Omega_n t)}{n}$$

where $\Omega_n = 2\pi n/T$. Notice that there is an n in the denominator of the terms in the series. Because the magnitude of the sine function is one, as n increases the influence of the additional terms gets smaller. That is one reason it is possible to terminate the series after a finite number of terms. Because of the abrupt discontinuity, the convergence of the series is slow. For continuous functions it is common to have at least n^2 in the denominator (giving much faster convergence).

Figure 3.12 shows the results for the loading function and the steady-state response. Observe how the sinusoids approximate the square wave. Notice, in particular, the tendency to overshoot due to the discontinuity in the function—the so-called *Gibbs phenomenon*. Taking more terms in the series improves the approximation (i.e., reducing the influence of the oscillations seen in the function $p(t)$ in Fig. 3.12), but the magnitude of the overshoot remains. The displacement is not sensitive to this phenomenon.

The displacement is derived from the loading by integration, which is inherently a smoothing process. As n gets large, $\sqrt{D_n} \to n^2\beta^2$. Hence, the convergence for $u(t)$ is much faster than the convergence of the loading function itself. The value of $\beta = 2\pi/T\omega = 0.393$ is well away from resonance.

3.5   Earthquake Ground Motion

In an earthquake, the reference frame for the structural elements moves relative to the inertial reference frame. Consider the situation illustrated in Fig. 3.13. In the deformed position we can locate the center of the mass at

$$\mathbf{x}(t) = u_g(t)\mathbf{e}_1 + x_o\mathbf{e}_1 + u(t)\mathbf{e}_1 \tag{3.41}$$

where x_o measures the distance from the origin to the center of mass in a position where the spring is unstretched (the reference position), $u(t)$ is the displacement relative to the moving frame, and $u_g(t)$ is the distance from the origin of the inertial reference frame to the moving structural support reference frame. Again, the motion is constrained by the rollers to be in the $\mathbf{e}_1$ direction.

The inertial force associated with the mass accrues in accord with the *absolute* motion while the internal resisting forces develop in accord with the *relative* motion. The acceleration is the second time derivative of the position

$$\ddot{\mathbf{x}}(t) = \ddot{u}_g(t)\mathbf{e}_1 + \ddot{u}(t)\mathbf{e}_1$$

Applying Newton's second law ($\mathbf{F} = m\mathbf{a}$) we get

$$-ku(t)\mathbf{e}_1 - c\dot{u}(t)\mathbf{e}_1 - mg\mathbf{e}_2 + N\mathbf{e}_2 = m\big(\ddot{u}_g(t)\mathbf{e}_1 + \ddot{u}(t)\mathbf{e}_1\big)$$

Taking the dot product of this equation with $\mathbf{e}_1$ gives the equation of motion:

$$m\ddot{u}(t) + c\dot{u}(t) + ku(t) = -m\ddot{u}_g(t) \tag{3.42}$$

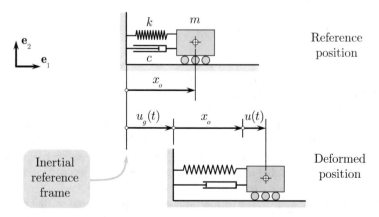

Fig. 3.13 *SDOF system with moving frame.* Earthquake ground motion can be modeled by considering the frame on which the SDOF system sits to be moving. The springs and dampers respond to the relative motion, but the inertial force must be reckoned relative to the inertial frame

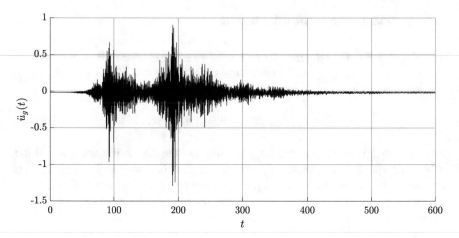

Fig. 3.14 *Earthquake ground motion.* Acceleration record: E–W component of the March 11, 2011 Japan earthquake. Accelerations are given in fractions of the acceleration of gravity *g*. Time is in seconds

From this equation we can observe that an effective loading results from the earthquake ground motion as mass times the ground acceleration. We have not considered any other applied load source $f(t)$ here, but it would be simple to add it. In general, we will have other loads during an earthquake, but we will ignore those for the time being to focus on the forces induced by the earthquake motion.

Ground Motions The ground acceleration can be measured with an accelerometer. An example of an earthquake ground motion is shown in Fig. 3.14. We have these records for many historical earthquakes. Generally, three components of motion are recorded at a point in space. The figure shows only one of those components as a function of time. Each earthquake is a little different, depending on the nature of the faulting and the local geology of the region through which the earthquake waves travel.

What actually happens in an earthquake is that a fault ruptures below the ground surface, releasing energy that has been stored through elastic deformation of the subsurface material, which has deformed as the two sides of the fault have attempted to move in different directions under tectonic drift. When the fault ruptures, waves propagate in all directions. If there is an accelerometer on the surface at a certain location, then that instrument measures the acceleration of the ground at that point caused by the passing waves. One accelerogram is not a complete picture of the earthquake, but it does give a good indication of what a structure located at that geographic location would experience.

There are some key features of the ground motion acceleration that affect the structural response. One parameter that was long thought to be the most important was the peak acceleration. It is important, but we now know that it is not the only important characteristic. Another very important feature is the duration—i.e., how long the ground shakes. The longer it goes, the more likely the structure will be damaged by the motion.

The frequency content of the earthquake is also important. In fact, in the 1985 Mexico City earthquake, the ground motions at the surface were fairly regular with a dominant period of about 2 s. The reason this occurred was because Mexico City sits on an ancient lakebed. As the waves propagated from the source (the epicenter was near the west coast of Mexico), they shook the base of the lakebed. The clay in the lakebed then responded to that excitation with waves that propagated to the surface. By the time the waves reached the surface, they had organized into motions that were nearly sinusoidal. Buildings with natural periods in the neighborhood of 2 s were severely damaged, and many collapsed. For most earthquakes, the frequency content of the motion is not readily observed from the accelerogram but might be discerned from some sort of spectral analysis (e.g., a Fourier transform).

Artificial Earthquakes   Earthquake-resistant design of structures is often based upon measured accelerograms. In fact, in the early days of earthquake engineering, the 1940 El Centro (California) earthquake acceleration record played a key role in forming ideas about the earthquake hazard because earthquake records were rare. As knowledge about earthquakes progressed, seismologists and engineers realized that the nature of the earthquake felt at a particular geographic location depends upon factors specific to that location.

As more information on recorded earthquakes was collected, the possibility of developing synthetic earthquake ground motions emerged. We will not go very deep in that field but will (for the sake of simplicity) conceptualize artificial motions as randomly phased sinusoids shaped by a time function. Thus, the form of the synthetic ground acceleration function is

$$\ddot{u}_g(t) = s(t) \sum_{i=1}^{N} A_i \sin(\Omega_i t + \varphi_i) \tag{3.43}$$

where A_i, Ω_i, and φ_i are the amplitude, frequency, and phase angles of the ith of N components of the motion. An example motion is shown in Fig. 3.15. The time shape function $s(t)$ is included to allow for a buildup and tapering off of the motion. Observed earthquakes generally build up rather abruptly (but not instantaneously) and they tend to die out more gradually. The time shape function used to generate the artificial earthquake shown is

$$s(t) = s_o t^a e^{-bt} \tag{3.44}$$

where s_o, a, and b are the time-function parameters. The earthquake characteristics can be altered by changing those parameter values (along with the amplitudes and phase angles inside the sum). Of course, other time-shaping functions are possible.

The velocity and displacement associated with an earthquake accelerogram can be computed by numerical integration. It is important to realize that the specification

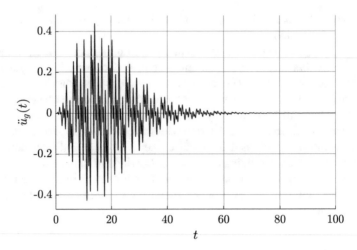

Fig. 3.15 *Artificial earthquake ground motion.* The duration is $t_D = 100$ and the time-shaping parameters are $s_o = 0.0125$, $a = 2$, and $b = 0.15$. There are four frequencies $[\,1, 2, 5, 10\,]$ with relative amplitudes $[\,2, 3, 4, 4\,]/\sqrt{45}$ and no phase shifts

of the acceleration function does not guarantee that the ground velocity will be zero at the end of the earthquake (which physically must be true). One simple way to adjust for this issue is to compute the terminal velocity v_f by integrating the acceleration

$$v_f = \int_0^{t_D} \ddot{u}_g(t)\, dt$$

Then adjust the velocity by making the initial velocity equal to the negative of the final velocity, i.e., $\dot{u}_g(0) = -v_f$, and adjusting the ground displacements by subtracting tv_f from the ground displacement obtained by direct integration of the acceleration. This adjustment will assure zero velocity of the ground at the end of the earthquake. It is, of course, possible to have a residual displacement. Although the response of the system to the earthquake does not depend on either the ground velocity or displacement, the latter value is needed to produce an animation of the motion of the structure. Code 3.1 computes an artificial earthquake ground motion. We will use this code throughout the book as a simple way to generate an earthquake.

Code 3.1 MATLAB code to compute a synthetic earthquake ground motion using phased sinusoids

```
function [EqA,EqV,EqD] = EQGroundMotion(dt,nSteps,EQOn,DF)
%   Compute artifical earthquake ground motion
%
%       dt : Analysis time increment
%   nSteps : Number of time steps for analysis
%     EQOn : =0 no EQ, =1 yes EQ
%       DF : Output control struct
```

```
%. Earthquake parameters
   tD = 100;                          % EQ Duration
   so = 0.0125; a = 2; b = 0.15;      % Time shape parameters
   Frq = [1, 2, 5, 10];              % Component Frequencies
   Phs = [0, 0, 0, 0];               % Component Phase shifts
   Amp = [2, 3, 4, 4];               % Component Amplitudes

%. Artificial earthquake ground motion parameters
   Amp = Amp/norm(Amp);              % Normalize amplitudes
   nEQSteps = ceil(tD/dt)+1;         % Compute number of EQ steps

%. Initialize variables
   EqA = zeros(nEQSteps+1,1);
   EqV = zeros(nEQSteps+1,1);
   EqD = zeros(nEQSteps+1,1);

%. Compute EQ acceleration, velocity, and displacement
   for i=1:nEQSteps
      t = tD*i/nEQSteps;
      Shape = so*t^a*exp(-b*t);
      if EQOn==0; Shape=0; end
      EqA(i+1) = Shape*dot(Amp,sin(Frq*t + Phs));
      EqV(i+1) = EqV(i) + dt*(EqA(i) + EqA(i+1))/2;
      EqD(i+1) = EqD(i) + dt*(EqV(i) + EqV(i+1))/2;
   end

%. Adjust so velocity goes to zero at tD
   EqVo = -EqV(end);
   Time = linspace(0,tD,nEQSteps+1)';
   EqV(:) = EqV(:) + EqVo;
   EqD(:) = EqD(:) + EqVo*Time;

%. Plot EQ and output to command window
   EQOut(dt,tD,Amp,Frq,Phs,so,a,b,EqA,EqV,EqD,DF,EQOn);

%. Trim to number of steps needed for the analysis
   EqA = EqA(1:nSteps);
   EqV = EqV(1:nSteps);
   EqD = EqD(1:nSteps);

end
```

This code executes Eq. 3.43 using the time shaping function given in Eq. 3.44, and adjusts the velocity and displacement so that the final ground velocity is zero. The arrays EqA, EqV, and EqD are earthquake ground acceleration, velocity, and displacement, respectively. The record is trimmed to the duration of the analysis, which is not necessarily the same as the duration of the earthquake. The function EQOut takes care of echoing the earthquake parameters to the command window and plotting the earthquake ground motion functions.

Response Spectrum Design of an SDOF system to resist an earthquake amounts to specifying the mass m, damping c, and stiffness k of the system. Usually, we only have a rough estimate of what the damping ratio should be for certain systems (e.g., $\xi = 0.03-0.04$ for steel frames, $\xi = 0.07-0.10$ for concrete structures). Hence, we might consider that parameter to be fixed for the purposes of design. In many cases, the mass does not change much during the process of design. Therefore, the design problem boils down to finding a suitable stiffness for the system.

We know that as we change the stiffness, the natural frequency changes. And the response to the earthquake can depend strongly upon the natural frequency. To get a picture of the design space, we compute the *response spectrum*—the maximum response to a given earthquake ground motion as a function of frequency. Once we know the response spectrum, we can make sensible design decisions (e.g., which frequencies to avoid).

Generation of the response spectrum is computationally intensive but straight-forward. Assume that the mass and damping ratio are fixed. We simply loop over natural periods T within a certain range. For each period, we compute the stiffness k and the damping coefficient c associated with the given values of m and ξ. With those physical properties, we compute the response of the system to the earthquake ground motion. From that response we select the maximum displacement of the system. We plot that point on the spectrum (maximum displacement vs. period) and move on to the next period.

Figure 3.16 shows how a single point on the response spectrum curve gets generated. For a given natural period (in this case $T = 5.97$), we do a complete analysis of the time history response of the system (the plot on the left side of the figure). Because we are holding the damping ratio fixed (in this case at $\xi = 0.1$), the only quantity that distinguishes the physical properties of the system is the frequency ω (which is the square root of k divided by m). Now, for the fixed system properties we can find the complete response caused by the earthquake in question. Out of that entire response, we pick one value—the absolute maximum displacement (in this case $u = -0.3228$, which occurs at time $t = 21.4$ s). We take that one value and plot one point on the response spectrum curve. The point gets plotted at $T = 5.97$ and $u_{max} = 0.3228$).

In each case we are subjecting the structure to the *same* earthquake ground motion. Figure 3.16 was constructed by doing an analysis for 500 different

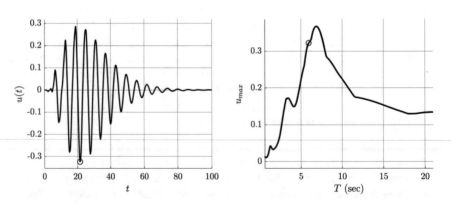

Fig. 3.16 *Construction of a response spectrum.* At left is the displacement history of the SDOF system with natural frequency $\omega = 1.0527$ (period $T = 5.97$) and $\xi = 0.1$ subjected to the earthquake of Fig. 3.15. The maximum displacement gives one point on the response spectrum at right. The one point is shown as the circle on the chart

frequencies between $\omega = 0.3$ to $\omega = 10$ (from $T = 0.638$ to $T = 20.94$), each one responding to the same earthquake.

The response spectrum is a different way of looking at the accelerogram. It is simply a summary of what that earthquake will do to all linear SDOF oscillators with a specified damping ratio. The response spectrum representation of the earthquake ground motion is quite different from the original time series. One reason it is valuable is because it directly indicates what the earthquake will do to a specific system, which is hard to discern from the earthquake ground motion time series itself. Unlike a Fourier transform, it is not possible to invert the response spectrum to get a time series because the response spectrum is created by taking the maximum response value for each frequency. These maxima do not necessarily occur at the same time and the time of occurrence of the maxima is not a continuous function of the frequency.

Often, the response spectra for a small selection of different damping ratios are shown on the same plot to give a sense of the effect of damping on the response.

Different Ways of Plotting the Response Spectrum   There are several ways that the response spectrum is typically presented. Figure 3.17 shows that we can plot the maximum displacement against either the natural frequency ω or the natural period $T = 2\pi/\omega$. These are simply two views of the exact same information.

There are two other important ways to present the response spectrum. The *pseudo-velocity* response spectrum plots $S_v = \omega\, u_{max}$ against either ω or T. The *pseudo-acceleration* response spectrum plots $S_a = \omega^2 u_{max}$ against either ω or T. The value of having these other variations of the response spectrum is as follows. The maximum force in the spring can be computed as $f_{max} = k u_{max}$. The response spectrum is a quick means of finding the maximum force (which is a key design quantity). Likewise, we can compute

$$f_{max} = k u_{max} = k\left(\frac{S_a}{\omega^2}\right) = m S_a \tag{3.45}$$

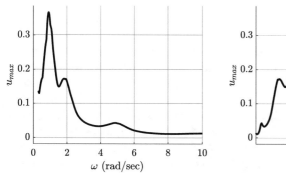
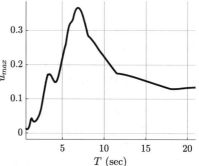

Fig. 3.17 *Different representations of the response spectrum.* We can plot the response spectrum (maximum displacement) against either the natural frequency ω or the natural period T

since $\omega^2 = k/m$. So, the maximum force can also be computed in a "mass times acceleration" sort of sense, but it is the same force—the maximum force in the spring—computed from a different, but very related, representation of the maximum displacement.

3.6 Nonlinear Response

While linear systems are very nice to work with mathematically, most real systems exhibit some sort of nonlinear response. In this section we will modify our analysis to take that into account.

The inertial force is always linear, but the restoring forces may not be. In particular, let us consider nonlinear restoring force in the spring element. For the time being, we will consider the damping mechanism to be linear, but there are many viscous models that are not linear (e.g., there are popular fluid drag resistance models where the force is proportional to v^2). The equation of motion for the nonlinear system is

$$m\ddot{u}(t) + c\dot{u}(t) + \sigma(\varepsilon(u(t))) = f(t) \tag{3.46}$$

where $\sigma(\varepsilon(u(t)))$ is the resisting force, which is a function of the deformation ε, which is a function of displacement, which is a function of time. For the SDOF system $\varepsilon(u) = u$, because the deformation is the same as the relative displacement $u(t)$. For more complex systems that will not be the case. The nature of the nonlinear response depends upon the specific characteristics of the function $\sigma(\varepsilon)$. We will consider two basic forms: (1) nonlinear elastic and (2) inelastic response.

Types of Nonlinear Response There are many different models of nonlinear response, each one designed to capture some observed behavior of some structural component or system. There are two very important types of response, though. We will study how those models affect the dynamic response of the system.

The first model is *nonlinear elasticity*. The main feature of nonlinear elasticity is that the system loads and unloads along the same curve. If it did not have this characteristic, then it would dissipate energy and, hence, it would not be elastic. The second model is *inelasticity*. We will consider the simplest model, elastoplasticity, wherein the system loads along an elastic line, yields at a load level we call the *yield force*, deforms plastically without increasing the resistance, unloads along a line parallel to the initial elastic line, and repeats the same behavior when it hits the yield point on the other side. The area enclosed by the cycle is the energy dissipated.

What we need to do is create mathematical models of these two response types so that we can incorporate them into our equation of motion. For elasticity, we can write a unique function. For inelasticity, we will implement the response as an algorithm.

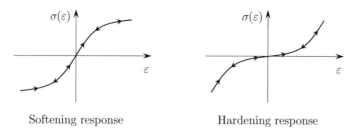

Softening response   Hardening response

Fig. 3.18 *Nonlinear elastic response models.* We consider two simple models of nonlinear elastic response. On the left is the softening model. On the right is the hardening model

Nonlinear Elasticity There are two models of nonlinear elasticity that we will consider: (1) softening response and (2) hardening response. The two models are shown schematically in Fig. 3.18. In a softening response, the stiffness of the system decreases with increasing deformation. In a hardening response, the stiffness of the system increases with increasing deformation.

A function that captures the softening response is

$$\sigma(\varepsilon) = \frac{k\,\varepsilon}{\sqrt{1 + C\varepsilon^2}} \tag{3.47}$$

where k and C are model constants. For very small values of ε, the system responds linearly with stiffness k. For larger values of ε the system softens, with the constant C determining the rate of softening.

A function that captures the hardening response is

$$\sigma(\varepsilon) = k\,\varepsilon\left(1 + C\varepsilon^2\right) \tag{3.48}$$

where k and C are model constants (not necessarily the same as the other model). Again, for very small values of ε, the system responds linearly with stiffness k. For larger values of ε the system hardens, with the constant C determining the rate of stiffening.

Inelastic Response The inelastic model is a little more complicated. The evolution of the response is not independent of the history. Therefore, we will describe the incremental process. The model is pictured in Fig. 3.19.

Consider the situation in the sketch on the left side of the figure. Assume that we have gotten to the state $(\sigma_n, \varepsilon_n)$ at time step n. That is what we will call the current state. Let us further assume that the motion causes an increase in the deformation ε in going from state n to state $n + 1$ (the arrow points to the right). We compute a trial force as if the system were going to behave elastically

$$\sigma^{tr} = k\left(\varepsilon_{n+1} - \varepsilon_p\right) \tag{3.49}$$

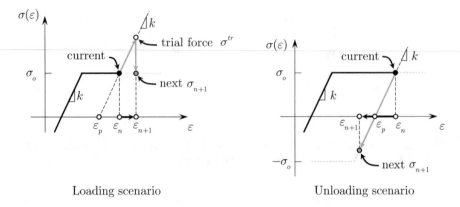

Fig. 3.19 *Inelastic response model.* Elastoplasticity is a model in which the internal force in the system is limited by the *yield force*. Unloading is along the elastic line

where ε_p is the *plastic deformation*, defined as the deformation that would be left if the system unloaded elastically from the current state. If the absolute value of the trial force is greater than the yield force σ_o, then we set the force σ_{n+1} equal to the yield force (and the tangent stiffness equal to zero). If it does not exceed the yield force, then the force is equal to the trial force (and the tangent stiffness is equal to k).

The plastic deformation ε_p is what we call an *internal variable*. It is an additional physical quantity that we must track to characterize the response. Physically, it represents the internal state of the resisting system (e.g., dislocations of the lattice structure of the material). The initial value of the internal variable is zero, and it gets updated each step. This feature is what gives the inelastic model its history dependence. More sophisticated models often have more internal variables.

Observe that in the sketch on the right side of the figure we have a situation where the next deformation is less than the current deformation. Now, when we compute the trial force, we predict the value in accord with the gray arrow with the exact same equation. This approach handles loading and unloading gracefully. The process is summarized in the following pseudocode:

Elastoplastic Response Model

1. Current state is σ_n and ε_n, with plastic deformation ε_p.
2. Compute new deformation ε_{n+1} from displacement u_{n+1}.
3. Compute the trial force $\sigma^{tr} = k\left(\varepsilon_{n+1} - \varepsilon_p\right)$.
4. If $|\sigma^{tr}| > \sigma_o$ then $\sigma_{n+1} = \sigma_o \operatorname{sign}(\sigma^{tr})$, tangent $k_T = 0$.
5. If $|\sigma^{tr}| < \sigma_o$ then $\sigma_{n+1} = \sigma^{tr}$, tangent $k_T = k$.
6. Update plastic deformation $\varepsilon_p = \varepsilon_{n+1} - \sigma_{n+1}/k$.

This algorithm works fine for going from an elastic state to an inelastic state in one step, but it will cut off the corner at the sharp bend in the curve. In reality, we should detect the exact point of yielding or unloading, which would be associated with a strain rate of zero. In the discrete setting, that instant will generally fall between two time points. This error can be reduced by taking smaller steps. It is, of course, possible to compute the time of the transition from elastic to plastic and break the time step into two pieces. That is a good strategy for the SDOF system, but when we extend to systems with many elements, tracking those discrete transitions is not very practical.

Implementation The single-component constitutive models are straightforward to implement. The following code gives a single function that contains a library of constitutive models.

Code 3.2 MATLAB code to compute the internal force, tangent, and update the internal variables given the element deformation. The function has four constitutive models: linear, nonlinear elastic (softening and hardening), and inelastic

```
function [Force,Tangent,iv] = NonlinearModels(Def,d,iv)
%   Library of constitutive models
%
%
%     Def : Deformation
%       d : Material props. [Type, k, C, So]
%      iv : Internal variable (plastic strain)

%. Extract constitutive model type
   Type = d(1);

%. Select element type
   switch Type

      case 1  % linear, elastic
         k = d(2);
         Force = k*Def;
         Tangent = k;

      case 2  % nonlinear elastic, softening
         k = d(2);
         C = d(3);
         denom = sqrt(1+C*Def^2);
         Force = k*Def/denom;
         Tangent = k/(denom^3);

      case 3  % nonlinear elastic, hardening
         k = d(2);
         C = d(3);
         Force = k*Def*(1+C*Def^2);
         Tangent = k*(1+3*C*Def^2);

      case 4  % elasto-plastic
         k  = d(2);
         So = d(4);
         Force = k*(Def-iv);
         Tangent = k;
         if abs(Force) > So
             Force = So*sign(Force);
             Tangent = 0;
             iv = Def - Force/k;
         end

   end % switch
end
```

The way we convey the type of model to use is through the material property array d. The first element of that array is the Type. That array also contains the parameters needed to drive the model (as defined in this section). The incoming Def is the deformation. In the case of the SDOF system, that is just the displacement $u(t)$, because in this context the displacement is the same as the deformation. In subsequent chapters there will be a clearer distinction between displacement and deformation. This function will be useful for the NDOF system and the truss, too.

3.7 Integrating the Equation of Motion

To solve the nonlinear equation of motion, we will integrate numerically with Newmark's method. Assume that we know the state at time t_n, (i.e., u_n, v_n, a_n). We have two Newmark equations relating displacement, velocity, and acceleration and we have the equation of motion, which we will aim to satisfy at time t_{n+1}. If we combine these three equations, we arrive at a system of equations for the three unknown state variables at the next state:

$$ma_{n+1} + cv_{n+1} + \sigma\left(\varepsilon\left(u_{n+1}\right)\right) = f(t_{n+1})$$

$$v_{n+1} = c_n + \eta a_{n+1} \tag{3.50}$$

$$u_{n+1} = b_n + \zeta a_{n+1}$$

where $b_n = u_n + hv_n + h^2\beta a_n$ and $c_n = v_n + h\gamma a_n$ are the parts of the new displacement and velocity, respectively, that can be computed from information at the previous time step. The parameters $\eta = h(1-\gamma)$ and $\zeta = h^2(0.5-\beta)$ are from the Newmark integrator. The equation of motion is nonlinear, but it is a nonlinear *algebraic* equation in terms of $(u_{n+1}, v_{n+1}, a_{n+1})$. We can solve these equations by Newton's method (see Appendix A).

Our equations to determine the next state of the system are given in Eq. 3.50. Because the Newmark equations are both linear, we could easily substitute them into the equation of motion and eliminate the displacement and velocity in favor of acceleration at the next state as the primary unknown. Denote $x = a_{n+1}$ as the unknown acceleration at time step $n+1$ that we seek. Let us apply Newton's method to the equation $g(x) = 0$ where

$$g(x) = mx + cv_{n+1}(x) + \sigma\left(\varepsilon\left(u_{n+1}(x)\right)\right) - f(t_{n+1}) \tag{3.51}$$

The displacement and velocity are functions of the unknown acceleration x through the Newmark relations

$$u_{n+1}(x) = b_n + \zeta x, \qquad v_{n+1}(x) = c_n + \eta x \tag{3.52}$$

The derivative of g is (using the chain rule)

$$g'(x) = m + c\eta + \frac{d\sigma}{d\varepsilon}\frac{d\varepsilon}{du_{n+1}}\zeta \qquad (3.53)$$

since $du_{n+1}/dx = \zeta$ and $dv_{n+1}/dx = \eta$. Note that in the present case of the SDOF system, $d\varepsilon/du = 1$ because the deformation in the spring is equal to the displacement. Both the function $g(x)$ and its derivative $g'(x)$ are evaluated for a given x in the routine NonlinearModels, presented in the previous section.

Knowing the constitutive function and its derivative, we can apply Newton's method as long as we have a starting estimate for x. We start the iteration with $x^0 = a_n$ and iterate by Newton's method

$$x^{i+1} = x^i - g(x^i)/g'(x^i)$$

At each step of the iteration we compute $|g(x^i)|$ to see if we have converged to within a specified tolerance. Upon convergence, x becomes the new acceleration. Once the Newton iteration has converged, the state is updated to prepare for the next time step.

An Approach to Computation The basic framework of the approach is to use a numerical time-stepping algorithm (e.g., Newmark's method) to deal with the temporal part of the problem. At each time step, we solve the equations of motion, which are often nonlinear. To solve the nonlinear algebraic equations, we use Newton's method. The algorithm can be summarized in the following pseudocode:

Algorithm for Nonlinear Dynamic Response

1. Initialize state. Set the plastic deformation ε_p to zero.
2. Compute the state for the initial displacement u_o

   - Compute the internal force and test for yielding.
   - If yielding has occurred, update the plastic deformation.

3. Loop over time steps:

   - Evaluate external force f_{n+1}. Evaluate b_n and c_n and initialize the acceleration $x^0 = a_n$ (the primary iteration variable) to the acceleration at the previous time step.
   - Newton iteration to solve equation of motion.

     - Compute displacement u_{n+1} and velocity v_{n+1} from the Newmark integrator with the current estimate of the acceleration.
     - Compute new deformation ε_{n+1} from displacement u_{n+1}.
     - Compute σ_{n+1} and tangent $k_T = d\sigma/d\varepsilon$.

(continued)

- Compute residual $g = mx^i + cv_{n+1} + \sigma_{n+1} - f_{n+1}$.
- Compute tangent $A = m + \eta c + \zeta k_T$.
- Update acceleration by Newton's method $x^{i+1} = x^i - g/A$.

- Once converged, set $a_{n+1} = x^i$. Update the state variables. If the new state is plastic, update ε_p.

4. Graphical output of results.

A systematic approach for doing nonlinear dynamic computations is evident even in the simple SDOF system. This framework will be useful for all of the computations that we present throughout the book, extended to suitably accommodate the different contexts and theories. As such, we can view the SDOF system as a computational template for much that is to come. For each particular case we will highlight the differences. The framework for the code will extend naturally to systems with multiple degrees of freedom. To see how to code this algorithm in MATLAB, take a look at the following script:

Code 3.3 MATLAB code for computing the dynamic response of the SDOF system

```
%  SDOF nonlinear dynamic response by Newmark's method
   clear; clc; DF = SDOFManageFigures;

%. Time parameters
   tf = 10.0;                    % Final time (duration of analysis)
   h = 0.01;                     % Analysis time step

%. Physical problem data
   LoadType = 2;                 % 0 none, 1 const, 2 sine, 3 blast, etc.
   EQon = 1;                     % =0 No EQ, =1 Include EQ
   m = 1.0;                      % Mass
   xi = 0.02;                    % Damping ratio
   d = [1, 5.0, 4.0, 2.0];       % Constitutive props. [Type, k, C, So]

%. Compute (linear) natural frequency and damping coefficient
   omega = sqrt(d(2)/m);
   c = 2*m*xi*omega;

%. Initial position and velocity
   uo = 0.0;
   vo = 0.0;

%. Numerical analysis parameters
   [beta,gamma,eta,zeta,nSteps,tol,itmax,tf] = NewmarkParams(tf,h);

%. Get earthquake ground motion
   [EqA,~,~] = EQGroundMotion(h,nSteps,EQon,DF);

%. Initialize
   t = 0;                        % Time
   Hist = zeros(nSteps,5);       % Storage for output
   iv = 0;                       % Internal variable (plastic strain)

%. Initialize displacement, velocity. Compute initial acceleration.
   uold = uo;   vold = vo;
   force = TimeFunction(t,LoadType,1);
```

```
    [sigma,~,iv]  =  NonlinearModels(uo,d,iv);
    aold = (force - c*vo - sigma)/m;

%. Compute motion by numerical integration. Loop over time steps.
    for i=1:nSteps

%... Store values for plotting
        Hist(i,:) = [t,uold,vold,force,sigma];

%... Compute the values of the state for the next time step.
        t = t + h;

%... Compute new acceleration from equation of motion.
        err = 1; it = 0;
        force = TimeFunction(t,LoadType,0);
        bn = uold + h*vold + beta*h^2*aold;
        cn = vold + gamma*h*aold;
        anew = aold;

        while (err > tol) && (it < itmax)
          unew = bn + zeta*anew;
          vnew = cn + eta*anew;
          [sigma,tang,~] = NonlinearModels(unew,d,iv);
          g = m*anew + c*vnew + sigma - force + m*EqA(i);
          A = m + eta*c + zeta*tang;
          anew  = anew - A\g;
          err = abs(g);
          it = it + 1;
        end

%... Compute new position and velocity
        unew = bn + zeta*anew;
        vnew = cn + eta*anew;

%... Update internal variables
        [sigma,~,iv] = NonlinearModels(unew,d,iv);

%... Update state to get ready for next step
        aold = anew;
        vold = vnew;
        uold = unew;

    end % loop over time steps

%. Output results
    SDOFOut(d,m,xi,uo,vo,Hist)
```

There are a few important things to observe in this code. First, the function
SDOFManageFigures establishes which plots to produce and sizes and positions
the figure windows. That information is stored in the 'struct' DF. Next, we provide
all values relevant both to describing the physical system and for executing the
numerical analysis. In particular, we specify the duration of the analysis tf,
the time step h, LoadType (which is the key to the type of loading used in
TimeFunction given in Code 3.4), and EQon (which turns on or off the earthquake ground motion). The mass m, damping ratio xi, and the spring constitutive
properties (which we specify through the array d, as described earlier) control the
physical properties of the system. Since we describe damping through the damping
ratio, we must then compute the actual damping coefficient c, which requires the
evaluation of the natural frequency omega. We also specify the initial displacement

uo and velocity vo. The initial acceleration is determined from the equation of motion.

The specification of the numerical analysis parameters include both the Newmark parameters β and γ and the Newton parameters (i.e., the convergence tolerance tol and the maximum number of iterations allowed itmax, which is a safeguard for the possibility of divergence of Newton's method). These parameters are set in the function NewmarkParams, which is contained in Code 3.5, at the end of this section.

We keep track of time with the variable t and record the history of anything we wish to plot later in the array Hist. The internal variable iv (plastic strain) is initialized to zero at the start of the analysis.

The time-stepping algorithm uses variables labeled as "old" with counterparts labeled as "new." The "old" values (i.e., uold, vold, and aold) represent the state at the converged previous time step n. The "new" values (i.e., unew, vnew, and anew) represent the state at the next time step $n+1$, which we seek. Newmark's method only requires the "old" state to compute the "new" ones. Thus, to start the computation, we put the initial values (uo and vo) into uold and vold and compute aold from the equation of motion.

Observe that it is possible for the initial displacement to cause the system to yield (for the inelastic model). Thus, we must compute the internal variable and internal force that corresponds with that initial state by calling NonlinearModels:

$$[\text{sigma},\sim,\text{iv}] = \text{NonlinearModels}(\text{u, d, iv}).$$

The $\sim$ in the second slot of the function output acknowledges that the tangent stiffness will be computed, but it is not needed at this time and therefore is not passed back.

Now we are ready to step through time. The first thing we do in each time step is write the results we wish to eventually plot to the array Hist. Since we are only storing the "old" and "new" states for the computation, this provides a way to record the response. Here we store every point that we compute. In the codes later in the book we will adopt a strategy of storing only enough points needed to make a nice plot. That is the main reason we avoid simply indexing the response variables in the code. Next, we increment the time by h, which is the size of the time step.

We solve the nonlinear equation of motion using Newton's method but note that some of the quantities we need do not change during that iteration (i.e., they are different for each time step, but not different for iterations within that time step). The applied force is retrieved from the function TimeFunction for the current time t using LoadType to indicate which load form to select from the library of possibilities. Note that the parameters associated with the loading are set inside that function. The third slot of the input to this function is a flag which, if set to 1, will print the features of the function to the command window. Otherwise, it simply returns the value of the function. The main reason for having TimeFunction is that additional functions can be added to the library very simply without changing the analysis code.

To prepare for the Newton iteration, we need to give a starting value. We do not use x in the code for the acceleration while we are iterating because we will simply set `anew` equal to that value upon convergence. So, we just use `anew` directly as the iteration variable. The best value to use to start the iteration is the converged acceleration from the previous time step `aold`. That is generally a good choice, even for situations where the acceleration changes quickly (like an earthquake). If the time step is small enough to give an accurate solution, then the old acceleration is a pretty reliable starting point for the iteration.

The Newton iteration is done with a "while" loop because we do not know how many iterations will be required. Note that we first compute estimates of the new displacement and velocity from the new acceleration, as required by Newton's method. We then call `NonlinearModels` again, this time in search of the internal force `sigma` (for the residual) and the derivative `tang` (for the tangent). In these intermediate states, the internal variable should *not* be updated. We can accomplish this with the call

$$[\texttt{sigma}, \texttt{tang}, \sim] = \texttt{NonlinearModels}(\texttt{u}, \texttt{d}, \texttt{iv}).$$

Even though we compute the internal variable update inside the function `NonlinearModels`, if we do not pass it back, then that update will not stick.

Once the iteration is converged, the acceleration satisfies the equation of motion (including the nonlinear constitutive equation). We compute `unew` and `vnew` one more time because the acceleration has been changed since we last computed them. Now, with the *converged* new displacement, we update the internal variables. We can accomplish this with the call

$$[\texttt{sigma}, \sim, \texttt{iv}] = \texttt{NonlinearModels}(\texttt{u}, \texttt{d}, \texttt{iv}).$$

We also compute the internal force at this point because we will need to store it in `Hist` for plotting later. Finally, we put the new state into the old storage locations to get ready for the next time step.

Once the final time has been reached, we proceed to the output. We hide the details of this last step in the function `SDOFOut`, but note that this function echoes all of the input quantities (for reference) and creates all of the plots that are set up for the problem. We will use this strategy throughout the book to avoid dispensing code that is not directly associated with the computation of response. For each theory and hence each main code, the outputs will need to be tailored to the application. Those will be visible mainly through the results presented.

Additional Functions   One of the features of dynamic analysis is that there can be many temporal variations of the loading, each one resulting in a different response. For convenience, we have created the function `TimeFunction` that we will use repeatedly in the codes presented in this book. The function gives a simple way to add and modify load forms to suit the purpose of the analysis. The code for this function is given in Code 3.4.

Code 3.4 MATLAB code for modeling the temporal form of dynamic loading

```
function [f] = TimeFunction(t,Type,flag)
%  Functions for time dependent loads
%
%       t : Time at which function is evaluated
%    Type : Flag for type of function
%    flag : Echo output flag ( =1 print, =0 don't )

   switch Type

     case 0   % No load
       f = 0;
       LoadName = {'none'};
       ParaName = []; Values = [];

     case 1   % Constant load
       Fo = 1;
       f = Fo;
       LoadName = {'Fo'};
       ParaName = {'Fo'}; Values = Fo;

     case 2   % Sinusoidal load
       Fo = 1;  Wo = 1.7;
       f = Fo*sin(Wo*t);
       LoadName = {'Fo*sin(Wo*t)'};
       ParaName = {'Fo','Wo'}; Values = [Fo, Wo];

     case 3   % Blast-type load
       Fo = 1.5;  a = 1;   b = 9.2;
       f = Fo*(exp(-a*t) - exp(-b*t));
       LoadName = {'Fo*(exp(-a*t)-exp(-b*t))'};
       ParaName = {'Fo','a','b'}; Values = [Fo, a, b];

     case 4   % smooth ramp up in time To
       Fo = 1;  To = 2;
       if t<To; f = Fo*(1-cos(pi*t/To))/2; else; f = Fo;  end
       LoadName = {'Fo*(1-cos(pi*t/To))/2, then constant'};
       ParaName = {'Fo','To'}; Values = [Fo,To];

     case 5   % linear ramp up in time To
       Fo = 1;  To = 2;
       if t<To; f = Fo*t/To; else; f = Fo; end
       LoadName = {'Fo*t/To, then constant'};
       ParaName = {'Fo','To'}; Values = [Fo,To];

     case 6   % wind up rotation to frequency Wo by To
       Fo = 1;  To = 12;   Wo = 8.0;
       if t<To; omega = t^2/(2*To); else; omega = (t-To/2); end
       f = Fo*sin(Wo*omega);
       LoadName = {'Fo*sin(Wo*t^2/(2*To)), then Fo*sin(Wo*(t-To/2))'};
       ParaName = {'Fo','To','Wo'}; Values = [Fo,To,Wo];

   end

%. Echo function name and values to command window
   TimeFunctionOut(Type,LoadName,ParaName,Values,flag)

end
```

This function is set up to do two things. First, it computes the value of $f(t)$ given the input t. Second, it records the name of the load form and the names of the parameters associated with it for output purposes. If `flag` is set to 1, the values of the loading parameters are printed to the command window. This way, the function does not get disassociated from its labels. Example time functions generated by this

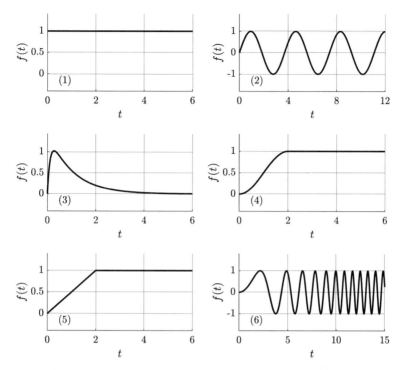

Fig. 3.20 *Time functions.* Results generated by `TimeFunction`. (1) constant, (2) sinusoidal, (3) blast, (4) smooth ramp up, (5) linear ramp up, (6) wind up to specified frequency in specified time

routine are shown in Fig. 3.20. The values of the parameters are the same as those in Code 3.4. The output function at the end simply echos the values to the command window to verify the function type and parameter values.

As mentioned earlier, the numerical analysis parameters are set in the function `NewmarkParams`, given in Code 3.5.

Code 3.5 MATLAB code for to set up the parameters for Newmark integration

```
function [beta,gamma,eta,zeta,nSteps,tol,itmax,tf] = NewmarkParams(tf,h)
% Establish parameters for Newmark integration

    beta = 0.25;                    % Newmark parameter
    gamma = 0.5;                    % Newmark parameter
    eta = (1-gamma)*h;              % Time integration parameter
    zeta = (0.5-beta)*h^2;          % Time integration parameter
    nSteps = ceil(tf/h)+1;          % Number of time steps in analysis
    tf = (nSteps-1)*h;              % Adjust final time to discrete times
    tol = 1.e-8;                    % Newton iteration tolerance
    itmax = 10;                     % Maximum allowable Newton iterations

%. Echo input values to command window
    fprintf('\n\n%s\n', ' Newmark integration of equations  ')
    fprintf('%s%8.3f\n','    Duration (tf)                   ',tf)
    fprintf('%s%8.3f\n','    Time increment (h)              ',h)
    fprintf('%s%8.3f\n','    Newmark (beta)                  ',beta)
```

```
    fprintf('%s%8.3f\n','    Newmark (gamma)            ',gamma)
    fprintf('%s%8.2e\n','    Newton tolerance           ',tol)
    fprintf('%s%8i\n',  '    Newton max iterations      ',itmax)
    fprintf('%s%8i\n',  '    Number of time steps       ',nSteps)

end
```

This code takes in the final time tf and the time step size h and computes the number of steps required for the analysis nSteps. The values of the Newmark parameters β and γ are set along with the values of η and ζ. The tolerance tol and maximum number of iterations itmax for the Newton iteration are also set in this function. The values are echoed to the command window. This function will be used in all codes that use numerical integration throughout the book.

3.8   Example

As an example consider a system that starts from rest, subjected to an applied load of $f(t) = p_o \sin \Omega t$, with $p_o = 4$ and $\Omega = 2$. The mass is $m = 1$, the elastic stiffness is $k = 5$. For the nonlinear elastic models, the parameter C is set to 4. And for the inelastic model the yield level is set to $\sigma_o = 2$. First consider the response of the linear elastic model, shown in Fig. 3.21.

The natural frequency of this system is $\sqrt{k/m} = 2.236$, which is slightly different from the driving frequency $\Omega = 2$. It is evident that the system is near resonance as the response shows *beating*, which is a buildup of motion followed by the decline as the response goes out of phase with the loading. The buildup and decline sequence repeats giving the beating response.

The results for this same example with all four constitutive models are shown in Fig. 3.22. Each model disrupts the near resonance in a different way. None of the nonlinear models shows a response as large as the linear model, but they are all quite different. This example illustrates how the constitutive nonlinearity can significantly change the dynamic response of the system.

Dynamic response is often counterintuitive because of the interplay between the restoring mechanism and the inertial force. Nonlinearity of the restoring force adds quite a bit of variety to the mix. Keep in mind that these four nonlinear models are only a small sample of the many possibilities available to represent the response of real materials. The nature of the loading also adds variety to the mix.

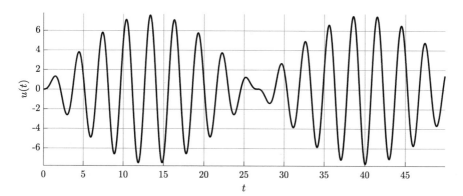

Fig. 3.21 *Dynamic response of an SDOF system (part 1).* Response of a linear system with $u_o = 0$, $v_o = 0$, $\xi = 0$, subjected to the applied load $f(t) = 4 \sin(2t)$. The mass is $m = 1$, the stiffness is $k = 5$. The natural frequency is $\omega = 2.236$

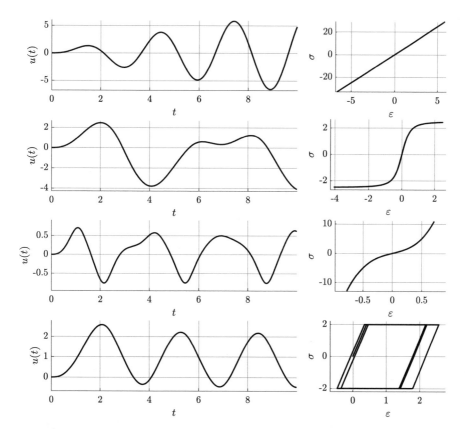

Fig. 3.22 *Dynamic response of an SDOF system (part 2).* Response of a system with $u_o = 0$, $v_o = 0$, $\xi = 0$, subjected to the applied load $f(t) = 4 \sin(2t)$. The mass is $m = 1$, the initial stiffness is $k = 5$, the nonlinear elastic parameter is $C = 4$, and the yield level is $\sigma_o = 2$. The top result is the linear elastic model, the second is the nonlinear elastic (softening) model, the third is the nonlinear elastic (hardening) model, and the bottom is the inelastic model

Chapter 4
Systems with Multiple Degrees of Freedom

In this chapter we consider structures with multiple degrees of freedom. Since we often label the number of degrees of freedom N, we call them NDOF systems. In this and the next several chapters we will analyze idealized *discrete* structures, wherein the parts associated with mass are considered rigid and the parts associated with restoring forces have no mass. This simplification will result in equations of motion that are coupled *ordinary* differential equations. While we will relax the assumptions that decouple mass and stiffness in the later chapters of the book, this idealization gives a great deal of insight into the dynamic response of structural systems.

In NDOF systems, the motions of the masses interact with each other because the restoring force elements connect them together. The additional complexity of this interaction creates new dynamic response features not exhibited by the SDOF system. In particular, we will find that an NDOF system possesses multiple natural frequencies, not just one. And each one of those frequencies is associated with a *mode shape*. The aim of this chapter is to see how those features come about, what they imply about the dynamic response of structures, and how they can be used to support the analysis of those structures.

While most of the results derived in this chapter apply to any discrete NDOF system (e.g., trusses and frames with lumped masses), we will focus on the particular case of the *shear building* to help make the concepts concrete. Through this simple model we can begin to understand the dynamic response of building structures.

Electronic Supplementary Material The online version of this chapter (https://doi.org/10.1007/978-3-030-89944-8_4) contains supplementary material, which is available to authorized users.

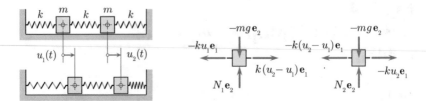

Fig. 4.1 *Two DOF system.* The 2-DOF system comprises two bodies of equal mass m connected with three springs of equal modulus k. The masses slide without friction and the springs are fixed at the two ends. The kinematic picture defining the displacements is shown at left. The free body diagrams of the masses are shown at right

4.1   The 2-DOF System as a Warm-up Problem

To gain an understanding of what is at stake in the dynamic analysis of an NDOF system, we will warm up with the very simple system with two degrees of freedom shown in Fig. 4.1. The system has two bodies, each of mass m, connected to the supports and to each other with three springs, each of stiffness k. The masses slide without friction and the force in each spring is proportional to the change in length of the spring, i.e., $f = k\Delta$.

Each mass requires a position vector to track its motion, constrained to move in the horizontal plane.[1] Let us write the position vectors as

$$\mathbf{x}_1(t) = u_1(t)\mathbf{e}_1, \quad \mathbf{x}_2(t) = u_2(t)\mathbf{e}_1$$

where the scalar time-dependent variables $u_1(t)$ and $u_2(t)$ characterize the motion relative to the reference configuration where the springs are unstretched. With the position vectors established, it is simple to compute the velocities and accelerations as

$$\dot{\mathbf{x}}_1(t) = \dot{u}_1(t)\mathbf{e}_1, \quad \dot{\mathbf{x}}_2(t) = \dot{u}_2(t)\mathbf{e}_1$$

$$\ddot{\mathbf{x}}_1(t) = \ddot{u}_1(t)\mathbf{e}_1, \quad \ddot{\mathbf{x}}_2(t) = \ddot{u}_2(t)\mathbf{e}_1$$

Balance of linear momentum comes from the free body diagrams of the masses shown in Fig. 4.1. We can determine the spring forces by considering the amount of stretch of each spring due to the motion of the masses. The first spring changes length by the amount $\Delta_1 = u_1$, the second by $\Delta_2 = u_2 - u_1$, and the third by $\Delta_3 = -u_2$. Balance of linear momentum for the masses gives two equations of motion (one for each mass)

---

[1] Note that the concept of *degree of freedom*, which we often abbreviate DOF, is associated with the independent motion variables. The number of independent motion variables required to describe the motion *is* the number of degrees of freedom of the system.

$$-k\Delta_1\mathbf{e}_1 + k\Delta_2\mathbf{e}_1 - mg\mathbf{e}_2 + N_1\mathbf{e}_2 = m\ddot{u}_1\mathbf{e}_1$$

$$-k\Delta_2\mathbf{e}_1 + k\Delta_3\mathbf{e}_1 - mg\mathbf{e}_2 + N_2\mathbf{e}_2 = m\ddot{u}_2\mathbf{e}_1$$

Taking the dot product of each equation with $\mathbf{e}_1$, substituting the expressions for the spring deformations in terms of displacements, and moving all terms to one side gives the system of equations that governs the motion of the system in free vibration:

$$m\ddot{u}_1 + 2ku_1 - ku_2 = 0$$
$$m\ddot{u}_2 - ku_1 + 2ku_2 = 0 \tag{4.1}$$

These equations represent a coupled system of second order ordinary differential equations with constant coefficients. The coupling of the equations of motion is caused by the elastic elements and creates some interesting new possibilities in response that are not present in systems with one degree of freedom.

The initial displacement and velocity must be specified for each degree of freedom. Hence, the initial conditions can be stated as

$$\begin{Bmatrix} u_1(0) \\ u_2(0) \end{Bmatrix} = \begin{Bmatrix} u_1^o \\ u_2^o \end{Bmatrix} \qquad \begin{Bmatrix} \dot{u}_1(0) \\ \dot{u}_2(0) \end{Bmatrix} = \begin{Bmatrix} v_1^o \\ v_2^o \end{Bmatrix}$$

where u_1^o, u_2^o, v_1^o, and v_2^o are given values of the state variables at $t = 0$.

Solution by Exponential Functions We can put the equations of motion into matrix form as follows:

$$\begin{bmatrix} m & 0 \\ 0 & m \end{bmatrix} \begin{Bmatrix} \ddot{u}_1 \\ \ddot{u}_2 \end{Bmatrix} + \begin{bmatrix} 2k & -k \\ -k & 2k \end{bmatrix} \begin{Bmatrix} u_1 \\ u_2 \end{Bmatrix} = \begin{Bmatrix} 0 \\ 0 \end{Bmatrix} \tag{4.2}$$

We refer to the coefficient matrix in front of the first term as the *mass matrix* and the coefficient matrix in front of the second term as the *stiffness matrix*. Similar to what we did for the SDOF system, let us assume a solution in the form of exponential functions as

$$\begin{Bmatrix} u_1(t) \\ u_2(t) \end{Bmatrix} = \begin{Bmatrix} \varphi_1 \\ \varphi_2 \end{Bmatrix} e^{\lambda t}$$

where the constants φ_1, φ_2, and λ are yet to be determined. Taking time derivatives, we can find the velocities and accelerations as

$$\begin{Bmatrix} \dot{u}_1(t) \\ \dot{u}_2(t) \end{Bmatrix} = \lambda \begin{Bmatrix} \varphi_1 \\ \varphi_2 \end{Bmatrix} e^{\lambda t} \qquad \begin{Bmatrix} \ddot{u}_1(t) \\ \ddot{u}_2(t) \end{Bmatrix} = \lambda^2 \begin{Bmatrix} \varphi_1 \\ \varphi_2 \end{Bmatrix} e^{\lambda t}$$

Substituting these back into Eq. 4.2, we get

$$\begin{bmatrix} 2k + \lambda^2 m & -k \\ -k & 2k + \lambda^2 m \end{bmatrix} \begin{Bmatrix} \varphi_1 \\ \varphi_2 \end{Bmatrix} e^{\lambda t} = \begin{Bmatrix} 0 \\ 0 \end{Bmatrix}$$

The situation is a little different from the single degree of freedom system, but it is also very similar. We want a solution that is not the zero solution (i.e., $u_1 = u_2 = 0$). Note that $e^{\lambda t}$ can never be zero. Therefore, the only hope that we can arrive at a nontrivial solution is if

$$\begin{bmatrix} 2k + \lambda^2 m & -k \\ -k & 2k + \lambda^2 m \end{bmatrix} \begin{Bmatrix} \varphi_1 \\ \varphi_2 \end{Bmatrix} = \begin{Bmatrix} 0 \\ 0 \end{Bmatrix} \tag{4.3}$$

Equation 4.3 is called an *eigenvalue problem*. It is a homogeneous linear system of equations. From a theorem of linear algebra, we know that this system has a solution other than $\varphi_1 = 0$ and $\varphi_2 = 0$ *if and only if* the determinant of the coefficient matrix is zero. Taking the determinant and setting it equal to zero gives

$$\det \begin{bmatrix} 2k + \lambda^2 m & -k \\ -k & 2k + \lambda^2 m \end{bmatrix} = \left(2k + \lambda^2 m \right)^2 - k^2 = 0$$

This equation is called the *characteristic equation*. In essence, it is an equation to determine the value of λ. If we can find a value of λ that works, then we have a general solution to the governing differential equation. If not, then our original assumption on the form of the solution was not a good one.

The characteristic equation is of fourth order in λ, but only even powers are present, so you can view it as being quadratic in λ^2. This equation happens to be pretty simple to factor in this particular case. Move the k^2 to the right side, take the square root of both sides and solve for λ^2 to get

$$2k + \lambda^2 m = \pm k, \quad \rightarrow \quad \lambda^2 = \frac{-2k \pm k}{m}$$

One thing is evident right away. There are two values of λ^2 and both are negative. To help clarify the solution, let us *define* two constants

$$\omega_1 = \sqrt{\frac{k}{m}}, \qquad \omega_2 = \sqrt{\frac{3k}{m}}$$

With these definitions, we can see that the four values of λ that satisfy the characteristic equation are

$$\lambda_1 = i\,\omega_1, \quad \lambda_2 = -i\,\omega_1, \quad \lambda_3 = i\,\omega_2, \quad \lambda_4 = -i\,\omega_2$$

Observe that the roots come in complex conjugate pairs.

We still need to determine the values of φ_1 and φ_2. To get them we will return to Eq. 4.3 with the two unique values of λ^2 that we found from the characteristic equation. If we substitute the first value, $\lambda^2 = -k/m$, we get

$$\begin{bmatrix} k & -k \\ -k & k \end{bmatrix} \begin{Bmatrix} \varphi_1^{(1)} \\ \varphi_2^{(1)} \end{Bmatrix} = \begin{Bmatrix} 0 \\ 0 \end{Bmatrix} \tag{4.4}$$

where we have added the superscript $\bullet^{(1)}$ to indicate that these values are associated with the first value of λ^2. Since Eq. 4.4 is singular, the only thing we can tell is that

$$\varphi_1^{(1)} = \varphi_2^{(1)}$$

We cannot uniquely determine their respective values. Both equations give the same result. Hence, we can say that the solution to the system of equations is

$$\begin{Bmatrix} \varphi_1^{(1)} \\ \varphi_2^{(1)} \end{Bmatrix} = A \begin{Bmatrix} 1 \\ 1 \end{Bmatrix}$$

where A is an arbitrary constant that cannot be determined from the eigenvalue problem. Because these values are known relative to each other, but not absolutely, we call them *mode shapes*. The shape is known, but the magnitude is not. For each value of λ^2 there is an associated mode shape.

We can find the second mode shape by plugging in the second value of $\lambda^2 = -3k/m$ into Eq. 4.4. If we do that, then we get

$$\begin{bmatrix} -k & -k \\ -k & -k \end{bmatrix} \begin{Bmatrix} \varphi_1^{(2)} \\ \varphi_2^{(2)} \end{Bmatrix} = \begin{Bmatrix} 0 \\ 0 \end{Bmatrix}$$

Following the same logic as we did for the first one, we find $\varphi_1^{(2)} = -\varphi_2^{(2)}$, or

$$\begin{Bmatrix} \varphi_1^{(2)} \\ \varphi_2^{(2)} \end{Bmatrix} = B \begin{Bmatrix} 1 \\ -1 \end{Bmatrix}$$

where, again, the constant B is arbitrary and cannot be determined from the eigenvalue problem.

Now that we have completed the solution of the eigenvalue problem, let us get back to finding the general solution to the differential equation. We found two mode shapes that correspond to the two unique values of λ^2. Note also that we got four values of λ, which appeared as two pairs of complex conjugate values. Hence, we can write the general solution to the differential equation as

$$\begin{Bmatrix} u_1(t) \\ u_2(t) \end{Bmatrix} = \begin{Bmatrix} 1 \\ 1 \end{Bmatrix} \left(A_1 e^{i\omega_1 t} + A_2 e^{-i\omega_1 t} \right) + \begin{Bmatrix} 1 \\ -1 \end{Bmatrix} \left(B_1 e^{i\omega_2 t} + B_2 e^{-i\omega_2 t} \right)$$

Noting Euler's identities $e^{ix} = \cos x + i \sin x$ and $e^{-ix} = \cos x - i \sin x$, we can rewrite this equation in terms of sines and cosines as

$$\begin{Bmatrix} u_1(t) \\ u_2(t) \end{Bmatrix} = \begin{Bmatrix} 1 \\ 1 \end{Bmatrix} (C_1 \cos \omega_1 t + D_1 \sin \omega_1 t)$$

$$+ \begin{Bmatrix} 1 \\ -1 \end{Bmatrix} (C_2 \cos \omega_2 t + D_2 \sin \omega_2 t) \tag{4.5}$$

where the new constants can be expressed in terms of the old ones as

$$C_1 = (A_1 + A_2), \quad D_1 = i(A_1 - A_2)$$
$$C_2 = (B_1 + B_2), \quad D_2 = i(B_1 - B_2)$$

We now have the complete solution to the original differential equation.

Initial Conditions To get the values of the constants, we implement the initial conditions. We know the displacement and velocity of each mass at time $t = 0$. Hence, we can write

$$\begin{Bmatrix} u_1(0) \\ u_2(0) \end{Bmatrix} = \begin{Bmatrix} 1 \\ 1 \end{Bmatrix} C_1 + \begin{Bmatrix} 1 \\ -1 \end{Bmatrix} C_2 = \begin{bmatrix} 1 & 1 \\ 1 & -1 \end{bmatrix} \begin{Bmatrix} C_1 \\ C_2 \end{Bmatrix} = \begin{Bmatrix} u_1^o \\ u_2^o \end{Bmatrix}$$

and

$$\begin{Bmatrix} \dot{u}_1(0) \\ \dot{u}_2(0) \end{Bmatrix} = \begin{Bmatrix} 1 \\ 1 \end{Bmatrix} \omega_1 D_1 + \begin{Bmatrix} 1 \\ -1 \end{Bmatrix} \omega_2 D_2 = \begin{bmatrix} \omega_1 & \omega_2 \\ \omega_1 & -\omega_2 \end{bmatrix} \begin{Bmatrix} D_1 \\ D_2 \end{Bmatrix} = \begin{Bmatrix} v_1^o \\ v_2^o \end{Bmatrix}$$

Solving these two equations for the constants gives

$$\begin{Bmatrix} C_1 \\ C_2 \end{Bmatrix} = \frac{1}{2} \begin{Bmatrix} u_1^o + u_2^o \\ u_1^o - u_2^o \end{Bmatrix} \qquad \begin{Bmatrix} D_1 \\ D_2 \end{Bmatrix} = \frac{1}{2\omega_1\omega_2} \begin{Bmatrix} \omega_2(v_1^o + v_2^o) \\ \omega_1(v_1^o - v_2^o) \end{Bmatrix}$$

It should not be too surprising that the initial conditions determine all of the constants of integration of the governing differential equation. What might be more surprising is that there is no intuitive relationship between the constants and the values of the initial displacement and velocity components. However, we will see that certain values of the initial conditions give rise to some interesting special cases and that those cases are related to the mode shapes.

Modal Vibration Let's take a look at two special cases of initial conditions. First, consider a case with initial displacement but no initial velocity (i.e., the system starts from rest in a displaced position). First, take the special case where $u_1^o = u_2^o = u_o$. In other words, the initial displacement is proportional to the first mode shape. In this case, the constants come out as $C_1 = u_o$, $C_2 = 0$, $D_1 = 0$, and $D_2 = 0$. The response is, then,

$$\begin{Bmatrix} u_1(t) \\ u_2(t) \end{Bmatrix} = u_o \begin{Bmatrix} 1 \\ 1 \end{Bmatrix} \cos \omega_1 t$$

For these initial conditions, the system oscillates at exactly the frequency ω_1. There is no influence of ω_2. The two masses remain completely synchronized as they move, crossing zero at exactly the same time. In essence, the system is vibrating exactly in the first mode. And when it does that, it does so at the first natural frequency.

Now consider initial conditions with $u_1^o = -u_2^o = u_o$. In other words, the initial displacement is proportional to the second mode shape. In this case, the constants come out to be $C_1 = 0$, $C_2 = u_o$, $D_1 = 0$, and $D_2 = 0$. The response is, then,

$$\begin{Bmatrix} u_1(t) \\ u_2(t) \end{Bmatrix} = u_o \begin{Bmatrix} 1 \\ -1 \end{Bmatrix} \cos \omega_2 t$$

In this case the system vibrates in the second mode at the second natural frequency ω_2.

These results shed light on what *natural frequencies* mean for systems with two degrees of freedom and what role the mode shapes play in the dynamic response. The system will respond to any given initial conditions, but it will only respond with synchronized motion when the initial conditions are proportional to one of the mode shapes. It is straightforward to follow the same approach with initial velocities to show that if the initial velocity is proportional to a mode shape then the system vibrates in that mode at the natural frequency associated with that mode.

In the remainder of this chapter we will seek to generalize the results that we have found in this simple example. The process will reveal some of the beauty of the eigenvalue problem as a mathematical tool and establish a notation built to go the distance with these problems. To get more information about the eigenvalue problem, see Appendix C. We only considered the free vibration problem for the system with two degrees of freedom. When we generalize the problem, we will also consider forced vibration of the system.

4.2 The Shear Building

One of the simplest and most useful idealizations of a building structure with N degrees of freedom is the shear building. The structure comprises N rigid (floor) masses supported by (massless) columns.[2] The elastic stiffness of the columns is

---

[2] The reason we assume that the columns are massless and the masses are rigid is that it allows us to treat the system as discrete rather than continuous. This assumption avoids some complex phenomena that we will look at later in the book.

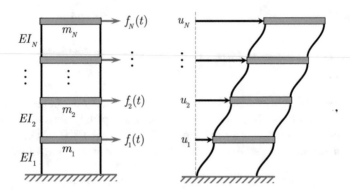

Fig. 4.2 *The shear building.* A shear building is a NDOF system with rigid floors with mass connected by flexible (massless) columns. The mass of level i is m_i and the bending rigidity of the columns below level i is EI_i. The columns are assumed to be axially rigid. The displacement of level i is u_i

characterized by the bending stiffness EI, where E is Young's modulus and I is the moment of inertia of the cross section. Each story might have a different mass and a different stiffness.

The geometry of the shear building is shown in Fig. 4.2. This model provides a fairly general framework for studying NDOF systems, in which we can implement features like damping, inelasticity, and earthquake ground motion. We will be able to find the classical solution to the differential equation for free and forced vibration for the linear case. We will also see how Newmark's method for numerical integration generalizes to the NDOF system.

Motion of the Shear Building Because the floors of the shear building are rigid and the columns are inextensible, the structure is constrained to move as shown in Fig. 4.2. Each level is associated with a horizontal displacement, which is a function of time. We can put those displacements into a single array, which we will call the *displacement vector*, as follows:

$$\mathbf{u}(t) = \begin{Bmatrix} u_1(t) \\ \vdots \\ u_N(t) \end{Bmatrix} \tag{4.6}$$

The *velocity vector*, $\dot{\mathbf{u}}(t)$, and the *acceleration vector*, $\ddot{\mathbf{u}}(t)$, can be simply computed as the first and second derivatives of the displacement vector, with respect to time.

The deformation of a column is shown in Fig. 4.3. For an individual column between levels $i-1$ and i, the deformation is simply the difference between the displacements at those levels. To wit,

$$\Delta_i(t) = u_i(t) - u_{i-1}(t) \tag{4.7}$$

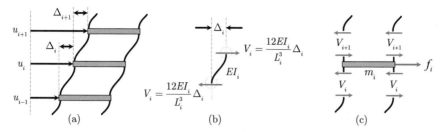

Fig. 4.3 *The shear building deformations and forces.* (**a**) Three adjacent stories showing relative displacements. (**b**) Forces required on a single column to maintain deformation. (**c**) Shear forces acting on mass at level i (moments and axial forces are not shown)

which works for the first level if we take $u_0 = 0$ (i.e., the motion of the base is zero). We do not include u_0 in the vector $\mathbf{u}(t)$. So, we will need to remember that $\Delta_1 = u_1$. The resisting force at each level will depend only on the relative motion at that level. We will consider various constitutive models, the first of which has linear elastic columns.

Equations of Motion Each mass can be taken as a free body diagram to establish the equations of motion. In Fig. 4.3a we show levels $i+1$, i, and $i-1$, along with the deformed columns between them. Figure 4.3b also shows the forces required to displace one end of the column relative to the other by an amount Δ_i without rotation at the ends. From beam theory, we know that the amount of shear force required to do that is

$$V_i = \frac{12E I_i}{L_i^3} \Delta_i$$

Moments in the amount $M_i = 6E I_i \Delta_i / L_i^2$ also develop at the ends of each column at top and bottom along with axial forces required to equilibrate the system. The forces acting on a story mass are shown in Fig. 4.3c. The moments and axial forces are not shown on the FBDs. Balance of linear momentum of the mass at level i in the horizontal direction gives

$$\left(\frac{24E I_{i+1}}{L_{i+1}^3} \right) \Delta_{i+1} - \left(\frac{24E I_i}{L_i^3} \right) \Delta_i + f_i = m_i \ddot{u}_i \qquad (4.8)$$

We get a similar expression at each level. The top level does not have the contribution from columns $i+1$ because there is no column above the top story.

To simplify the formulation, let $k_i = 24E I_i / L_i^3$ represent the *story stiffness*. Now the equations can be written as (specialized to four degrees of freedom for purposes of illustration)

$$k_2\Delta_2 - k_1\Delta_1 + f_1 = m_1\ddot{u}_1$$
$$k_3\Delta_3 - k_2\Delta_2 + f_2 = m_2\ddot{u}_2$$
$$k_4\Delta_4 - k_3\Delta_3 + f_3 = m_3\ddot{u}_3$$
$$-k_4\Delta_4 + f_4 = m_4\ddot{u}_4$$

Note the difference in the last equation that happens because there is no story above the top one. Substituting the deformation–displacement relationships, the equations of motion become

$$k_2\,(u_2 - u_1) - k_1 u_1 + f_1 = m_1\ddot{u}_1$$
$$k_3\,(u_3 - u_2) - k_2\,(u_2 - u_1) + f_2 = m_2\ddot{u}_2$$
$$k_4\,(u_4 - u_3) - k_3\,(u_3 - u_2) + f_3 = m_3\ddot{u}_3$$
$$-k_4\,(u_4 - u_3) + f_4 = m_4\ddot{u}_4$$

which can be rearranged to read

$$m_1\ddot{u}_1 + (k_1 + k_2)\,u_1 - k_2 u_2 = f_1$$
$$m_2\ddot{u}_2 - k_2 u_1 + (k_2 + k_3)\,u_2 - k_3 u_3 = f_2$$
$$m_3\ddot{u}_3 - k_3 u_2 + (k_3 + k_4)\,u_3 - k_4 u_4 = f_3$$
$$m_4\ddot{u}_4 - k_4 u_3 + k_4 u_4 = f_4$$

We can put the equations of motion into matrix form. Create 4×1 arrays for the displacement, acceleration, and applied force as

$$\mathbf{u}(t) = \begin{Bmatrix} u_1(t) \\ \vdots \\ u_4(t) \end{Bmatrix}, \quad \ddot{\mathbf{u}}(t) = \begin{Bmatrix} \ddot{u}_1(t) \\ \vdots \\ \ddot{u}_4(t) \end{Bmatrix}, \quad \mathbf{f}(t) = \begin{Bmatrix} f_1(t) \\ \vdots \\ f_4(t) \end{Bmatrix} \tag{4.9}$$

In addition, define the 4×4 mass and stiffness matrices, respectively, as

$$\mathbf{M} = \begin{bmatrix} m_1 & 0 & 0 & 0 \\ 0 & m_2 & 0 & 0 \\ 0 & 0 & m_3 & 0 \\ 0 & 0 & 0 & m_4 \end{bmatrix}, \quad \mathbf{K} = \begin{bmatrix} k_1 + k_2 & -k_2 & 0 & 0 \\ -k_2 & k_2 + k_3 & -k_3 & 0 \\ 0 & -k_3 & k_3 + k_4 & -k_4 \\ 0 & 0 & -k_4 & k_4 \end{bmatrix}$$

Now we can write the equations of motion in a compact matrix form as

$$\mathbf{M}\ddot{\mathbf{u}}(t) + \mathbf{K}\mathbf{u}(t) = \mathbf{f}(t) \tag{4.10}$$

This process shows how the mass and stiffness values are assembled into the equations, and ultimately into the matrices that represent those equations. The patterns of appearance of the mass and stiffness coefficients are evident in the matrix form. It should be pretty obvious how to extend these equations to the $N \times N$ case. These equations will form the basis of our *classical solution* to the equations of motion. We will first consider the undamped free vibration problem. Next, we will look at ways of specifying damping. Finally, we will consider the forced vibration problem for a few important loading cases.

There is a more systematic way of forming the mass and stiffness matrices for the shear building. In the next chapter we will reformulate the problem with a notation and strategy that will serve us well for the remainder of the book.

4.3 Free Vibration of the NDOF System

For a system with N degrees of freedom, we get an equation of motion for undamped free vibration of the form

$$\mathbf{M}\ddot{\mathbf{u}}(t) + \mathbf{K}\mathbf{u}(t) = \mathbf{0} \tag{4.11}$$

Motivated by the solution to the system with two degrees of freedom, let us assume that the displacement vector is a constant vector $\boldsymbol{\varphi}$ multiplying a sinusoidal function.[3] To wit,

$$\mathbf{u}(t) = \boldsymbol{\varphi}\,(C \cos \omega t + D \sin \omega t) \tag{4.12}$$

The acceleration is $\ddot{\mathbf{u}}(t) = -\omega^2 \boldsymbol{\varphi}\,(C \cos \omega t + D \sin \omega t)$. Substituting these expressions into the equation of motion gives

$$\left[\mathbf{K} - \omega^2 \mathbf{M}\right] \boldsymbol{\varphi}\,(C \cos \omega t + D \sin \omega t) = \mathbf{0}$$

The term $(C \cos \omega t + D \sin \omega t)$ is not zero at all times unless C and D are both zero (the trivial solution). So, the condition for a nonzero solution is the eigenvalue problem

$$\left[\mathbf{K} - \omega^2 \mathbf{M}\right] \boldsymbol{\varphi} = \mathbf{0} \tag{4.13}$$

---

[3] To be consistent with what we did with the SDOF system, we should assume $\mathbf{u} = \boldsymbol{\varphi} e^{\lambda t}$. If we do so, then the eigenvalue equation that results is $\mathbf{K}\boldsymbol{\varphi} = -\lambda^2 \mathbf{M}\boldsymbol{\varphi}$. Since both $\mathbf{K}$ and $\mathbf{M}$ are positive definite, we can premultiply that equation with $\boldsymbol{\varphi}^T$ to show that $-\lambda^2$ must be a positive real number. Therefore, λ must be purely imaginary. That is tantamount to assuming that the solution is $\mathbf{u} = \boldsymbol{\varphi} e^{i\omega t}$, which is the same as assuming the sinusoidal form we have selected here.

Solving the eigenvalue problem yields N natural frequencies $\{\omega_1, \ldots, \omega_N\}$ and N mode shapes $\{\boldsymbol{\varphi}_1, \ldots, \boldsymbol{\varphi}_N\}$ (see Appendix C). Thus, the general solution of the equation of motion is

$$\mathbf{u}(t) = \sum_{i=1}^{N} \boldsymbol{\varphi}_i \left(C_i \cos \omega_i t + D_i \sin \omega_i t \right) \tag{4.14}$$

For reference (we will need it for initial conditions), the velocity can be computed as

$$\dot{\mathbf{u}}(t) = \sum_{i=1}^{N} \omega_i \boldsymbol{\varphi}_i \left(-C_i \sin \omega_i t + D_i \cos \omega_i t \right) \tag{4.15}$$

For large problems (i.e., many degrees of freedom), we usually do not solve the eigenvalue problem the way we did for the two DOF system (i.e., setting up and solving the characteristic equation to find the eigenvalues and then computing the eigenvectors by solving the original Eq. 4.13 with those specific eigenvalues). MATLAB has a built-in function to solve the generalized eigenvalue problem

$$[\,\boldsymbol{\Phi}, \boldsymbol{\Omega}\,] = \text{eig}(\mathbf{K}, \mathbf{M})$$

The ith column of $\boldsymbol{\Phi}$ is the eigenvector $\boldsymbol{\varphi}_i$ and the ith diagonal element of the matrix $\boldsymbol{\Omega}$ is the square of the ith frequency ω_i^2. Because the mass and stiffness matrices are positive-definite and symmetric, the eigenvectors and eigenvalues are guaranteed to be real-valued (i.e., not complex numbers).

The constants C_i and D_i can be determined from the initial conditions. You might recall that the process we followed for the two-degree-of-freedom system involved rather tedious algebra. It turns out that the eigenvectors have a property called *orthogonality* that will greatly simplify the computations associated with the initial conditions.

Example—Mode Shapes The *mode shapes* $\boldsymbol{\varphi}_n$ are configurations of the structure associated with undamped free vibration. To visualize what a mode shape looks like, consider the following example. The five-story shear building shown in Fig. 4.4 has the stiffness and mass properties given in the caption of the figure.

These mode shapes are typical of shear buildings. They are in order from the smallest natural frequency to the largest. In this case we have $\omega_1 = 2.083$, $\omega_2 = 5.777$, $\omega_3 = 9.057$, $\omega_4 = 11.623$, and $\omega_5 = 13.608$. The *fundamental mode* is the mode shape with the lowest natural frequency. All of the displacements are in the same direction for this mode. As the mode number gets higher, the tortuosity of the mode shape increases. For the same maximum displacement, the higher modes store more potential energy and vibrate at a faster rate.

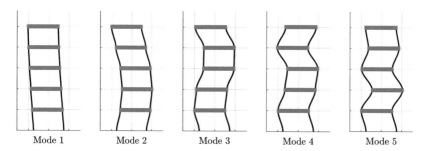

Mode 1   Mode 2   Mode 3   Mode 4   Mode 5

Fig. 4.4 *Mode shapes.* A five-story shear building with k going linearly from 60 at the first level to 40 at the top level, with mass of one for all levels

4.3.1 Orthogonality of the Eigenvectors

In a vector space, we use the inner (dot) product to measure the degree to which two vectors point in the same direction. In fact, the cosine of the angle between two vectors in three-dimensional space is

$$\cos \theta (\mathbf{v}, \mathbf{w}) = \frac{\mathbf{v} \cdot \mathbf{w}}{\|\mathbf{v}\| \|\mathbf{w}\|} \tag{4.16}$$

where $\| \bullet \|$ is the norm, or length, of $\bullet$. If $\cos \theta = 1$, that implies that the two vectors point in exactly the same direction (i.e., the angle between them is zero). If $\cos \theta = 0$, then the vectors are orthogonal. The idea of orthogonality is clear in three-dimensional space. It means that the vectors are perpendicular to each other. In N-dimensional space we do not have the physical notion of perpendicularity, but we can think of orthogonality as meaning that the vectors point in the most different directions possible. We can also include a *metric tensor* in our definition of the inner product as

$$\mathbf{v} \cdot \mathbf{w} = \mathbf{v}^T \mathbf{G} \mathbf{w} \tag{4.17}$$

where $\mathbf{G}$ is a given (positive-definite) matrix that defines the metric. Now, the condition $\mathbf{v}^T \mathbf{G} \mathbf{w} = 0$ implies that the vectors $\mathbf{v}$ and $\mathbf{w}$ are orthogonal with respect to the metric $\mathbf{G}$ or, equivalently, the vectors are $\mathbf{G}$–orthogonal.

The eigenvectors associated with Eq. 4.13 are *orthogonal* with respect to the mass matrix, i.e.,

$$\boldsymbol{\varphi}_i^T \mathbf{M} \boldsymbol{\varphi}_j = 0 \quad \text{if } \omega_i^2 \neq \omega_j^2 \tag{4.18}$$

The proof goes as follows. First, note that $\boldsymbol{\varphi}_i$ and ω_i satisfy the eigenvalue problem, as do $\boldsymbol{\varphi}_j$ and ω_j. Hence, we can write

$$\left[\mathbf{K} - \omega_i^2 \mathbf{M}\right] \boldsymbol{\varphi}_i = \mathbf{0}$$

$$\left[\mathbf{K} - \omega_j^2 \mathbf{M}\right] \boldsymbol{\varphi}_j = \mathbf{0}$$

Premultiply the first equation with $\boldsymbol{\varphi}_j^T$ and the second equation by $\boldsymbol{\varphi}_i^T$ and expand them to get

$$\boldsymbol{\varphi}_j^T \mathbf{K} \boldsymbol{\varphi}_i - \omega_i^2 \boldsymbol{\varphi}_j^T \mathbf{M} \boldsymbol{\varphi}_i = 0$$
$$\boldsymbol{\varphi}_i^T \mathbf{K} \boldsymbol{\varphi}_j - \omega_j^2 \boldsymbol{\varphi}_i^T \mathbf{M} \boldsymbol{\varphi}_j = 0$$

(4.19)

because the stiffness and mass matrices are symmetric,

$$\boldsymbol{\varphi}_j^T \mathbf{K} \boldsymbol{\varphi}_i = \boldsymbol{\varphi}_i^T \mathbf{K} \boldsymbol{\varphi}_j, \qquad \boldsymbol{\varphi}_j^T \mathbf{M} \boldsymbol{\varphi}_i = \boldsymbol{\varphi}_i^T \mathbf{M} \boldsymbol{\varphi}_j$$

Now, subtract the first equation from the second in Eqs. 4.19 to get

$$\left(\omega_i^2 - \omega_j^2\right) \boldsymbol{\varphi}_i^T \mathbf{M} \boldsymbol{\varphi}_j = 0$$

If the frequencies are distinct, then the eigenvectors must be $\mathbf{M}$–orthogonal. It is important to note that the orthogonality only holds with the mass matrix included as a metric in the product $\boldsymbol{\varphi}_i^T \mathbf{M} \boldsymbol{\varphi}_j$. That is what we mean by *mass orthogonal*.

If the frequencies are equal, then the eigenvectors are not necessarily orthogonal, but rather they live in a plane such that any vector in that plane is an eigenvector associated with the repeated eigenvalue. Consider, for example, eigenvectors $\boldsymbol{\varphi}_m$ and $\boldsymbol{\varphi}_n$ that are associated with the same eigenvalue, i.e., $\omega_m = \omega_n = \hat{\omega}$. Now, consider a vector that is a linear combination of the two eigenvectors, $\hat{\boldsymbol{\varphi}} = a\boldsymbol{\varphi}_m + b\boldsymbol{\varphi}_n$. Let us plug $\hat{\boldsymbol{\varphi}}$ into the eigenvalue problem to see what happens. To wit,

$$\left[\mathbf{K} - \hat{\omega}\mathbf{M}\right]\hat{\boldsymbol{\varphi}} = \left[\mathbf{K} - \hat{\omega}\mathbf{M}\right]\left(a\boldsymbol{\varphi}_m + b\boldsymbol{\varphi}_n\right)$$
$$= a\left[\mathbf{K} - \hat{\omega}\mathbf{M}\right]\boldsymbol{\varphi}_m + b\left[\mathbf{K} - \hat{\omega}\mathbf{M}\right]\boldsymbol{\varphi}_n = \mathbf{0}$$

which is true for *any* values of a and b because each of the two individual vectors satisfy the eigenvalue problem. Observe how this proof depends crucially on $\hat{\omega}$ appearing in both terms. What this result implies is that $\boldsymbol{\varphi}_m$ and $\boldsymbol{\varphi}_n$ span a two-dimensional subspace and that any vector within that subspace satisfies the eigenvalue problem (and is therefore an eigenvector).

In such a case, it is possible to pick two orthogonal vectors to represent that subspace, so we can always think of the eigenvectors as a set of mass-orthogonal vectors. In practice, you can pick two random vectors (i.e., arbitrarily chosen pairs of a and b) and then orthogonalize them. It is straightforward to extend this proof to show that n repeated eigenvalues are associated with an n-dimensional subspace of vectors in which all vectors are eigenvectors. While this issue does not come up in the context of the shear building, it is common in other kinds of structures (especially three-dimensional trusses and frames).

Note that the eigenvectors associated with Eq. 4.13 are also stiffness-orthogonal, i.e.,

$$\boldsymbol{\varphi}_i^T \mathbf{K} \boldsymbol{\varphi}_j = 0 \quad \text{if } \omega_i^2 \neq \omega_j^2$$

The proof of this assertion is simple. Premultiply the eigenvalue problem for $\boldsymbol{\varphi}_i$ and ω_i by $\boldsymbol{\varphi}_j$ to get

$$\boldsymbol{\varphi}_j^T \mathbf{K} \boldsymbol{\varphi}_i - \omega_i^2 \boldsymbol{\varphi}_j^T \mathbf{M} \boldsymbol{\varphi}_i = 0 \tag{4.20}$$

Since the eigenvectors are mass-orthogonal, then this equation proves that they must also be stiffness-orthogonal. We can also observe that, if we let $i = j$ in Eq. 4.20, we have

$$\omega_i^2 = \frac{\boldsymbol{\varphi}_i^T \mathbf{K} \boldsymbol{\varphi}_i}{\boldsymbol{\varphi}_i^T \mathbf{M} \boldsymbol{\varphi}_i} \tag{4.21}$$

This ratio is often called the *Rayleigh quotient*.

The Rayleigh Quotient Rayleigh[4] observed that what we now call the Rayleigh quotient could be formed with *any* vector $\mathbf{w}$ as

$$\mathcal{R}(\mathbf{w}) = \frac{\mathbf{w}^T \mathbf{K} \mathbf{w}}{\mathbf{w}^T \mathbf{M} \mathbf{w}} \tag{4.22}$$

He showed that, over all possible vectors $\mathbf{w}$, there is at least one that yields the minimum value of $\mathcal{R}(\mathbf{w})$, and the value of $\mathcal{R}$ for that vector is equal to the square of the fundamental natural frequency, ω_1.

The proof is instructive. Let us expand the vector $\mathbf{w}$ in terms of the eigenvectors as

$$\mathbf{w} = \sum_{i=1}^{N} a_i \boldsymbol{\varphi}_i \tag{4.23}$$

which is possible because the eigenvectors span N–space. For simplicity, and without loss of generality, we will assume that the eigenvectors have been scaled to unity modal mass, i.e., $\boldsymbol{\varphi}_i^T \mathbf{M} \boldsymbol{\varphi}_i = 1$. We can now think of the Rayleigh quotient

---

[4] John William Strutt, Lord Rayleigh, 1842–1919, made numerous contributions to dynamics, including his book *The Theory of Sound* (1877).

as being a function of $a_1, \ldots, a_N$. Specifically, substituting the expansion into the quotient, and noting orthogonality of the eigenvectors with respect to both $\mathbf{M}$ and $\mathbf{K}$ gives

$$\mathcal{R}(a_1, \ldots, a_N) = \frac{\sum_{i=1}^{N} a_i^2 \omega_i^2}{\sum_{i=1}^{N} a_i^2} \tag{4.24}$$

To find the minimum of $\mathcal{R}$, take the derivative with respect to a_j and set the result equal to zero to get (using the quotient rule for differentiation)

$$\frac{\partial \mathcal{R}}{\partial a_j} = \frac{2a_j \omega_j^2 \sum_{i=1}^{N} a_i^2 - 2a_j \sum_{i=1}^{N} a_i^2 \omega_i^2}{\left(\sum_{i=1}^{N} a_i^2\right)^2} = 0$$

which is satisfied if

$$\sum_{i=1}^{N} a_i^2 \omega_i^2 = \omega_j^2 \sum_{i=1}^{N} a_i^2 \tag{4.25}$$

We do not need to solve for the individual a_j values because we can substitute this result back into Eq. 4.24 to find that the minimum of the function relative to a_j is $\mathcal{R}_{min}^j = \omega_j^2$. Thus, the smallest value among all of these local minima is $\mathcal{R}_{min} = \omega_1^2$. It is also evident that if we could select among vectors $\mathbf{w}$ that have $a_1 = 0$ (i.e., no contribution from the first mode), then the minimization of the Rayleigh quotient would give a value of ω_2^2.

Inverse Iteration   One way of using the Rayleigh quotient is to randomly select a set of vectors $\mathbf{w}_i$ and compute the Rayleigh quotient. The smallest value computed, then, must be the closest to the actual fundamental eigenvalue. A more systematic approach is the algorithm called *inverse iteration*, which forms a more orderly way to generate the vectors $\mathbf{w}_i$. One historically significant approach is suggested by the sequence

$$\mathbf{w}_{i+1} = \alpha_i \left[\mathbf{K}^{-1} \mathbf{M} \right] \mathbf{w}_i \tag{4.26}$$

where α_i is a scaling factor. If you don't scale the vectors at each iteration, they get smaller and smaller, eventually causing underflow. One simple way to scale is to make one of the elements of the vector equal to one. From a single specified vector $\mathbf{w}_0$, the remaining vectors can be determined by applying Eq. 4.26. Following a similar approach to the one used for the Rayleigh quotient, one can prove that this sequence of vectors converges to the eigenvector associated with the fundamental frequency. The fundamental frequency can then be computed from the Rayleigh quotient with the last vector in the sequence. This algorithm is called *inverse iteration*.

To prove convergence, let us represent the vectors $\mathbf{w}_i$ as a modal expansion in terms of the eigenvectors $\boldsymbol{\varphi}_n$. To wit,

$$\mathbf{w}_i = \sum_{n=1}^{N} \boldsymbol{\varphi}_n a_n^i \tag{4.27}$$

where a_n^i is the contribution of the nth eigenvector at the ith iteration. Premultiplying Eq. 4.26 by $\boldsymbol{\varphi}_m^T \mathbf{M}$ gives

$$\boldsymbol{\varphi}_m^T \mathbf{M} \sum_{n=1}^{N} \boldsymbol{\varphi}_n a_n^{i+1} = \alpha_i \, \boldsymbol{\varphi}_m^T \mathbf{M} \sum_{n=1}^{N} \mathbf{K}^{-1} \mathbf{M} \boldsymbol{\varphi}_n a_n^i$$

From the eigenvalue problem we know that $\mathbf{M}\boldsymbol{\varphi}_n = \omega_n^2 \mathbf{M} \mathbf{K}^{-1} \mathbf{M} \boldsymbol{\varphi}_n$. Noting the orthogonality of the eigenvectors, we find that

$$a_m^{i+1} = \alpha_i \left(\frac{1}{\omega_m^2} \right) a_m^i \tag{4.28}$$

Specializing this equation to $m = 1$, we can solve for a convenient value of the scaling factor $\alpha_i = \omega_1^2 a_1^{i+1} / a_1^i$. Substituting this result back into Eq. 4.28 we get

$$\frac{a_m^{i+1}}{a_1^{i+1}} = \left(\frac{\omega_1^2}{\omega_m^2} \right) \frac{a_m^i}{a_1^i}$$

What this equation implies is that the relative value of a_m/a_1, i.e., the ratio of the mth coefficient to the first, decreases in each iteration by a factor equal to the square of the ratio of ω_1 to ω_m. That ratio is one for $m = 1$ and less than one for all of the rest because ω_1 is the smallest eigenvalue. Hence, the first term remains constant for each iteration and the remaining terms decrease, which is the reason the specific choice we made for α_i is useful. After enough iterations, the only term that remains is a_1 and therefore the initial vector $\mathbf{w}_0$ has converged to $\boldsymbol{\varphi}_1$. The speed of convergence depends upon how the eigenvalues are spaced. It should be evident that the influence of the higher modes gets washed out very quickly.

If the second frequency is close to the first, then convergence will be slow. If the initial vector was fortuitously selected such that $a_1^0 = 0$, then the iteration would converge to the second eigenvector. Once the first eigenvector is found, it is possible to project it out of the starting vector to converge to the second one. Specifically, if the starting vector is $\mathbf{w}_0$, a new starting vector can be computed as

$$\bar{\mathbf{w}}_0 = \mathbf{w}_0 - \left(\frac{\boldsymbol{\varphi}_1^T \mathbf{M} \mathbf{w}_0}{\boldsymbol{\varphi}_1^T \mathbf{M} \boldsymbol{\varphi}_1} \right) \boldsymbol{\varphi}_1$$

It is easy to show that $\boldsymbol{\varphi}_1^T \mathbf{M} \bar{\mathbf{w}}_0 = 0$. In other words, the starting vector $\bar{\mathbf{w}}_0$ is $\mathbf{M}$-orthgonal to the first eigenvector $\boldsymbol{\varphi}_1$. This process is called *deflation* and can be continued to find additional eigenvectors by projecting out the ones already found. Numerical errors tend to creep in, so it is advisable to project out the known components periodically during the iteration process.

It is also possible to revise the iteration formula to include a *shift* as

$$\mathbf{w}_{i+1} = \alpha_i \left[\mathbf{K} - c^2 \mathbf{M} \right]^{-1} \mathbf{M} \, \mathbf{w}_i \tag{4.29}$$

where c is a given number. This iteration will converge to the frequency ω_m that is closest to c. The proof is a little more complicated than the algorithm without shifts, but it is instructive. To set up the proof, first observe that we can write the spectral representation of the inverse of the matrix $\mathbf{K} - c^2 \mathbf{M}$ as

$$\left[\mathbf{K} - c^2 \mathbf{M} \right]^{-1} = \sum_{n=1}^{N} \frac{\boldsymbol{\varphi}_n \boldsymbol{\varphi}_n^T}{K_n - c^2 M_n} \tag{4.30}$$

where $K_n = \boldsymbol{\varphi}_n^T \mathbf{K} \boldsymbol{\varphi}_n$ and $M_n = \boldsymbol{\varphi}_n^T \mathbf{M} \boldsymbol{\varphi}_n$ are the modal stiffness and mass, respectively. To prove that this expression is the inverse, all we need to do is show that the product of the matrix and inverse is the identity. Compute

$$\left[\mathbf{K} - c^2 \mathbf{M} \right]^{-1} \left[\mathbf{K} - c^2 \mathbf{M} \right] = \sum_{n=1}^{N} \frac{\boldsymbol{\varphi}_n \boldsymbol{\varphi}_n^T \left[\mathbf{K} - c^2 \mathbf{M} \right]}{K_n - c^2 M_n} = \mathbf{I}$$

So, is it really the inverse? If $\mathbf{I} \boldsymbol{\varphi}_m = \boldsymbol{\varphi}_m$ for any $\boldsymbol{\varphi}_m$ then it is, since the eigenvectors span the N-dimensional space. Using the expression just derived for the identity we can compute

$$\mathbf{I} \boldsymbol{\varphi}_m = \sum_{n=1}^{N} \frac{\boldsymbol{\varphi}_n \boldsymbol{\varphi}_n^T \left[\mathbf{K} - c^2 \mathbf{M} \right] \boldsymbol{\varphi}_m}{K_n - c^2 M_n} = \frac{\boldsymbol{\varphi}_m \left(K_m - c^2 M_m \right)}{K_m - c^2 M_m} = \boldsymbol{\varphi}_m$$

Orthogonality of the eigenvectors with respect to both $\mathbf{K}$ and $\mathbf{M}$ means that only the mth term in the sum survives. Now, taking Eq. 4.29, premultiplying both sides by $\boldsymbol{\varphi}_m^T \mathbf{M}$, using Eq. 4.30, and substituting the eigenvector expansions of the vectors $\mathbf{w}_i$ and $\mathbf{w}_{i+1}$ from Eq. 4.27, we can show that

$$a_m^{i+1} = \alpha_i \left(\frac{1}{\omega_m^2 - c^2} \right) a_m^i$$

where we have used the fact that $K_m / M_m = \omega_m^2$. Now, specialize the equation to the case $m = n$, such that ω_n is the eigenvalue closest to c, to solve for α_i. Doing so and

Fig. 4.5 *Spacing of eigenvalues.* The distance from ω_n^2 to c^2 is smaller than the distance from c^2 to any other eigenvalue

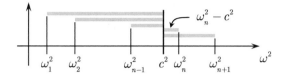

substituting that value back gives the result

$$\frac{a_m^{i+1}}{a_n^{i+1}} = \left(\frac{\omega_n^2 - c^2}{\omega_m^2 - c^2}\right)\frac{a_m^i}{a_n^i} \tag{4.31}$$

This relationship proves that the ratio a_m/a_n decreases in proportion to the ratio of the distance $\omega_n^2 - c^2$ to the distance $\omega_m^2 - c^2$, as illustrated in Fig. 4.5. Since we selected n such that ω_n is closest to c, the ratio is one for $n = m$ and it is less than one (in absolute value) for all of the others. That proves that the iteration in Eq. 4.29 converges to φ_n from any starting vector $\mathbf{w}_0$. Obviously, when $c = 0$ we get the result from the unshifted inverse iteration, and Fig. 4.5 makes it clear why $n = 1$ in that case.

With several great algorithms available to solve the eigenvalue problem, inverse iteration with shifting and deflation is mostly a historical footnote at this point. However, the concept can be found at the heart of the *subspace iteration* algorithm (see Appendix C) for finding a subset of eigenvalues and eigenvectors of a system.

4.3.2 Initial Conditions

Equation 4.14 contains unknown constants C_i and D_i for $i = 1, \ldots, N$. To determine these constants, we need initial conditions. Let us assume that the initial displacement is $\mathbf{u}_o$ and the initial velocity is $\mathbf{v}_o$ (i.e., each component has a specified value). These values are sufficient to find the values of the C_i and D_i. At time $t = 0$ we can set the displacement from Eq. 4.14 and velocity from Eq. 4.15 equal to the initial values

$$\mathbf{u}(0) = \sum_{i=1}^{N} \varphi_i C_i = \mathbf{u}_o, \qquad \dot{\mathbf{u}}(0) = \sum_{i=1}^{N} \omega_i \varphi_i D_i = \mathbf{v}_o$$

Premultiply both of those equations by $\varphi_j^T \mathbf{M}$ and note that, due to orthogonality, only one of the terms in the sum survives. Therefore,

$$C_j = \frac{\varphi_j^T \mathbf{M} \mathbf{u}_o}{\varphi_j^T \mathbf{M} \varphi_j} \qquad D_j = \frac{1}{\omega_j} \frac{\varphi_j^T \mathbf{M} \mathbf{v}_o}{\varphi_j^T \mathbf{M} \varphi_j}$$

Finally, substitute the values of the constants back into the general solution to get the complete solution to the equations of motion

$$\mathbf{u}(t) = \sum_{i=1}^{N} \boldsymbol{\varphi}_i \left\{ \left(\frac{\boldsymbol{\varphi}_i^T \mathbf{M} \mathbf{u}_o}{\boldsymbol{\varphi}_i^T \mathbf{M} \boldsymbol{\varphi}_i} \right) \cos \omega_i t + \frac{1}{\omega_i} \left(\frac{\boldsymbol{\varphi}_i^T \mathbf{M} \mathbf{v}_o}{\boldsymbol{\varphi}_i^T \mathbf{M} \boldsymbol{\varphi}_i} \right) \sin \omega_i t \right\} \qquad (4.32)$$

If the initial displacements or velocities are in the same ratios as one of the eigenvectors, then the response will be in that shape and will oscillate in the associated natural frequency (because of orthogonality of the eigenvectors), as we demonstrate in the next section.

Pure Modal Vibration Consider the case of an initial displacement in the nth mode shape $\mathbf{u}_o = u_o \boldsymbol{\varphi}_n$ with zero initial velocity $\mathbf{v}_o = \mathbf{0}$. Then

$$C_j = \frac{\boldsymbol{\varphi}_j^T \mathbf{M} \mathbf{u}_o}{\boldsymbol{\varphi}_j^T \mathbf{M} \boldsymbol{\varphi}_j} = \begin{cases} u_o & \text{if } j = n \\ 0 & \text{if } j \neq n \end{cases}$$

Thus, the response for this initial condition is

$$\mathbf{u}(t) = u_o \boldsymbol{\varphi}_n \cos \omega_n t$$

This result generalizes what we found for the two-degree-of-freedom system. The system vibrates in a pure mode (with mode shape $\boldsymbol{\varphi}_n$) at the natural frequency ω_n. Again, this observation gives the physical meaning to the notion of natural frequency for the NDOF system.

Observe that we can induce pure modal vibration by specifying that the initial velocity is proportional to a pure mode as $\mathbf{v}_o = v_o \boldsymbol{\varphi}_n$, with zero initial displacement $\mathbf{u}_o = \mathbf{0}$. Now,

$$D_j = \frac{1}{\omega_j} \frac{\boldsymbol{\varphi}_j^T \mathbf{M} \mathbf{v}_o}{\boldsymbol{\varphi}_j^T \mathbf{M} \boldsymbol{\varphi}_j} = \begin{cases} v_o/\omega_n & \text{if } j = n \\ 0 & \text{if } j \neq n \end{cases}$$

The response for this initial condition is

$$\mathbf{u}(t) = \frac{v_o}{\omega_n} \boldsymbol{\varphi}_n \sin \omega_n t$$

It is also straightforward to show that any combination of initial displacement and velocity in a single pure mode will result in response in that mode. These simple cases provide a physical context for understanding the concept of *natural frequencies* and *mode shapes*, which appeared in our development as purely mathematical notions.

Forced Vibration of the NDOF System Now let us take a look at an undamped system that is subjected to a forcing function $\mathbf{f}(t)$, as defined in Eq. 4.9. The equations of motion for this case are

$$\mathbf{M}\ddot{\mathbf{u}}(t) + \mathbf{K}\mathbf{u}(t) = \mathbf{f}(t) \tag{4.33}$$

To solve this equation we will use an *eigenvector expansion* of the unknown displacement. Let

$$\mathbf{u}(t) = \sum_{i=1}^{N} \boldsymbol{\varphi}_i a_i(t) \tag{4.34}$$

where $a_i(t)$ is an as yet unknown function of time. There are N of these functions and we will build the actual displacement $\mathbf{u}(t)$ from them. Differentiate Eq. 4.34 twice to get the acceleration (noting that $\boldsymbol{\varphi}_i$ is constant) and substitute back into the equation of motion to get

$$\mathbf{M}\sum_{i=1}^{N} \boldsymbol{\varphi}_i \ddot{a}_i(t) + \mathbf{K}\sum_{i=1}^{N} \boldsymbol{\varphi}_i a_i(t) = \mathbf{f}(t)$$

Because the mass $\mathbf{M}$ and stiffness $\mathbf{K}$ do not depend upon the summation index i, they can be moved into the summation. If we premultiply the equation by $\boldsymbol{\varphi}_j^T$ (which can also be moved under the sum because j is distinct from i) and note mass and stiffness orthogonality, we have

$$\left(\boldsymbol{\varphi}_j^T \mathbf{M}\boldsymbol{\varphi}_j\right)\ddot{a}_j(t) + \left(\boldsymbol{\varphi}_j^T \mathbf{K}\boldsymbol{\varphi}_j\right)a_j(t) = \boldsymbol{\varphi}_j^T \mathbf{f}(t)$$

If we divide this equation through by the leading coefficient on the acceleration term, noting the result from Eq. 4.21, we have

$$\ddot{a}_j(t) + \omega_j^2 a_j(t) = \frac{\boldsymbol{\varphi}_j^T \mathbf{f}(t)}{\boldsymbol{\varphi}_j^T \mathbf{M}\boldsymbol{\varphi}_j} = p_j(t), \qquad j = 1, \ldots, N$$

where $p_j(t)$ is the amount of loading associated with the jth mode. What we have is a completely uncoupled set of differential equations for each $a_j(t)$. Each one can be solved using the techniques we developed for forced vibration of the SDOF system. Hence, we can quickly find the response, for example, for the suddenly applied constant load and the sinusoidal load.

One common loading scenario is for there to be a pattern of loads that is driven by a common scalar function of time

$$\mathbf{f}(t) = f(t)\mathbf{p}_o$$

where $\mathbf{p}_o$ is a prescribed constant vector and $f(t)$ is a scalar function of time (e.g., $f(t) = 1$ or $f(t) = \sin \Omega t$). In this circumstance, all of the N uncoupled equations would have a particular solution of the same form, just scaled to the fraction of the vector $\mathbf{p}_o$ that participates in that mode. Let

$$p_n = \frac{\boldsymbol{\varphi}_n^T \mathbf{p}_o}{\boldsymbol{\varphi}_n^T \mathbf{M} \boldsymbol{\varphi}_n}$$

and let $a_n(t)$ be the solution to the equation

$$\ddot{a}_n(t) + \omega_n^2 a_n(t) = p_n\, f(t)$$

The complete solution can then be constructed from Eq. 4.34.

Example Consider a shear building with a constant load suddenly at the top level, starting from rest. In this case $f(t) = 1$ and $\mathbf{p}_o = [\,0, 0, \ldots, 1\,]^T$ (a one in the last DOF with zeros everywhere else). Now, the solution to each modal equation is (from the SDOF solution)

$$a_n(t) = \frac{p_n}{\omega_n^2}\left(1 - \cos \omega_n t\right)$$

and the complete solution is

$$\mathbf{u}(t) = \sum_{n=1}^{N} \boldsymbol{\varphi}_n \frac{p_n}{\omega_n^2}\left(1 - \cos \omega_n t\right)$$

Figure 4.6 shows the response computed for a five-story building with story stiffness k varying linearly from 60 at the bottom level to 40 at the top level, a mass of one at each level, and a suddenly applied force of one at the top. It is evident from the response that it takes about 0.5 s for the effects of the load to reach the

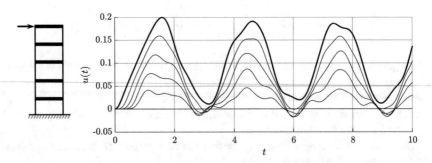

Fig. 4.6 *Response of shear building example.* Displacement response of a five-story shear building with a load suddenly applied at the top level and held constant. The structure is at rest at $t = 0$. The darker line is the response of the top level

bottom level. The primary period of response appears to be about 3 s, which matches the fundamental period of vibration of the system. While the fundamental mode dominates the response, the extra wiggles indicate that higher modes contribute, too.

This example illustrates how the classical solutions to the SDOF problems are directly applicable to NDOF problems. We will revisit this approach later in our study of resonance of NDOF systems, where we use the classical solution to the SDOF system subjected to a sinusoidal load.

4.4 Structural Damping

If you excite a structural system, the response eventually dies out. Deciding the best way to model that phenomenon mathematically is complicated. There are a number of physical mechanisms that might be involved in the energy dissipation that causes the decay of motion. One that has been particularly successful is a linear velocity-dependent mechanism, which leads to exponential decay of motion with time (as we saw in the SDOF system). The damping mechanism in complex structures is not well understood. Therefore, we will develop an approach to damping that serves the purpose of giving reasonable phenomenological response.

What Is Damping? In the study of the SDOF system, we introduced damping through a specific device (the dashpot) which develops force in proportion to the velocity. The relationship between force and velocity was taken to be linear, and the damping constant c was the physical parameter that we built into our model that resulted in the equations of motion. Linear viscous damping yields an exponential decay of the amplitude of motion. One might also imagine that the decay of motion is caused by a mechanism like friction. In the SDOF model we can show that friction causes a linear decay of the amplitude of motion. Unfortunately, the friction model does not preserve the nature of the differential equation as being linear. We will adopt the viscous damping model as most representative, but we will create our damping matrix for the NDOF system guided by a desire for simplicity in modeling.

The concept of *classical damping* goes back at least to Rayleigh and is based upon the diagonalization properties of the eigenvectors of the undamped system. Recall that the undamped system in free vibration was governed by the equation of motion

$$\mathbf{M}\ddot{\mathbf{u}}(t) + \mathbf{K}\mathbf{u}(t) = \mathbf{0}$$

which gave rise to the eigenvalue problem

$$\left[\mathbf{K} - \omega^2 \mathbf{M}\right] \boldsymbol{\varphi} = \mathbf{0}$$

that yielded N pairs of eigenvalues and eigenvectors $(\omega_i, \boldsymbol{\varphi}_i)$ for $i = 1, \ldots, N$. We proved that the eigenvectors were mass- and stiffness-orthogonal. In fact, if we normalize the eigenvectors so that $\boldsymbol{\varphi}_i^T \mathbf{M} \boldsymbol{\varphi}_i = 1$ (i.e., unity modal mass), then $\boldsymbol{\varphi}_i^T \mathbf{K} \boldsymbol{\varphi}_i = \omega_i^2$. We can store the eigenvectors as columns of an array as

$$\boldsymbol{\Phi} = \left[\boldsymbol{\varphi}_1, \ldots, \boldsymbol{\varphi}_N \right] \tag{4.35}$$

Now, unity modal mass implies that $\boldsymbol{\Phi}^T \mathbf{M} \boldsymbol{\Phi} = \mathbf{I}$ and

$$\Omega = \boldsymbol{\Phi}^T \mathbf{K} \boldsymbol{\Phi} = \begin{bmatrix} \omega_1^2 & & \\ & \ddots & \\ & & \omega_N^2 \end{bmatrix}$$

This property is called *diagonalization* of the mass and stiffness matrices.

The equation of motion for a system with linear viscous damping in free vibration is

$$\mathbf{M}\ddot{\mathbf{u}}(t) + \mathbf{C}\dot{\mathbf{u}}(t) + \mathbf{K}\mathbf{u}(t) = \mathbf{0} \tag{4.36}$$

where $\mathbf{M}$, $\mathbf{C}$, and $\mathbf{K}$ are $N \times N$ matrices. The stiffness and mass matrices are the ones derived previously from first principles. Our goal is to create a damping matrix $\mathbf{C}$ that is diagonalized by the eigenvectors of the undamped system.

4.4.1   Modal Damping

Assume that we want the *undamped* eigenvectors $\boldsymbol{\Phi}$ to diagonalize the damping matrix $\mathbf{C}$, just like it does the mass and stiffness matrices. We will write this outcome as follows:

$$\mathbf{D} = \boldsymbol{\Phi}^T \mathbf{C} \boldsymbol{\Phi} = \begin{bmatrix} 2\xi_1\omega_1 & & \\ & \ddots & \\ & & 2\xi_N\omega_N \end{bmatrix}$$

where ξ_i is the *damping ratio* associated with the ith mode. Let

$$\mathbf{a}(t) = [a_1(t), \ldots, a_N(t)]^T$$

be the unknown, time-dependent coefficients of a modal expansion of the displacement. Now we can write

$$\mathbf{u}(t) = \boldsymbol{\Phi}\mathbf{a}(t) \tag{4.37}$$

where $\mathbf{\Phi}$ is given by Eq. 4.35. Substituting Eq. 4.37 into Eq. 4.36 and premultiplying by the transpose of $\mathbf{\Phi}$ gives

$$\left[\mathbf{\Phi}^T\mathbf{M}\mathbf{\Phi}\right]\ddot{\mathbf{a}}(t) + \left[\mathbf{\Phi}^T\mathbf{C}\mathbf{\Phi}\right]\dot{\mathbf{a}}(t) + \left[\mathbf{\Phi}^T\mathbf{K}\mathbf{\Phi}\right]\mathbf{a}(t) = \mathbf{0}$$

Because the (undamped) eigenvectors diagonalize $\mathbf{M}$, $\mathbf{C}$, and $\mathbf{K}$, these equations are uncoupled and have the form (assuming that the eigenvectors have been scaled to unity modal mass)

$$\ddot{a}_i(t) + 2\xi_i\omega_i\dot{a}_i(t) + \omega_i^2 a_i(t) = 0, \qquad i = 1, \ldots, N \qquad (4.38)$$

It should be evident now why we selected the elements of $\mathbf{D}$ as we did. The uncoupled equations of motion are identical to the SDOF equations of motion. Hence, the interpretation of ξ as the damping ratio is appropriate.

One common misconception appears to arise because of the coincidence that the shear building has N stories and therefore N degrees of freedom and therefore N eigenvalues and eigenvectors. In the model described by Eq. 4.38, we get N damping ratios ξ_i, and it is tempting to associate those with the stories of the shear building as we do the masses m_i, and the story stiffness values k_i. However, the damping ratios are associated with the *mode shapes* and not the degrees of freedom of the system. As such, the N SDOF systems of Eq. 4.38 are *not* a collection of SDOF shear buildings stacked one on top of the other. The equations do not map onto N physical SDOF oscillators, but rather N mathematically independent differential equations that arise from the modal decomposition.

We will complete the solution of the damped equations of motion later but let us first complete the computation of the damping matrix $\mathbf{C}$ *given* the modal damping ratios ξ_i. Note that $\mathbf{D} = \mathbf{\Phi}^T\mathbf{C}\mathbf{\Phi}$. If we premultiply this equation by $\mathbf{\Phi}^{-T}$ and post-multiply it by $\mathbf{\Phi}^{-1}$, where the superscript $\bullet^{-1}$ means inverse and the superscript $\bullet^{-T}$ means inverse transpose (the order of those two operations does not matter), then we can solve for $\mathbf{C}$ as

$$\mathbf{C} = \mathbf{\Phi}^{-T}\mathbf{D}\mathbf{\Phi}^{-1}$$

The inverse operation might seem like an impractical and computationally intensive step, but we can simplify it by noting that $\mathbf{\Phi}^T\mathbf{M}\mathbf{\Phi} = \mathbf{I}$. We can do two things to this equation: (1) premultiply by $\mathbf{\Phi}^{-T}$, which gives $\mathbf{M}\mathbf{\Phi} = \mathbf{\Phi}^{-T}$ and (2) post-multiply by $\mathbf{\Phi}^{-1}$, which gives $\mathbf{\Phi}^T\mathbf{M} = \mathbf{\Phi}^{-1}$. Using these results, we can write the damping matrix as

$$\mathbf{C} = \mathbf{M}\mathbf{\Phi}\mathbf{D}\mathbf{\Phi}^T\mathbf{M} \qquad (4.39)$$

This damping matrix is called the *modal damping* matrix. It is very simple to compute if the eigenvectors are known. This approach requires the specification of damping in *all* N modes, and it requires the complete solution of the $N \times N$ eigenvalue problem. For large values of N, this approach may not be practical.

Another practical downside to this approach is that the damping matrix that comes from Eq. 4.39 is full, not banded like the mass and stiffness matrices. That can have an impact on numerical methods like Newmark's method, where we must solve a system of equations. For nonlinear problems, that happens at every time step.

4.4.2   Rayleigh Damping

Rayleigh was among the first to solve NDOF problems with damping. One of his ideas was to create a damping matrix $\mathbf{C}$ that had the same orthogonality features of $\mathbf{M}$ and $\mathbf{K}$. The simplest candidate is to take $\mathbf{C}$ as a linear combination of $\mathbf{M}$ and $\mathbf{K}$. To wit,

$$\mathbf{C} = \mu_1 \mathbf{M} + \mu_2 \mathbf{K}$$

where μ_1 and μ_2 are, as yet, unknown constants. It is straightforward to demonstrate that the undamped mode shapes are orthogonal with respect to $\mathbf{C}$

$$\boldsymbol{\varphi}_j^T \mathbf{C} \boldsymbol{\varphi}_i = \boldsymbol{\varphi}_j^T \left(\mu_1 \mathbf{M} + \mu_2 \mathbf{K} \right) \boldsymbol{\varphi}_i = \mu_1 \boldsymbol{\varphi}_j^T \mathbf{M} \boldsymbol{\varphi}_i + \mu_2 \boldsymbol{\varphi}_j^T \mathbf{K} \boldsymbol{\varphi}_i$$

Using orthogonality of $\mathbf{M}$ and $\mathbf{K}$, we find that

$$\boldsymbol{\varphi}_j^T \mathbf{C} \boldsymbol{\varphi}_i = 0 \quad \text{if} \quad \omega_i^2 \neq \omega_j^2$$

for *any* values of the constants μ_1 and μ_2. This model is called *Rayleigh damping*. We can find the damping coefficients by noting

$$\boldsymbol{\varphi}_j^T \mathbf{C} \boldsymbol{\varphi}_j = \mu_1 \left(\boldsymbol{\varphi}_j^T \mathbf{M} \boldsymbol{\varphi}_j \right) + \mu_2 \left(\boldsymbol{\varphi}_j^T \mathbf{K} \boldsymbol{\varphi}_j \right)$$

Dividing this equation through by $\boldsymbol{\varphi}_j^T \mathbf{M} \boldsymbol{\varphi}_j$, noting that

$$\frac{\boldsymbol{\varphi}_j^T \mathbf{K} \boldsymbol{\varphi}_j}{\boldsymbol{\varphi}_j^T \mathbf{M} \boldsymbol{\varphi}_j} = \omega_j^2 \quad \text{and} \quad \frac{\boldsymbol{\varphi}_j^T \mathbf{C} \boldsymbol{\varphi}_j}{\boldsymbol{\varphi}_j^T \mathbf{M} \boldsymbol{\varphi}_j} = 2\xi_j \omega_j$$

we can generate equations for μ_1 and μ_2. Of course, we can do this for any j. So, we will pick two distinct values, $j = m$ and $j = n$, to get two equations in two unknowns (again assuming that we are specifying the damping ratios)

$$\begin{aligned}
2\xi_m \omega_m &= \mu_1 + \mu_2 \omega_m^2 \\
2\xi_n \omega_n &= \mu_1 + \mu_2 \omega_n^2
\end{aligned} \tag{4.40}$$

We can find the damping coefficients by solving the system of equations

$$\begin{bmatrix} 1 & \omega_m^2 \\ 1 & \omega_n^2 \end{bmatrix} \begin{Bmatrix} \mu_1 \\ \mu_2 \end{Bmatrix} = \begin{Bmatrix} 2\xi_m\omega_m \\ 2\xi_n\omega_n \end{Bmatrix}$$

We can select *any* mode numbers m and n, but we can *only* select two. That means we must make a choice, and that choice will affect the nature of damping in the system. In fact, once μ_1 and μ_2 are established, we can compute the damping in any mode k as

$$\xi_k = \tfrac{1}{2}\left(\frac{\mu_1}{\omega_k} + \mu_2\,\omega_k\right) \tag{4.41}$$

This result shows that the damping associated with mass (i.e., μ_1) is inversely proportional to frequency, and the damping associated with the stiffness (i.e., μ_2) is proportional to frequency. The stiffness-proportional damping forces the damping of the higher modes to be fairly high. This may be a welcome feature in finite element analysis of continuous systems where the higher modes might not be very accurate.

Example (Rayleigh Damping) Consider a shear building with 40 levels, each one with mass equal to one and with story stiffness values varying linearly from 900 at the bottom to 600 at the top. We specify the damping ratio to be 0.1 in the first and 20th modes. Figure 4.7 shows what that choice implies about the damping in all the other modes. The higher modes have damping higher than the 20th mode, while all of the modes between 1 and 20 have lower damping. You can also see that the higher damping ratios are asymptotic to a straight line when damping ratio is plotted against frequency.

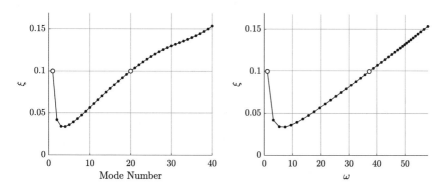

Fig. 4.7 *Example—Rayleigh damping.* A forty-story shear building with $\xi = 0.1$ specified in mode 1 and mode 20. The mass is one at all levels and the stiffness decreases linearly from 900 at the bottom level to 600 at the top level. The graph shows the damping implied in all of the other modes of the system

4.4.3 Caughey Damping

We can generalize the idea of Rayleigh damping to allow the specification of the damping ratio in more than two modes. To do that we need the following lemma:

Lemma *If* (φ_j, ω_j) *are the jth mode shape and frequency associated with mass matrix* $\mathbf{M}$ *and stiffness matrix* $\mathbf{K}$, *then*

$$\mathbf{M}\left(\mathbf{M}^{-1}\mathbf{K}\right)^n \varphi_j = \omega_j^{2n}\mathbf{M}\varphi_j \tag{4.42}$$

where the definition of a matrix raised to the power n is the matrix multiplied by itself, in accord with the rules of matrix multiplication, n times (for example, $\mathbf{A}^2 = \mathbf{AA}$, $\mathbf{A}^3 = \mathbf{AAA}$). *With this convention, then, we can write*

$$\left(\mathbf{M}^{-1}\mathbf{K}\right)^2 = \left(\mathbf{M}^{-1}\mathbf{K}\right)\left(\mathbf{M}^{-1}\mathbf{K}\right) = \mathbf{M}^{-1}\mathbf{K}\mathbf{M}^{-1}\mathbf{K}$$

with similar results for higher powers.

Proof First note the property of the eigenvalue problem

$$\mathbf{K}\varphi_j = \omega_j^2\mathbf{M}\varphi_j \quad \rightarrow \quad \mathbf{M}^{-1}\mathbf{K}\varphi_j = \omega_j^2\varphi_j$$

Operating recursively, then, we can write

$$\begin{aligned}
\mathbf{M}\left(\mathbf{M}^{-1}\mathbf{K}\right)^n \varphi_j &= \mathbf{M}\left(\mathbf{M}^{-1}\mathbf{K}\right)^{n-1}\left(\mathbf{M}^{-1}\mathbf{K}\right)\varphi_j \\
&= \mathbf{M}\left(\mathbf{M}^{-1}\mathbf{K}\right)^{n-1}\omega_j^2\varphi_j \\
&= \mathbf{M}\left(\mathbf{M}^{-1}\mathbf{K}\right)^{n-2}\left(\mathbf{M}^{-1}\mathbf{K}\right)\omega_j^2\varphi_j \\
&= \mathbf{M}\left(\mathbf{M}^{-1}\mathbf{K}\right)^{n-2}\omega_j^4\varphi_j \\
&\quad\vdots \\
&= \omega_j^{2n}\mathbf{M}\varphi_j
\end{aligned}$$

Now we can write the *Caughey damping matrix* as

$$\mathbf{C} = \sum_{k=1}^{n} \mu_k \mathbf{M}\left(\mathbf{M}^{-1}\mathbf{K}\right)^{k-1} \tag{4.43}$$

The first two terms give Rayleigh damping. As before, we will let the damping ratio in any mode j be defined through the relationship

$$2\xi_j\omega_j = \frac{\varphi_j^T C \varphi_j}{\varphi_j^T M \varphi_j}$$

Now consider n (distinct) modes $m(i)$ for $i = 1, 2, \ldots, n$. Premultiplying Eq. 4.43 by $\varphi_{m(i)}^T$ and post-multiplying by $\varphi_{m(i)}$, we get n equations for the coefficients $\mu_1, \ldots, \mu_n$

$$2\xi_{m(i)}\omega_{m(i)} = \sum_{k=1}^{n} \mu_k \left(\omega_{m(i)}^2\right)^{k-1}, \quad i = 1, 2, \ldots, n \tag{4.44}$$

This is an $n \times n$ linear system of equations, which can be solved as the following examples suggest.

Example (Caughey Damping, Three Terms) To show how the computation of the damping coefficients works for Caughey damping, consider the three-term Caughey expansion

$$C = \mu_1 M + \mu_2 K + \mu_3 K M^{-1} K \tag{4.45}$$

We will specify damping in modes k, m, and n. For this case, Eq. 4.44 manifests as a system of three equations in three unknowns

$$\begin{bmatrix} 1 & \omega_k^2 & \omega_k^4 \\ 1 & \omega_m^2 & \omega_m^4 \\ 1 & \omega_n^2 & \omega_n^4 \end{bmatrix} \begin{Bmatrix} \mu_1 \\ \mu_2 \\ \mu_3 \end{Bmatrix} = \begin{Bmatrix} 2\xi_k\omega_k \\ 2\xi_m\omega_m \\ 2\xi_n\omega_n \end{Bmatrix}$$

Finding the values of μ_1, μ_2, and μ_3 comes from the solution of this system of equations. The pattern of the coefficient matrix and right side of the linear system of equations is evident, even for this simple example.

Orthogonality Orthogonality of the Caughey damping matrix is simple to prove. From Eq. 4.42 in conjunction with the definition of the damping matrix from Eq. 4.43, we have

$$\varphi_i^T C \varphi_j = \varphi_i^T \sum_{k=1}^{n} \mu_k M \left(M^{-1} K\right)^{k-1} \varphi_j$$

$$= \sum_{k=1}^{n} \mu_k \omega_j^{2(k-1)} \varphi_i^T M \varphi_j$$

The orthogonality of **C** is inherited from the orthogonality of **M**.

Setting $i = j$ in the above equation and dividing through by $\boldsymbol{\varphi}_j^T \mathbf{M} \boldsymbol{\varphi}_j$ gives

$$2\xi_j \omega_j = \sum_{k=1}^{n} \mu_k \omega_j^{2(k-1)}$$

We can, therefore, solve for the damping in *any* mode j as

$$\xi_j = \tfrac{1}{2} \sum_{k=1}^{n} \mu_k \omega_j^{(2k-3)}$$

It is possible to have some very strange damping ratios result from Caughey damping. Be careful!

Example (Caughey Damping, Four Terms) With Caughey damping we can specify the damping ratio in any number of modes. They do not have to be successive. For example, we could specify the damping in modes 1, 10, 20, and 30 in the forty-story shear building with the same mass and stiffness properties as the Rayleigh damping example. In this case, damping is set to $\xi = 0.1$ in each of those four modes. We can compute the damping in the other modes, and it comes out in accord with Fig. 4.8.

It is evident from these examples that one of the main challenges with Rayleigh and Caughey damping is controlling the damping in the modes in which the damping ratio is *not* specified. In both cases, the damping decreases in the lower modes if you specify the damping in the first mode. That leaves either the first mode overly damped or the other lower modes insufficiently damped. A little experimentation shows that it is essential to specify the damping of the first mode because, if you do not, the first mode damping will be very high.

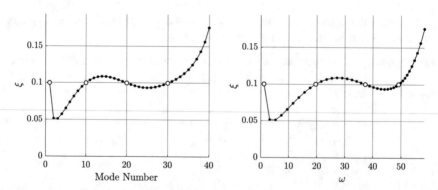

Fig. 4.8 *Example—Caughey damping.* The same forty-story shear building as Fig. 4.7 with $\xi = 0.1$ specified in modes 1, 10, 20, and 30. The chart shows the damping implied in all of the other modes of the system

In Caughey damping the variation of damping is a polynomial with order equal to the number of modes with specified damping. Note that for odd numbers the tail past the last specified mode will be downward, opening up the possibility of negative damping.

The three damping models are implemented in Code 4.1.

Code 4.1 MATLAB code to create a damping matrix **C** by Rayleigh, Caughey, or modal damping

```matlab
function [C,V,freq,Damp] = DampingMatrix(M,K,Damp,DF)
%  Compute damping matrix from mass and stiffness matrices
%
%           M : Mass Matrix
%           K : Stiffness Matrix
%        Damp : Struct with damping information
%   Damp.type : 1 Rayleigh damping
%               2 Caughey damping
%               3 Full Modal
%     Damp.xi : Damping ratios in specified modes
%  Damp.modes : Modes where damping ratios are specified

%. Compute undamped eigenvalues and eigenvectors
   [V,D] = eig(K,M);
   freq = sqrt(abs(diag(D)));
   nFree = length(freq);

%. Compute damping matrix
   switch Damp.type

      case 1 % Rayleigh damping
        Damp.Name = 'Rayleigh Damping';
        mm = Damp.modes(1); nn = Damp.modes(2);
        AA = [1 ,freq(mm)^2; 1, freq(nn)^2];
        bb = [2*Damp.xi(1)*freq(mm); 2*Damp.xi(2)*freq(nn)];
        mu = AA\bb;
        C = mu(1)*M + mu(2)*K;

      case 2 % Caughey damping
        Damp.Name = 'Caughey Damping';
        nModes = size(Damp.xi,1);      % number of modes to include
        AA = zeros(nModes,nModes); bb = zeros(nModes,1);
        for i=1:nModes
          mm = Damp.modes(i);
          for j=1:nModes
            nn = (j-1)*2;
            AA(i,j) = freq(mm)^nn;
          end
          bb(i) = 2*Damp.xi(i)*freq(mm);
        end
        mu = AA\bb;

%..... Compute damping matrix from mass and stiffness matrices
        MK = M\K; C = zeros(nFree,nFree);
        for i=1:nModes
          C = C + mu(i)*M;
          M = M*MK;
        end

      case 3 % Full modal damping
        Damp.Name = 'Modal Damping';
        mu = zeros(nFree,1);
        CC = 2*diag(Damp.xi.*freq);
        C = M*V*CC*V'*M;

   end
```

```
%. Output for damping model
    Damp = DampingOut(Damp,freq,mu,DF);

end
```

This routine takes in the mass and stiffness matrices, along with the struct Damp (specified in the input file for the structure) that specifies which model to use and provides the associated parameters (e.g., which modes are specified for Rayleigh damping and the values of damping in those modes). The routine returns the damping matrix **C**, the undamped eigenvectors φ_n, and the undamped frequencies, ω_n in all modes (these are stored in arrays V and freq, respectively). The function DampingOut simply prints the results to the command window and produces the graphs shown in Figs. 4.7 and 4.8. The struct DF that is passed into the damping routine contains values that control the plots.

4.4.4 Non-classical Damping

We can apply the standard approach of taking the solution to the governing equation of motion to be in the form of exponential functions, but we must first convert the second order system to a first order differential equation. Define the $2N \times 1$ array (and its first derivative)

$$\mathbf{z}(t) = \begin{Bmatrix} \mathbf{u}(t) \\ \dot{\mathbf{u}}(t) \end{Bmatrix} \qquad \dot{\mathbf{z}}(t) = \begin{Bmatrix} \dot{\mathbf{u}}(t) \\ \ddot{\mathbf{u}}(t) \end{Bmatrix}$$

We can write the equations of motion and enforce the velocity constraint equation $\dot{\mathbf{u}}(t) = \dot{\mathbf{u}}(t)$ by writing the following first order differential equation

$$\mathbf{A}\dot{\mathbf{z}}(t) + \mathbf{B}\mathbf{z}(t) = \mathbf{0} \tag{4.46}$$

where the coefficient matrices have the form

$$\mathbf{A} = \begin{bmatrix} \mathbf{I} & \mathbf{0} \\ \mathbf{0} & \mathbf{M} \end{bmatrix} \qquad \mathbf{B} = \begin{bmatrix} \mathbf{0} & -\mathbf{I} \\ \mathbf{K} & \mathbf{C} \end{bmatrix}$$

Note that the damping matrix **C** does not need to have orthogonal properties relative to the undamped mode shapes. Now, take the solution in the form of exponential functions as

$$\mathbf{z}(t) = \psi e^{\lambda t} \qquad \dot{\mathbf{z}}(t) = \lambda \psi e^{\lambda t}$$

Substituting these back into the governing equations, we get

$$\left[\mathbf{B} + \lambda \mathbf{A}\right]\boldsymbol{\psi} e^{\lambda t} = \mathbf{0}$$

We know that the exponential cannot be zero, so in order to have a solution to this problem, we must have

$$\left[\mathbf{B} + \lambda \mathbf{A}\right]\boldsymbol{\psi} = \mathbf{0}$$

This is an eigenvalue problem, and we can solve it in the usual way (e.g., ask MATLAB to do it). The main consequence of non-classical damping is that we should expect both our eigenvalues and eigenvectors to be complex valued. Once the eigenvalues and eigenvectors are known, the solution to Eq. 4.46 can be carried out using a modal decomposition, as before. We will not pursue this avenue further here but leave it as an exercise for the reader.

4.5 Damped Forced Vibration of the NDOF System

For a system with N degrees of freedom, we will get an equation of motion for damped forced vibration of the form

$$\mathbf{M}\ddot{\mathbf{u}}(t) + \mathbf{C}\dot{\mathbf{u}}(t) + \mathbf{K}\mathbf{u}(t) = \mathbf{f}(t)$$

Assume orthogonal damping and take a modal expansion of the displacement, $\mathbf{u}(t) = \boldsymbol{\Phi}\mathbf{a}(t)$. Substituting the modal expansion into the equations of motion, premultiplying the resulting equation by $\boldsymbol{\varphi}_i^T$, and dividing through by the modal mass $\boldsymbol{\varphi}_i^T \mathbf{M}\boldsymbol{\varphi}_i$ gives

$$\ddot{a}_i(t) + 2\xi_i \omega_i \dot{a}_i(t) + \omega_i^2 a_i(t) = \frac{\boldsymbol{\varphi}_i^T \mathbf{f}(t)}{\boldsymbol{\varphi}_i^T \mathbf{M}\boldsymbol{\varphi}_i} = p_i(t) \tag{4.47}$$

for $i = 1, \ldots, N$. Each equation is an SDOF equation and can be solved classically using the methods of Chap. 3. The general solution to the free vibration problem is

$$\mathbf{u}_h(t) = \sum_{i=1}^{N} \boldsymbol{\varphi}_i e^{-\xi_i \omega_i t} \left(C_i \cos \bar{\omega}_i t + D_i \sin \bar{\omega}_i t\right) \tag{4.48}$$

where $\bar{\omega}_i = \omega_i \sqrt{1 - \xi_i^2}$ is the damped natural frequency for the ith mode. The particular solution $\mathbf{u}_p(t)$ will depend upon the precise nature of the loading function $\mathbf{f}(t)$.

Free Vibration by Modal Analysis Consider free vibration of the NDOF system. In this case the homogeneous solution is the complete solution. The displacement is

$$\mathbf{u}(t) = \sum_{i=1}^{N} \boldsymbol{\varphi}_i e^{-\xi_i \omega_i t} \left(C_i \cos \bar{\omega}_i t + D_i \sin \bar{\omega}_i t \right) \tag{4.49}$$

and, by differentiation, the velocity is

$$\dot{\mathbf{u}}(t) = \sum_{i=1}^{N} \boldsymbol{\varphi}_i e^{-\xi_i \omega_i t} \left[-\xi_i \omega_i \left(C_i \cos \bar{\omega}_i t + D_i \sin \bar{\omega}_i t \right) \right.$$
$$\left. + \bar{\omega}_i \left(-C_i \sin \bar{\omega}_i t + D_i \cos \bar{\omega}_i t \right) \right] \tag{4.50}$$

We can find the values of the constants C_i and D_i by using the initial conditions $\mathbf{u}(0) = \mathbf{u}_o$ and $\dot{\mathbf{u}}(0) = \mathbf{v}_o$. To wit,

$$\mathbf{u}_o = \sum_{i=1}^{N} \boldsymbol{\varphi}_i C_i, \qquad \mathbf{v}_o = \sum_{i=1}^{N} \boldsymbol{\varphi}_i \left(\bar{\omega}_i D_i - \xi_i \omega_i C_i \right) \tag{4.51}$$

Premultiply both equations by $\boldsymbol{\varphi}_n^T \mathbf{M}$, noting orthogonality of the mode shapes, to get

$$C_n = \frac{\boldsymbol{\varphi}_n^T \mathbf{M} \mathbf{u}_o}{\boldsymbol{\varphi}_n^T \mathbf{M} \boldsymbol{\varphi}_n}, \qquad D_n = \frac{1}{\bar{\omega}_n} \left(\frac{\boldsymbol{\varphi}_n^T \mathbf{M} \mathbf{v}_o}{\boldsymbol{\varphi}_n^T \mathbf{M} \boldsymbol{\varphi}_n} + \xi_n \omega_n \frac{\boldsymbol{\varphi}_n^T \mathbf{M} \mathbf{u}_o}{\boldsymbol{\varphi}_n^T \mathbf{M} \boldsymbol{\varphi}_n} \right) \tag{4.52}$$

These values can be substituted back into Eqs. 4.49 to give a complete solution to the free vibration problem for the NDOF system.

Implementation This computation is straightforward to implement into a dynamics code. We will hold off on showing how to implement the NDOF code more fully until next chapter, where we will delve into the computation of the stiffness matrix (actually, the tangent stiffness for the nonlinear problem), and the following chapter, where we will delve into the implementation of the earthquake analysis. To see how to organize the modal analysis computation, consider the following code:

Code 4.2 MATLAB code to perform modal analysis of the NDOF system

```
%  NDOF Dynamics (Modal Analysis)
   clear; clc;

%. Analysis features
   tf = 10.0;                        % Duration of anaysis
   nSteps = 300;                     % Number of time steps to plot

%. Input structure data
   [x,ix,id,iv,d,mass,Damp,force,u,S,DF] = ShearBldgInputModal;
```

```
%. Compute initial stiffness, mass, and damping matrices
   [K,~,~,~,~] = NDOFAssemble(u,iv,id,ix,d,S);
   [M,~] = NDOFMassMatrix(mass,id,S);
   [C,V,freq,Damp] = DampingMatrix(M,K,Damp,DF);

%. Establish initial conditions
   [uo,vo] = NDOFICs(V,S);

%. Compute coefficients for classical solution
   xi = Damp.XI;
   freqD = freq.*sqrt(1-xi.^2);
   Cn = V'*M*uo;
   Dn = V'*M*vo;
   Dn = Cn.*(xi.*freq./freqD) + Dn./freqD;

%. Initialize storage arrays
   Time = linspace(0,tf,nSteps)';
   Hist = zeros(nSteps,S.nDOF);

%. Compute response by modal analysis
   for n=1:nSteps
     t = Time(n);
     ee = exp(-xi.*freq.*t);
     u = V*(ee.*(Cn.*cos(freqD.*t) + Dn.*sin(freqD.*t)));
     Hist(n,:) = [u',0];
   end

%. Output
   NDOFEchoInputModal(S,tf,d,mass,xi,uo,vo,freq)
   NDOFGraphsModal(Time,Hist,S,DF)
```

To start, we must specify the duration of analysis (tf) and how many steps to use to create the plots (nSteps). We get the physical specification of the model from the MATLAB function called ShearBldgInput. There is quite a bit coming out of this function. We will not get into details at this point (we will use the same input for the codes in the next two chapters). Suffice it to say that the information provided is adequate to compute the mass **M** and stiffness **K** matrices, needed for the computation. The function that creates **M** (NDOFMassMatrix, see Code 6.2) and the function that creates **K** (NDOFAssemble, see Code 5.2) will be presented in subsequent chapters.

The initial conditions are specified in the input function. If we set S.Mode to 0, then we will get whatever initial conditions we specify in the function NDOFICs. That requires giving a specific value to the initial displacement and velocity for every degree of freedom in the system. If we set S.Mode to n, then the function NDOFICs will set the initial displacement and velocity equal to φ_n. The values S.Uo and S.Vo give the magnitude of the initial displacement, U_o, and velocity, V_o, respectively. The initial conditions are established with the following function:

Code 4.3 MATLAB code to establish the initial conditions for the NDOF system

```
function [uo,vo] = NDOFICs(V,S)
%     Establish initial conditions
%
%         V : Eigenvectors
%         S : Struct with problem size parameters
%             S.Mode =0 specify ICs, >0 IC equal mode shape S.Mode
%             S.Uo = magnitude of initial displacement
```

```
%              S.Vo = magnitude of initial velocity
%. Specify each component
   if S.Mode==0
      uo = S.Uo*ones(S.nFree,1);
      vo = S.Vo*ones(S.nFree,1);
   elseif (S.Mode > 0)
      uo = S.Uo*V(:,S.Mode);
      vo = S.Vo*V(:,S.Mode);
   end

end
```

You can see that the code is set to give all ones to the degrees of freedom if S.Mode is zero. To investigate more elaborate initial conditions would require modification of this function.

Example Consider the five-story shear building shown in Fig. 4.9, with the properties given in the caption of the figure. The initial conditions are such that all five levels are given the same displacement and same velocity. The heavier line is the response of the top level.

Due to the initial velocity, all of the masses embark on a positive displacement trajectory right from the start. It is evident that the first level departs from that trajectory immediately. It is the only level that has any deformation and therefore internal force at $t = 0$. Once the first level starts to depart from the others, the second level is engaged. The others follow with some time delay. This response shows that the initial displacement disturbance propagates a wave that travels up the building. The top story departs from the original trajectory at about 0.6 s.

As the motion progresses it is obvious that the system is oscillating, but each mass responds a little differently. It is also evident that there is a dominant oscillation that settles into a fairly regular period very close to the fundamental period of

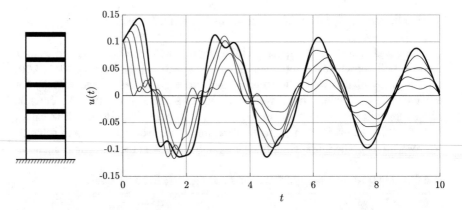

Fig. 4.9 *NDOF damped free vibration 1.* A five-story shear building with k going linearly from 60 at the first level to 40 at the top level, with mass of one for all levels. Modal damping is used with damping ratios of $\xi = 0.03$ in all modes. Initial displacement is $\mathbf{u}_o = 0.1[1, 1, 1, 1, 1]$ and the initial velocity is $\mathbf{v}_o = 0.1[1, 1, 1, 1, 1]$

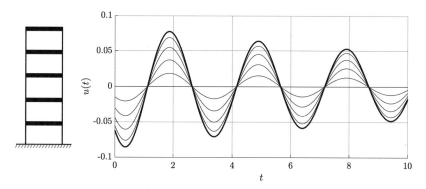

Fig. 4.10 *NDOF damped free vibration 2.* The same five-story shear building as Fig. 4.9, but the initial displacement is $\mathbf{u}_o = 0.1\varphi_1$ and the initial velocity is $\mathbf{v}_o = 0.2\varphi_1$

3.017 s. This is common for NDOF systems. Of course, the decay in the response amplitude is due to damping.

Now consider the same structure subjected to different initial conditions. In this case we set the initial displacement and velocity to be proportional to the first mode shape, as indicated in Fig. 4.10. As the theory would predict, the response is synchronized as the vibration is in a pure mode. The initial values can be gleaned from the plot at $t = 0$. The synchronized motion has all five masses passing through zero displacement at the same time. Also, the observed period of the motion exactly corresponds with the fundamental period of the structure $T_1 = 3.017$, which corresponds to the fundamental natural frequency of $\omega_1 = 2.083$.

The same experiment can be conducted by setting the initial conditions proportional to any of the other five modes. Doing so would show that the system vibrates synchronously at the specified natural frequency.

4.6   Resonance in NDOF Systems

One of the phenomena that we encountered in the single-degree-of-freedom system was *resonance*, which occurred when the system was subjected to a sinusoidally varying loading function $p_o \sin \Omega t$ and the driving frequency Ω was exactly equal to the natural frequency ω of the system. In a resonating system, the response builds up steadily until it reaches an asymptotic limit defined by the damping ratio, ξ. For the SDOF system, the maximum response was equal to $u_{max} \approx 1/2\xi$. If the system is undamped, then the amplitude of vibration increases linearly with time without bound. An illustration of resonant response is shown in Fig. 3.8.

It is reasonable to ask whether the NDOF system exhibits resonance and, if so, what is the nature of the phenomenon in this context. For example, the NDOF system has N natural frequencies, not just one. Does the system resonate at all of the natural frequencies? The NDOF system also has N degrees of freedom where loads can be applied. Does the location of the sinusoidally varying load have an influence on the phenomenon?

The equations of motion for the system with a sinusoidal load are

$$\mathbf{M\ddot{u}}(t) + \mathbf{C\dot{u}}(t) + \mathbf{Ku}(t) = \mathbf{p}_o \sin \Omega t$$

where $\mathbf{p}_o$ is a constant pattern of loads, all of which vary in time sinusoidally at the driving frequency Ω. Consider the modal decomposition of the equations of motion given in Eq. 4.47. Specialized for this case, those equations take the form

$$\ddot{a}_n(t) + 2\xi_n \omega_n \dot{a}_n(t) + \omega_n^2 a_n(t) = p_n \sin \Omega t$$

where $p_n = \boldsymbol{\varphi}_n^T \mathbf{p}_o / \boldsymbol{\varphi}_n^T \mathbf{M} \boldsymbol{\varphi}_n$ is the contribution of the load $\mathbf{p}_o$ to the nth modal equation. The steady-state solution for each modal component can be found using the method described for the SDOF system. The result is

$$a_n(t) = \frac{p_n}{\omega_n^2 \sqrt{D_n}} \sin\left(\Omega t - \alpha_n\right)$$

where

$$D_n = \left(1 - \beta_n^2\right)^2 + \left(2\xi_n \beta_n\right)^2, \qquad \tan \alpha_n = \frac{2\xi_n \beta_n}{1 - \beta_n^2}$$

and $\beta_n = \Omega/\omega_n$ is the ratio of the driving frequency Ω to the nth (undamped) natural frequency ω_n. We will not consider the transient part of the solution because resonance is a process that builds over time and the transient part of the solution dies out relatively quickly. With the modal components determined, the complete solution can be reconstructed as

$$\mathbf{u}(t) = \sum_{n=1}^{N} \boldsymbol{\varphi}_n \frac{p_n}{\omega_n^2 \sqrt{D_n}} \sin\left(\Omega t - \alpha_n\right)$$

We can look at the displacement at one degree of freedom, say DOF m. To extract that, we premultiply the equation by $\mathbf{e}_m^T$ (the mth base vector that has a one in the mth slot and a zero in all of the others). To wit,

$$u_m(t) = \sum_{n=1}^{N} \frac{q_{mn}\, p_n}{\omega_n^2 \sqrt{D_n}} \sin\left(\Omega t - \alpha_n\right)$$

where $q_{mn} = \mathbf{e}_m^T \boldsymbol{\varphi}_n$ is the mth term of mode shape $\boldsymbol{\varphi}_n$.

There are a few things to notice about this solution. First, each modal component oscillates at the frequency Ω. Each sinusoidal component has a different phase angle, but that angle remains constant for each mode throughout time. Since the sine function oscillates between -1 and 1, the magnitude of the response comes from the other terms. Each term is inversely proportional to the square of its respective natural frequency. That means that the higher frequencies contribute significantly less than the lower frequencies. Second, p_n is the dot product between the shape of the force distribution and the nth mode shape. Thus, the contribution of each mode depends upon how much excitation happens in the nth mode. Third, the factor q_{mn} measures how much response comes out at DOF m. For the shear building we would typically expect the top level to have the greatest response. Finally, the response is inversely proportional to the square root of D_n.

Let's take a closer look at D_n. In the case where there is no damping, D_n goes to zero at each β_n. Therefore, the solution blows up each time the driving frequency matches a natural frequency. That implies that resonance happens at all natural frequencies. Two cases are shown in Fig. 4.11. The first case has the driving frequency equal to the first natural frequency, while the second has the driving frequency equal to the second natural frequency. You can see the linear increase in the amplitude of vibration with time in each case, but it is clear that the rate of increase is not the same, being much larger in the first case than the second.

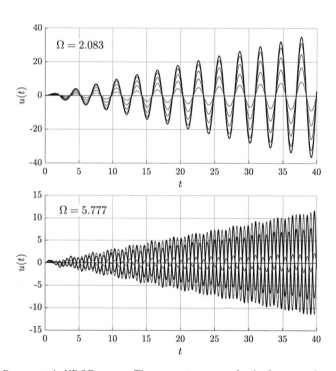

Fig. 4.11 *Resonance in NDOF systems.* The resonant response for the five-story shear building in Fig. 4.9, subject to a load $p_o = 10 \sin \Omega t$ at the top level. The system is undamped. The plot on top has $\Omega = \omega_1$. The plot on the bottom has $\Omega = \omega_2$

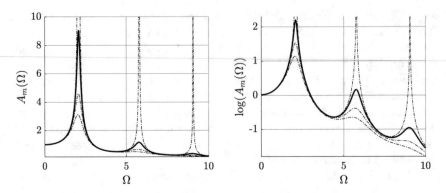

Fig. 4.12 *Resonance response spectrum.* The response spectrum for the five-story shear building in Fig. 4.9, subject to a load $p_o = 10 \sin \Omega t$ at the top. The dark curve is for $\xi = 0.05$. The curve above is undamped (i.e., $\xi = 0$). The two curves below are for $\xi = 0.1$ and $\xi = 0.2$

Figure 4.12 shows what happens for nonzero damping. The figure is a plot of the resonance function

$$A_m(\Omega) = \sum_{n=1}^{N} \frac{|q_{mn}| \, |p_n|}{\omega_n^2 \sqrt{D_n}}$$

for different values of the damping ratio, ξ and m is the top level. The dark solid line is the value $\xi = 0.05$. The dotted line above that is the curve for the undamped case (i.e., $\xi = 0$). The two dotted lines below the solid curve are for values $\xi = 0.1$ and $\xi = 0.2$. There are two views of the same function. The left plot is the function and the right plot is the natural logarithm of the function. This particular plot is for the structure shown in Fig. 4.9. The only salient part of the structure for this function is the natural frequency spectrum, in this case with $\omega_1 = 2.083$, $\omega_2 = 5.777$, and $\omega_3 = 9.057$.

In the first view it is evident how quickly the height of the peaks of the function diminishes due to the inverse proportionality to ω_n^2. The peak around each natural frequency gets narrower for the higher modes. That has two implications. First, the amplification of the motion due to resonance is greatly diminished if the driving frequency is slightly different from the natural frequency. Second, since the damped curves live under the undamped one, the amplitude increase due to resonance is dramatically influenced by damping. The plot of the logarithm of the function affords a better look at the nature of the peaks for all of the natural frequencies.

The resonance function supports the observations one can make by doing the analysis of the different cases with damping. For example, if we drive the system at $\Omega = 2.083$ (with $\xi = 0.05$), the amplitude of the motion is asymptotic to approximately $u_{max} = 8.8$. If we drive the system at $\Omega = 5.777$, the amplitude is asymptotic to approximately $u_{max} = 1.0$. The response spectrum predicts these values pretty well. If you drive the system at a higher frequency, it is not uncommon for the early transient phase to show a larger amplitude than the asymptotic steady-state prediction.

4.7 Numerical Integration of the NDOF Equations

It is straightforward to implement numerical integration of the equations of motion for the NDOF system because the integrators (e.g., Newmark or the generalized trapezoidal rule) apply to each displacement component separately. Only the equation of motion is coupled.

As usual, we will denote displacement, velocity, and acceleration at time t_n as $\mathbf{u}_n$, $\mathbf{v}_n$ and $\mathbf{a}_n$, respectively.[5] Also, let $\mathbf{f}_n = \mathbf{f}(t_n)$ be the value of the force vector at time t_n. If we define $\eta = h(1 - \gamma)$ and $\zeta = h^2 (0.5 - \beta)$ and let

$$\mathbf{b}_n = \mathbf{u}_n + h\mathbf{v}_n + h^2 \beta\, \mathbf{a}_n, \qquad \mathbf{c}_n = \mathbf{v}_n + h\gamma\, \mathbf{a}_n \qquad (4.53)$$

be the portion of the displacement and velocity at step $n+1$ that can be computed from information at step n, then we can write Newmark's equations as

$$\mathbf{u}_{n+1} = \mathbf{b}_n + \zeta\, \mathbf{a}_{n+1}, \qquad \mathbf{v}_{n+1} = \mathbf{c}_n + \eta\, \mathbf{a}_{n+1} \qquad (4.54)$$

and the equations of motion in discrete form at time step $n+1$ as

$$\mathbf{M}\mathbf{a}_{n+1} + \mathbf{C}\mathbf{v}_{n+1} + \mathbf{K}\mathbf{u}_{n+1} = \mathbf{f}_{n+1} \qquad (4.55)$$

With these definitions, we can rewrite the discrete equation of motion as

$$\mathbf{M}\mathbf{a}_{n+1} + \mathbf{C}(\mathbf{c}_n + \eta\, \mathbf{a}_{n+1}) + \mathbf{K}(\mathbf{b}_n + \zeta\, \mathbf{a}_{n+1}) = \mathbf{f}_{n+1}$$

Rearranging terms we can solve for the new acceleration as

$$\mathbf{a}_{n+1} = [\mathbf{M} + \eta\, \mathbf{C} + \zeta\, \mathbf{K}]^{-1} (\mathbf{f}_{n+1} - \mathbf{C}\mathbf{c}_n - \mathbf{K}\mathbf{b}_n)$$

With the acceleration determined at the next time step, we can go back to Eqs. 4.54 and compute the velocity and displacement and the new time step. Once complete, we can move on to the next time step. One of the obvious upsides of numerical integration is that it is trivial to specify different loading functions $\mathbf{f}(t)$. One downside is that computation of the new acceleration requires the solution of a system of equations. For a linear problem the so-called *effective mass matrix*

---

[5] The acceleration $\mathbf{a}_n$ is not to be confused with the use of $\mathbf{a}(t)$ for the vector of modal components in modal analysis. If there is a subscript, then it is the discrete acceleration in the numerical integration scheme. If it is a function of time, then it is a vector of modal components.

$\mathbf{M} + \eta\mathbf{C} + \zeta\mathbf{K}$ can be factored once, so that the solution only requires forward reduction and back substitution at each time step.

In the numerical integration approach, we do not rely in any way on the orthogonality properties of the eigenvectors. There is no need to construct an orthogonal damping matrix. That said, the efficiency of the equation solution will depend upon the structure of the matrices. A full damping matrix would increase the storage requirements and solution times.

We will defer deeper exploration of the numerical solution of the NDOF system to the following chapters wherein we lay out a more systematic formulation of the problem and give consideration to nonlinear problems and earthquake excitation.

Chapter 5
Nonlinear Response of NDOF Systems

In this chapter, we lay out an approach to nonlinear analysis of the NDOF system with two objectives: (1) connect with the nonlinear models we introduced for the SDOF systems in Chap. 3 and (2) set up an approach to formulating the equations of motion in a way that will extend gracefully to trusses, frames, and other complex structures. In particular, we will establish a strategy based upon the *principle of virtual work* that leads to an algorithm for directly assembling the equations of motion. This approach will serve as a foundation for formulating equations of motion throughout the rest of the book. Once formulated, we will solve the equations of motion by extending the numerical tools already developed for the SDOF system.

In the history of structural analysis and design, static analysis long reigned as *the* approach to modeling the response of structures. That changed when earthquake engineering emerged as an important area of structural engineering. There is no way to deny the dynamic nature of earthquake excitation or the response of structures to that excitation. But beyond that, strong ground motions present huge, albeit rare, demands on structures and it is not economical to meet those demands without allowing some nonlinear response. Engineering of structures to withstand strong ground motions has provided the impetus for development of nonlinear analysis tools. Design methods based upon performance of the structure have these tools at their core.

There are *many* nonlinear models available in the literature, and it is a nontrivial extension to move from the single-component models (scalar force is a function of scalar deformation) to multi-component models (which we will develop in Chap. 14). While we only scratch the surface on modeling, this introduction should serve as a solid foundation on which to build. The codes presented in this chapter and

Electronic Supplementary Material The online version of this chapter (https://doi.org/10.1007/978-3-030-89944-8_5) contains supplementary material, which is available to authorized users.

K. D. Hjelmstad, *Fundamentals of Structural Dynamics*,
https://doi.org/10.1007/978-3-030-89944-8_5

the next will provide a route to investigating how nonlinearity affects the response of complex structures.

One of the biggest challenges in establishing a computational framework that is reliable and understandable is in the description of the structure—the labeling of physical features that lends itself to generalization. As such, we will start by looking at the shear building through a different lens. It should be obvious from the previous chapter that the shear building is simple enough that a reformulation seems unnecessary. On the other hand, the simplicity of the shear building makes it a great candidate for working through this task in preparation for more geometrically complicated structures (like trusses and frames). We will also see that this approach is valuable for addressing nonlinear analysis in a systematic way.

5.1   A Point of Departure

In our formulation of the equations of motion for the shear building in the last chapter, we isolated the story masses as free body diagrams and established the equations of motion using Newton's second law. Each story had an applied force and was subject to the restoring forces provided by the shears at the ends of the columns. Through a few algebraic substitutions, we were able to reorganize the equations of motion into the form

$$\mathbf{M}\ddot{\mathbf{u}}(t) + \mathbf{C}\dot{\mathbf{u}}(t) + \mathbf{K}\mathbf{u}(t) = \mathbf{f}(t)$$

That approach is valid but does not extend easily to more general types of structures with more complicated structural geometries, and it is not ideal for considering nonlinearity.

To formulate the equations of motion in this chapter, we will use a few strategies that are the essence of structural analysis. First, we solve the problem of establishing continuity of motion from one element to the next by introducing *global displacement* functions $\mathbf{u}(t)$ that are shared by the elements. Second, we introduce Boolean matrices $\mathbf{B}_e$ whose only job is to pick out the global displacements associated with element e from the global displacement vector $\mathbf{u}(t)$. The $\mathbf{B}_e$ matrices play an important role in the assembly of the equations of motion. Third, we will employ the *principle of virtual work* to reframe the equations of motion in a way that lends itself to direct assembly of equations from element contributions.

We will formulate the equations of motion assuming that the constitutive response of the columns can be nonlinear and use the Newmark time integrator to discretize the differential equation. The resulting equations will be nonlinear algebraic equations, which can be solved by Newton's method (see Appendix A.2). These algorithmic components will combine to form a reliable framework for writing a computer code to carry out the computations associated with nonlinear structural dynamics.

5.2   The Shear Building, Revisited

Let us conceptualize the shear building as a set of *nodes* with associated mass (i.e., the rigid floors) connected together by *elements* that provide the restoring forces (i.e., the columns). For mathematical simplicity, consider the columns to be massless and the masses to be rigid (and constrained to move in a horizontal direction, which implies that the columns are axially rigid).

The key to describing the geometry and topology of the structure is to give the position of the nodes and to indicate which nodes each element is attached to. A simplified version of the shear building is shown as an example in Fig. 5.1. Number the nodes sequentially from 1 to 5, starting at the bottom. Note that we include the bottom node, even though its motion is constrained by a boundary condition. Number the elements from 1 to 4 starting at the bottom. The element numbers are independent from the node numbers, even though for the shear building there is a clearly discernible pattern.

Nodal Coordinates To describe the location of the nodes, we need only give one coordinate, its vertical distance from the base of the structure (which we take as the origin). We call the coordinate locations $\mathbf{x} = [x_1, x_2, \ldots, x_N]^T$, where each subscript indicates a node number and N is the number of nodes. For the structure in Fig. 5.1 (assuming a story height of 10), the nodal coordinate array is

$$\mathbf{x} = \begin{bmatrix} 0 & 10 & 20 & 30 & 40 \end{bmatrix}^T$$

This definition establishes the *nodal numbering scheme*. Setting $x_1 = 0$ establishes that node 1 is the node located at $x = 0$. Likewise, setting $x_2 = 10$ establishes that node 2 is the node located at $x = 10$. The creation of the $\mathbf{x}$ array corresponds with the numbering of the nodes, and vice versa. The node number is the row index of the array $\mathbf{x}$.

Fig. 5.1 *Description of shear building.* The structure comprises a set of nodes connected together by elements. The nodes and elements are identified by their number, which can be assigned arbitrarily but will be sequential starting with 1. In the sketch on the left, the element numbers are circled, the node numbers are not. In the sketch on the right, we number the displacements and masses in accord with their DOF number

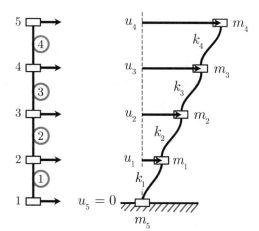

Boundary Conditions The way we will take care of the boundary conditions in our code is through the id array. In the input we will enter a "1" if a node is fixed and a "0" if it is free. For our example structure, then, id $= [\, 1, 0, 0, 0, 0\,]$. To establish the global DOF numbering, we work our way through the id array and each time we see a "0" we increment the positive DOF counter by one. Each time we encounter a "1" we decrement the negative DOF counter by one.

After this sweep we have id $= [-1, 1, 2, 3, 4]$. The largest positive number in the array at this point is equal to the number of *free* degrees of freedom we have (4 in this case). We will call this number nFree. We sweep through the array one last time and for each negative value we replace it with its absolute value plus nFree. After the final sweep, id $= [\, 5, 1, 2, 3, 4\,]$. That becomes the global degree of freedom assignments for the nodes. While this might seem like overkill for the shear building, it is essential for more complex structures like trusses and frames.

There are two things that are important about the id array. First, it represents the equation numbers for our system of equations. Those need to be unique and sequentially numbered, starting with one. Second, we number our restrained degrees of freedom *after* the free degrees of freedom. This allows us to partition the equations very simply as we assemble them. The only thing that matters in this equation numbering scheme is whether or not the motion associated with the degree of freedom is restrained.

Element Connections We also establish an *element connection* array ix, where each row of the array is associated with an element. The first column is called the *start* node, and the second column is called the *end* node. When we describe which nodes an element *e* is connected to, we will do so in an oriented way—from start node to end node. While this labeling is arbitrary, once you establish it, then that is the convention for that structure. The ix array for the structure in Fig. 5.1 is

$$\mathtt{ix} = \begin{bmatrix} 1 & 2 \\ 2 & 3 \\ 3 & 4 \\ 4 & 5 \end{bmatrix} \tag{5.1}$$

Notice that the first row corresponds to element $e = 1$, and that element is connected to nodes 1 and 2. That establishes the *start* node to be 1 and the *end* node to be 2. The ix array establishes which element we mean when we refer to element *e*.

Nodal Displacements and Forces To track the displacement of the structure, we define the (global) displacement array $\mathbf{u}(t)$. Similarly, we define a nodal force array $\mathbf{f}(t)$. These arrays are organized in degree-of-freedom order (*not* node order) as

$$\mathbf{u}(t) = \begin{Bmatrix} u_1(t) \\ \vdots \\ u_N(t) \end{Bmatrix} \qquad \mathbf{f}(t) = \begin{Bmatrix} f_1(t) \\ \vdots \\ f_N(t) \end{Bmatrix} \tag{5.2}$$

where each $u_i(t)$ represents the displacement of DOF i. Each $f_i(t)$ represents the externally applied force at DOF i. Note that the DOF number is *not* the same as the node number. For the shear building the node at the base of the structure is numbered last and is associated with no motion. The force $f_N(t)$ is the force associated with DOF N. Since DOF N is restrained, the associated force is a *reaction force*, which means that it is unknown at the outset of the analysis (and is something that we will seek as a response quantity). In the code we will input the force vector in node order and then convert it to DOF order before starting the solution.

Element Deformation The key kinematic quantity that we need to track for the elements of the shear building is the relative deformation of the columns, which is the difference between their end displacements. For element e, it is the difference between the displacement at the end node $j(e)$ and the start node $i(e)$. In other words,

$$\Delta_e = u_{j(e)} - u_{i(e)} \tag{5.3}$$

The *only* thing that affects the force in element e is the deformation Δ_e. Let us define a $2 \times N$ matrix $\mathbf{B}_e$ as

$$\mathbf{B}_e = \begin{bmatrix} 0 \dots 1 & 0 \dots 0 \\ 0 \dots 0 & 1 \dots 0 \end{bmatrix} \tag{5.4}$$

where the 1 in the first row of matrix $\mathbf{B}_e$ is in the column with the DOF number associated with the *start* node $i(e) = \text{id}(\text{ix}(e, 1))$, and the 1 in the second row of matrix $\mathbf{B}_e$ is in the column with the DOF number associated with the *end* node $j(e) = \text{id}(\text{ix}(e, 2))$ of element e. There is one $\mathbf{B}_e$ array for each element e. What this matrix does is pick out the two displacements associated with element e

$$\mathbf{u}_e = \begin{Bmatrix} u_{i(e)} \\ u_{j(e)} \end{Bmatrix} = \mathbf{B}_e \mathbf{u}$$

These Boolean matrices are very helpful in formulating our equations, performing a task that we call *localization*. For our example structure we have

$$\mathbf{B}_1 = \begin{bmatrix} 0 & 0 & 0 & 0 & 1 \\ 1 & 0 & 0 & 0 & 0 \end{bmatrix} \quad \mathbf{B}_2 = \begin{bmatrix} 1 & 0 & 0 & 0 & 0 \\ 0 & 1 & 0 & 0 & 0 \end{bmatrix}$$
$$\mathbf{B}_3 = \begin{bmatrix} 0 & 1 & 0 & 0 & 0 \\ 0 & 0 & 1 & 0 & 0 \end{bmatrix} \quad \mathbf{B}_4 = \begin{bmatrix} 0 & 0 & 1 & 0 & 0 \\ 0 & 0 & 0 & 1 & 0 \end{bmatrix} \tag{5.5}$$

Because $\mathbf{B}_e$ is a Boolean matrix and because we know explicitly from the ix and id arrays where the 1 and 0 entries are, we will not need to actually form this matrix in a computer implementation of the theory. For derivation, however, it will be very useful. In Sect. 5.5 we demonstrate how this matrix helps us to assemble the equations.

Now the deformation in the element can be computed as

$$\Delta_e = \boldsymbol{\phi}^T \mathbf{u}_e = \boldsymbol{\phi}^T \mathbf{B}_e \mathbf{u} \tag{5.6}$$

where $\boldsymbol{\phi}^T = [-1, 1]$. Basically, the function of the $\boldsymbol{\phi}$ matrix is to operate on the element displacement array $\mathbf{u}_e$ and produce the relative displacement between the two ends (i.e., the difference between $u_{j(e)}$ and $u_{i(e)}$).

5.3 The Principle of Virtual Work

To begin the formulation of the equations of motion, note that the equation of motion for degree of freedom i is given by

$$m_i \ddot{u}_i - V_{a(i)} + V_{b(i)} - f_i = 0 \tag{5.7}$$

where $V_{a(i)}$ is the element shear force from the members *above* the node with DOF number i, and $V_{b(i)}$ is the element force from the members *below* the node with DOF number i. The forces $V_e(\Delta_e)$ are, strictly speaking, associated with the elements, and they are functions of the deformation within the element they come from. The applied and internal forces for our example structure are shown in Fig. 5.2.

The node associated with the fixed boundary at the base deserves comment. For the shear building, degree of freedom N is associated with that boundary point. We can associate a mass with that node (and do so in the code), but that mass is irrelevant because $u_N(t) = 0$ and hence $\ddot{u}_N(t) = 0$. Because of the constraint of motion, the force V_b for the equation of motion for that node is equal to the unknown reaction

Fig. 5.2 *Nodal forces.* The structure degrees of freedom and element numbers are shown at left. The free body diagrams of each node are shown at right. The force V_e is the force in element e. The force f_i is the applied force at DOF i. Note that R is a reaction force because the motion is constrained to be zero at the boundary (i.e., $u_5 = 0$)

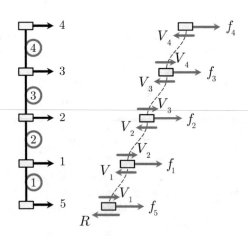

force R. The applied load at that node (f_5 in Fig. 5.2) is irrelevant because a reaction force has the property that it will equilibrate all other forces exerted at that point. In practice, we can eliminate R from the free body diagram for the base node and simply interpret f_N as being the reaction force.

Imagine that each degree of freedom is given a virtual displacement w_i, where i is the DOF number associated with that node. Define the *virtual work function* by taking the residual for each degree of freedom, multiplying by the virtual displacement, and summing over all DOF in the structure. To wit,

$$\mathcal{G} = \sum_{i=1}^{N} \left(m_i \ddot{u}_i - V_{a(i)} + V_{b(i)} - f_i \right) w_i \tag{5.8}$$

If $\mathcal{G} = 0$ for all w_i, then the equations of motion are satisfied for all i. That is the *principle of virtual work*. The proof of the assertion is simple. Assume that the residual (i.e., the term in parentheses) is *not* zero. Now, $\mathcal{G}$ must be zero for *any* w_i, including the case where it is equal to the term in parenthesis. A quantity squared is always positive. The only way positive values can sum to zero is if all of them are individually equal to zero. This is the *fundamental theorem of the calculus of variations*. Thus, the principle of virtual work is an alternative, but equivalent, way of representing Eq. 5.7.

We will do a little manipulation of the virtual work function to put it in a more useful form for computing. To avoid the equation involving the reaction R at the base, we will take the virtual displacement $w_5 = 0$. In general, at a support point where the real displacement is constrained (usually to zero), the virtual displacement will be taken as zero. In essence, that simply omits the contribution of the equation associated with the constrained node from the sum in Eq. 5.8. The cost is that the principle of virtual work will not give us back the equation for the reaction force at the support, which will not be a problem if we can find it some other way (in this case by recognizing that it must be equal to the shear in the first element).

We can convert the sum over nodes to a sum over elements as

$$\sum_{i=1}^{N} \left(V_{b(i)} - V_{a(i)} \right) w_i = \sum_{e=1}^{M} V_e \left(w_{j(e)} - w_{i(e)} \right) \tag{5.9}$$

where M is the number of elements, $w_{i(e)}$ is the virtual displacement of global DOF $i(e)$, and $w_{j(e)}$ is the virtual displacement of global DOF $j(e)$. In the present example $M = 4$ and $N = 5$. Write out the terms to see how the sum works (noting that $w_5 = 0$):

$$\sum_{i=1}^{N} \left(V_{b(i)} - V_{a(i)} \right) w_i = \begin{array}{l} (V_1 - V_2)w_1 + (V_2 - V_3)w_2 \\ + (V_3 - V_4)w_3 + (V_4 - 0)w_4 \end{array}$$

Similarly, we can write out the sum over elements as

$$\sum_{e=1}^{M} V_e\left(w_{j(e)} - w_{i(e)}\right) = \begin{aligned} & V_1(w_1 - 0) + V_2(w_2 - w_1) \\ & + V_3(w_3 - w_2) + V_4(w_4 - w_3) \end{aligned}$$

Observe that, term by term, these two sums are identical. Equation 5.8 can, therefore, be rewritten in the form

$$\sum_{i=1}^{N} \left(m_i \ddot{u}_i - f_i\right)w_i + \sum_{e=1}^{M} V_e\left(w_{j(e)} - w_{i(e)}\right) = 0 \qquad (5.10)$$

Further, since the virtual displacements $\mathbf{w} = [\, w_1, \ldots, w_N \,]^T$ are stored in exactly the same order as the real displacements $\mathbf{u}$, note that

$$w_{j(e)} - w_{i(e)} = \boldsymbol{\phi}^T \mathbf{B}_e \mathbf{w}$$

Hence, we can write the virtual work function as

$$\mathcal{G} = \sum_{i=1}^{N} \left(m_i \ddot{u}_i - f_i\right)w_i + \sum_{e=1}^{M} V_e\left(\Delta_e\right) \boldsymbol{\phi}^T \mathbf{B}_e \mathbf{w}$$

If we define the *mass matrix* $\mathbf{M}$ so that $M_{ii} = m_i$, where the index i is associated with the DOF number, and note that the force vector $\mathbf{f}(t)$ is stored with the same index ordering as $\mathbf{u}$ and $\mathbf{w}$, then

$$\sum_{i=1}^{N} \left(m_i \ddot{u}_i - f_i\right)w_i = \mathbf{w}^T(\mathbf{M}\ddot{u} - \mathbf{f})$$

To prove it, just write out the terms of the matrix products. With this notation, we can write the virtual work function in the final form

$$\mathcal{G} = \mathbf{w}^T \left(\mathbf{M}\ddot{u} + \sum_{e=1}^{M} \mathbf{B}_e^T \boldsymbol{\phi}\, V_e\left(\Delta_e\right) - \mathbf{f}\right) \qquad (5.11)$$

If $\mathcal{G} = 0$ is to be true for any choice of the virtual displacements $\mathbf{w}$, then

$$\mathbf{M}\ddot{u} + \sum_{e=1}^{M} \mathbf{B}_e^T \boldsymbol{\phi}\, V_e\left(\Delta_e\right) - \mathbf{f} = \mathbf{0} \qquad (5.12)$$

Equation 5.12 is the *equation of motion* of the nonlinear shear building. The first term is the inertial resistance, the second term is the resistance provided by the elements, and the third term is the applied force. The restoring forces provided by the columns can be *assembled* from the element forces V_e, using the $\mathbf{B}_e$ operator. Note that each element contributes forces to two nodes. Equation 5.12 is a *nodal* equation (i.e., it still represents the equation of motion of each node), but we are adding the contributions of the element forces two at a time, element by element, rather than adding in the forces associated with each node, as we did in Eq. 5.7.

Equation 5.7 is important not only because it shows the way to automatic assembly of the equations of motion (which is essential for programming) but also because the element forces $V_e(\Delta_e)$ show up explicitly in the equation. That puts us in a better position to implement nonlinear element response.

The Linear Shear Building Revisited Before we tackle the nonlinear problem, it is instructive to revisit the linear one. Note that for the linear problem the column forces are

$$V_e = k_e \Delta_e = k_e \boldsymbol{\phi}^T \mathbf{B}_e \mathbf{u} \tag{5.13}$$

Substituting this expression into Eq. 5.12, we get

$$\mathbf{M}\ddot{\mathbf{u}} + \sum_{e=1}^{M} \mathbf{B}_e^T \boldsymbol{\phi} \, k_e \boldsymbol{\phi}^T \mathbf{B}_e \mathbf{u} - \mathbf{f} = \mathbf{0} \tag{5.14}$$

If we define the *element stiffness matrix* $\mathbf{k}_e = \boldsymbol{\phi} \, k_e \boldsymbol{\phi}^T$, then the second term, which involves a sum over elements, is simply $\mathbf{Ku}$, where the global stiffness matrix is assembled from element stiffnesses as

$$\mathbf{K} = \sum_{e=1}^{M} \mathbf{B}_e^T \mathbf{k}_e \mathbf{B}_e$$

We will discuss the assembly process in more detail in Sect. 5.5. This process of assembling global matrices from element contributions is one of the most important innovations in structural analysis. It should be evident that it is the result of the principle of virtual work.

5.4 Nonlinear Dynamic Computations

Equation 5.12 is a nonlinear second order differential equation. The only route available to solve this equation is numerical integration. We will couple Newmark's method, which will take care of integrating the differential equation in time, and Newton's method, which will take care of solving the nonlinear algebraic equations that result from the temporal discretization.

The state of motion will be characterized by discrete arrays that approximate the unknown function $\mathbf{u}(t)$ and its derivatives: $\mathbf{u}_n = \mathbf{u}(t_n)$, $\mathbf{v}_n = \dot{\mathbf{u}}(t_n)$, and $\mathbf{a}_n = \ddot{\mathbf{u}}(t_n)$. Recall that the Newmark integrator can be written as

$$\mathbf{u}_{n+1} = \mathbf{b}_n + \zeta\, \mathbf{a}_{n+1}$$
$$\mathbf{v}_{n+1} = \mathbf{c}_n + \eta\, \mathbf{a}_{n+1}$$

(5.15)

where $\mathbf{b}_n$, $\mathbf{c}_n$, ζ, and η are given by Eqs. 4.53. We can think of these equations plus the equation of motion as three vector equations in three vector unknowns—the new state $\mathbf{u}_{n+1}$, $\mathbf{v}_{n+1}$, and $\mathbf{a}_{n+1}$. Notice that the Newmark equations for the displacement and velocity are very simple linear functions of the new acceleration. We will think of displacement and velocity as being a function of acceleration, as in $\mathbf{u}_{n+1}(\mathbf{x})$ and $\mathbf{v}_{n+1}(\mathbf{x})$, where $\mathbf{x} = \mathbf{a}_{n+1}$ is the unknown acceleration at the next time step. The Newton iteration, then, involves the single unknown $\mathbf{x}$.

We can define the *residual* as being a function of the unknown new acceleration $\mathbf{x}$. Let

$$\mathbf{g}(\mathbf{x}) = \mathbf{M}\mathbf{x} + \sum_{e=1}^{M} \mathbf{B}_e^T\, \boldsymbol{\phi}\, V_e\, (\Delta_e\, (\mathbf{b}_n + \zeta\mathbf{x})) - \mathbf{f}_{n+1}$$

(5.16)

where the function $V_e(\Delta_e)$ is expressed as it is because Δ_e depends upon the displacement $\mathbf{u}$, which depends upon the acceleration $\mathbf{x}$ (because of Newmark). Now we can think of the solution process through the lens of Newton's method. If we can find an acceleration $\mathbf{x}$ that satisfies $\mathbf{g}(\mathbf{x}) = \mathbf{0}$, then that acceleration is the new acceleration we seek, i.e., $\mathbf{a}_{n+1}$.

To execute Newton's method, we need to compute the tangent matrix for the residual $\mathbf{g}$. We will use the directional derivative (see Appendix B). The first term is easy:

$$\mathcal{D}(\mathbf{M}\mathbf{x}) \cdot \mathbf{h} = \frac{d}{d\varepsilon}\Big[\mathbf{M}(\mathbf{x} + \varepsilon\mathbf{h})\Big]_{\varepsilon=0} = \mathbf{M}\mathbf{h}$$

To compute the directional derivative of the second term, let us first compute the directional derivative of $V_e(\Delta_e(\mathbf{u}(\mathbf{x})))$. We will start from the inside and work outward. The directional derivative of $\mathbf{u}(\mathbf{x})$, from the Newmark equations, is

$$\mathcal{D}(\mathbf{u}(\mathbf{x})) \cdot \mathbf{h} = \frac{d}{d\varepsilon}\Big[\mathbf{b}_n + \zeta(\mathbf{x} + \varepsilon\mathbf{h})\Big]_{\varepsilon=0} = \zeta\mathbf{h}$$

From the relationship $\Delta_e = \boldsymbol{\phi}^T \mathbf{B}_e \mathbf{u}$, noting that the first two terms are constant, we can compute the directional derivative

$$\mathcal{D}(\Delta_e(\mathbf{u})) \cdot \mathbf{h} = \boldsymbol{\phi}^T \mathbf{B}_e(\mathcal{D}\mathbf{u} \cdot \mathbf{h}) = \zeta\boldsymbol{\phi}^T \mathbf{B}_e\mathbf{h}$$

Now, by the chain rule we have

$$\mathcal{D}\big(V_e(\Delta_e)\big) \cdot \mathbf{h} = \frac{\partial V_e}{\partial \Delta_e}\big(\mathcal{D}\Delta_e \cdot \mathbf{h}\big) = \zeta\left(\frac{\partial V_e}{\partial \Delta_e}\right)\boldsymbol{\phi}^T \mathbf{B}_e \mathbf{h}$$

Finally, putting all of this together, we can compute the directional derivative of $\mathbf{g}$ as

$$\mathcal{D}\mathbf{g} \cdot \mathbf{h} = \mathbf{M}\mathbf{h} + \zeta \sum_{e=1}^{M} \mathbf{B}_e^T \boldsymbol{\phi}\left(\frac{\partial V_e}{\partial \Delta_e}\right)\boldsymbol{\phi}^T \mathbf{B}_e \mathbf{h}$$

Since $\mathcal{D}\mathbf{g} \cdot \mathbf{h} = \mathbf{A}\mathbf{h}$, the tangent matrix for the nonlinear shear building is

$$\mathbf{A} = \mathbf{M} + \zeta \sum_{e=1}^{M} \mathbf{B}_e^T \boldsymbol{\phi}\left(\frac{\partial V_e}{\partial \Delta_e}\right)\boldsymbol{\phi}^T \mathbf{B}_e \tag{5.17}$$

There is quite a bit to unpack from Eq. 5.17. First, note that the tangent is basically the mass matrix plus a little more because ζ is a small number for a small time step h. This is good news because the mass matrix is always positive definite, so the equation solving in Newton's method should be generally well conditioned.

Let us take a closer look at the second term. Note that $\partial V_e / \partial \Delta_e$ is a scalar and represents the slope of the V_e versus Δ_e curve. In other words, it is the tangent to the force/deformation relationship of element e. This quantity is simple to compute for all of the models we have considered. In fact, it is identical to what we used for the nonlinear SDOF model. For example, the nonlinear, elastic softening model has

$$V_e(\Delta_e) = \frac{k_e \Delta_e}{\sqrt{1 + C_e \Delta_e^2}}$$

where k_e and C_e are material constants for element e. The derivative of this function is

$$\frac{\partial V_e}{\partial \Delta_e} = \frac{k_e}{\left(1 + C_e \Delta_e^2\right)^{3/2}}$$

It is also interesting to note that for a linear model $V_e(\Delta_e) = k_e \Delta_e$ and the derivative is $\partial V_e / \partial \Delta_e = k_e$. In the next section we explore the assembly of the residual and the tangent matrix.

5.5   Assembly of Equations

To execute Newton's method, we need to form the residual and tangent arrays through summation over elements as

$$\mathbf{r} = \sum_{e=1}^{M} \mathbf{B}_e^T \mathbf{r}_e \qquad \mathbf{K} = \sum_{e=1}^{M} \mathbf{B}_e^T \mathbf{k}_e \mathbf{B}_e$$

where the element residual and tangent are defined as

$$\mathbf{r}_e = \boldsymbol{\phi}\, V_e = \begin{Bmatrix} -V_e \\ V_e \end{Bmatrix} \qquad \mathbf{k}_e = \boldsymbol{\phi}\, k_e\, \boldsymbol{\phi}^T = \begin{bmatrix} k_e & -k_e \\ -k_e & k_e \end{bmatrix} \qquad (5.18)$$

and $k_e = \partial V_e / \partial \Delta_e$.

To see how this calculation goes, let's look at one of the contributions to that sum. For our example structure, for element $e = 3$ we can compute the matrix product $\mathbf{B}_3^T \mathbf{k}_3 \mathbf{B}_3$ as (where $\mathbf{B}_3$ is given in Eq. 5.5)

$$\begin{bmatrix} 0 & 0 \\ 1 & 0 \\ 0 & 1 \\ 0 & 0 \\ 0 & 0 \end{bmatrix} \begin{bmatrix} k_3 & -k_3 \\ -k_3 & k_3 \end{bmatrix} \begin{bmatrix} 0 & 1 & 0 & 0 & 0 \\ 0 & 0 & 1 & 0 & 0 \end{bmatrix} = \begin{bmatrix} 0 & 0 & 0 & 0 & 0 \\ 0 & k_3 & -k_3 & 0 & 0 \\ 0 & -k_3 & k_3 & 0 & 0 \\ 0 & 0 & 0 & 0 & 0 \\ 0 & 0 & 0 & 0 & 0 \end{bmatrix}$$

and $\mathbf{B}_3^T \boldsymbol{\varphi}\, V_3$ as

$$\begin{bmatrix} 0 & 0 \\ 1 & 0 \\ 0 & 1 \\ 0 & 0 \\ 0 & 0 \end{bmatrix} \begin{Bmatrix} -V_3 \\ V_3 \end{Bmatrix} = \begin{Bmatrix} 0 \\ -V_3 \\ V_3 \\ 0 \\ 0 \end{Bmatrix}$$

The most interesting observation about this operation is that the element stiffness $\mathbf{k}_e$ simply gets placed in the $i(e)$ and $j(e)$ rows and columns of the global stiffness matrix and the element residual $\mathbf{r}_e$ gets placed in the $i(e)$ and $j(e)$ rows of the global residual. We know $i(e)$ and $j(e)$ because they are stored implicitly in the arrays id and ix. In the process of summing over the elements, some of the locations in the global matrices get contributions from more than one element. For example, the element $e = 4$ contributes to the (3,3) location just as element $e = 3$ did. Thus, the final global tangent matrix has $k_3 + k_4$ in the (3,3) slot.

To see how to code this in MATLAB, let us consider a simple example. We will eventually show the general case, but to see the basic structure of the code, take a look at the following script:

Code 5.1 MATLAB code for assembly for the example structure

```
%. Set initial problem data
   M = 4; N = 5;
   id = [5,1,2,3,4];
   ix = [1,2; 2,3; 3,4; 4,5];
   ke = [5,4,3,2];  % element stiffness
   Ve = [5,7,9,3];  % element internal force

%. Initialize residual and tangent
   A = zeros(N,N);
   r = zeros(N,1);

%. Assemble the matrices by elements
   for e=1:M
      inode = ix(e,1); jnode = ix(e,2);
      ii = [id(inode); id(jnode)];
      Ae = ke(e)*[1,-1;-1,1];
      re = Ve(e)*[-1;1];
      A(ii,ii) = A(ii,ii) + Ae;
      r(ii) = r(ii) + re;
   end
```

There are a few things to notice about the above code. First, we established values of V_e and k_e arbitrarily for the purposes of illustration. In a structural dynamics code, those would generally be computed based upon the results of the Newton iteration.

The most important observation is that we do not form $\mathbf{B}_e$ or execute the matrix multiplications shown above. The algorithm of direct assembly is equivalent. In the loop over elements, we first look up $i(e)$ and $j(e)$ (called inode and jnode in the code) from the ix array. Recall that this array points to *node* numbers. We then look up the global DOF number in the id array. Assembly is done in accord with DOF numbers.

This code makes use of MATLAB's approach to indexing arrays. We build the 2×1 array ii based upon which DOF the element connects to. Then the 2×2 element stiffness $\mathbf{A}_e$ and the 2×1 element residual $\mathbf{r}_e$ gets assembled into the rows (and columns for $\mathbf{A}_e$) associated with the values in ii.

For reference, the result of running Code 5.1 gives

$$
\mathbf{A} = \begin{bmatrix} 9 & -4 & 0 & 0 & -5 \\ -4 & 7 & -3 & 0 & 0 \\ 0 & -3 & 5 & -2 & 0 \\ 0 & 0 & -2 & 2 & 0 \\ -5 & 0 & 0 & 0 & 5 \end{bmatrix}, \quad \mathbf{r} = \begin{Bmatrix} -2 \\ -2 \\ 6 \\ 4 \\ -5 \end{Bmatrix}
$$

These results can be easily verified by a hand calculation. Draw free body diagrams of the nodes and compute the net force on the nodes from the element shears to show that $\mathbf{r}$ is correct. As an exercise, it is also instructive to form $\mathbf{A}$ explicitly using the $\mathbf{B}_e$ matrices directly.

The fact that there is no need to form the $\mathbf{B}_e$ matrices or to execute the matrix multiplications that show up in Eq. 5.17 is a practical difference between derivation and implementation. The $\mathbf{B}_e$ matrices are crucial to the mathematical *formulation* of the tangent stiffness and residual. We will use this same approach to formulate all of the theories in this book. Doing so will allow the very simple assembly process to be identical for all applications.

5.6 Adding Damping to the Equations of Motion

We have derived the nonlinear equations of motion for the undamped case. We can easily include damping by incorporating the linear viscous model discussed in Chap. 4. If we do so, then the discrete equations of motion are

$$\mathbf{M}\ddot{\mathbf{u}}(t) + \mathbf{C}\dot{\mathbf{u}}(t) + \sum_{e=1}^{M} \mathbf{B}_e^T \, \boldsymbol{\phi} \, V_e\,(\Delta_e(\mathbf{u}(t))) = \mathbf{f}(t) \tag{5.19}$$

where $\mathbf{C}$ is the damping matrix, which we can construct as shown in Sect. 4.4 as either Rayleigh, Caughey, or modal damping. The damping matrix depends upon the natural frequencies and mode shapes, and those only exist for the linear system. For nonlinear problems it is usually the case that in the neighborhood of the undeformed configuration, there exists a range of behavior that is essentially linear and elastic. We can build the damping matrix from the mass and linear stiffness matrix. The eigenvectors will orthogonalize the mass and damping matrices, but as the response gets into the nonlinear range, the tangent stiffness will depart from the linear stiffness. Thus, the eigenvectors will *not* diagonalize the tangent stiffness matrix. When integrating the equations of motion numerically with the Newmark integrator, we do not depend upon the modal properties to advance the solution.

We can modify the residual given in Eq. 5.16 to include damping. To wit,

$$\mathbf{g}\,(\mathbf{x}) = \mathbf{M}\mathbf{x} + \mathbf{C}\mathbf{v}_{n+1}(\mathbf{x}) + \sum_{e=1}^{M} \mathbf{B}_e^T \, \boldsymbol{\phi} \, V_e\,(\Delta_e\,(\mathbf{u}_{n+1}(\mathbf{x}))) - \mathbf{f}_{n+1} \tag{5.20}$$

where $\mathbf{v}_{n+1}(\mathbf{x}) = \mathbf{c}_n + \eta\,\mathbf{x}$ and $\mathbf{u}_{n+1}(\mathbf{x}) = \mathbf{b}_n + \zeta\mathbf{x}$ are the velocity and displacement at the next time step, as estimated by the Newmark integrator. The tangent matrix associated with this residual is

$$\mathbf{A} = \mathbf{M} + \eta\,\mathbf{C} + \zeta\,\sum_{e=1}^{M} \mathbf{B}_e^T \, \boldsymbol{\phi}\,\left(\frac{\partial V_e}{\partial \Delta_e}\right) \boldsymbol{\phi}^T \mathbf{B}_e \tag{5.21}$$

It should be clear that adding damping to the model is fairly straightforward. It is possible to have more complex models (e.g., a velocity-squared fluid drag model) for which the velocity term would also be nonlinear. For such a case, we would use the directional derivative approach to find the appropriate contribution to the tangent matrix.

5.7 The Structure of the NDOF Code

The pseudocode to carry out the nonlinear dynamic analysis of the NDOF shear building is outlined in the box that follows.

Nonlinear Dynamic Analysis of NDOF Systems

1. Input problem data (physical, numerical analysis, etc.).

   - Create the global DOF array id from boundary conditions.
   - Create the assembly array ix.
   - Initialize the internal variables iv to zero.
   - Put forces **f** in global DOF order.

2. Compute elastic stiffness matrix **K** and mass matrix **M**.
3. Compute damping matrix **C** from **K** and **M** (e.g., modal damping).
4. Initialize displacement and velocity, $\mathbf{u}_o$ and $\mathbf{v}_o$.
5. Compute initial acceleration from equation of motion, $\mathbf{u}_o$, and $\mathbf{v}_o$.
6. Loop over time steps:

   - Evaluate external force **f** for current time step. Evaluate $\mathbf{b}_n$, $\mathbf{c}_n$, and initialize $\mathbf{x} = \mathbf{a}_n$ to the acceleration at the previous time step.
   - Newton iteration to solve equation of motion.

      - Compute displacement and velocity from Newmark integrator.
      - Form tangent and residual: **A** and **g**.
      - Solve for new acceleration by Newton update.

   - Finalize velocity and displacement with Newmark integrator.
   - Update internal variables.
   - Update state for next time step ("new" values → "old" locations).

7. Output graphics.

The initialization of the internal variables only applies to models that need internal variables (e.g., elastoplasticity, where the plastic strain is required). Since the nonlinearities are local, each member needs its own internal variable.

5.8 Implementation

There are a few things introduced in this chapter that deserve mention by way
of implementation of the concepts into a computer code. First, we have defined
more specifically the data structures needed to code structural dynamics (or statics,
for that matter). Second, we have formulated an automatic assembly process for
creating the equations of motion. Finally, we have extended the numerical solution
scheme introduced for the SDOF problem to the NDOF setting.

The conceptualization of a structure as a set of nodes connected by elements is
essential to progress as we move forward. Hence, we must incorporate the nodal
coordinate array x, the element connection array ix, and the global DOF array id
more formally into the code. For the shear building, these are still very simple, with

$$
\texttt{ix} =
\begin{bmatrix}
1 & 2 \\
2 & 3 \\
3 & 4 \\
\vdots & \vdots \\
M & M+1
\end{bmatrix}
, \qquad
\texttt{id} =
\begin{bmatrix}
M+1 \\
1 \\
2 \\
\vdots \\
M
\end{bmatrix}
$$

where M is the number of elements and $N = M + 1$ is the number of nodes. These
arrays are easily generated in the input function. The nodal coordinate array x is
not really needed for the shear building because we specify the story stiffnesses
k directly. However, the nodal coordinates can be used to compute the element
lengths, which could be used to determine k if the flexural stiffness EI values were
specified instead of k. That is minor implementation detail. The node locations are
also important for graphical output, e.g., animation of the response.

The main new level of complication introduced in this chapter is the computation
of the tangent stiffness and residual. Code 5.2 accomplishes this task.

Code 5.2 MATLAB code to create the tangent **A** and the part of the residual **g** associated with
the internal element forces for the shear building

```
function [A,g,reac,ElmFD,iv] = NDOFAssemble(u,iv,id,ix,d,S)
%   Compute the residual and tangent for the NDOF shear building
%
%        u : Current displacement (for nonlinear models
%       iv : Internal variables
%       ix : Element connection matrix
%       id : Global DOF numbers
%        d : Element properties [Type, k, C, So]
%        S : Struct with problem size parameters

%. Initialize arrays for residual, tangent, and element force/def
    g = zeros(S.nDOF,1);
    A = zeros(S.nDOF,S.nDOF);
    ElmFD = zeros(S.nElem,2);

    for e=1:S.nElem

%... Localize element quantities
        ii = id(ix(e,:));                    % global DOF for element e
```

```
      ive = iv(e);                              % iv for element e
      de = d(e,:);                              % properties for element e

%... Compute deformation and associated force
      Def = u(ii(2)) - u(ii(1));
      [Force,Tang,ive]  = NonlinearModels(Def,de,ive);

%... Compute residual and tangent
      re = Force*[-1; 1];
      ke = Tang*[1,-1;-1,1];
      g(ii) = g(ii) + re;
      A(ii,ii) = A(ii,ii) + ke;
      ElmFD(e,:) = [Force,Def];
      iv(e) = ive;

    end % loop over elements

%. Trim residual and tangent
    reac = g(S.nDOF);                           % grab reaction force
    g = g(1:S.nFree);                           % trim to free DOF
    A = A(1:S.nFree,1:S.nFree);                 % trim to free DOF

end
```

The information passed into this function includes the displacement u, the internal variables iv (the inelastic strains for each element), and the information about the structure needed to carry out the assembly process. For the nonlinear problem, we need the current displacement (which is the complete global displacement vector, including the restrained DOF at the bottom) because that will determine the deformations in the members, which will determine the forces in those members, which will affect the tangent stiffness and will sum to be the residual.

There are five output quantities passed out of NDOFAssemble. The first two are the tangent matrix and the residual (at least the parts associated with the internal resistance). The third is an array of reaction forces (just one value for the shear building). The reactions are naturally computed when you assemble the residual. In fact, the residual is simply the net force that elements contribute to the nodes. For the node at the base of the structure, the sum of element contributions is simply equal to the reaction force, by Newton's third law. The reaction force is passed back only for output purposes—it is not used to advance the computation. The fourth item, the array ElmFD is an array containing the element forces and deformations. Like the reaction force, these are not needed to advance the solution, but we pass them back in case we want to store them to plot them later. Finally, the fifth item is the *internal variable* iv. This is the only array that is both passed in and passed out of the function. That opens the possibility that those values could be changed and retained after exiting the function (recall that all local variables are forgotten upon exit of a function in MATLAB).

We will use this function in several places in our main program, depending on what we currently need. In those cases, we will take care to return only those items required at the time. However, we will compute all of them each time we enter the function.

This function loops over the elements, computes the element contributions to the tangent and residual, and assembles them into the global tangent and residual.

To do that, the deformations in each element must be computed (i.e., `Def`) and that requires finding the DOF associated with that element. For element e, the associated nodes are `ix(e,1)` and `ix(e,2)`. The associated global degrees of freedom are the `id` values associated with those nodes. Thus, the two values stored in array `ii` are the global DOF numbers associated with element e. Those values are used to assemble the element contributions into the global arrays *and* to pick the two end displacements associated with that element from the global displacements array **u**. Once these quantities have been found, the element deformation is easily computed as the difference between them. In essence, the `ii` array is doing the work of the matrix $\mathbf{B}_e$.

With the element deformation known, we can find the force that goes along with that deformation. To compute the force, we need the element constitutive properties in the array `d` and the internal variable associated with that element from `iv`. The function that computes the force $V_e(\Delta_e)$ and the element tangent stiffness $k_e = \partial V_e / \partial \Delta_e$ is `NonlinearModels` presented in Sect. 3.6 in Chap. 3.

We will hold off on giving the main program until the next chapter because it will have all of the features to solve the complete NDOF problem at that point. However, let's take a look at an example of what this code can do.

Example Consider the same five-story shear building used for the examples from Figs. 4.9 and 4.10. The stiffness values vary from $k = 60$ at the bottom to $k = 40$ at the top. All of the masses are $m_i = 1$. In this case we will drive the system with sinusoidal loads applied at DOF 3 and 5. The time-dependent forcing function is $\mathbf{f}(t) = \mathbf{f}_o \sin \Omega t$, where $\mathbf{f}_o = [\,0, 0, 1, 0, 1\,]^T$ gives the spatial pattern of the loads. We will use $\Omega = 2.083$, which is the first natural frequency of the linear system.

The results for the linear constitutive model are shown in Fig. 5.3. As expected, all of the elements respond linearly, albeit with different slopes because they each have a different stiffness value. The heavier line is the bottom level. The amplitude of the displacement response shows a linear growth with time, while oscillating at the driving frequency. This is the expected resonance phenomenon that we first saw

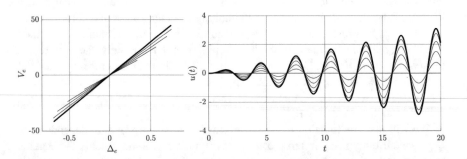

Fig. 5.3 *Linear elastic constitutive model.* Response of the shear building from Fig. 4.9 with a sinusoidal driving force at nodes 3 and 5. The sinusoidal forcing function is $\mathbf{f}_o \sin \Omega t$, with $\Omega = \omega_1 = 2.083$. The plot on the left shows the internal forces vs. their associated deformations. The plot on the right shows the nodal displacements vs. time

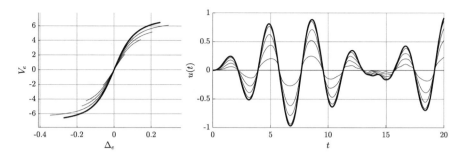

Fig. 5.4 *Nonlinear elastic (softening) constitutive model.* Response of the shear building from Fig. 4.9 with the same loading as Fig. 5.3. The nonlinear parameter in the model (see Eq. 5.22) is $C_e = 70$ for all elements

in the SDOF system when the driving force was equal to the natural frequency of the system. The heavier line is the top level. It is evident that the top level displaces the most. After 20 s, the amplitude is nearly 2. The damping ratio in all modes is $\xi = 0.03$.

The results for the nonlinear elastic model with a softening force/deformation relationship given by

$$V_e = \frac{k_e \Delta_e}{\sqrt{1 + C_e \Delta_e^2}} \qquad (5.22)$$

is shown in Fig. 5.4. The extent of the nonlinearity is evident in the plot of force vs. deformation on the left. The most obvious difference between this model and the linear one is that the nonlinearity has disrupted the resonance. One way to think about it is that resonance depends upon the frequency, which depends upon the ratio of stiffness to mass. For the nonlinear model there isn't a unique stiffness, but one might think of the secant line that goes from peak to peak on the element response as an *effective* stiffness. For this model, that effective stiffness is not equal to the linear stiffness value k_e. Hence, the resonant frequency shifts accordingly. The maximum displacement is around 0.8 for this model.

The results for the nonlinear elastic model with a hardening force/deformation relationship given by

$$V_e = k_e \Delta_e \left(1 + C_e \Delta_e^2\right) \qquad (5.23)$$

is shown in Fig. 5.5. The observations for this nonlinear elastic model are similar to those for the softening model. The nonlinearity, again, disrupts resonance. In this case, since the element stiffens with increasing deformation, the response is reduced from the softening model. In this case the maximum displacement is less than 0.6.

Finally, let us consider the response of the inelastic model, shown in Fig. 5.6. The yield value of $V_o = 5$ is easily observed on the force vs. deformation plot on the left. When the internal force reaches the yield value, inelastic deformation ensues. Recall

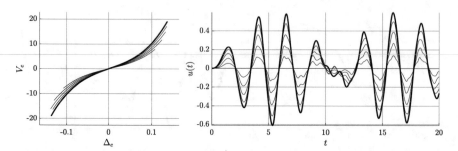

Fig. 5.5 *Nonlinear elastic (hardening) constitutive model.* Response of the shear building from Fig. 4.9 with the same loading as Fig. 5.3. The nonlinear parameter in the model (see Eq. 5.23) is $C_e = 70$ for all elements

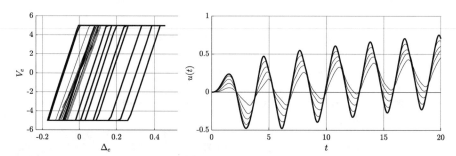

Fig. 5.6 *Inelastic constitutive model.* Response of the shear building from Fig. 4.9 with the same loading as Fig. 5.3. The yield value for the model is $V_o = 5$ for all elements

that the internal variable is the plastic strain for this model. Again, the nonlinearity disrupts resonance. The nature of the displacement vs. time curves are noticeably different from the previous models, and there appears to be a net drift in the response caused by the asymmetry of the loading process. Yielding happens at a certain point in the loading phase and unloading happens at a different point. How that unfolds in a particular case will determine if the drift is toward the positive or the negative (or if none happens).

These examples give a little flavor of what nonlinearity does to the dynamic response of a structural system. It should be evident that with so many parameters available, the variety is endless. One goal of the structural engineer is to develop insights about the dynamic response of complex structures. The examples show that the dynamic response features of linear systems (e.g., vibration in natural modes and resonance) are dramatically altered when the elements respond nonlinearly. Sometimes dynamic analysis results seem counterintuitive, at least initially. The computational tools provide a laboratory to carry out investigations of structural response. Seeing how a system responds dynamically to applied loads is an avenue to developing insight about the system. Having the ability to make changes to the system and observe how that affects the response is a powerful way to gain experience with dynamic response of structures.

Chapter 6
Earthquake Response of NDOF Systems

The motion of the NDOF system in an earthquake includes the motion at the base, which we will take as a *prescribed motion*. In essence, we assume that the ground motion just happens to the structure and that the structure does not influence the ground motion (which is not completely true, but it is a good place to start with the problem). In this chapter, we add earthquake ground motions as a source of excitation, show how they impart dynamic forces in the shear building, and analyze the response those forces induce.

Figure 6.1 shows the kinematics of motion for the shear building subjected to an earthquake ground motion. Now the absolute motion of the ith degree of freedom is $u_i^a(t) = u_g(t) + u_i(t)$, where $u_i(t)$ is the relative motion. The elements respond to the relative motion while the inertial resistance develops in accord with the absolution motion. The total motion can be put into a vector as

$$\mathbf{u}^a(t) = u_g(t)\mathbf{1} + \mathbf{u}(t)$$

where $\mathbf{1} = [\, 1, \ldots, 1\,]^T$ is a vector of ones. With this convention we can write the equation of motion as

$$\mathbf{M}\big(\ddot{u}_g(t)\mathbf{1} + \ddot{\mathbf{u}}(t)\big) + \mathbf{C}\dot{\mathbf{u}}(t) + \mathbf{r}(\mathbf{u}) = \mathbf{f}(t)$$

where $\mathbf{M}$ is the mass matrix, $\mathbf{C}$ is the damping matrix, $\mathbf{r}(\mathbf{u})$ is the assembled internal resistance of the elements, and $\mathbf{f}(t)$ is the applied load. Note that the only difference from the non-earthquake problem is the presence of the ground motion term. The mathematical form of the internal resistance is exactly the same. We can rewrite this equation in the form

---

Electronic Supplementary Material   The online version of this chapter (https://doi.org/10. 1007/978-3-030-89944-8_6) contains supplementary material, which is available to authorized users.

159
K. D. Hjelmstad, *Fundamentals of Structural Dynamics*,
https://doi.org/10.1007/978-3-030-89944-8_6

Fig. 6.1 *Kinematics of earthquake ground motion for the shear building.* For earthquake excitation of the shear building, each floor has a common inertial reference frame. The relative displacements **u**(t) are added to the ground motions $u_g(t)$

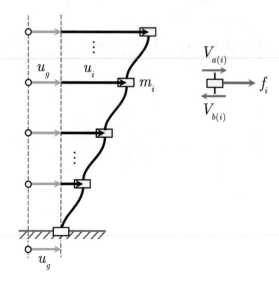

$$\mathbf{M}\ddot{\mathbf{u}}(t) + \mathbf{C}\dot{\mathbf{u}}(t) + \mathbf{r}(\mathbf{u}) = \mathbf{f}(t) - \ddot{u}_g(t)\mathbf{M1} \tag{6.1}$$

From this perspective, the earthquake ground motion acts like an equivalent lateral force on the system. Most importantly, the force is mobilized by the mass of the structure through the mass matrix **M** and the vector **1**. Note that the ground motion causes loading in *all* degrees of freedom of the structure.

Because the effect of the ground motion is like a force, there is very little modification needed to derive the equations of motion for the NDOF system. Following the argument in the principle of virtual work, substituting the absolute acceleration, we arrive at the equations of motion for the nonlinear system

$$\mathbf{M}\ddot{\mathbf{u}}(t) + \mathbf{C}\dot{\mathbf{u}}(t) + \sum_{e=1}^{M} \mathbf{B}_e^T \boldsymbol{\phi} \, V_e\left(\Delta_e(\mathbf{u})\right) = \mathbf{f}(t) - \ddot{u}_g(t)\mathbf{M1} \tag{6.2}$$

The discrete equations are similar to what we had before, with the addition of the earthquake loading term. The residual is

$$\mathbf{g}(\mathbf{x}) = \mathbf{M}\mathbf{x} + \mathbf{C}\mathbf{v}_{n+1}(\mathbf{x}) + \sum_{e=1}^{M} \mathbf{B}_e^T \boldsymbol{\phi} \, V_e(\mathbf{u}_{n+1}(\mathbf{x})) - \mathbf{f}_{n+1} + \ddot{u}_g(t_{n+1})\mathbf{M1}$$

where $\mathbf{x} = \mathbf{a}_{n+1}$ is the acceleration at step $n+1$ that we seek. The tangent matrix is exactly the same as before (Eq. 5.21):

$$\mathbf{A} = \mathbf{M} + \eta \, \mathbf{C} + \zeta \sum_{e=1}^{M} \mathbf{B}_e^T \boldsymbol{\phi} \left(\frac{\partial V_e}{\partial \Delta_e}\right) \boldsymbol{\phi}^T \mathbf{B}_e$$

Thus, from the perspective of numerical integration of the equations of motion, and the code to do it, there is almost no additional complication created by adding the earthquake excitation. The response, of course, will have many features that are unique to the earthquake problem.

The central issue in the modeling of the shear building subjected to earthquake ground motion is the selection of the motion. However, that issue is not different from what we discussed in Chap. 3. The earthquake motion is generally given as an accelerogram $\ddot{u}_g(t)$. This function can be generated artificially, or it can be taken from a recorded motion.

6.1 Special Case of the Elastic System

We can get some insight from considering the response of a linearly elastic system subjected to earthquake ground motion with no additional applied load. For an elastic NDOF system the equations of motion take the form:

$$\mathbf{M}\ddot{\mathbf{u}}(t) + \mathbf{C}\dot{\mathbf{u}}(t) + \mathbf{K}\mathbf{u}(t) = -\ddot{u}_g(t)\mathbf{M1}$$

If our system remains linear and the damping is classical, then it is possible to do a modal analysis. Let

$$\mathbf{u}(t) = \sum_{i=1}^{N} \boldsymbol{\varphi}_i a_i(t) \tag{6.3}$$

where $\boldsymbol{\varphi}_i$ is the eigenvector associated with the ith mode and $a_i(t)$ is an as yet unknown function of time.[1] With this modal expansion, the equations of motion become (assuming no external force besides the earthquake)

$$\mathbf{M}\sum_{i=1}^{N} \boldsymbol{\varphi}_i \ddot{a}_i(t) + \mathbf{C}\sum_{i=1}^{N} \boldsymbol{\varphi}_i \dot{a}_i(t) + \mathbf{K}\sum_{i=1}^{N} \boldsymbol{\varphi}_i a_i(t) = -\ddot{u}_g \mathbf{M1}$$

Premultiply by $\boldsymbol{\varphi}_j^T$, note orthogonality, and divide through by the modal mass to get

$$\ddot{a}_j(t) + 2\xi_j \omega_j \dot{a}_j(t) + \omega_j^2 a_j(t) = -\ddot{u}_g(t)\left(\frac{\boldsymbol{\varphi}_j^T \mathbf{M1}}{\boldsymbol{\varphi}_j^T \mathbf{M}\boldsymbol{\varphi}_j}\right) \tag{6.4}$$

---

[1] The letter a gets a pretty hard workout in this book. In this case $a_i(t)$ is a continuous function of time, not a discrete acceleration. The subscript i indexes the mode number, not a time step number (which we usually designate with the letter n).

This equation holds for all $j = 1, \ldots, N$ modes. Observe that the natural frequency and damping ratio are different for each mode, as is the contribution of the earthquake excitation for each mode. The term on the right side of the modal equation of motion includes the factor

$$\vartheta_j = \frac{\boldsymbol{\varphi}_j^T \mathbf{M} \mathbf{1}}{\boldsymbol{\varphi}_j^T \mathbf{M} \boldsymbol{\varphi}_j} \tag{6.5}$$

which does not depend upon the ground motion. We call ϑ_j the *modal participation factor* or *mass participation factor*. In essence, it is the amount of the earthquake excitation associated with the jth mode. You can get a lot of insight about earthquake response by looking at how these participation factors trend with mode number j. In general, there is a steady decrease in magnitude of the participation factor, which has led to the conventional wisdom that most earthquake response is in the lowest modes of the structure.

Once the solutions for the individual modes are computed, they can be reassembled to create the entire solution. Because there are no classical solutions for the earthquake excitation, the modal equations generally need to be solved numerically.

6.2 Modal Recombination

The participation factor ϑ_j given in Eq. 6.5 depends upon how the mode shapes are scaled. There is one $\boldsymbol{\varphi}_j$ in the numerator and two in the denominator. Once, we recombine the modal contributions in Eq. 6.3, the scaling issue is resolved because we multiply each response term $a_j(t)$ by $\boldsymbol{\varphi}_j$.

We can think of constructing other response quantities from modal contributions. Let us assume that we solve the equations

$$\ddot{s}_j(t) + 2\xi_j \omega_j \dot{s}_j(t) + \omega_j^2 s_j(t) = -\ddot{u}_g(t) \tag{6.6}$$

to find $s_j(t)$ for $j = 1, \ldots, N$. Each one of these responses is an SDOF response to the full earthquake ground motion $\ddot{u}_g(t)$. We can construct the displacement vector as

$$\mathbf{u}(t) = \sum_{j=1}^{N} \boldsymbol{\varphi}_j \vartheta_j s_j(t) \tag{6.7}$$

We can also observe that

$$\mathbf{K}\mathbf{u}(t) = \sum_{j=1}^{N} \mathbf{K}\boldsymbol{\varphi}_j \vartheta_j s_j(t) = \sum_{j=1}^{N} \omega_j^2 \mathbf{M}\boldsymbol{\varphi}_j \vartheta_j s_j(t) \tag{6.8}$$

where we have made use of the eigenvalue problem property $\mathbf{K}\boldsymbol{\varphi}_j = \omega_j^2\mathbf{M}\boldsymbol{\varphi}_j$. The quantity $\mathbf{Ku}$ represents the net force generated by the internal resisting elements. We can compute the *base shear*, V_{base}, by adding up all of those forces. We can do that by premultiplying the net force $\mathbf{Ku}$ with the vector $\mathbf{1}^T$. To wit,

$$V_{base}(t) = \mathbf{1}^T\mathbf{Ku}(t) = \sum_{j=1}^{N}\omega_j^2\left(\mathbf{1}^T\mathbf{M}\boldsymbol{\varphi}_j\right)\vartheta_j s_j(t) \tag{6.9}$$

To see that the base shear is also equal to the force in the bottom element, notice that due to the tridiagonal structure of the stiffness matrix (see, for example, Sect. 4.2),

$$\mathbf{1}^T\mathbf{K} = \left[k_1, 0, \dots, 0\right] \quad \rightarrow \quad \mathbf{1}^T\mathbf{Ku} = k_1 u_1$$

Now, define

$$\mathcal{L}_j = \left(\mathbf{1}^T\mathbf{M}\boldsymbol{\varphi}_j\right)\vartheta_j = \frac{\left(\boldsymbol{\varphi}_j^T\mathbf{M1}\right)^2}{\boldsymbol{\varphi}_j^T\mathbf{M}\boldsymbol{\varphi}_j} \tag{6.10}$$

and Eq. 6.11 can be rewritten as

$$V_{base}(t) = \sum_{j=1}^{N}\omega_j^2\mathcal{L}_j s_j(t) \tag{6.11}$$

The beauty of the *effective modal mass* $\mathcal{L}_j$ is that it is unitless, and therefore provides a reliable measure of modal participation that is not dependent on the scaling of the eigenvectors.

The effective modal masses, $\mathcal{L}_j$, sum to the total mass of the structure. To prove this, let us start by doing a modal expansion of the vector $\mathbf{1}$ as

$$\mathbf{1} = \sum_{i=1}^{N}\boldsymbol{\varphi}_i b_i$$

where b_i is the ith modal coordinate of the vector $\mathbf{1}$. Premultiply this equation by $\boldsymbol{\varphi}_j^T\mathbf{M}$ and note that, due to orthogonality, only one term in the sum survives to give

$$b_j = \frac{\boldsymbol{\varphi}_j^T\mathbf{M1}}{\boldsymbol{\varphi}_j^T\mathbf{M}\boldsymbol{\varphi}_j}$$

Plugging this result back into the original modal expansion gives

$$1 = \sum_{i=1}^{N} \varphi_i \left(\frac{\varphi_i^T \mathbf{M1}}{\varphi_i^T \mathbf{M} \varphi_i} \right) = \sum_{i=1}^{N} \frac{\varphi_i \varphi_i^T}{\varphi_i^T \mathbf{M} \varphi_i} \mathbf{M1}$$

which implies that

$$\sum_{i=1}^{N} \frac{\varphi_i \varphi_i^T}{\varphi_i^T \mathbf{M} \varphi_i} = \mathbf{M}^{-1}$$

We can use this result along with Eq. 6.10 to compute the sum of the effective modal masses. To wit,

$$\sum_{i=1}^{N} \mathcal{L}_i = \sum_{i=1}^{N} \frac{\left(\mathbf{1}^T \mathbf{M} \varphi_i \right) \left(\varphi_i^T \mathbf{M1} \right)}{\varphi_i^T \mathbf{M} \varphi_i} = \mathbf{1}^T \mathbf{M} \mathbf{M}^{-1} \mathbf{M1} = \mathbf{1}^T \mathbf{M1}$$

For the shear building, the mass matrix is diagonal. Hence, $\mathbf{1}^T \mathbf{M1}$ is simply the sum of the diagonal element of $\mathbf{M}$, which is the sum of the masses in the system, which is the total mass. Thus, the sum of the effective modal masses is the total mass of the structure.

It is worth noting that $\mathcal{L}_j$ only appears in the expression for the base shear. We can compute, for example, the force in element e as

$$V_e = k_e \sum_{j=1}^{N} \left(\phi^T \mathbf{B}_e \varphi_j \right) \vartheta_j s_j(t) \tag{6.12}$$

where ϕ and $\mathbf{B}_e$ were defined previously. Or, we could compute the displacement at DOF i as

$$u_i = \mathbf{e}_i^T \mathbf{u} = \sum_{j=1}^{N} \left(\mathbf{e}_i^T \varphi_j \right) \vartheta_j s_j(t) \tag{6.13}$$

where $\mathbf{e}_i$ is an N–vector with a one in slot i and zeros in all other slots. In each case, we can identify a quantity that is unitless—$(\phi^T \mathbf{B}_e \varphi_j) \vartheta_j$ in the case of the element force and $(\mathbf{e}_i^T \varphi_j) \vartheta_j$ in the case of the displacement at level i. However, only $\mathcal{L}_j$ has the property that it is purely positive. Hence, it can be used to see how participation varies with mode number.

6.3 Response Spectrum Methods

The *earthquake response spectrum* was defined in Chap. 3 to be a plot of the maximum displacements versus natural frequency ω (at a fixed damping ratio ξ) for an SDOF oscillator for a specific earthquake ground motion. The response spectrum is just a different way of looking at that specific earthquake and is a concept that strictly makes sense only for a linear SDOF system. One might think of an earthquake response spectrum as a *structural transform* of the time series $\ddot{u}_g(t)$ into the frequency domain (i.e., a function of ω). It represents a summary of the maximum response of all SDOF systems to the given earthquake ground motion.

To get the maximum response for a particular SDOF system, all one needs to do is look up the ordinate of the response spectrum at the natural frequency or period of that system. There is an advantage to this approach in earthquake resistant design because the aim is to iteratively home in on the physical properties of the system (e.g., the stiffness k and mass m) that meet the various design requirements.

In our analysis of the NDOF system, it was evident that we could use the orthogonality properties of the mode shapes to decouple the equations of motion for the NDOF system into N equations that looked like the SDOF equations. Each equation differed in the frequency ω_i, possibly the damping ratio ξ_i, and the participation factor ϑ_i. Certainly, we could use the response spectrum to find the maximum displacement for each of those N SDOF oscillators. The problem lies in how to combine those responses to find the maximum response of the NDOF system because the individual maxima do not necessarily occur at the same time. Hence, just adding them together does not really make mathematical sense.

The three most common modal combination methods for response spectrum analysis are the *absolute sum* method, the *square root of the sum of the squares* (SRSS) method, and the *complete quadratic combination* (CQC) method. These methods can be applied to any response quantity. For example, let us apply them to the base shear from Eq. 6.11. Let s_j^{max} be the maximum absolute value of $s_j(t)$ over time, where $s_j(t)$ is the solution to Eq. 6.6. We can compute the maximum base shear associated with the jth mode as

$$V_j = \omega_j^2 \mathcal{L}_j s_j^{max} \tag{6.14}$$

The s_j^{max} value can be picked directly from the response spectrum at the frequency ω_j and damping ratio ξ_j. The absolute sum method would estimate the maximum base shear as

$$V_{base}^{max} = V_1 + V_2 + \ldots + V_N \tag{6.15}$$

The SRSS method would estimate the maximum base shear as

$$V_{base}^{max} = \sqrt{V_1^2 + V_2^2 + \ldots + V_N^2} \tag{6.16}$$

A more general modal combination rule would estimate the maximum base shear as

$$V_{base}^{max} = \sqrt{\sum_{i=1}^{N} \sum_{j=1}^{N} V_i \, C_{ij} \, V_j} \qquad (6.17)$$

where the C_{ij} values determine how the components combine. The SRSS method, for example, is a special case of this combination rule where $C_{ii} = 1$ and $C_{ij} = 0$ if $i \neq j$. The CQC method (assuming that all modal damping ratios are the same and equal to ξ) defines the combination matrix as

$$C_{ij} = \frac{8\xi^2 \left(1+r_{ij}\right) r_{ij}^{3/2}}{\left(1-r_{ij}^2\right)^2 + 4\xi^2 r_{ij} \left(1+r_{ij}\right)^2} \qquad r_{ij} = \min\left(\frac{T_i}{T_j}, \frac{T_j}{T_i}\right) \qquad (6.18)$$

where $T_i = 2\pi/\omega_i$ is the period associated with mode i. The CQC was designed to be a more rational approach for closely spaced natural frequencies. There are several other combination rules available in the literature.

These methods form the basis of response spectrum methods for NDOF systems. They are all approximate ways to generate the maximum response of the system. Because they combine maximum values that occur at different times, the combined effect generally gives a conservative estimate of the actual maximum of the quantity (e.g., base shear) in question.

It is important to note that each physical response quantity requires its own response expression. For example, one can compute the maximum force in element e by taking the maximum response for each mode and substituting into Eq. 6.12. As we did for the base shear, we can define individual modal components for the maximum force in element e as

$$V_e^j = k_e \left(\boldsymbol{\phi}^T \mathbf{B}_e \boldsymbol{\varphi}_j\right) \vartheta_j s_j^{max} \qquad (6.19)$$

where, again, the s_j^{max} value can be picked from the response spectrum. The maximum force in element e can then be estimated using, say, SSRS to be

$$V_e^{max} = \sqrt{V_e^1 + V_e^2 + \cdots + V_e^N} \qquad (6.20)$$

To see how the response spectrum calculation goes, let's look at an example.

Example Consider a three-story shear building with mass $m_n = 1$ at all three levels and stiffness values $k_1 = 60$, $k_2 = 50$, and $k_3 = 40$. The structure is subjected to the earthquake ground motion shown in Fig. 6.2. The response spectrum is also shown in the figure, with the dots being at the natural frequencies of the structure. The mode shapes are also shown. The damping ratio for the response spectrum and the damping for all three modes is $\xi = 0.1$.

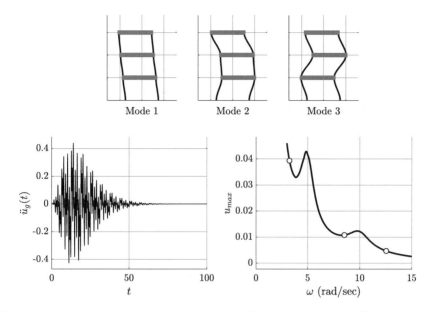

Fig. 6.2 *Response spectrum example.* Mode shapes of the three-story shear building, the earthquake ground acceleration, and the response spectrum. Note that the open circles on the response spectrum correspond to the natural frequencies of the structure

Table 6.1 *Results for response spectrum example.* Values for the computation of the maximum base shear for the three-story shear building example. Note that $\mathcal{L}_n^u = (\mathbf{e}_3^T \boldsymbol{\varphi}_n)\vartheta_n$ is the modal multiplier for the maximum displacement

Mode n	ω_n	s_n^{max}	$\mathcal{L}_n$	V_n	$\mathcal{L}_n^u$	u_n
1	3.247	0.03934	2.6426	1.096	1.2544	0.0493
2	8.516	0.01083	0.2838	0.223	0.3163	0.0034
3	12.528	0.00475	0.0737	0.055	0.0619	0.0003

The natural frequencies of the structure are given in Table 6.1 along with the value s_n^{max} from the response spectrum, $\mathcal{L}_n$ from Eq. 6.10, and the modal base shear from Eq. 6.14. The modal base shears can be combined with either the absolute sum value of $V_b^{ABS} = 1.37$ or SRSS value of $V_b^{SRSS} = 1.12$. These values can be compared to the actual maximum base shear computed from a time-history analysis of $V_b^{max} = 1.17$. Observe that the absolute sum method overestimates the actual value and the SRSS method underestimates it slightly.

The maximum displacement at level 3 can be estimated, too. The absolute sum combination gives $u_{max}^{ABS} = 0.0530$ and the SRSS combination gives $u_{max}^{SRSS} = 0.0494$. The actual maximum displacement at the top level from the time-history analysis is $u_{max} = 0.058$.

6.4   Implementation

This chapter ends our exploration of the dynamics of the shear building. It is worthwhile to take a look at the structure of the code to do the computations, now that all of the features have been presented. The organization of the computations for the NDOF system is contained in the following code:

Code 6.1 MATLAB code to analyze the dynamic response of the NDOF system

```
%. NDOF Structural Dynamics (Nonlinear, EQ)
   clear; clc;

%. Analysis features
   tf = 50.0;                        % Duration of anaysis
   h = 0.005;                        % Analysis time step
   nOutFreq = 1;                     % Output reporting frequency

%. Numerical analysis parameters
   [beta,gamma,eta,zeta,nSteps,tol,itmax,tf] = NewmarkParams(tf,h);

%. Input structure data
   [x,ix,id,iv,d,mass,Damp,force,u,S,DF] = ShearBldgInput;

%. Initialize ouput storage arrays
   [nS,Hist] = NDOFCreateHist(nOutFreq,nSteps,S);

%. Compute initial stiffness, mass, EQ force, and damping matrices
   [K,~,~,~,~] = NDOFAssemble(u,iv,id,ix,d,S);
   [M,Q] = NDOFMassMatrix(mass,id,S);
   [C,V,freq,Damp] = DampingMatrix(M,K,Damp,DF);

%. Establish initial conditions
   [uo,vo] = NDOFICs(V,S);

%. Get earthquake ground motion
   [EQaccel,~,EQdispl] = EQGroundMotion(h,nSteps,S.EQon,DF);

%. Set up initial parameters for first time step
   t = 0;

%. Initialize position, velocity, and acceleration
   uold = uo; vold = vo;
   Ft = TimeFunction(t,S.TimeType,1);
   F = Ft*force;
   u = [uold; 0];
   [~,g,Vb,EF,iv] = NDOFAssemble(u,iv,id,ix,d,S);
   g = g + C*vo - F;
   aold = -M\g;

%. Compute motion by numerical integration
   for n=1:nSteps

%... Take care of the storage of the time history
      Eq = EQdispl(n);  u = [uold; 0];
      [nS,Hist] = NDOFStoreHist(nS,Hist,t,u,iv,Eq,EF,F,Vb);

%.... Compute the values for the next time step
      t = t + h; err = 1; its = 0;
      Feq = Q*EQaccel(n);
      F = force*TimeFunction(t,S.TimeType,0);
      bn = uold + h*vold + beta*h^2*aold;
      cn = vold + gamma*h*aold;
      anew = aold;
```

```
%.... Newton iteration to determine the acceleration at time step n+1
      while (err>tol) && (its<itmax)
        its = its + 1;
        unew = bn + zeta*anew;
        vnew = cn + eta*anew;
        u = [unew; 0];
        [A,g,~,~,~] = NDOFAssemble(u,iv,id,ix,d,S);
        g = M*anew + C*vnew + g - F + Feq;
        A = M + eta*C + zeta*A ;
        anew = anew - A\g;
        err = norm(g);
      end

%.... Update velocity and displacement
      vnew = cn + eta*anew;
      unew = bn + zeta*anew;

%.... Update internal variables
      [~,g,Vb,EF,iv] = NDOFAssemble(u,iv,id,ix,d,S);

%.... Update state for next iteration
      aold = anew;
      vold = vnew;
      uold = unew;

  end

%. Echo input and produce output graphics
   NDOFEchoInput(S,d,mass,uo,vo,V,freq,force,Q)
   NDOFGraphs(x,Hist,S,nS,Q,V,freq,force,DF)
```

The code starts with setting the duration tf, the time step h, and the output frequency nOutFreq. The variable nOutFreq is included because it is not necessary to save the results of every time step to get good graphical output. The numerical analysis parameters include the Newmark parameters β and γ, and the Newton iteration tolerance and maximum allowable number of iterations. These values are obtained from the function NewmarkParams.

With this preparation, we call the function ShearBldgInput to describe the specific geometry and properties of the system. An example of that function will be presented after we finish going through the main program. One aspect to note is the use of the MATLAB 'struct' S, which contains all of the problem size parameters (e.g., number of nodes, number of elements). Using S simply makes it easier to pass those parameters to the various functions that need them. We also create a 'struct' DF that helps with the management of the graphical output.

The mechanism for storing results for later plotting is essential, but otherwise unimportant. All results are stored in a 'struct' called Hist. For example, the displacements are in Hist.u, the element forces are in Hist.N, and the base shear is in Hist.V. Using the 'struct' makes it very simple to add more fields and it is nice to "hide" the I/O in the function NDOFStoreHist because we need to take care of the fact that we do not need to store every step computed. The information in Hist is passed to the plotting routines at the end. Of course, storage must be set up before starting the time-stepping part. The function NDOFCreateHist takes care of setting up Hist.

Now, we can compute the initial stiffness, the mass, and the damping matrices. The linear stiffness matrix is needed because we use it to find the eigenvectors and eigenvalues, which form the basis of our damping matrix. If not for that, we would not need to form a stiffness matrix at this point. We get the linear stiffness matrix because the array u has been initialized to zero in the input function. The formation of the mass matrix is straightforward, but the function NDOFMassMatrix also takes care of the formation of $\mathbf{Q} = \mathbf{M1}$ for the earthquake excitation. Code 6.2 computes these matrices. The formation of the damping matrix has been discussed previously.

Code 6.2 MATLAB code to form the mass matrix **M** and the effective earthquake force vector **Q** for the NDOF system

```
function [M,Q] = NDOFMassMatrix(Mass,id,S)
%    Assemble global mass and Q matrices
%
%         Mass : Nodal masses
%           id : Global DOF numbers by node
%            S : Struct with problem size parameters

%. Zero global mass matrix M, and Q
   M = zeros(S.nDOF,S.nDOF);
   Q = zeros(S.nDOF,1);

%. Assemble M and Q = M*1
   for n = 1:S.nNodes
     ii = id(n,:);
     M(ii,ii) = M(ii,ii) + Mass(n);
     Q(ii) = Q(ii) + Mass(n);
   end

%. Trim to free DOF
   M = M(1:S.nFree,1:S.nFree);
   Q = Q(1:S.nFree);

end
```

The code for the initial conditions has already been discussed. The call to EQGroundMotion brings back the accelerations and displacements at every time step. We need the accelerations to compute the effective earthquake force. We need the displacements for the output graphics, specifically, the animation of the motion using the absolute displacement.

To start any dynamics problem, you must set the initial displacement and velocity to their known values and compute the initial acceleration from the equations of motion. In this case, we need to find the internal forces associated with the initial conditions, accounting for the fact that the initial displacements might induce some nonlinearity (including some yielding, hence the inclusion of iv in our call to NDOFAssemble). Note that we include the applied forces $\mathbf{f}(0)$, but we assume that the earthquake acceleration is zero at $t = 0$.

Finally, we arrive at the numerical integration of the differential equation. After we store the current results in Hist, we increment the time, compute the new applied force and earthquake effective force, compute the part of the new state $\mathbf{b}_n$ and $\mathbf{c}_n$ that is known from the previous state, and compute the starting value for the new acceleration to be the converged acceleration from the previous time step.

In the Newton "while" loop, we compute the new displacement and velocity based upon the current value of the new acceleration (which is changing with each iteration) using the Newmark equations. With the new displacement (which is an array of size nFree), we create a full displacement (including the restrained DOF at the base) and pass that into NDOFAssemble to get the tangent and residual. We then complete the tangent and residual to include the dynamic terms and compute the new estimate of the acceleration from Newton's method. Finally, we compute the norm of the residual, which we use to check convergence of Newton's method.

We update the displacement and velocity one more time after the iteration has converged because the acceleration was improved as the last computation that happened prior to dropping out of the "while" loop. With the converged state, we update the internal variables. Do not pass back iv while in the throes of the Newton's iterations because the intermediate values are erroneous estimates. Finally, the newly computed state is put in the arrays from the old state to get ready for the next time step.

Once the time steps have been exhausted, the output commences, through the functions NDOFEchoInput and NDOFGraphs. You can see the evidence of what is in those functions through the figures presented with results from the code.

Input Function The physical properties of the structure are established in the function ShearBldgInput, including the geometry, topology, mass, damping, loading, and material properties. The bookkeeping associated with element, node, and DOF numbering is performed in this function. The internal variables are initialized. In essence, everything that can change from one problem to the next is done in this function. An example of the input function is contained in Code 6.3.

Code 6.3 MATLAB code to prepare the input for the shear building for the NDOF code

```
function [x,ix,id,iv,d,mass,Damp,force,u,S,DF] = ShearBldgInput
%   Create model of shear building
%
%          x : Nodal coordinates
%         ix : Element Connectivity array
%         id : Global DOF numbers
%         iv : Internal variables (inelastic model needs this)
%          d : element properties [Type, k, C, So]
%              Type: 1 Linear, elastic
%                    2 Nonlinear, elastic, softening
%                    3 Nonlinear, elastic, hardening
%                    4 Elasto-plastic
%       mass : Nodal masses
%       Damp : Struct with damping information
%      force : spatial shape of forcing function
%          u : Displacement vector sized to nDOF (initialized to zero)
%          S : Struct for problem size, nNodes, nElem, nDOF, nFree
%         DF : Graphical output management

%. Problem size data
   S.nElem = 3;                 % Number of elements (stories)
   S.nNodes = S.nElem+1;        % Number of nodes
   S.nDOF = S.nNodes;           % Number of DOF
   S.nFree = S.nDOF-1;          % Number of free DOF

%. Mass, stiffness, damping
   Type = 4;                    % Constitutive model type
   L = 10;                      % Story height
   Mo = [1; 1];                 % Masses [Bottom, Top] story
```

```
Ko = [60,  40];              % Stiffness [Bottom, Top] story
Co = [70,  70];              % NL parameter [Bottom, Top] story
So = [ 1,   1];              % Yield force [Bottom, Top] story

%. Compute inputs by linear model
   x = linspace(0,S.nElem*L,S.nNodes)';        % Nodal coordinates
   mass = linspace(Mo(1),Mo(2),S.nNodes)';     % Nodal masses
   d(:,1) = linspace(Type,Type,S.nElem)';      % Model Type
   d(:,2) = linspace(Ko(1),Ko(2),S.nElem)';    % Elastic stiffness
   d(:,3) = linspace(Co(1),Co(2),S.nElem)';    % NL elastic parameter
   d(:,4) = linspace(So(1),So(2),S.nElem)';    % Yield force levels

%. Element connections. Number elements from bottom to top
   ix = zeros(S.nElem,2);
   for e=1:S.nElem; ix(e,:) = [e, e+1]; end

%. Global DOF assignments. Bottom node (Node 1) is fixed
   id = zeros(S.nNodes,1);
   for n=1:S.nNodes; id(n) = n-1; end
   id(1) = S.nNodes;

%. Establish the shape of the applied force vector
   force = zeros(S.nFree,1);
   force(3) = 1.0;

%. Damping
   Damp.type = 3;            % =1 Rayleigh, =2 Caughey, =3 Modal
   Xi = 0.1;                 % Reference damping ratio
   Damp.modes = [1,2,3];     % Modes for specified damping
   Damp.xi = Xi*ones(3,1);   % damping ratio in each specified mode

%. Initialize internal variables (plastic strains) and total displacement
   iv = zeros(S.nElem,1);
   u = zeros(S.nDOF,1);

%. Initial condition, Time function, and EQ data
   S.Mode = 0;               % Mode for initial condition
   S.Uo = 0;                 % Magnitude of initial displacement
   S.Vo = 0;                 % Magnitude of initial velocity
   S.TimeType = 0;           % =0 none =1 Constant, =2 Sinusoid, etc
   S.EQon = 1;               % =0 No EQ, =1 Yes EQ

%. Manage which figures to produce
   DF = NDOFManageFigures;
end
```

Output Storage Scheme The only remaining functions that have yet to be described are those associated with the storage scheme, i.e., NDOFCreateHist and NDOFStoreHist. These functions take care of storing the results of the analysis for later plotting. The storage scheme recognizes that we do not need to store every time step to generate good graphics. The functions are given in Codes 6.4 and 6.5.

Code 6.4 MATLAB code to create the struct Hist and the output counter array nS for storage of results

```
function [nS,Hist] = NDOFCreateHist(nOutFreq,nSteps,S)
%   Create struct to handle the storage of the response
%
%      nOutFreq : Frequency of output of time steps
%        nSteps : Number of time steps
%             S : Struct with problems size parameters, etc.
```

```
%. Set up struct Hist for storage
   nReport = ceil(nSteps/nOutFreq);      % Number of steps to report
   Hist.t = zeros(nReport,1);            % Time
   Hist.u = zeros(nReport,S.nDOF);       % Displacement
   Hist.D = zeros(nReport,S.nElem);      % Element deformations
   Hist.N = zeros(nReport,S.nElem);      % Element forces
   Hist.Y = zeros(nReport,S.nElem);      % Internal variables
   Hist.E = zeros(nReport,1);            % Ground displacement
   Hist.F = zeros(nReport,S.nFree);      % Applied forces
   Hist.V = zeros(nReport,1);            % Base Shear

%. Initialize output counter array
   nS = [0,0,nOutFreq,nReport,nSteps];

end
```

Code 6.5 MATLAB code to store computed results in Hist to be used later for graphical output

```
function [nS,Hist] = NDOFStoreHist(nS,Hist,t,u,iv,Eq,EF,F,Vb)
%  Storage response in Hist if this is a recorded time step
%
%        nS : [out,iOut,nOutput,nReport,nSteps]
%      Hist : Struct containing response history
%         t : time
%         u : Displacement (full DOF)
%        iv : Internal variables
%        Eq : Earthquake ground displacement
%        EF : [Force,Def] for elements
%         F : Applied forces (and free DOF)
%        Vb : Base shear

%. Write the values out to history, if appropriate
   if (nS(1)==0)
     nS(2) = nS(2) + 1;                  % Increment iOut
     iOut = nS(2);
     Hist.t(iOut,:) = t;                 % Time
     Hist.u(iOut,:) = u';                % Displacement
     Hist.D(iOut,:) = EF(:,2)';          % Element deformations
     Hist.N(iOut,:) = EF(:,1)';          % Element forces
     Hist.Y(iOut,:) = iv(:)';            % Internal variables
     Hist.E(iOut,:) = Eq;                % Ground displacement
     Hist.F(iOut,:) = F';                % Applied forces
     Hist.V(iOut,:) = Vb;                % Base shear
     nS(1) = nS(3);                      % Reset output counter
   end
   nS(1) = nS(1) - 1;                    % Decrement output counter

end
```

All of the codes in the book will use functions similar to these to manage the storage of results.

6.5 Example

To get a flavor of what the code can produce, consider a three-story shear building with $m = 1$ at all levels, $k_1 = 60$, $k_2 = 50$ and $k_3 = 40$. The element model is elastoplastic with yield value of $V_o = 1.0$ at all levels. The damping ratio is $\xi = 0.1$ for all modes. There are no applied loads, the structure starts from rest, and it is

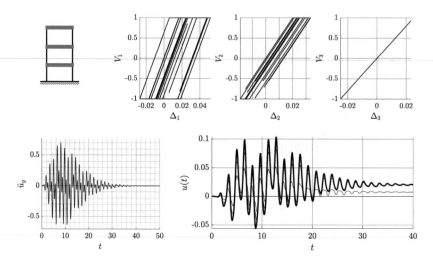

Fig. 6.3 *Earthquake response of an inelastic shear building.* Response of a three-story shear building subjected to the earthquake ground motion shown. Each element is elastoplastic with yield values $V_o = 1.0$. The duration of the earthquake is 51 s.

subjected to the earthquake ground motion, shown in Fig. 6.3. The response of the structure is also shown in this figure.

The duration of the earthquake is 51 s. It is a four-component earthquake defined by Eqs. 3.43 and 3.44 with $a = 2$, $b = 0.25$, and $s_o = 0.06$, which gives a peak acceleration of around 0.7. The frequencies are 1, 2, 4, and 8 with amplitudes of 0.1734, 0.0867, 0.6937, and 0.6937, respectively. The phase angles are all zero.

There are many interesting features to the response of this system. First, it is evident from the force versus deformation plots that there is significant yielding of the elements. The bottom level experiences the most yielding. The top level remains elastic throughout the earthquake. The permanent deformation is evident in the plot of displacement versus time. The solid horizontal line is the ground level DOF, confirming that the plot is of *relative* displacement. The maximum displacement is 0.103 and the maximum base shear is 1.00. At the end of the response, you see mostly free vibration until the motion dies out. The residual displacements are quite evident because each node damps out to its residual value. These results indicate that the final structure is damaged.

The variety of phenomena one can explore about a structure in an earthquake is enormous. This is just one result for one design for one earthquake. In earthquake-resistant design, one would want to subject the structure to a suite of possible earthquake ground motions to see if it comes through them in acceptable fashion.

Chapter 7
Special Methods for Large Systems

In this book we focus on computational approaches to structural dynamics because the calculations required to find the response of a system to dynamic loads involves tedious and repetitive steps that are ideally suited to the computer. To make the codes in the book as readable as possible we avoid clever strategies that might be used in commercial codes to speed up calculations. For the small problems we solve as examples in the book, speed of computation is seldom an issue. However, real-world systems can be very complex and the time it takes to execute structural dynamics computations grows with problem size, particularly for nonlinear problems. In this chapter, we examine two approaches to deal with large problems. Both methods involve projecting the problem onto a smaller subspace to improve the speed of calculation. One downside is that such methods are approximate and, therefore, require care in application.

The first approach we consider involves projecting the full problem onto a Ritz subspace. The approach is motivated by how the modal participation factors tend to be small for the higher frequency modes in earthquake analysis problems. The second approach, the *static correction method*, gives a way to improve results for modal analysis when using a truncated set of modes. This method can be executed without knowing all the eigenvectors of the system. Solving the eigenvalue problem for a large system can be computationally intensive. In Appendix C we show how the *subspace iteration* algorithm allows the computation of a subset of eigenvectors and eigenvalues.

We will develop the methods in this chapter in the context of the shear building, but these methods are applicable to any discrete problem with multiple degrees of freedom.

Electronic Supplementary Material The online version of this chapter (https://doi.org/10. 1007/978-3-030-89944-8_7) contains supplementary material, which is available to authorized users.

K. D. Hjelmstad, *Fundamentals of Structural Dynamics*,
https://doi.org/10.1007/978-3-030-89944-8_7

7.1   Ritz Projection Onto a Smaller Subspace

The governing equations of motion for a linear shear building subjected to earth-quake ground motion can be expressed as

$$\mathbf{M\ddot{u}}(t) + \mathbf{C\dot{u}}(t) + \mathbf{Ku}(t) = -\ddot{u}_g(t)\mathbf{Q}$$

where, for example, $\mathbf{Q} = \mathbf{M1}$ for the shear building (we will see how this array looks for other theories later in the book). As we have seen, this problem can be solved by forming the coefficient matrices and then doing one of two things: (1) solve the equations with a time-stepping algorithm like Newmark's method or (2) compute the natural modes and frequencies, use the modes to decouple the equations, solve N single-degree-of-freedom problems, and construct the complete solution from the modal components.

In this section we investigate the possibility of projecting the problem onto a smaller subspace of Ritz vectors (which could also be eigenvectors if we enhance the Ritz vectors using subspace iteration, as outlined in Appendix C). Consider the approximation

$$\mathbf{u}(t) = \mathbf{\Psi s}(t) \tag{7.1}$$

where $\mathbf{\Psi}$ is a constant $N \times n$ matrix whose columns are the Ritz vectors and $\mathbf{s}(t)$ is an $n \times 1$ matrix whose elements are the time-dependent coefficients of the Ritz basis that represents the evolving displacements $\mathbf{u}(t)$. In general, we will expect that $n \ll N$. The Ritz vectors are linearly independent and span the n–dimensional subspace. Is it possible to generate accurate results with this subspace approximation?

The concept of reduction of dimension in mechanics is a fairly well-worn path. For example, beam theory is a reduction of three-dimensional continuum mechanics to a one-dimensional ordinary differential equation. In the later chapters of this book we will introduce methods that reduce continuous systems to discrete ones, which is a reduction from infinite to finite dimensions. From linear algebra, we know that we need N independent vectors to span N-dimensional space. If most of the action is along only a few of those dimensions, then it seems reasonable to simply ignore the dimensions that contribute little to the solution.

Dwindling Participation Factors   The success of the Ritz projection really depends upon how much of the solution we are projecting out. One way of thinking about that is to examine the participation factors associated with the loading term, as mentioned in the previous section. For the earthquake the effective loading term is $-\ddot{u}_g(t)\mathbf{Q}$, which changes with time through the scalar ground acceleration function. The spatial pattern of loading is determined by the vector $\mathbf{Q}$ (the mass mobilization caused by the ground motion).

If we take the Ritz vectors to be the eigenvectors (mode shapes), then the motivation is clear because the equations are uncoupled. We get

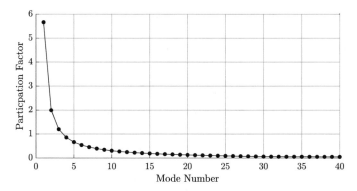

Fig. 7.1 *Modal participation factors for a forty-story shear building.* The participation factors ϑ_i plotted against mode number i for a forty-story shear building computed from Eq. 6.5. The story stiffness varies linearly from 900 at the base to 600 at the top, and the story masses all equal 2

$$\ddot{s}_i(t) + 2\xi_i\omega_i\dot{s}_i(t) + \omega_i^2 s_i(t) = -\ddot{u}_g(t)\vartheta_i$$

where ϑ_i is the participation factor associated with the ith mode, as defined in Eq. 6.5. It should be evident that the smaller the participation factor, the smaller the amplitude of $s_i(t)$ that results from the earthquake excitation (ignoring, of course, possible dynamic phenomena like resonance).

Figure 7.1 shows the variation of the modal participation factors for a forty-story shear building with uniform mass and linearly varying story stiffnesses. The dominance of the first mode is evident. The amplitudes of the participation factors decline fairly quickly with almost no participation from the higher modes.

Incomplete Basis For the linear problem we can project the problem to a smaller subspace using Ritz vectors. Substituting the approximation given by Eq. 7.1 into the equation of motion and then premultiplying by $\boldsymbol{\Psi}^T$, we get the reduced equations

$$\tilde{\mathbf{M}}\ddot{\mathbf{s}}(t) + \tilde{\mathbf{C}}\dot{\mathbf{s}}(t) + \tilde{\mathbf{K}}\mathbf{s}(t) = -\ddot{u}_g(t)\boldsymbol{\Psi}^T\mathbf{Q}$$

where

$$\tilde{\mathbf{M}} = \boldsymbol{\Psi}^T\mathbf{M}\boldsymbol{\Psi}, \qquad \tilde{\mathbf{C}} = \boldsymbol{\Psi}^T\mathbf{C}\boldsymbol{\Psi}, \qquad \tilde{\mathbf{K}} = \boldsymbol{\Psi}^T\mathbf{K}\boldsymbol{\Psi} \tag{7.2}$$

are $n \times n$ matrices. If the Ritz vectors are not eigenvectors, then these matrices might be full, but they are small if $n \ll N$. That allows the solution of a much smaller system of equations than the original system, which may be a significant computational savings if you are using Newmark's method, for example.

Let $\tilde{\mathbf{u}}_n = \mathbf{s}(t_n)$, $\tilde{\mathbf{v}}_n = \dot{\mathbf{s}}(t_n)$, and $\tilde{\mathbf{a}}_n = \ddot{\mathbf{s}}(t_n)$ be the *effective* displacement, velocity, and acceleration in the reduced subspace. At each time step we solve for the new acceleration using the Newmark integrator. To wit,

$$\tilde{\mathbf{a}}_{n+1} = \tilde{\mathbf{A}}^{-1}\left(-\ddot{u}_g(t_{n+1})\mathbf{\Psi}^T\mathbf{Q} - \tilde{\mathbf{C}}\tilde{\mathbf{c}}_n - \tilde{\mathbf{K}}\tilde{\mathbf{b}}_n\right)$$

The reduced effective mass matrix for Newmark's method is

$$\tilde{\mathbf{A}} = \tilde{\mathbf{M}} + \eta\tilde{\mathbf{C}} + \zeta\tilde{\mathbf{K}}$$

where $\eta = (1-\gamma)h$ and $\zeta = (0.5 - \beta)h^2$ are Newmark coefficients, and

$$\tilde{\mathbf{b}}_n = \tilde{\mathbf{u}}_n + h\tilde{\mathbf{v}}_n + \beta h^2\tilde{\mathbf{a}}_n, \qquad \tilde{\mathbf{c}}_n = \tilde{\mathbf{v}}_n + \gamma h\tilde{\mathbf{a}}_n$$

are the part of the new effective displacement and velocity that can be computed from the state at time n. Since the coefficient matrix is constant, it can be formed and factored once and then used at every time step. This is a very fast operation. We will see that it is possible to project the nonlinear problem, but the coefficient matrix will change at each iteration of each time step.

The Projected Damping Matrix In the previous section, we implied that the projected damping matrix is computed from the full damping matrix as

$$\tilde{\mathbf{C}} = \mathbf{\Psi}^T\mathbf{C}\mathbf{\Psi}$$

However, note that we generally form $\mathbf{C}$ from a model like Rayleigh damping, modal damping, or Caughey damping, rather than from first principles. With that in mind, we might simply create the reduced damping matrix directly from the reduced mass and stiffness matrices. First, solve the eigenvalue problem

$$\tilde{\mathbf{K}}\tilde{\mathbf{\Phi}} = \tilde{\mathbf{M}}\tilde{\mathbf{\Phi}}\tilde{\mathbf{\Omega}} \quad \rightarrow \quad \tilde{\mathbf{\Phi}}, \tilde{\mathbf{\Omega}}$$

to get the reduced eigenvalues $\tilde{\mathbf{\Omega}}$ (squares of natural frequencies on the diagonals) and eigenvectors $\tilde{\mathbf{\Phi}}$ (mode shapes). Now we can compute the modal damping matrix as

$$\tilde{\mathbf{C}} = \tilde{\mathbf{M}}\tilde{\mathbf{\Phi}}\tilde{\mathbf{D}}\tilde{\mathbf{\Phi}}^T\tilde{\mathbf{M}} \quad \text{where} \quad \tilde{\mathbf{D}} = \begin{bmatrix} 2\xi_1\omega_1 & & \\ & \ddots & \\ & & 2\xi_n\omega_n \end{bmatrix}$$

Or, alternatively, we could compute Rayleigh damping as

$$\tilde{\mathbf{C}} = \mu_1\tilde{\mathbf{M}} + \mu_2\tilde{\mathbf{K}}$$

and compute the coefficients μ_1 and μ_2 as we have previously.

The reduced damping matrix $\tilde{\mathbf{C}}$ does not include any damping in the neglected modes. We could specify zero damping in the higher modes when solving the full dynamics problem, but in that case, there would be undamped response in the higher

modes. In the reduced case we have entirely suppressed the response in the higher
modes so there is nothing left to damp out.

Projecting the Nonlinear Problem   The discrete version of the nonlinear problem
can be written as

$$\mathbf{M}\mathbf{a}_{n+1} + \mathbf{C}\mathbf{v}_{n+1} + \mathbf{r}(\mathbf{u}_{n+1}) = -\ddot{u}_g(t_{n+1})\mathbf{Q}$$

where the internal resistance $\mathbf{r}(\mathbf{u}_{n+1})$ is generally assembled from element contributions, as we have seen previously. We can project this problem by noting that

$$\mathbf{u}_{n+1} = \mathbf{\Psi}\tilde{\mathbf{u}}_{n+1}, \qquad \mathbf{v}_{n+1} = \mathbf{\Psi}\tilde{\mathbf{v}}_{n+1}, \qquad \mathbf{a}_{n+1} = \mathbf{\Psi}\tilde{\mathbf{a}}_{n+1}$$

In other words, the real discrete displacement, velocity, and acceleration can be
obtained from the effective displacement, velocity, and acceleration by multiplying
by the projection matrix $\mathbf{\Psi}$. Substituting these expressions and projecting the
equation by premultiplying by $\mathbf{\Psi}^T$, we get

$$\tilde{\mathbf{M}}\tilde{\mathbf{a}}_{n+1} + \tilde{\mathbf{C}}\tilde{\mathbf{v}}_{n+1} + \mathbf{\Psi}^T\mathbf{r}\left(\mathbf{\Psi}\tilde{\mathbf{u}}_{n+1}\right) = -\ddot{u}_g(t_{n+1})\mathbf{\Psi}^T\mathbf{Q}$$

where the reduced mass and damping matrices are given by Eq. 7.2. The residual
needed for Newton's method is

$$\tilde{\mathbf{g}} = \tilde{\mathbf{M}}\tilde{\mathbf{a}}_{n+1} + \tilde{\mathbf{C}}\tilde{\mathbf{v}}_{n+1} + \mathbf{\Psi}^T\mathbf{r}\left(\mathbf{\Psi}\tilde{\mathbf{u}}_{n+1}\right) + \ddot{u}_g(t_{n+1})\mathbf{\Psi}^T\mathbf{Q}$$

The tangent can be computed as

$$\tilde{\mathbf{A}} = \tilde{\mathbf{M}} + \eta\,\tilde{\mathbf{C}} + \zeta\,\mathbf{\Psi}^T\mathbf{K}_T\mathbf{\Psi}$$

In the nonlinear problem we need to compute $\mathbf{r}(\mathbf{u})$ and $\mathbf{K}_T = \partial\mathbf{r}/\partial\mathbf{u}$, both of
which require the *full* displacement vector $\mathbf{u} = \mathbf{\Psi}\tilde{\mathbf{u}}$. All the remaining parts of the
computation require only the projected arrays $\tilde{\mathbf{a}}$ and $\tilde{\mathbf{v}}$. So, the strategy will be to do
Newmark's method on the reduced system, but project up to the full displacement
when we need to (e.g., to compute the internal forces, check for and resolve yielding,
and to assemble $\mathbf{r}$ and the tangent stiffness matrix).

Projection of Initial Conditions   The initial conditions $\mathbf{u}_o$ and $\mathbf{v}_o$ (the full vectors)
project to the reduced space as

$$\mathbf{u}_o = \mathbf{\Psi}\tilde{\mathbf{u}}_o, \qquad \mathbf{v}_o = \mathbf{\Psi}\tilde{\mathbf{v}}_o$$

To solve for the reduced initial conditions, we premultiply the equations by $\mathbf{\Psi}^T\mathbf{M}$
to get

$$\mathbf{\Psi}^T\mathbf{M}\mathbf{u}_o = \mathbf{\Psi}^T\mathbf{M}\mathbf{\Psi}\tilde{\mathbf{u}}_o = \tilde{\mathbf{M}}\tilde{\mathbf{u}}_o$$

and

$$\mathbf{\Psi}^T \mathbf{M} \mathbf{v}_o = \mathbf{\Psi}^T \mathbf{M} \mathbf{\Psi} \tilde{\mathbf{v}}_o = \tilde{\mathbf{M}} \tilde{\mathbf{v}}_o$$

Solve these equations to get the projected initial displacement and velocity

$$\tilde{\mathbf{u}}_o = \tilde{\mathbf{M}}^{-1} \mathbf{\Psi}^T \mathbf{M} \mathbf{u}_o, \qquad \tilde{\mathbf{v}}_o = \tilde{\mathbf{M}}^{-1} \mathbf{\Psi}^T \mathbf{M} \mathbf{v}_o$$

The algorithm for doing nonlinear Ritz projected dynamics is summarized in the following box:

Ritz Projected Dynamic Analysis

1. Initialize the problem geometry and set up arrays $\mathbf{M}$ and $\mathbf{K}$.
2. Establish the Ritz projection matrix $\mathbf{\Psi}$.
3. Compute projected matrices $\tilde{\mathbf{M}}$, $\tilde{\mathbf{K}}$, $\tilde{\mathbf{Q}} = \mathbf{\Psi}^T \mathbf{Q}$.
4. Compute the projected damping matrix $\tilde{\mathbf{C}}$.
5. Initialize $\mathbf{u}_o$ and $\mathbf{v}_o$ and project to $\tilde{\mathbf{u}}_o$ and $\tilde{\mathbf{v}}_o$.
6. Compute initial acceleration $\tilde{\mathbf{a}}_o$ from the equation of motion.
7. Loop over time steps:

   - Evaluate external force $\tilde{\mathbf{f}}_{n+1} = -\ddot{u}_g(t_{n+1}) \tilde{\mathbf{Q}}$.
   - Evaluate $\tilde{\mathbf{b}}_n$, and $\tilde{\mathbf{c}}_n$ and initialize the acceleration $\tilde{\mathbf{x}}^0 = \tilde{\mathbf{a}}_n$ to the acceleration at the previous time step.
   - Compute the new acceleration by Newton's method.

     – Compute $\tilde{\mathbf{u}}_{n+1}^i = \tilde{\mathbf{b}}_n + \zeta \tilde{\mathbf{x}}^i$ and $\tilde{\mathbf{v}}_{n+1}^i = \tilde{\mathbf{c}}_n + \eta \tilde{\mathbf{x}}^i$.
     – Project to full displacement $\mathbf{u} = \mathbf{\Psi} \tilde{\mathbf{u}}_{n+1}^i$
     – Form tangent and residual $\mathbf{K}_T$ and $\mathbf{r}$ in the full space.
     – Project tangent and the residual $\tilde{\mathbf{r}} = \mathbf{\Psi}^T \mathbf{r}$, $\tilde{\mathbf{K}} = \mathbf{\Psi}^T \mathbf{K}_T \mathbf{\Psi}$
     – Complete the residual as $\tilde{\mathbf{g}} = \tilde{\mathbf{M}} \tilde{\mathbf{a}}_{n+1}^i + \tilde{\mathbf{C}} \tilde{\mathbf{v}}_{n+1}^i + \tilde{\mathbf{r}} - \tilde{\mathbf{f}}_{n+1}$
     – Complete the tangent as $\tilde{\mathbf{A}} = \tilde{\mathbf{M}} + \eta \tilde{\mathbf{C}} + \zeta \tilde{\mathbf{K}}$
     – Compute new acceleration by Newton $\tilde{\mathbf{x}}^{i+1} = \tilde{\mathbf{x}}^i - \tilde{\mathbf{A}}^{-1} \tilde{\mathbf{g}}$.

   - Converged solution $\tilde{\mathbf{x}}^m$ is the new acceleration $\tilde{\mathbf{a}}_{n+1}$.
   - Finalize velocity and displacement with Newmark integrator.
   - Project to full displacement to update internal variables.
   - Update state for next time step ("new" values → "old" locations).

8. Output graphics.

It is very simple to convert the original full dynamics code to one that uses the Ritz projection. There are only a few places that need projection. The first part of the computation is exactly the same as the standard NDOF problem. We gather the inputs and create the `id` and `ix` matrices as before. We assemble the full system matrices as before. The first point where we use the projection is to compute the projected mass, stiffness, and damping matrices. Next, we compute the projected initial displacements and velocities. Then, at each time step, we need to project up to the full displacement in order to compute the full internal residual and full internal part of the tangent. Once we have those, we project them down to the subspace to do the time integration with the Newmark integrator. At the end of the time step, the internal variables must be updated, which requires a projection back to the full space.

Note that since the displacement changes with each Newton iteration, we need to project up and back down for each iteration. That can be a costly operation. There are several practical strategies (like not recomputing the internal tangent each iteration, unless the number of Newton iterations gets too large).

7.2   Static Correction Method

The mode shapes of a linear system allow the equations of motion to be uncoupled. Consider the equations of motion of a damped linear system with N degrees of freedom subjected to an applied force

$$\mathbf{M\ddot{u}}(t) + \mathbf{C\dot{u}}(t) + \mathbf{Ku}(t) = \mathbf{f}(t) \tag{7.3}$$

where $\mathbf{M}$, $\mathbf{C}$, and $\mathbf{K}$ are the $N \times N$ mass, damping, and stiffness matrices. The vector $\mathbf{u}(t)$ is the $N \times 1$ response and the vector $\mathbf{f}(t)$ is the $N \times 1$ applied force. The N natural frequencies and mode shapes, $(\omega_n, \boldsymbol{\varphi}_n)$, emanate from the eigenvalue problem $\mathbf{K\varphi} = \omega^2 \mathbf{M\varphi}$. The vectors $\boldsymbol{\varphi}_n$, $n = 1, ..., N$ constitute a basis for N-dimensional space.

The eigenvectors have the property that they are orthogonal relative to $\mathbf{M}$, $\mathbf{K}$, and $\mathbf{C}$ (assuming classical damping), i.e.,

$$\boldsymbol{\varphi}_m^T \mathbf{M} \boldsymbol{\varphi}_n = 0, \quad \boldsymbol{\varphi}_m^T \mathbf{C} \boldsymbol{\varphi}_n = 0, \quad \boldsymbol{\varphi}_m^T \mathbf{K} \boldsymbol{\varphi}_n = 0 \qquad m \neq n$$

We can compute the modal mass, damping, and stiffness, respectively, as

$$M_n = \boldsymbol{\varphi}_n^T \mathbf{M} \boldsymbol{\varphi}_n, \quad C_n = \boldsymbol{\varphi}_n^T \mathbf{C} \boldsymbol{\varphi}_n, \quad \text{and} \quad K_n = \boldsymbol{\varphi}_n^T \mathbf{K} \boldsymbol{\varphi}_n$$

and note that

$$K_n = \omega_n^2 M_n, \qquad C_n = 2\xi_n \omega_n M_n \tag{7.4}$$

where ξ_n is the damping ratio associated with the nth mode.

Expand the solution vector $\mathbf{u}(t)$ in terms of the eigenvector basis as

$$\mathbf{u}(t) = \sum_{n=1}^{N} \boldsymbol{\varphi}_n a_n(t) \tag{7.5}$$

where $a_n(t)$ is the contribution of $\boldsymbol{\varphi}_n$ to the solution. Plugging this expression for the displacement into Eq. 7.3, premultiplying by $\boldsymbol{\varphi}_m^T$, and noting orthogonality gives

$$M_m \ddot{a}_m(t) + C_m \dot{a}_m(t) + K_m a_m(t) = \boldsymbol{\varphi}_m^T \mathbf{f}(t) \tag{7.6}$$

The resulting N equations are uncoupled and can be solved by any of the methods developed previously for single-degree-of-freedom systems (including numerical methods). Once the solution is found for each component $a_m(t)$, the results can be combined through Eq. 7.5 to reconstruct the displacement array $\mathbf{u}(t)$. This solution is exact if the integration of the individual modal equations of motion is accurate.

For systems with a large number of degrees of freedom, one can consider using a truncated modal expansion where, instead of summing over all N eigenvectors in Eq. 7.5, we sum over a subset $\mathcal{A}$ (which is usually a set that starts with 1 and indexes sequentially up to $M < N$). This idea is attractive because it is possible to compute a subset of the eigenvectors using an algorithm like subspace iteration, as described in Appendix C. The solution of the eigenvalue problem can be a time-consuming computational task for a large structure. In such a circumstance, the remaining eigenvalues and eigenvectors would not be known and, therefore, would not be available to support the modal expansion of $\mathbf{u}(t)$.

We have seen that the participation of the modes often decreases as the mode number increases, suggesting that the lower modes are more important to the overall response of the system than the higher modes. While that is true for many types of loads, it is not always the case. The *static correction method* provides a way to improve the solution obtained with a truncated basis.

The main idea of the static correction method is that we segregate the modes into two sets: $\mathcal{A}$ and $\mathcal{B}$. The union of these two sets is the complete set of eigenvectors. We can write the modal expansion in Eq. 7.5 as

$$\mathbf{u}(t) = \sum_{n \in \mathcal{A}} \boldsymbol{\varphi}_n a_n(t) + \sum_{n \in \mathcal{B}} \boldsymbol{\varphi}_n b_n(t) \tag{7.7}$$

The crux of the static correction method (and probably the origin of the appellation "static") comes from recognizing that for the modes with higher frequencies, the modal mass and modal damping are much smaller than the modal stiffness (i.e., $M_n \ll K_n$ and $C_n \ll K_n$), the first being inversely proportional to the square of the frequency and the second being inversely proportional to the frequency, according to Eq. 7.4. Hence, the first two terms on the left side of Eq. 7.6 are negligible compared to the third term for the higher modes. Neglecting those terms is similar to assuming that $\ddot{b}_n(t) = 0$ and $\dot{b}_n(t) = 0$, which would be the very definition of *static* response in those modes. Technically, we should use the term *quasi-static* because $b_n(t)$ is still a function of time since the right side of the equation is time dependent.

Neglecting the first two terms of Eq. 7.6, we can write the equations of motion for the components associated with index set $\mathcal{B}$ as

$$K_m b_m(t) = \boldsymbol{\varphi}_m^T \mathbf{f}(t), \qquad m \in \mathcal{B}$$

which we can easily solve to get

$$b_m(t) = \frac{\boldsymbol{\varphi}_m^T \mathbf{f}(t)}{K_m}, \qquad m \in \mathcal{B}$$

Substituting this result back into Eq. 7.7 gives

$$\mathbf{u}(t) = \sum_{n \in \mathcal{A}} \boldsymbol{\varphi}_n a_n(t) + \sum_{n \in \mathcal{B}} \frac{\boldsymbol{\varphi}_n \boldsymbol{\varphi}_n^T}{K_n} \mathbf{f}(t) \tag{7.8}$$

Note that the expansion still requires the eigenvectors in index set $\mathcal{B}$. We can eliminate those by noting that the inverse of the stiffness matrix can be represented by its spectral decomposition as

$$\mathbf{K}^{-1} = \sum_{n=1}^{N} \frac{\boldsymbol{\varphi}_n \boldsymbol{\varphi}_n^T}{K_n} = \sum_{n \in \mathcal{A}} \frac{\boldsymbol{\varphi}_n \boldsymbol{\varphi}_n^T}{K_n} + \sum_{n \in \mathcal{B}} \frac{\boldsymbol{\varphi}_n \boldsymbol{\varphi}_n^T}{K_n} \tag{7.9}$$

The proof that the inverse of the stiffness matrix can be written this way is simple. First, note that

$$\mathbf{K}^{-1}\mathbf{K} = \mathbf{I} = \sum_{n=1}^{N} \frac{\boldsymbol{\varphi}_n \boldsymbol{\varphi}_n^T \mathbf{K}}{K_n}$$

The identity matrix operates on any vector and gives back the same vector. For the vector $\boldsymbol{\varphi}_m$ we can compute

$$\mathbf{I}\boldsymbol{\varphi}_m = \sum_{n=1}^{N} \frac{\boldsymbol{\varphi}_n \boldsymbol{\varphi}_n^T \mathbf{K} \boldsymbol{\varphi}_m}{K_n} = \frac{\boldsymbol{\varphi}_m K_m}{K_m} = \boldsymbol{\varphi}_m$$

where we have noted the orthogonality of the eigenvectors and the definition of the modal stiffness K_m. Since the result holds for any eigenvector $\boldsymbol{\varphi}_m$ it holds for any vector because the eigenvectors form a basis for N-dimensional space.

With the modal expansion of the inverse of the stiffness matrix, we can write

$$\sum_{n \in \mathcal{B}} \frac{\boldsymbol{\varphi}_n \boldsymbol{\varphi}_n^T}{K_n} = \mathbf{K}^{-1} - \sum_{n \in \mathcal{A}} \frac{\boldsymbol{\varphi}_n \boldsymbol{\varphi}_n^T}{K_n}$$

which we can substitute back into Eq. 7.8 to give

$$\mathbf{u}(t) = \mathbf{K}^{-1}\mathbf{f}(t) + \sum_{n \in \mathcal{A}} \boldsymbol{\varphi}_n \left(a_n(t) - \frac{\boldsymbol{\varphi}_n^T}{K_n}\mathbf{f}(t) \right) \tag{7.10}$$

This is the *static correction method*. Observe that this method only requires the eigenvectors associated with the index set $\mathcal{A}$.

For the special case where the spatial distribution of the loading is fixed and driven by a scalar time-dependent function, i.e., $\mathbf{f}(t) = f(t)\mathbf{p}_o$, the static correction equation takes the form

$$\mathbf{u}(t) = \left(\mathbf{K}^{-1}\mathbf{p}_o \right) f(t) + \sum_{n \in \mathcal{A}} \boldsymbol{\varphi}_n \left(a_n(t) - \frac{\boldsymbol{\varphi}_n^T \mathbf{p}_o}{K_n} f(t) \right) \tag{7.11}$$

The significance of this form is that the solution of "static" equations implied by $\mathbf{K}^{-1}\mathbf{p}_o$ is done only once. The constants $\boldsymbol{\varphi}_n^T \mathbf{p}_o / K_n$ can also be computed once.

It is hard to make definitive statements about when this approach yields the most benefit because it depends upon the specific nature of the loading, the frequency spectrum of the structure in question, and the number of modes included in index set $\mathcal{A}$. The static correction method should generally be better than the solution obtained by simply truncating the solution to include only the vectors in $\mathcal{A}$. The difference between simply truncating and doing the static correction can be significant in some cases, as the following example shows.

Example To illustrate what is at stake with the static correction approach, consider a 30-story shear building with all story masses equal to 5 and the story stiffness varying linearly from a value of 6200 at the bottom story to 5200 at the top (in a consistent set of units). The fundamental natural period of the structure is 3.55 s, which is reasonable for a structure of this size. The damping ratio is $\xi = 0.02$ in all modes. A load of $f(t) = 10\sin(3t)$ is applied at the fifth level (from the bottom). The structure starts from rest. The lowest five frequencies, along with the highest frequency, are given in Table 7.1. It is evident that the driving frequency is between the first and second natural frequencies.

Table 7.1 *Example*. Natural frequencies and periods associated with the 30-story shear building. The first five modes and the last mode are included

Mode	Frequency	Period
1	1.768	3.553
2	5.213	1.205
3	8.661	0.725
4	12.089	0.520
5	15.485	0.406
...	...	...
30	68.894	0.091

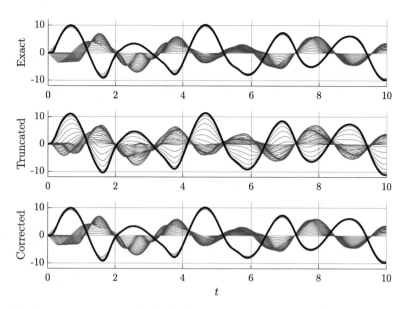

Fig. 7.2 Element forces vs. time for the 30-story shear building. The top plot is the exact solution with all 30 modes included. The middle plot is the truncated solution using only the first five modes. The bottom plot is the solution using the first five modes with the static correction. The heavy line is the force in the first story

The response of the structure is shown in Fig. 7.2. The results compare the element forces using the full modal (exact) solution, the truncated solution (first five modes), and the truncated solution (first five modes) with the static correction. The element forces in all 30 stories are plotted against time. The force in the bottom level is the heavier line, which is equal to the base shear. It is evident that while the base shear is well represented by the truncated solution, several elements give dramatically different values when compared to the exact solution. The static-correction solution resolves those issues almost entirely.

The maximum absolute errors in the element forces are shown in Fig. 7.3. The errors are computed as the difference between the approximate and exact solutions, normalized by the largest absolute value of the exact solution over time. The errors for the truncated solution get as high as 40%. The static correction holds the errors in force to be less than 3%. It is interesting to note that the maximum error does not decline with time for the truncated solution, but it does with the static correction.

Figure 7.4 shows how the number of modes included in the truncated and static-correction solutions affect the outcome. The maximum absolute error in element forces over all forces and all time are plotted against the number of modes included in the truncated solutions. Using a single mode results in large errors for both cases, although the error is smaller with the static correction. The errors for the solution with static correction decreases monotonically and drops to a very low value with few modes. The truncated solution does not decrease monotonically, and the errors

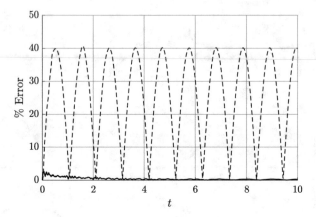

Fig. 7.3 Plot error in element forces as a function of time for the two solutions truncated after the first five modes. The dotted line is the truncated solution without correction and the solid line is the truncated solution with static correction

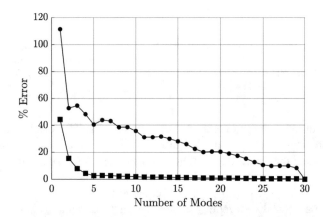

Fig. 7.4 Plot of the maximum error in element forces vs. number of modes used for the truncated solution without correction (circles) and the truncated solution with static correction (squares)

drop much more slowly as more modes are added. Observe that the error drops to zero when all 30 modes are included, as expected. This plot puts the results for the case with five modes into perspective.

7.3 Summary

Even though the theory of structural dynamics does not change much with time, the capabilities of computers do. Structural dynamics will probably always have problems so complex that they tax whatever current computers can do. There was period of time where significant research efforts were devoted to finding algorithms and strategies that would make structural dynamics computations feasible. Some approaches have mostly historical interest, others have been woven into the very fabric of structural dynamics.

The ideas presented in this chapter only hint at strategies that can be used to speed up the solution of large structures. There are many others. It is worth pondering such questions as, "Does the tangent need to be recomputed for every iteration in Newton's method?" or "Can localized nonlinearities be handled more efficiently?" These questions always come with a burden of decision about the algorithm that the user must make or that the code itself must resolve. Also, implementation is often complex.

Chapter 8
Dynamic Analysis of Truss Structures

The shear building gives a good introduction to discrete NDOF structural systems, encompassing many of the phenomena that are important to structural dynamics. As a structural model, however, it is very limited. There are several directions to branch out from the shear building to get to more practical structures. For our first branch, we will consider the analysis of truss structures. Trusses represent an important class of structures used widely in practice. The main reason for considering them next is because the mechanics of truss elements is very simple, and that simplicity will allow us to focus on the geometry and topology of multiply connected structures. Truss analysis will afford us an opportunity to expand on ideas presented in the previous chapters.

8.1 What Is a Truss?

A truss is a collection of points called *nodes* connected together by *elements* (which are also often called *members*). A sketch of a typical truss structure is shown in Fig. 8.1. The members are pinned together where they join and, hence, individually exert no net moment on the node. Certain nodes are restrained against motion, denoted on the sketch with a pin or roller. The task of truss analysis is to compute the nodal displacements, reaction forces, and element forces caused by the applied loads or any other source of excitation (like earthquake ground motion).

A load can be applied at any node (for nodes that are restrained, the force is a reaction force). In the sketch we have a force $\mathbf{p}_i$ applied at node i and a force $\mathbf{p}_j$ applied at node j. It is possible to have a reaction force and an applied force at the

Electronic Supplementary Material   The online version of this chapter (https://doi.org/10.1007/978-3-030-89944-8_8) contains supplementary material, which is available to authorized users.

Fig. 8.1 *Typical truss structure* A truss is a collection of nodes connected together by elements, restrained at certain points, and subjected to load at the nodes

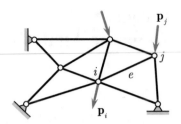

same node, e.g., at a node with a roller boundary condition. Any component of the applied force in a direction that is restrained from motion is directly absorbed by the reaction force there. In our formulation we will assume that external forces can only be applied in a direction that is free to move.

In the sketch you can see that element e is connected to nodes i and j. These nodes will be endowed with *node numbers* to distinguish one from another. The ordering of the nodes is arbitrary but is defined when we create the nodal coordinate array x, in which the row order of the array defines the nodal order. We will also number the elements. Again, the ordering is arbitrary, but is defined when we create the element connectivity array ix, in which the row order of the array defines the element order.

There are three degrees of freedom (DOF) at each node for a structure occupying three-dimensional space and two DOF for structures that can be idealized as planar.[1]

8.2   Element Kinematics

Consider the motion of element e, which is connected to nodes $i(e)$ and $j(e)$, as shown in Fig. 8.2. The element has an undeformed length of L_e and is oriented in the direction of the unit vector $\mathbf{n}_e$. Under the actions applied to the truss, the nodes move to new positions labeled as $i'(e)$ and $j'(e)$. The motion is such that node $i(e)$ experiences a displacement of $\mathbf{u}_{i(e)}$ and node $j(e)$ experiences a displacement of $\mathbf{u}_{j(e)}$, as indicated in the figure.

The member must remain connected to the two nodes and moves accordingly. The deformed element has length $\lambda_e L_e$ and is oriented in the direction of the unit vector $\mathbf{m}_e$. The element has stretched and rotated. The stretch λ_e is the ratio of the deformed length to the original length. By vector addition we can get from $i(e)$ to $j'(e)$ by two routes. They must be equal. Hence,

$$\mathbf{u}_{i(e)} + \lambda_e L_e \mathbf{m}_e = L_e \mathbf{n}_e + \mathbf{u}_{j(e)}$$

which we can rearrange to get

---

[1] Throughout this chapter we will consider three-dimensional structures. In the implementation it is very straightforward to consider structures restricted to a plane.

Fig. 8.2 *Displacement of an element e*. Element *e* is attached to node $i(e)$ and node $j(e)$. The nodes displace to locations $i'(e)$ and $j'(e)$. The unit vectors $\mathbf{n}_e$ and $\mathbf{m}_e$ point along the undeformed and deformed axis of the bar, respectively. The stretch of the element is λ_e

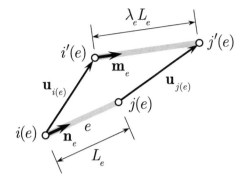

$$\lambda_e \mathbf{m}_e = \mathbf{n}_e + \frac{1}{L_e} \left(\mathbf{u}_{j(e)} - \mathbf{u}_{i(e)} \right) \tag{8.1}$$

It is evident that the nodal displacement difference $\mathbf{u}_{j(e)} - \mathbf{u}_{i(e)}$ plays a central role in the expression for the stretch of element e.

It will be advantageous to define this displacement difference in terms of the global displacement array. Let us assume that the global displacement array[2] $\mathbf{u} = [u_1, \ldots, u_{\mathcal{N}}]^T$ is stored in DOF order, where $\mathcal{N}$ is the total number of degrees of freedom ($\mathcal{N} = 3N$ for a three-dimensional truss and $\mathcal{N} = 2N$ for a planar truss, where N is the number of nodes). We will describe how to organize the global DOF for the code in Sect. 8.10. Our strategy is to number the DOF sequentially in x, y, z order starting at node 1 and moving through the structure in node order. We will, however, deal with restrained degrees of freedom in a way that will allow us to partition the global arrays in the code, numbering them at the end of the list after the free degrees of freedom.

The *displacement difference* for element e can be computed as

$$\boldsymbol{\Delta}_e = \mathbf{u}_{j(e)} - \mathbf{u}_{i(e)} = \boldsymbol{\phi}^T \mathbf{B}_e \mathbf{u} \tag{8.2}$$

The matrices $\boldsymbol{\phi}$ and $\mathbf{B}_e$ are similar to the ones we introduced for the shear building but tailored to the truss. The $\mathbf{B}_e$ matrix has six rows (four for 2D) and $\mathcal{N}$ columns. Its purpose is to pick out displacements associated with element e and has the following structure:

$$\mathbf{B}_e = \begin{bmatrix} 0 & 1 & 0 & 0 & \cdots & 0 & 0 & \cdots & 0 & \cdots & 0 \\ 0 & 0 & 1 & 0 & \cdots & 0 & 0 & \cdots & 0 & \cdots & 0 \\ 0 & 0 & 0 & 1 & \cdots & 0 & 0 & \cdots & 0 & \cdots & 0 \\ 0 & 0 & 0 & 0 & \cdots & 0 & 0 & \cdots & 1 & \cdots & 0 \\ 0 & 0 & 0 & 0 & \cdots & 1 & 0 & \cdots & 0 & \cdots & 0 \\ 0 & 0 & 0 & 0 & \cdots & 0 & 1 & \cdots & 0 & \cdots & 0 \end{bmatrix} \tag{8.3}$$

---

[2] We will use $\mathbf{u}$ to represent the global displacement array and $\mathbf{u}_i$ as the displacement vector at node i. When we refer to the ith component of $\mathbf{u}$ it will be u_i. This notation is a bit subtle. We want to use the same symbol for displacement, but it has some nuanced particular usages.

The 1 entries in the first three rows are associated with the degrees of freedom associated with node $i(e)$ and the 1 entries in the last three rows are associated with the degrees of freedom associated with node $j(e)$. The purpose of the matrix is to select the vectors $\mathbf{u}_{i(e)}$ and $\mathbf{u}_{j(e)}$ from the global displacement array $\mathbf{u}$ (which is in DOF order). The matrix $\boldsymbol{\phi}$ is defined as

$$\boldsymbol{\phi}^T = \begin{bmatrix} -\mathbf{I} & \mathbf{I} \end{bmatrix} \tag{8.4}$$

where $\mathbf{I}$ is the 3×3 identity matrix (2×2 for 2D). Hence, $\boldsymbol{\phi}$ is a 3×6 matrix (2×4 for 2D). Its role is to compute the difference between $\mathbf{u}_{j(e)}$ and $\mathbf{u}_{i(e)}$.

If we take the dot product of each side of Eq. 8.1 with itself, we get

$$\lambda_e^2 = \left(\mathbf{n}_e + \frac{1}{L_e} \boldsymbol{\Delta}_e \right) \cdot \left(\mathbf{n}_e + \frac{1}{L_e} \boldsymbol{\Delta}_e \right) \tag{8.5}$$

Distributing the dot product gives

$$\lambda_e^2 = 1 + \frac{2}{L_e} (\boldsymbol{\Delta}_e \cdot \mathbf{n}_e) + \frac{1}{L_e^2} (\boldsymbol{\Delta}_e \cdot \boldsymbol{\Delta}_e) \tag{8.6}$$

We will use the stretch to define the element strain.

The Element Strain To characterize the constitutive response of an element, we need to define strain. There are many valid ways to measure strain. Some of the most popular strain measures are given in Table 8.1. Any function of λ_e is a valid measure of strain. All of the strain measures listed in Table 8.1 have roughly the same values when strains are small. Note that small values of strain correspond to stretches in the neighborhood of $\lambda_e = 1$.

To formulate our theory, we will use the *Lagrangian strain*. The reason for that should be obvious soon. The Lagrangian strain for element e is defined as

$$\varepsilon_e = \tfrac{1}{2} \left(\lambda_e^2 - 1 \right) \tag{8.7}$$

From Eqs. 8.6 and 8.7, we have the truss bar strain–displacement relationship

$$\varepsilon_e = \frac{1}{L_e} (\boldsymbol{\Delta}_e \cdot \mathbf{n}_e) + \frac{1}{2L_e^2} (\boldsymbol{\Delta}_e \cdot \boldsymbol{\Delta}_e) \tag{8.8}$$

Table 8.1 Different measures of uniaxial strain, expressed in terms of the stretch λ, which is the ratio of deformed length to reference length

Strain	Common name
$\lambda - 1$	Engineering strain
$\frac{1}{2}(\lambda^2 - 1)$	Lagrangian strain
$1 - 1/\lambda$	Natural or "true" strain
$\frac{1}{2}(1 - 1/\lambda^2)$	Eulerian strain
$\log \lambda$	Logarithmic strain

If we neglect the second term in this expression, we get the strain–displacement relationship for a linearized strain $\varepsilon_e^{lin} = (\mathbf{\Delta}_e \cdot \mathbf{n}_e) / L_e$, which is commonly used in linear truss analysis. When the strain is a nonlinear function of the displacement, we often refer to it as a *geometric nonlinearity* to distinguish it from a *material nonlinearity*, which is when stress is a nonlinear function of strain (more on that later).

8.3 Element and Nodal Static Equilibrium

For simplicity, we will first consider static equilibrium of the truss. To develop the picture of equilibrium, we will consider the deformed element shown in Fig. 8.3. The force vector $\mathbf{N}_{i(e)}$ acts at end $i(e)$ and the force vector $\mathbf{N}_{j(e)}$ acts at end $j(e)$. The end forces on element e must be equal and opposite to be in equilibrium, as shown in Fig. 8.3. Hence, we can write

$$\mathbf{N}_{j(e)} = \bar{N}_e \mathbf{m}_e, \quad \mathbf{N}_{i(e)} = -\bar{N}_e \mathbf{m}_e \tag{8.9}$$

where the scalar magnitude of the axial force $\bar{N}_e$ is positive in tension and negative in compression. The direction of the force vector is determined by the unit vector $\mathbf{m}_e$, which points from the *start* node $i(e)$ to the *end* node $j(e)$ in the deformed configuration.

By Newton's third law, the element end forces act in an equal and opposite manner on the nodes. Hence, if the force on the member is $\mathbf{N}_{i(e)}$ the force on the node must be $-\mathbf{N}_{i(e)}$. The nodes, of course, have forces acting on them from the other elements attached to that node and from the applied nodal force (e.g., $\mathbf{p}_i$ acting on node i). The equation of equilibrium for node i is

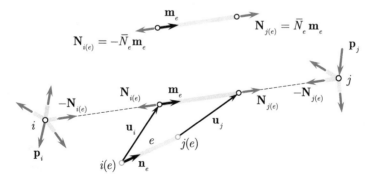

Fig. 8.3 *Element and node equilibrium.* The forces acting on the ends of the element are transferred to the nodes. The vector end force must have the same magnitude $\bar{N}_e$ and opposite directions to be in equilibrium

$$\sum_{k \in \mathcal{K}(i)} (-\mathbf{N}_k) + \mathbf{p}_i = \mathbf{0} \tag{8.10}$$

This equation holds for $i = 1, \ldots, N$, where N is the number of nodes in the structure. This is nothing more than the *method of joints*. The notation on the summation means that we are summing on the index $k \in \mathcal{K}(i)$, which is the set of all elements that attach to node i. This equation establishes equilibrium in the *deformed* configuration.

8.4   The Principle of Virtual Work

We can take our equilibrium equation for node i, multiply it (dot product) by an arbitrary virtual displacement vector $\mathbf{w}_i$, and sum over all N nodes to get the *virtual work function*

$$\mathcal{G} = \sum_{i=1}^{N} \left(\sum_{k \in \mathcal{K}(i)} (-\mathbf{N}_k) + \mathbf{p}_i \right) \cdot \mathbf{w}_i \tag{8.11}$$

if $\mathcal{G} = 0$ for all $\mathbf{w}_i$ then the fundamental theorem of the calculus of variations guarantees that Eq. 8.10 is satisfied individually for each node i. We call this the *principle of virtual work* for the static truss.

The double sum on the internal forces is a sum over nodes of a sum over all elements framing into those nodes. As we did with the shear building, we can show that this sum is exactly the same as a sum over all elements of the contributions from the two ends of each element. Consider the example shown in Fig. 8.4. The sketch of the structure shows the node and member numbers. There are four nodes and five members. The free body diagrams of the structure are shown in Fig. 8.5. The member forces are drawn in the positive (tensile) sense. The sketch on the left shows the free body diagrams of the nodes. The sketch on the right shows the free body diagrams of the elements. The sketches implicitly define the positive directions of $\mathbf{m}_1, \ldots, \mathbf{m}_5$.

Fig. 8.4 *Simple example truss.* The structure is shown with member and node numbers. Loads are applied at nodes 1 and 2. Motions are restrained at nodes 3 and 4 (hence, the forces are reaction forces)

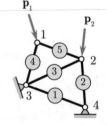

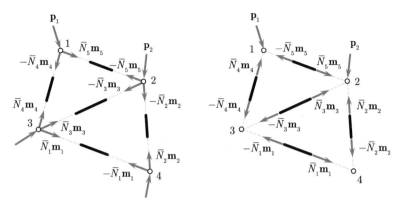

Fig. 8.5 *Nodal and element free body diagrams.* The sketch on the left shows the free body diagrams of nodes. The sketch on the right shows the free body diagrams of members

Let us compute the sum of virtual displacements dotted with element forces in two different ways. First, execute the sum over *nodes* as written in Eq. 8.11. To wit,

$$
\sum_{i=1}^{4} \mathbf{w}_i \cdot \sum_{k \in \mathcal{K}(i)} \mathbf{N}_k =
\begin{aligned}
& \mathbf{w}_1 \cdot \left(-\bar{N}_4 \mathbf{m}_4 + \bar{N}_5 \mathbf{m}_5 \right) + \\
& \mathbf{w}_2 \cdot \left(-\bar{N}_2 \mathbf{m}_2 - \bar{N}_3 \mathbf{m}_3 - \bar{N}_5 \mathbf{m}_5 \right) + \\
& \mathbf{w}_3 \cdot \left(\bar{N}_1 \mathbf{m}_1 + \bar{N}_3 \mathbf{m}_3 + \bar{N}_4 \mathbf{m}_4 \right) + \\
& \mathbf{w}_4 \cdot \left(-\bar{N}_1 \mathbf{m}_1 + \bar{N}_2 \mathbf{m}_2 \right)
\end{aligned}
$$

Evidently, the members framing into node 1 belong to set $\mathcal{K}(1) = \{4, 5\}$, the members framing into node 2 belong to set $\mathcal{K}(2) = \{2, 3, 5\}$, the members framing into node 3 belong to set $\mathcal{K}(3) = \{1, 3, 4\}$, and the members framing into node 4 belong to set $\mathcal{K}(4) = \{1, 2\}$. The organization of the calculation in this equation has virtual displacements multiplying the sum of forces at each node.

Conversely, we can write the same expression as a sum over *elements* as

$$
\sum_{e=1}^{5} \bar{N}_e \mathbf{m}_e \cdot \left(\mathbf{w}_{j(e)} - \mathbf{w}_{i(e)} \right) =
\begin{aligned}
& \bar{N}_1 \mathbf{m}_1 \cdot (\mathbf{w}_3 - \mathbf{w}_4) + \\
& \bar{N}_2 \mathbf{m}_2 \cdot (\mathbf{w}_4 - \mathbf{w}_2) + \\
& \bar{N}_3 \mathbf{m}_3 \cdot (\mathbf{w}_3 - \mathbf{w}_2) + \\
& \bar{N}_4 \mathbf{m}_4 \cdot (\mathbf{w}_3 - \mathbf{w}_1) + \\
& \bar{N}_5 \mathbf{m}_5 \cdot (\mathbf{w}_1 - \mathbf{w}_2)
\end{aligned}
$$

This example shows how the crucial conversion from the sum over nodes of the sum of members framing into the nodes is equivalent to a sum over elements. This conversion is only possible because of how we set up the virtual work function (that is where we introduced the sum over nodes in the first place). This conversion will be key in defining how we assemble our equations.

With this observation, we can write the virtual work done by the internal forces in the form

$$\sum_{i=1}^{N} \mathbf{w}_i \cdot \sum_{k \in \mathcal{K}(i)} \mathbf{N}_k = \sum_{e=1}^{M} \left(\mathbf{N}_{j(e)} \cdot \mathbf{w}_{j(e)} + \mathbf{N}_{i(e)} \cdot \mathbf{w}_{i(e)} \right) \tag{8.12}$$

The end forces on element e must be oriented along the member and must be in equilibrium. Hence, we can write $\mathbf{N}_{j(e)} = \bar{N}_e \mathbf{m}_e$ and $\mathbf{N}_{i(e)} = -\bar{N}_e \mathbf{m}_e$. Substituting these expressions into Eq. 8.11 gives

$$\mathcal{G} = -\sum_{e=1}^{M} \bar{N}_e \mathbf{m}_e \cdot \left(\mathbf{w}_{j(e)} - \mathbf{w}_{i(e)} \right) + \sum_{i=1}^{N} \mathbf{p}_i \cdot \mathbf{w}_i \tag{8.13}$$

We can write $\mathbf{w}_{j(e)} - \mathbf{w}_{i(e)} = \boldsymbol{\phi}^T \mathbf{B}_e \mathbf{w}$, where $\mathbf{w}$ is the *global* virtual displacement array in the same DOF order as $\mathbf{u}$. Also, observe that the sum in the second term of Eq. 8.13 can be written as $\mathbf{w}^T \mathbf{p}$ if we define the *global* force vector as $\mathbf{p}$ (also in global DOF order). Now, we can write the virtual work function in the final form

$$\mathcal{G} = \mathbf{w}^T \left(\mathbf{p} - \sum_{e=1}^{M} \mathbf{B}_e^T \boldsymbol{\phi} \, \mathbf{m}_e \bar{N}_e \right) \tag{8.14}$$

If we assert that $\mathcal{G} = 0$ must hold for *any* choice of the arbitrary virtual displacement $\mathbf{w}$, then the fundamental theorem of the calculus of variations demands that

$$\mathbf{p} - \sum_{e=1}^{M} \mathbf{B}_e^T \boldsymbol{\phi} \, \mathbf{m}_e \bar{N}_e = \mathbf{0} \tag{8.15}$$

which is the equation of static equilibrium for the truss.

The proof is simple. Assume that the quantity in parentheses $(\bullet)$ in Eq. 8.14 is *not* zero. Since $\mathbf{w}$ is arbitrary, we could pick $\mathbf{w} = (\bullet)$, i.e., pick the virtual displacement to be equal to the quantity in parentheses. Now, $\mathcal{G}$ reduces to $(\bullet)^T (\bullet)$. The inner product of a vector with itself is the length of the vector. The only way for the length of a vector to be zero is for that vector to be the zero vector (i.e., every element is zero). That proves the assertion.

The principle of virtual work is a very powerful way of thinking about the equations of equilibrium because it gives us a way to convert the method of joints to a sum over elements. The equilibrium equation expressed in Eq. 8.15 opens up a canonical algorithm to assembling the equations.

8.5 Constitutive Models for Axial Force

The equations we have derived so far are valid for large deformations, as the representation of deformation was built on a geometrically exact model and the equations of equilibrium were written in the deformed configuration. This generality opens up the important question of how to model the internal response of the elements to loading. We are using the *stress resultant* $\bar{N}_e$, the net axial force, to represent the stress acting on a cross section of the truss bar. However, finite deformation mechanics leaves us with a few questions to resolve. First, which measure of stress is most appropriate? And second, once the stress measure is established, how do we characterize the constitutive response? In this section, we take a closer look at those questions.

Measures of Stress Consider the situation illustrated in Fig. 8.6. There are three main ways to measure stress in continuum mechanics, as illustrated in the figure. The *Cauchy* stress σ_e is the actual stress acting on the truss cross section in the deformed configuration. While we will not go into details here, the Lagrangian strain is the natural dual to the second *Piola–Kirchhoff* stress Σ_e, which represents the same state of stress, but represented on the reference configuration.

If you imagine that the deformed bar has gotten longer, the cross section will contract due to Poisson's effect. For simplicity, we will assume that the volume of material does not change. The deformed area a_e relates to the undeformed area A_e as $\lambda_e L_e a_e = L_e A_e$. The first Piola–Kirchhoff stress, P_e, gives the same net force as the Cauchy stress, σ_e. The second Piola–Kirchhoff, Σ_e, refers the stress to the reference configuration, as shown in the sketch at right. These are related as

$$\sigma_e a_e = P_e A_e = \bar{N}_e, \qquad \lambda_e \Sigma_e = P_e \qquad (8.16)$$

We will use the second Piola–Kirchhoff stress as the most natural companion for Lagrangian strain. If we define the second Piola–Kirchhoff stress resultant $N_e = \Sigma_e A_e$, then

Deformed configuration   Reference configuration

Fig. 8.6 *Representing the state of stress. The Cauchy stress σ_e acts normal to the cross section in the deformed configuration (sketch at left). The first Piola–Kirchhoff stress P_e acts in the same vector direction as the Cauchy stress, but on the reference configuration. The second Piola–Kirchhoff stress Σ_e acts in the axial direction in the reference configuration*

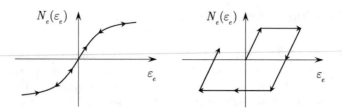

Fig. 8.7 *Axial force constitutive function.* The axial force constitutive function $N_e(\varepsilon_e)$ is the model of how the material behaves. The axial force used in the equilibrium equations is $\bar{N}_e = \lambda_e N_e$

$$\bar{N}_e = \lambda_e \Sigma_e A_e = \lambda_e N_e. \tag{8.17}$$

The details of why this association makes sense can be found in any textbook on continuum mechanics (e.g., Hjelmstad, *Fundamentals of Structural Mechanics*, Springer, 2005). In a nutshell, the product of Lagrangian strain and second Piola–Kirchhoff stress, integrated over the domain in the reference configuration, completely accounts for the internal energy possessed by the body. One advantage of working with the second Piola–Kirchhoff stress is that we can refer to the cross-sectional area A_e in the reference configuration in our constitutive models.

Constitutive Function The total axial force in the member is $\bar{N}_e$. We will take $N_e = \bar{N}_e/\lambda_e$ to be the *constitutive function*. In other words, when we seek the force induced in the material caused by the strain ε_e, we will get it from the constitutive function $N_e(\varepsilon_e)$. Two of the models we will use (nonlinear elasticity and elastoplasticity) are shown in Fig. 8.7.

We can define the response functions as we have previously. For example, the softening nonlinear elastic model for the truss element has the model

$$N_e\left(\varepsilon_e\right) = \frac{EA_e\varepsilon_e}{\sqrt{1 + C_e\varepsilon_e^2}} \tag{8.18}$$

where EA_e and C_e are model parameters. Note that we can compute the *constitutive tangent* as

$$\frac{\partial N_e}{\partial \varepsilon_e} = \frac{EA_e}{\left(1 + C_e\varepsilon_e^2\right)^{\frac{3}{2}}} \tag{8.19}$$

It should be evident that everything we have done previously (for any constitutive model) is identical from this point forward if we take N_e as the constitutive function with independent variable ε_e.

A *linear* constitutive model is $N(\varepsilon_e) = EA_e\varepsilon_e$. That is a linear relationship between the Lagrangian strain and the force N_e, but that does not make the model linear because the strain–displacement relationship, Eq. 8.8, is not linear. For small deformations, $\lambda_e \approx 1$ and the distinction between $\bar{N}_e$ and N_e disappears. In that case, the linear theory would suggest that E is Young's modulus of the material and A_e is the cross-sectional area. For larger strains those interpretations would require verification, and the interpretation of E in, for example, Eq. 8.18 might be slightly different.

Back to the Equations of Equilibrium Define the vector $\mathbf{F}_e(\mathbf{u})$ (see Eq. 8.1) as

$$\mathbf{F}_e(\mathbf{u}) = \mathbf{n}_e + \frac{1}{L_e}\mathbf{\Delta}_e = \lambda_e\mathbf{m}_e \tag{8.20}$$

where $\mathbf{\Delta}_e = \mathbf{u}_{j(e)} - \mathbf{u}_{i(e)}$. If the displacements are very small, then $\mathbf{F}_e(\mathbf{u})$ reduces to $\mathbf{n}_e$. In essence, $\mathbf{F}_e(\mathbf{u})$ is a vector that points in the direction of the deformed axis of the element, but it is not a unit vector (in fact, it has length λ_e). With these definitions, we can write the static equation of equilibrium of the structure, Eq. 8.15, as

$$\sum_{e=1}^{M} \mathbf{B}_e^T \boldsymbol{\phi} \, \mathbf{F}_e(\mathbf{u}) N_e(\varepsilon_e(\mathbf{u})) - \mathbf{p} = \mathbf{0} \tag{8.21}$$

where we have noted that $\bar{N}_e = \lambda_e N_e(\varepsilon_e)$.

8.6 Solving the Static Equations of Equilibrium

Equation 8.21 is a *displacement-based* statement of equilibrium because $\mathbf{u}$ is the primary variable. In essence, you can look at the solution of this equation as a search for a vector $\mathbf{u}$ that, when plugged into Eq. 8.21, satisfies the equation. As this is a nonlinear equation, we will use Newton's method to execute the search (see Appendix A). To set that up, the *residual* for Eq. 8.21 is defined by the function

$$\mathbf{g}(\mathbf{u}) = \sum_{e=1}^{M} \mathbf{B}_e^T \boldsymbol{\phi} \, \mathbf{F}_e(\mathbf{u}) N_e(\varepsilon_e(\mathbf{u})) - \mathbf{p} \tag{8.22}$$

Configurations $\mathbf{u}$ that satisfy $\mathbf{g}(\mathbf{u}) = \mathbf{0}$ are *equilibrium configurations*. Newton's method also requires the *tangent matrix* $\mathbf{A}$, which is the derivative of $\mathbf{g}(\mathbf{u})$ with respect to $\mathbf{u}$. The simplest way to compute the tangent is with the directional derivative (see Appendix B), noting that $\mathcal{D}\mathbf{g}(\mathbf{u}) \cdot \mathbf{h} = \mathbf{A}\mathbf{h}$, where $\mathbf{h}$ is an arbitrary variation of $\mathbf{u}$.

We can compute the derivative of $\mathbf{F}_e N_e$ (which are the only terms that are functions $\mathbf{u}$) using product rule:

$$\mathcal{D}\big[\mathbf{F}_e N_e\big] = \mathcal{D}\big[\mathbf{F}_e\big]N_e + \mathbf{F}_e\mathcal{D}\big[N_e\big]$$

where we are using the shorthand notation $\mathcal{D}[\bullet] = \mathcal{D}[\bullet] \cdot \mathbf{h}$. The directional derivative of N_e can be done with the chain rule. Let us first find the directional derivative of $\mathbf{F}_e(\mathbf{u})$, defined in Eq. 8.20. To compute this derivative, we need to know how to compute the directional derivative of the displacement difference $\mathbf{\Delta}_e = \mathbf{\phi}^T \mathbf{B}_e \mathbf{u}$, which can be done as follows:

$$\mathcal{D}[\mathbf{\Delta}_e] = \frac{d}{d\epsilon} \left[\mathbf{\phi}^T \mathbf{B}_e (\mathbf{u} + \epsilon \mathbf{h}) \right]_{\epsilon=0} = \mathbf{\phi}^T \mathbf{B}_e \mathbf{h} \tag{8.23}$$

Since $\mathbf{n}_e$ is constant, the directional derivative of $\mathbf{F}_e(\mathbf{u})$ is simply

$$\mathcal{D}[\mathbf{F}_e] = \frac{1}{L_e} \mathcal{D}[\mathbf{\Delta}_e] = \frac{1}{L_e} \mathbf{\phi}^T \mathbf{B}_e \mathbf{h}$$

Recall that the Lagrangian strain is

$$\varepsilon_e(\mathbf{u}) = \frac{1}{L_e} \mathbf{n}_e \cdot \mathbf{\Delta}_e + \frac{1}{2L_e^2} \mathbf{\Delta}_e \cdot \mathbf{\Delta}_e$$

So, the directional derivative of the Lagrangian strain is

$$\mathcal{D}[\varepsilon_e] = \frac{1}{L_e} \mathbf{n}_e \cdot \mathcal{D}[\mathbf{\Delta}_e] + \frac{1}{2L_e^2} \mathcal{D}[\mathbf{\Delta}_e \cdot \mathbf{\Delta}_e]$$

By the product rule, we can compute the derivative in the last term as

$$\mathcal{D}[\mathbf{\Delta}_e \cdot \mathbf{\Delta}_e] = 2\mathbf{\Delta}_e \cdot \mathcal{D}[\mathbf{\Delta}_e] = 2\mathbf{\Delta}_e \cdot \mathbf{\phi}^T \mathbf{B}_e \mathbf{h}$$

With the derivative of the displacement difference given by Eq. 8.23, we can put the pieces together to get the directional derivative of strain as

$$\mathcal{D}[\varepsilon_e] = \frac{1}{L_e} \mathbf{F}_e^T(\mathbf{u}) \mathbf{\phi}^T \mathbf{B}_e \mathbf{h}$$

Putting it all together, we get the directional derivative of $\mathbf{g}(\mathbf{u})$ as

$$\mathcal{D}\mathbf{g}(\mathbf{u}) \cdot \mathbf{h} = \sum_{e=1}^{M} \mathbf{B}_e^T \mathbf{\phi} \left(\mathbf{F}_e \frac{1}{L_e} \frac{\partial N_e}{\partial \varepsilon_e} \mathbf{F}_e^T + \frac{N_e}{L_e} \mathbf{I} \right) \mathbf{\phi}^T \mathbf{B}_e \mathbf{h} \tag{8.24}$$

where $\mathbf{I}$ is the identity matrix. This expression holds for any constitutive function $N_e(\varepsilon_e)$. Since $\mathcal{D}\mathbf{g} \cdot \mathbf{h} = \mathbf{A}\mathbf{h}$, we can factor out the arbitrary vector $\mathbf{h}$ and identify the *tangent matrix* as

$$\mathbf{A}(\mathbf{u}) = \sum_{e=1}^{M} \mathbf{B}_e^T \boldsymbol{\phi} \left(\mathbf{F}_e \frac{1}{L_e} \frac{\partial N_e}{\partial \varepsilon_e} \mathbf{F}_e^T + \frac{N_e}{L_e} \mathbf{I} \right) \boldsymbol{\phi}^T \mathbf{B}_e \qquad (8.25)$$

As was the case for the shear building, we will assemble the global tangent and residual from element contributions. Let us define the *element tangent* stiffness matrix as

$$\mathbf{k}_e = \mathbf{F}_e \frac{1}{L_e} \frac{\partial N_e}{\partial \varepsilon_e} \mathbf{F}_e^T + \frac{N_e}{L_e} \mathbf{I} \qquad (8.26)$$

where $\mathbf{F}_e(\mathbf{u})$ and $N_e(\varepsilon_e(\mathbf{u}))$ are functions of $\mathbf{u}$. Similarly, the *element residual* can be defined as

$$\mathbf{r}_e = \mathbf{F}_e N_e \qquad (8.27)$$

The implementation of the computation of the element tangent and residual is given in the following function:

Code 8.1 MATLAB code to compute the element residual and tangent for the truss element

```
function [A,g,Ne,E,iv] = TrussElem(x,u,d,iv,nDim)
%  Compute tangent and residual for 2D or 3D nonlinear truss element
%
%           x : Nodal coordinates for element
%           u : Nodal displacements for element
%           d : Material properties of element
%          iv : Internal variables for element
%        nDim : Dimension of space

%. Compute element length (Le) and direction cosines (ne)
   ne = (x(2,:)-x(1,:))';      % vector pointing from i-end to j-end
   Le = norm(ne);              % length of element
   ne = ne/Le;                 % unit vector in direction of element

%. Compute element strain
   du = (u(2,:)-u(1,:))'/Le;
   Fe = ne + du;
   E  = dot(du,ne+du/2);

%. Compute stress from strain. Compute material tangent
   [Ne,tg,iv] = NonlinearModels(E,d,iv);

%. Create element stiffness in global coordinates
   re = Ne*Fe;
   ke = tg*(Fe*Fe')/Le + (Ne/Le)*eye(nDim);

%. Create full element stiffness and residual for assembly
   A = [ke, -ke; -ke,  ke];
   g = [-re; re ];

%. Compute the actual axial bar force from N
   lambda = sqrt(1+2*E);
   Ne = Ne*lambda;

end
```

The first thing we compute in this function is the element length (Le) and the unit vector $\mathbf{n}_e$ that points from the i–node to the j–node (ne). Next, the element deformation is computed from Eq. 8.8. Note that du is actually Δ_e/L_e and the Lagrangian strain ε_e is called E.

It should be evident that the element tangent and element residual can be assembled into the global tangent and residual as

$$A(\mathbf{u}) = \sum_{e=1}^{M} \mathbf{B}_e^T \hat{\mathbf{k}}_e(\mathbf{u})\mathbf{B}_e, \qquad \mathbf{g}(\mathbf{u}) = \sum_{e=1}^{M} \mathbf{B}_e^T \hat{\mathbf{r}}_e(\mathbf{u}) - \mathbf{p} \qquad (8.28)$$

where

$$\hat{\mathbf{k}}_e = \boldsymbol{\phi}\,\mathbf{k}_e\,\boldsymbol{\phi}^T = \begin{bmatrix} \mathbf{k}_e & -\mathbf{k}_e \\ -\mathbf{k}_e & \mathbf{k}_e \end{bmatrix}, \qquad \hat{\mathbf{r}}_e = \boldsymbol{\phi}\,\mathbf{r}_e = \begin{Bmatrix} -\mathbf{r}_e \\ \mathbf{r}_e \end{Bmatrix} \qquad (8.29)$$

This process of assembly is exactly the same one we saw for the shear building, with a few slight modifications. With the global residual and tangent, we can iteratively compute the solution to the nonlinear static truss problem by Newton's method. The pseudocode for the algorithm is given in the following box:

Static Analysis of Truss Structures

1. Make initial guess of displacement $\mathbf{u}_0$.
2. While $err > tol$

   - Assemble tangent $A(\mathbf{u}_i)$ and residual $\mathbf{g}(\mathbf{u}_i)$
   - Update displacement $\mathbf{u}_{i+1} = \mathbf{u}_i - [A(\mathbf{u}_i)]^{-1}\mathbf{g}(\mathbf{u}_i)$
   - Compute norm of residual $err = \|\mathbf{g}(\mathbf{u}_{i+1})\|$
   - Increment counter $i \leftarrow i + 1$

3. Converged value of $\mathbf{u}_i$ is the solution

8.7   Dynamic Analysis of Truss Structures

In order to develop a discrete dynamic model, we need to make a few assumptions. First, all of the mass of the structure will be associated with the nodes, and the masses will behave as point masses. This assumption will avoid the complications that arise with rotations. Second, we will assume that all of the truss elements have negligible mass (in other words, their mass is small compared with the mass of the non-structural parts of the structure). We will eventually amend that assumption to one where the elements have mass, but are rigid in bending (see Sect. 8.8).

With these assumptions, the equations of motion come out to be discrete ordinary differential equations. Because they are discrete, we will be able to apply all of the numerical methods that we used for the SDOF and shear building models. In particular, we will use Newmark's method for time stepping and Newton's method to solve the nonlinear equations at each time step.

In this section we will derive the theory behind the dynamic analysis of a truss with nonlinear resisting elements and use these developments to create a code for truss dynamics. Virtually everything that we derived in the previous section for the static analysis of trusses will carry over. In particular, the description of the structure is the same (except that we will need to add mass and damping properties) and the element kinematics and constitutive behavior are identical (except that now everything will depend upon time). The point of departure is that we will need to replace equations of equilibrium with equations of motion.

Nodal Balance of Momentum Each node must satisfy balance of linear momentum. Hence, if we take node i as a free body diagram, the net force comes from exactly the same set of forces that we considered in the static problem in the previous section. For dynamics, these forces are equal to the mass associated with the node times the acceleration of that node. As we have observed for the shear building, the nodal accelerations are coupled only through the forces transmitted by the members that connect them. In a truss, this sharing of force is more complex than it was in the shear building.

Let us apply the method of joints to establish balance of linear momentum. Because the masses are considered point masses, we do not need balance of angular momentum. We pick up the derivation at Eq. 8.10, and rewrite it as an equation of motion using $\mathbf{F} = m\mathbf{a}$. To wit,

$$-\sum_{k \in \mathcal{K}(i)} \mathbf{N}_k(t) + \mathbf{p}_i(t) = m_i \ddot{\mathbf{u}}_i(t) \tag{8.30}$$

which holds for all nodes $i = 1, \ldots, N$. In this equation, m_i is the mass associated with node i and N is the number of nodes. Recall that the notation on the summation means that we are summing on the index $k \in \mathcal{K}(i)$, the set of all elements that attach to node i. Of course, all of the quantities depend upon time t.

We have not explicitly accounted for the forces that arise because the mass creates weight in a gravitational field. Those forces are implicitly contained in the applied force vector $\mathbf{p}_i(t)$. At some point it will be important to model the nature of the time dependence of the forces. The weight forces are constant and act in the direction of gravity. Additional forces that are time dependent will act in addition to the weight forces. However, in most cases we would not want the weight forces to appear to be suddenly applied (recall the response of the SDOF system under constant but suddenly applied loads), but rather find static equilibrium prior to the action of the time-dependent loads. This will be especially important if the weight forces cause compression in the elements that bring the structure close to a buckling limit state.

The Dynamic Principle of Virtual Work We can take our nodal equations of motion, multiply them by an arbitrary virtual displacement $\mathbf{w}_i$, and then sum over all of the nodes to get the virtual work function

$$\mathcal{G} = \sum_{i=1}^{N} \left(m_i \ddot{\mathbf{u}}_i(t) - \mathbf{p}_i(t) + \sum_{k \in \mathcal{K}(i)} \mathbf{N}_k(t) \right) \cdot \mathbf{w}_i \tag{8.31}$$

If $\mathcal{G} = 0$ for all $\mathbf{w}_i$, then, by the fundamental theorem of the calculus of variations, the equations of motion Eq. 8.30 are all individually satisfied. Note that $\mathbf{w}_i$ is *not* a function of time. This statement is called *the dynamic principle of virtual work*. This principle is not an *energy principle* like Hamilton's principle of least action, although there are some similarities in the terms that appear.

We can convert the sum over nodes of the sum over elements framing into those nodes to a sum over elements, as we did previously. Hence, we can write the virtual work function as (suppressing, for notational convenience, the dependence on time)

$$\mathcal{G} = \sum_{i=1}^{N} \left(m_i \ddot{\mathbf{u}}_i - \mathbf{p}_i \right) \cdot \mathbf{w}_i + \sum_{e=1}^{M} \left(\mathbf{N}_{j(e)} \cdot \mathbf{w}_{j(e)} + \mathbf{N}_{i(e)} \cdot \mathbf{w}_{i(e)} \right) \tag{8.32}$$

The parallels of this derivation with the static case should be evident. In fact, the term $m_i \ddot{\mathbf{u}}_i$ looks like an *effective applied load*, grouping naturally with $\mathbf{p}_i$. Noting that the end forces are $\mathbf{N}_{j(e)} = \lambda_e N_e \mathbf{m}_e$ and $\mathbf{N}_{i(e)} = -\lambda_e N_e \mathbf{m}_e$, and that $\mathbf{w}_{j(e)} - \mathbf{w}_{i(e)} = \boldsymbol{\phi}^T \mathbf{B}_e \mathbf{w}$, we can write

$$\mathcal{G} = \sum_{i=1}^{n} \left(m_i \ddot{\mathbf{u}}_i - \mathbf{p}_i \right) \cdot \mathbf{w}_i + \mathbf{w}^T \sum_{e=1}^{M} \mathbf{B}_e^T \boldsymbol{\phi} \mathbf{F}_e N_e(\varepsilon_e) \tag{8.33}$$

where, again, $N_e(\varepsilon_e)$ is the constitutive function and $\mathbf{w}$ is the representation of the nodal vectors $\mathbf{w}_i$ in DOF order as a global array. As we did before, we will define global displacement $\mathbf{u}$ and global loads $\mathbf{p}$ in the same DOF order.

In order to make headway with the acceleration term, we need to associate the nodal acceleration $\ddot{\mathbf{u}}_i(t)$ with the global acceleration array $\ddot{\mathbf{u}}(t)$. To do this, we will introduce the matrix

$$\mathbf{D}_i = \begin{bmatrix} 0 \cdots 1 \ 0 \ 0 \cdots 0 \ 0 \\ 0 \cdots 0 \ 1 \ 0 \cdots 0 \ 0 \\ 0 \cdots 0 \ 0 \ 0 \cdots 1 \ 0 \end{bmatrix} \tag{8.34}$$

that has three rows (for a 3D truss, two rows for a 2D truss) and a number of columns equal to the number of degrees of freedom $\mathcal{N}$ in the system (the length of the $\mathbf{u}$ array). The purpose of $\mathbf{D}_i$ is to pick out the displacements associated with node i from the global array. Hence, the matrix is all zeros except for a one in each row.

The columns of the array with ones correspond to the global DOF numbers of the degrees of freedom associated with that node. Those are stored in the ith row of the id array. So the one in the first row would be in column $\mathrm{id}(i, 1)$. The one in the second row would be in column $\mathrm{id}(i, 2)$, etc.

With this definition we can write the nodal displacement and acceleration vectors for node i as

$$\mathbf{u}_i(t) = \mathbf{D}_i \mathbf{u}(t), \qquad \ddot{\mathbf{u}}_i(t) = \mathbf{D}_i \ddot{\mathbf{u}}(t) \tag{8.35}$$

Now, we can write the virtual work of the applied load in the form

$$\sum_{i=1}^{N} \mathbf{p}_i \cdot \mathbf{w}_i = \sum_{i=1}^{N} (\mathbf{D}_i \mathbf{w})^T (\mathbf{D}_i \mathbf{p}) = \mathbf{w}^T \left(\sum_{i=1}^{N} \mathbf{D}_i^T \mathbf{D}_i \right) \mathbf{p} \tag{8.36}$$

It is easy to prove that

$$\sum_{i=1}^{N} \mathbf{D}_i^T \mathbf{D}_i = \mathbf{I} \tag{8.37}$$

the $\mathcal{N} \times \mathcal{N}$ identity matrix, because each $\mathbf{D}_i$ has only one entry in each row, no two matrices have a one in the same column, and there is no column that is zero for all matrices. Because of this fact, we can write

$$\sum_{i=1}^{N} \mathbf{p}_i \cdot \mathbf{w}_i = \mathbf{w}^T \mathbf{p} \tag{8.38}$$

which we had noted with apparent impunity earlier in the static case without defining $\mathbf{D}_i$. The reason we need $\mathbf{D}_i$ now is because dealing with the acceleration term is not as obvious. If we note that the nodal force can be picked from the global vector $\mathbf{p}$ as $\mathbf{p}_i = \mathbf{D}_i \mathbf{p}$, then we can premultiply by $\mathbf{D}_i^T$ and sum over all nodes to get

$$\sum_{i=1}^{N} \mathbf{D}_i^T \mathbf{p}_i = \sum_{i=1}^{N} \mathbf{D}_i^T \mathbf{D}_i \, \mathbf{p} = \mathbf{p} \tag{8.39}$$

which shows, formally, how to construct the global force vector from the nodal forces. As usual, we will accomplish this assembly using the id array, rather than forming and multiplying by the $\mathbf{D}_i$ matrices.

The Mass Matrix Let us examine the nodal sum of the inertial terms of the dynamic principle of virtual work:

$$\sum_{i=1}^{N} m_i \ddot{\mathbf{u}}_i \cdot \mathbf{w}_i = \sum_{i=1}^{N} (\mathbf{D}_i \mathbf{w})^T m_i (\mathbf{D}_i \ddot{\mathbf{u}}) = \mathbf{w}^T \left(\sum_{i=1}^{N} m_i \mathbf{D}_i^T \mathbf{D}_i \right) \ddot{\mathbf{u}}$$

We can define the global *mass matrix* as

$$\mathbf{M} = \sum_{i=1}^{N} m_i \mathbf{D}_i^T \mathbf{D}_i \tag{8.40}$$

By the definition of the $\mathbf{D}_i$ matrices, we know that the mass matrix is diagonal with the nodal masses on the diagonals. What this operation does is associate the correct mass with each degree of freedom. In the shear building, the mass m_i was a single value at node i. Now there are three degrees of freedom (two in 2D) associated with node i. We need to assemble the mass m_i into the appropriate degrees of freedom in accord with the DOF number. Figure 8.8 illustrates this case.

The mass matrix that results for the structure in Fig. 8.8 is

$$\mathbf{M} = \begin{bmatrix} m_1 & 0 & 0 & 0 & 0 & 0 & 0 & 0 \\ 0 & m_1 & 0 & 0 & 0 & 0 & 0 & 0 \\ 0 & 0 & m_2 & 0 & 0 & 0 & 0 & 0 \\ 0 & 0 & 0 & m_2 & 0 & 0 & 0 & 0 \\ 0 & 0 & 0 & 0 & m_4 & 0 & 0 & 0 \\ 0 & 0 & 0 & 0 & 0 & m_3 & 0 & 0 \\ 0 & 0 & 0 & 0 & 0 & 0 & m_3 & 0 \\ 0 & 0 & 0 & 0 & 0 & 0 & 0 & m_4 \end{bmatrix}$$

Observe that the appropriate masses are associated with the correct elements of the mass matrix.

Incorporating the definition of the mass matrix, we can write the virtual work function in the final form

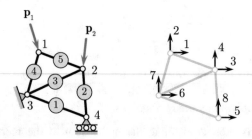

Fig. 8.8 *Example structure for definition of mass matrix.* A simple structure is shown with node and element numbers. The DOF numbers are shown, numbering the unrestrained degrees of freedom first and the restrained degrees of freedom last

$$\mathcal{G} = \mathbf{w}^T \left(\mathbf{M}\ddot{\mathbf{u}}(t) + \sum_{e=1}^{m} \mathbf{B}_e^T \, \boldsymbol{\phi} \, \mathbf{F}_e(\mathbf{u}(t)) N_e(\varepsilon_e(\mathbf{u}(t))) - \mathbf{p}(t) \right)$$

The virtual displacement $\mathbf{w}$ is arbitrary. Thus, if $\mathcal{G} = 0$ must hold for any virtual displacement, then we can conclude from the fundamental theorem of the calculus of variations that

$$\mathbf{M}\ddot{\mathbf{u}}(t) + \sum_{e=1}^{M} \mathbf{B}_e^T \, \boldsymbol{\phi} \, \mathbf{F}_e(\mathbf{u}(t)) N_e(\varepsilon_e(\mathbf{u}(t))) = \mathbf{p}(t) \tag{8.41}$$

which is the governing equation of motion.

Equation 8.41 is the *undamped* equation of motion. We can add damping to our system using the common notion of velocity dependent damping. We can construct a damping matrix using concepts like Rayleigh, Caughey, or modal damping, as before. If we build a damping matrix from mass and stiffness matrices, the best choice would be to use the initial stiffness (not the tangent stiffness, which changes from one step to the next). Damping can be added to give our final equation of motion for the nonlinear truss

$$\mathbf{M}\ddot{\mathbf{u}}(t) + \mathbf{C}\dot{\mathbf{u}}(t) + \sum_{e=1}^{M} \mathbf{B}_e^T \, \boldsymbol{\phi} \, \mathbf{F}_e(\mathbf{u}(t)) N_e(\varepsilon_e(\mathbf{u}(t))) = \mathbf{p}(t) \tag{8.42}$$

Solving the Equations of Motion We can solve the nonlinear equation of motion in a manner similar to what we did for the shear building, using Newmark's method for the time integrator and Newton's method to satisfy the equations of motion. Let us call $\mathbf{x} = \mathbf{a}_{n+1}$ the unknown acceleration that we seek. We can write the Newmark integrator as

$$\mathbf{u}_{n+1}(\mathbf{x}) = \mathbf{b}_n + \zeta \mathbf{x}$$
$$\mathbf{v}_{n+1}(\mathbf{x}) = \mathbf{c}_n + \eta \mathbf{x} \tag{8.43}$$

where $\mathbf{b}_n$, $\mathbf{c}_n$, ζ, and η are defined in Eqs. 4.53. We can define the discrete residual of Eq. 8.42 as

$$\mathbf{g}(\mathbf{x}) = \mathbf{M}\mathbf{x} + \mathbf{C}\mathbf{v}_{n+1}(\mathbf{x}) - \mathbf{p}_{n+1}$$
$$+ \sum_{e=1}^{M} \mathbf{B}_e^T \, \boldsymbol{\phi} \, \mathbf{F}_e(\mathbf{u}_{n+1}(\mathbf{x})) N_e(\varepsilon_e(\mathbf{u}_{n+1}(\mathbf{x}))) \tag{8.44}$$

The directional derivatives of displacement and velocity are simple to compute. Specifically, $\mathcal{D}(\mathbf{u}_{n+1}) \cdot \mathbf{h} = \zeta \mathbf{h}$ and $\mathcal{D}(\mathbf{v}_{n+1}) \cdot \mathbf{h} = \eta \mathbf{h}$. Using the directional derivative, we can find the dynamic tangent matrix to be

$$\mathbf{T} = \mathbf{M} + \eta\,\mathbf{C} + \zeta\mathbf{A}(\mathbf{x}) \tag{8.45}$$

where the static tangent is, as derived previously,

$$\mathbf{A}(\mathbf{x}) = \sum_{e=1}^{M} \mathbf{B}_e^T \boldsymbol{\phi} \left(\mathbf{F}_e \frac{1}{L_e} \frac{\partial N_e}{\partial \varepsilon_e} \mathbf{F}_e^T + \frac{N_e}{L_e} \mathbf{I} \right) \boldsymbol{\phi}^T \mathbf{B}_e \tag{8.46}$$

So, it turns out that the hard part of the equations of motion for dynamics was the part that also shows up in the static problem. The computation of the tangent stiffness matrix can be carried out element by element, noting that we can define an element tangent and residual as

$$\mathbf{k}_e = \mathbf{F}_e \frac{1}{L_e} \frac{\partial N_e}{\partial \varepsilon_e} \mathbf{F}_e^T + \frac{N_e}{L_e} \mathbf{I}, \qquad \mathbf{r}_e = \mathbf{F}_e N_e(\varepsilon_e)$$

which can be assembled using the standard assembly process as outlined previously. For those quantities that depend on the displacement, we will use the latest version of $\mathbf{u}_{n+1}(\mathbf{x})$ associated with the estimate of the acceleration $\mathbf{x}$ at the next time step, emanating from the Newton iteration.

The pseudocode for the dynamic analysis of truss structures is given in the following box.

Dynamic Analysis of Truss Structures

1. Input structure geometry and loads.
2. Establish initial displacement $\mathbf{u}_o$ and velocity $\mathbf{v}_o$.
3. Compute initial acceleration $\mathbf{a}_o$ from the equations of motion.
4. Loop over time steps

   - Compute new load $\mathbf{p}_{n+1}$.
   - Compute $\mathbf{b}_n$ and $\mathbf{c}_n$. Set $\mathbf{x}^0 \leftarrow \mathbf{a}_n$.
   - While $err > tol$

     - Compute $\mathbf{u}_{n+1}^i$ and $\mathbf{v}_{n+1}^i$ from Newmark.
     - Compute tangent $\mathbf{T}^i$ and residual $\mathbf{g}^i$.
     - Update acceleration $\mathbf{x}^{i+1} = \mathbf{x}^i - [\mathbf{T}^i]^{-1}\mathbf{g}^i$.
     - Compute norm of residual $err = \|\mathbf{g}^i\|$.
     - Increment counter $i \leftarrow i + 1$.

   - Converged value of $\mathbf{x}^i \rightarrow \mathbf{a}_{n+1}$ is the solution.
   - Compute final displacement and velocity $\mathbf{u}_{n+1}$ and $\mathbf{v}_{n+1}$.
   - Update internal variables (if applicable).
   - Update state $\mathbf{u}_n \leftarrow \mathbf{u}_{n+1}$, $\mathbf{v}_n \leftarrow \mathbf{v}_{n+1}$, $\mathbf{a}_n \leftarrow \mathbf{a}_{n+1}$.

(continued)

5. Loop over time steps complete.
6. Output graphics.

The dynamic solution process for the truss is exactly the same as the shear building. We loop over time steps with Newmark's integration method. At each time step we satisfy the equations of motion by Newton's method. The heart of the computation is to form the residual and tangent. Once we converge on a new acceleration, we update the variables and move on.

8.8   Distributed Element Mass

In our analysis of the truss in the previous section, we assumed that all of the mass of the structure was invested as nodal mass. We know that some of the mass comes from the elements themselves. In the next chapter we will start to take on the issue in a more complete context, but let's first consider an approximation that will allow us to include both the self-weight of the elements and the influence of the mass of the elements. To keep the idealization consistent with the context of this chapter we will assume that the bars are *flexurally rigid* (but still axially flexible).

Let us reconsider the kinematics of the truss bar, shown in Fig. 8.9 (see Fig. 8.2 for the original kinematics of motion). The displacement of a typical particle at a distance s from the left end can be expressed as

$$\mathbf{u}(s, t) = \mathbf{u}_{i(e)} + s\,\mathbf{F}_e \qquad (8.47)$$

where $s \in [\,0, L_e\,]$. The vector $\mathbf{u}_{i(e)}$ is the displacement at the left end of the bar and $\mathbf{F}_e = \lambda_e \mathbf{m}_e$ is, from Eq. 8.1,

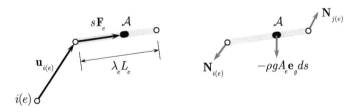

Fig. 8.9 *Truss element with distributed effects.* To include mass distributed along the truss element e, we assume that the bar is rigid in flexure, but flexible with respect to axial deformation. The position of the typical particle $\mathcal{A}$ is the position of the left end plus a distance $s\,\mathbf{F}_e$ along the bar. The sketch at right is a free body diagram of the element showing the end forces and the self-weight of the typical particle

$$\mathbf{F}_e = \mathbf{n}_e + \frac{1}{L_e} \boldsymbol{\Delta}_e$$

where $\mathbf{n}_e$ is a unit vector along the axis of the bar in the undeformed configuration, L_e is the undeformed length of the bar, and $\boldsymbol{\Delta}_e = \mathbf{u}_{j(e)} - \mathbf{u}_{i(e)}$ is the difference in displacements at the two ends of the bar (as originally defined in Eq. 8.2). It is straightforward to verify that the displacement at the right end is $\mathbf{u}(L_e, t) = \mathbf{u}_{j(e)}$, as illustrated in Fig. 8.2.

Now, the acceleration of the typical particle can be computed by differentiating Eq. 8.47 twice with respect to time with the result

$$\ddot{\mathbf{u}}(s, t) = \ddot{\mathbf{u}}_{i(e)} + s \, \ddot{\mathbf{F}}_e \tag{8.48}$$

where $\ddot{\mathbf{F}}_e = \ddot{\boldsymbol{\Delta}}_e / L_e = (\ddot{\mathbf{u}}_{j(e)} - \ddot{\mathbf{u}}_{i(e)})/L_e$, since the vector $\mathbf{n}_e$ is constant. We will use the acceleration of the typical particle to establish balance of momentum of the overall bar, as formulated in Chap. 1, especially the example in Sect. 1.6.

Consider the free body diagram shown on the right side of Fig. 8.9. Balance of linear momentum gives

$$\mathbf{N}_{i(e)} + \mathbf{N}_{j(e)} - \int_0^{L_e} \rho_e g A_e \mathbf{e}_g \, ds = \int_0^{L_e} \rho_e A_e \left(\ddot{\mathbf{u}}_{i(e)} + s \ddot{\mathbf{F}}_e \right) ds$$

where ρ_e is the density of the material, A_e is the cross-sectional area in the undeformed configuration, g is the acceleration of gravity, and $\mathbf{e}_g$ is a unit vector in the direction of gravity. We account for the contribution of all of the particles by integrating over the domain of the bar. Carrying out the integrals and defining the total mass of the bar as $\varrho_e = \rho_e A_e L_e$ and the total weight as $\varrho_e g \, \mathbf{e}_g$, this equation reduces to

$$\mathbf{N}_{i(e)} + \mathbf{N}_{j(e)} = \varrho_e \left(g \, \mathbf{e}_g + \tfrac{1}{2} \ddot{\mathbf{u}}_{i(e)} + \tfrac{1}{2} \ddot{\mathbf{u}}_{j(e)} \right) \tag{8.49}$$

A second equation of motion of the bar comes from balance of angular momentum about the left end of the bar, which can be written as

$$L_e \mathbf{F}_e \times \mathbf{N}_{j(e)} - \int_0^{L_e} s \, \mathbf{F}_e \times \rho_e g A_e \mathbf{e}_g \, ds = \int_0^{L_e} s \, \mathbf{F}_e \times \rho_e A_e \left(\ddot{\mathbf{u}}_{i(e)} + s \ddot{\mathbf{F}}_e \right) ds$$

Note that the moment arm for the force at the right end of the bar is $L_e \mathbf{F}_e$ and the moment arm for the typical particle is $s \, \mathbf{F}_e$. Carrying out the integrals, and dividing through by L_e, balance of angular momentum reduces to

$$\mathbf{F}_e \times \mathbf{N}_{j(e)} = \mathbf{F}_e \times \varrho_e \left(\tfrac{1}{2} g \, \mathbf{e}_g + \tfrac{1}{6} \ddot{\mathbf{u}}_{i(e)} + \tfrac{1}{3} \ddot{\mathbf{u}}_{j(e)} \right) \tag{8.50}$$

This equation does not give us the force $\mathbf{N}_{j(e)}$ directly, but it tells us about part of it. In fact, it tells us everything except the part of the force oriented in the axial direction. Let us write the end forces in the following form:

$$\mathbf{N}_{i(e)} = \varrho_e \left(\tfrac{1}{2} g \, \mathbf{e}_g + \tfrac{1}{3} \ddot{\mathbf{u}}_{i(e)} + \tfrac{1}{6} \ddot{\mathbf{u}}_{j(e)} \right) - N_e \mathbf{F}_e$$
$$\mathbf{N}_{j(e)} = \varrho_e \left(\tfrac{1}{2} g \, \mathbf{e}_g + \tfrac{1}{6} \ddot{\mathbf{u}}_{i(e)} + \tfrac{1}{3} \ddot{\mathbf{u}}_{j(e)} \right) + N_e \mathbf{F}_e \tag{8.51}$$

Taking the cross product of $\mathbf{F}_e$ with the second of these equations, gives Eq. 8.50 because any vector crossed with itself is zero. Adding the two equations together gives Eq. 8.49. Therefore, these expressions for the end forces are consistent with balance of momentum of the bar. Furthermore, if we neglect the distributed mass (i.e., set $\rho_e = 0$), these equations are exactly the same as our representation of the end forces without distributed effects where we assumed that the axial force had magnitude $\bar{N}_e = \lambda_e N_e$ oriented in the $\mathbf{m}_e$ direction, with the magnitude to be determined through the equations of motion in conjunction with the constitutive equations.

Recall that the virtual work functional for the truss structure, given in Eq. 8.32, is

$$\mathcal{G} = \sum_{i=1}^{N} \left(m_i \ddot{\mathbf{u}}_i - \mathbf{p}_i \right) \cdot \mathbf{w}_i + \sum_{e=1}^{M} \left(\mathbf{N}_{j(e)} \cdot \mathbf{w}_{j(e)} + \mathbf{N}_{i(e)} \cdot \mathbf{w}_{i(e)} \right)$$

where m_i is the mass of node i and $\mathbf{p}_i$ is the nodal load applied at node i (which is time dependent). The first sum is over all nodes in the structure; the second sum is over all elements of the structure. Let us examine the summand of the second term in this equation. Substitute the end forces Eqs. 8.51 to get

$$\mathbf{N}_{j(e)} \cdot \mathbf{w}_{j(e)} + \mathbf{N}_{i(e)} \cdot \mathbf{w}_{i(e)} = \mathbf{w}^T \left[\mathbf{B}_e^T \boldsymbol{\phi} \mathbf{F}_e N_e + \mathcal{D}_e^T \hat{\mathbf{m}}_e \mathcal{D}_e \ddot{\mathbf{u}} + \mathcal{D}_e^T \hat{\mathbf{g}}_e \right]$$

In this expression we have made the definition

$$\mathcal{D}_e = \begin{bmatrix} \mathbf{D}_{i(e)} \\ \mathbf{D}_{j(e)} \end{bmatrix} \tag{8.52}$$

where $\mathbf{D}_i$ is the $3 \times \mathcal{N}$ array ($2 \times \mathcal{N}$ for 2D) defined in Eq. 8.34 that picks out the global degrees of freedom for node i. The array $\mathcal{D}_e$ picks out the global degrees of freedom associated with the nodes at the two ends of element e. We have also defined the arrays

$$\hat{\mathbf{g}}_e = \tfrac{1}{2} \varrho_e \, g \begin{Bmatrix} \mathbf{e}_g \\ \mathbf{e}_g \end{Bmatrix}, \qquad \hat{\mathbf{m}}_e = \begin{bmatrix} \tfrac{1}{3} \varrho_e \mathbf{I} & \tfrac{1}{6} \varrho_e \mathbf{I} \\ \tfrac{1}{6} \varrho_e \mathbf{I} & \tfrac{1}{3} \varrho_e \mathbf{I} \end{bmatrix} \tag{8.53}$$

where $\mathbf{I}$ is the 3×3 identity matrix (2×2 in 2D). The matrix $\hat{\mathbf{m}}_e$ is called the *element distributed mass matrix*. With these definitions, we can write the structure mass matrix as

$$\mathbf{M} = \sum_{i=1}^{N} m_i \mathbf{D}_i^T \mathbf{D}_i + \sum_{e=1}^{M} \mathcal{D}_e^T \hat{\mathbf{m}}_e \mathcal{D}_e \qquad (8.54)$$

It should be evident that the gravity forces $\hat{\mathbf{g}}_e$ contribute the load vector. The total nodal force vector is

$$\mathbf{p} = \sum_{i=1}^{N} \mathbf{D}_i^T \mathbf{p}_i - \sum_{e=1}^{M} \mathcal{D}_e^T \hat{\mathbf{g}}_e \qquad (8.55)$$

It is important to realize that the forces $\mathbf{p}_i$ depend upon time, but the gravity forces $\hat{\mathbf{g}}_e$ are constant. In the previous section we mentioned that the static weight associated with nodal masses could be lumped in with the nodal forces in $\mathbf{p}_i$. However, to account for the fact that they are constant, they are better kept separate. To acknowledge that different dynamic loading mechanisms might be active at the same time, one might define the load vector as a combination of loads from different sources as

$$\mathbf{p}(t) = \sum_{j=1}^{J} \mathbf{p}^{(j)} f_j(t)$$

where J is the number of different loading mechanisms and each $\mathbf{p}^{(j)}$ is a specific pattern of loads and $f_j(t)$ determines how that pattern varies with time. In the code we assemble the gravity loads separately from the applied dynamic loads and provide the option to "turn off gravity" if we want to neglect the self-weight of the structure.

The assembly of the load vector and mass matrix now includes a nodal portion and an element portion to the assembly process, in contrast with the element stiffness, which is assembled only from element contributions. It is not surprising that the self-weight of the bar contributes equally to both nodes, with half of the total weight going to each node. Less intuitive is how the mass of the bar is apportioned in the element mass matrix, showing that it is not correct to simply lump half of the total mass at each end of the bar (although that approximation is sometimes used). The derivation proves that the distributed mass matrix, as simple as it turned out to be, is exact for large deformation.

The code segment for computing the mass matrix is Code 8.2, which is presented in the next section.

8.9   Earthquake Response of Truss Structures

The previous section outlined the formulation of the nonlinear dynamics problem for trusses. The loading envisioned in the formulation comprised time-dependent loads $\mathbf{p}(t)$ applied to the nodes. In this section we will formulate the dynamic equations of motion that include earthquake excitation.

In an earthquake, all of the points on the structure connected to the ground are subject to the ground motion. As we have done for the shear building problem, we will assume that the motion of the support points is prescribed to be the motion of the ground $\mathbf{u}_g(t)$. Therefore, we are not accounting for any effect the structure might have on the ground (i.e., soil–structure interaction). We will also assume that all support points feel exactly the same motion—a good assumption for a building structure and perhaps an assumption that needs to be revisited for a bridge. In a long bridge, for example, the support points might feel different motions because it takes time for the displacement waves associated with the earthquake ground motion to propagate. In our formulation we will envision the structure to be defined in a moving frame, and that frame will be subject to a uniform translation in accord with the ground motion, as shown in Fig. 8.10.

We can very easily modify our equations of motion to reflect the influence of a ground acceleration as input to the support points. The absolute motion of node i is

$$\mathbf{u}_i^a(t) = \mathbf{u}_g(t) + \mathbf{u}_i(t) \tag{8.56}$$

where $\mathbf{u}_i(t)$ is the relative displacement of node i (i.e., the displacement associated with deformation relative to a frame that is moving with the supports). The absolute acceleration of node i is

$$\ddot{\mathbf{u}}_i^a(t) = \ddot{\mathbf{u}}_g(t) + \ddot{\mathbf{u}}_i(t) \tag{8.57}$$

Using the virtual work argument from the previous section, the inertial term is

Fig. 8.10 *Truss subjected to ground motion.* For the earthquake problem, we assume that all support points experience the same ground motion $\mathbf{u}_g(t)$

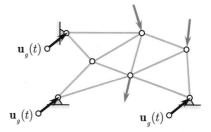

$$\sum_{i=1}^{N} m_i \left(\ddot{\mathbf{u}}_g + \ddot{\mathbf{u}}_i \right) \cdot \mathbf{w}_i = \sum_{i=1}^{N} m_i \left(\ddot{\mathbf{u}}_g + \mathbf{D}_i \ddot{\mathbf{u}} \right) \cdot \mathbf{D}_i \mathbf{w}$$

$$= \mathbf{w}^T \left(\sum_{i=1}^{N} m_i \mathbf{D}_i^T \ddot{\mathbf{u}}_g + m_i \mathbf{D}_i^T \mathbf{D}_i \ddot{\mathbf{u}} \right)$$

(8.58)

Similarly, we can add the ground acceleration to Eq. 8.48 to get

$$\ddot{\mathbf{u}}(s, t) = \ddot{\mathbf{u}}_g + \ddot{\mathbf{u}}_{i(e)} + s \ddot{\mathbf{F}}_e$$

(8.59)

This accounts for the effects of ground motion on the distributed mass of the elements. The expansion of $\mathbf{N}_{j(e)} \cdot \mathbf{w}_{j(e)} + \mathbf{N}_{i(e)} \cdot \mathbf{w}_{i(e)}$ then adds the new term

$$\tfrac{1}{2} \varrho_e \ddot{\mathbf{u}}_g \cdot \left(\mathbf{w}_{j(e)} + \mathbf{w}_{i(e)} \right) = \mathbf{w}^T \left(\tfrac{1}{2} \varrho_e \mathcal{D}_e^T \left\{ \begin{matrix} \ddot{\mathbf{u}}_g \\ \ddot{\mathbf{u}}_g \end{matrix} \right\} \right)$$

which accounts for the acceleration of the ground. We leave the details of the derivation to the reader.

Incorporating the nodal and element terms, the virtual work associated with acceleration can be written $\mathbf{w}^T \left(\mathbf{Q} \ddot{\mathbf{u}}_g + \mathbf{M} \ddot{\mathbf{u}} \right)$, where the mass matrix $\mathbf{M}$ is defined in Eq. 8.54 and we have defined

$$\mathbf{Q} = \sum_{i=1}^{N} m_i \mathbf{D}_i^T + \sum_{e=1}^{M} \tfrac{1}{2} \varrho_e \mathcal{D}_e^T \begin{bmatrix} \mathbf{I} \\ \mathbf{I} \end{bmatrix}$$

(8.60)

The number of columns in $\mathcal{D}_e$, which is defined in Eq. 8.52, is equal to twice the dimension of space, which is the dimension of the identity matrix $\mathbf{I}$. The quantity $\mathbf{Q}$ takes the place of $\mathbf{M1}$ in the shear building earthquake formulation.

Now we can write our equations of motion to reflect the influence of ground acceleration as an input to the support points. We will assume that our damping forces accrue in proportion to the relative velocities, as usual. Hence, our equation of motion for the structure subjected to earthquake ground motion is

$$\mathbf{M} \ddot{\mathbf{u}}(t) + \mathbf{C} \dot{\mathbf{u}}(t) + \mathbf{r}(\mathbf{u}(t)) = \mathbf{p}(t) - \mathbf{Q} \ddot{\mathbf{u}}_g(t)$$

(8.61)

where $\mathbf{u}(t)$ is the global relative displacement vector (i.e., the displacement relative to a frame that is moving with the supports) and the internal resistance is (as before)

$$\mathbf{r}(\mathbf{u}) = \sum_{e=1}^{M} \mathbf{B}_e^T \boldsymbol{\phi} \, \mathbf{F}_e(\mathbf{u}(t)) N_e(\varepsilon_e(\mathbf{u}(t)))$$

(8.62)

If one could idealize the earthquake as acting in a fixed direction $\mathbf{n}$, then we could write $\ddot{\mathbf{u}}_g(t) = \ddot{u}_g(t)\mathbf{n}$.

Example of Q   Let us construct the $\mathbf{Q}$ array for the structure shown in Fig. 8.8. We will also assume that the nodal masses for this structure are $m_1 = 2$, $m_2 = 5$, $m_3 = 1$, and $m_4 = 4$. Neglect the mass of the elements. The $\mathbf{D}$ matrices for the four nodes are

$$\mathbf{D}_1 = \begin{bmatrix} 1 & 0 & 0 & 0 & 0 & 0 & 0 & 0 \\ 0 & 1 & 0 & 0 & 0 & 0 & 0 & 0 \end{bmatrix} \qquad \mathbf{D}_2 = \begin{bmatrix} 0 & 0 & 1 & 0 & 0 & 0 & 0 & 0 \\ 0 & 0 & 0 & 1 & 0 & 0 & 0 & 0 \end{bmatrix}$$

$$\mathbf{D}_3 = \begin{bmatrix} 0 & 0 & 0 & 0 & 0 & 1 & 0 & 0 \\ 0 & 0 & 0 & 0 & 0 & 0 & 1 & 0 \end{bmatrix} \qquad \mathbf{D}_4 = \begin{bmatrix} 0 & 0 & 0 & 0 & 1 & 0 & 0 & 0 \\ 0 & 0 & 0 & 0 & 0 & 0 & 0 & 1 \end{bmatrix}$$

We will also assume that the earthquake acts in the fixed direction $\mathbf{n} = [\,0.6, 0.8\,]^T$. With these values we get

$$\mathbf{Q}\ddot{u}_g = \begin{bmatrix} m_1 & 0 \\ 0 & m_1 \\ m_2 & 0 \\ 0 & m_2 \\ m_4 & 0 \\ m_3 & 0 \\ 0 & m_3 \\ 0 & m_4 \end{bmatrix} \ddot{u}_g = \begin{bmatrix} 2 & 0 \\ 0 & 2 \\ 5 & 0 \\ 0 & 5 \\ 4 & 0 \\ 1 & 0 \\ 0 & 1 \\ 0 & 4 \end{bmatrix} \begin{Bmatrix} 0.6 \\ 0.8 \end{Bmatrix} \ddot{u}_g = \begin{bmatrix} 1.2 \\ 1.6 \\ 3.0 \\ 4.0 \\ 2.4 \\ 0.6 \\ 0.6 \\ 3.2 \end{bmatrix} \ddot{u}_g \qquad (8.63)$$

Note that the result needs to be trimmed to the free degrees of freedom by lopping off the last three rows, which are associated with restrained degrees of freedom.

Implementation of Mass Matrix   All the information needed to form the mass matrix and the $\mathbf{Q}$ matrix is contained in the `id` and `ix` arrays, along with nodal masses, which are part of the input specification as `mass`, and the element mass, with the density specified as an element material property in `d`. The code to compute the arrays associated with mass is provided in the following listing.

Code 8.2   MATLAB code to compute the arrays associated with mass in the truss structure

```
function [M,Q,P,D] = TrussMassMatrix(mass,x,d,ix,id,S)
%   Assemble global mass matrix M, EQ matrices Q and D, gravity load P
%
%     mass : Nodal masses
%        x : Nodal coordinates
%        d : Material properties, rho is d(5)
%       ix : Element connectivities
%       id : Global DOF numbers
%        S : Struct with problem size parameters

%. Initialize arrays
    I = eye(S.nDim);                    % Identity matrix of size nDim
    M = zeros(S.nDOF,S.nDOF);           % Mass matrix
    Q = zeros(S.nDOF,1);                % Effective EQ force matrix
    P = zeros(S.nDOF,1);                % Self-weight gravity forces
```

```
   D = zeros(S.nDOF,1);                    % EQ displacement array
%. Assemble M, Q, and D
   for n = 1:S.nNodes
      ii = id(n,:);
      M(ii,ii) = M(ii,ii) + mass(n)*I;
      Q(ii) = Q(ii) + mass(n)*S.EQdir;
      P(ii) = P(ii) - mass(n)*S.GRvec;
      D(ii) = D(ii) + S.EQdir;
   end % loop on n

%. Add contribution of element mass
   for e = 1:S.nElem
      inode = ix(e,1);                     % element start node
      jnode = ix(e,2);                     % element end node
      mat = ix(e,3);                       % element material set
      Le = norm(x(jnode,:) - x(inode,:));  % element length
      rhoe = d(mat,5)*Le;                  % total mass of element
      me = rhoe*[I/3, I/6; I/6, I/3];      % element mass matrix
      qe = (rhoe/2)*[S.EQdir; S.EQdir];    % element EQ force vector
      pe = (rhoe/2)*[S.GRvec; S.GRvec];    % element self-weight
      ii = [id(inode,:),id(jnode,:)];      % global node numbers
      M(ii,ii) = M(ii,ii) + me;            % assemble element mass
      Q(ii) = Q(ii) + qe;                  % assemble Q
      P(ii) = P(ii) - pe;                  % assemble P
   end

%. Trim to free DOF
   M = M(1:S.nFree,1:S.nFree);
   Q = Q(1:S.nFree,:);
   P = P(1:S.nFree,:);

end
```

The information going into this function is the nodal mass (mass), the nodal coordinates (x), the constitutive properties (d), the element connections (ix), the global DOF numbers (id), and the 'struct' S, which contains information about the problem size, the direction of action of the earthquake (S.EQdir), and the gravity vector (S.GRdir), which has magnitude g and points in the direction of gravity (e.g., up for gravity acting in the vertical direction). Gravity forces can be turned off by setting this value to zero.

The first task is to initialize all arrays. The mass matrix **M** and the effective earthquake force vector **Q** are defined as indicated in Eqs. 8.54 and 8.60. Then, we assemble the arrays. The array **P** is the self-weight associated with the elements, i.e., the second term in Eq. 8.55. These forces are combined with the nodal applied loads in the main program. The array **D** is an assembled version of the earthquake direction vector. With this array, we can compute the absolute displacement of all nodes as $\mathbf{u}^a = \mathbf{u} + u_g \mathbf{D}$. The next task is to assemble the nodal contributions to these arrays followed by assembly of the element contributions. Finally, we trim the arrays to the free degrees of freedom (except we do not trim D because we use it only to convert to absolute displacement for plotting an animation of the truss, in which case we need the displacement at all degrees of freedom).

8.10 Implementation

The implementation of the truss analysis code is very similar to the NDOF code.
The first task in the dynamic analysis of any structure is to describe the structure
itself. We relegate that task to the function `TrussInput`. This function establishes
the nodal coordinates, element connections, element material properties, boundary
conditions, applied forces, nodal masses, and the damping characteristics. Similar
to the NDOF code, those properties are stored in the arrays `x`, `ix`, `d`, `id`, `f`, `mass`,
and `Damp` respectively. For the truss, the `x`, `id`, and `f` arrays have multiple columns,
one for each spatial dimension of the structure. The code that generates the structural
properties for the example that we explore in Sect. 8.11 is:

Code 8.3 MATLAB code to establish the physical properties and dimensions of the example
structure shown in Fig. 8.12

```
function [x,ix,idG,d,f,mass,Damp,iv,S,DF] = TrussInput
% Create model of truss
%
%        x : Nodal coordinates
%       ix : Element connections and material set
%       id : Boundary conditions, then global DOF assignments
%        d : Material properties [Model Type, EA, C, No, rhoA]
%        f : Nodal forces (in node order)
%     mass : Nodal masses
%     Damp : Struct with information on damping
%       iv : Internal variables (for inelastic model)
%        S : Struct with problem size parameters
%       DF : Struct to manage graphic output

%. Nodal coordinates
   x = [  8.0,   16.0 ;
         14.0,   16.0 ;
         20.0,   16.0 ;
         26.0,   16.0 ;
          0.0,    0.0 ;
          8.0,    0.0 ;
         14.0,    8.0 ;
         20.0,   12.5 ];

%. Element Connectivity, [i-node, j-node, material set]
   ix = [ 5,   1,   1 ;
          1,   2,   1 ;
          2,   3,   1 ;
          3,   4,   1 ;
          5,   6,   1 ;
          6,   7,   1 ;
          7,   8,   1 ;
          8,   4,   1 ;
          6,   1,   1 ;
          7,   2,   1 ;
          8,   3,   1 ;
          1,   7,   1 ;
          2,   8,   1 ];

%. Set [Type, EA, C, No, rhoA] for each material set.
   d = [4, 30000, 0,  4,  0.1];

%. Extract problem size parameters from x, ix, and d arrays
   S.nNodes = size(x,1);            % Number of nodes
   S.nDim = size(x,2);              % Dimension of space
```

```
S.nDOF = S.nNodes*S.nDim;          % Total number of DOF
S.nElem = size(ix,1);             % Number of elements
S.nMat = size(d,1);               % Number of material sets

%. Boundary Condition, 1=fixed, 0=free
  id = zeros(S.nNodes,S.nDim);
  id(5,:) = [1, 1];
  id(6,:) = [0, 1];

%. Nodal forces
  f = zeros(S.nNodes,S.nDim);
  f(4,:) = [0,-1];
  f(7,:) = [0,-0.8];
  f(8,:) = [0,-0.9];

%. Change id array to global DOF numbers, nFree is # of free DOF
  [idG,S.nFree] = Bound(id,S.nDim,S.nNodes);

%. Establish mass for each node
  Mo = 0.0;
  mass = Mo*ones(S.nNodes,1);

%. Damping
  Damp.type = 1;                   % =1 Rayleigh, =2 Caughey, =3 Modal
  Damp.modes = [1,3];              % Modes for specified damping
  Damp.xi = [0.05; 0.05];          % damping ratio in each specified mode

%. Initialize internal variables to zero
  iv = zeros(S.nElem,1);

%. Initial condition, Time function, and EQ data
  g = 0;                           % Acceleration of gravity
  S.Mode = 0;                      % Mode for initial condition
  S.Uo = 0;                        % Magnitude of initial displacement
  S.Vo = 0;                        % Magnitude of initial velocity
  S.TimeType = 0;                  % =0 none =1 Constant, =2 Sinusoid, etc
  S.EQon = 1;                      % =0 No EQ, =1 Yes EQ
  S.EQdir = [1.0; 0.0];            % EQ direction
  S.GRvec = [0.0; 1.0]*g;          % Gravity vector

%. Set which figures to include
  DF = TrussManageFigures;
  PlotInitialTruss(x,f,ix,id,S,DF);

end
```

The function `PlotInitialTruss` was written to produce a figure like Fig. 8.12 that shows node and element numbers as well as the boundary restraints (in light gray) and the applied nodal loads (in darker gray). In a commercial code, the input would likely be done through a graphical user interface and the data structure would be invisible to the analyst. Note that we call the function `Bound` to convert the `id` array as input for the boundary conditions (1 if fixed, 0 if free) to global DOF numbers, as detailed in Appendix E (Code E.1), which also outlines how the nodal forces `f` are converted to global DOF order with Code E.3.

Main Program Code 8.4 is the main program for the dynamic truss analysis code. The structure of the truss code is very similar to the SDOF and NDOF codes.

Code 8.4 MATLAB code to carry out the dynamic structural analysis of two- and three-dimensional trusses with nonlinear elements

```
%. Dynamic Nonlinear 2D or 3D Truss Analysis with Earthquake
   clear; clc;

%. Anaysis features
   tf = 50;                    % Final time of analysis
   h = 0.01;                   % Analysis time step
   nOutFreq = 1;               % Output reporting frequency

%. Numerical analysis parameters
   [beta,gamma,eta,zeta,nSteps,tol,itmax,tf] = NewmarkParams(tf,h);

%. Input problem data
   [x,ix,id,d,f,mass,Damp,iv,S,DF] = TrussInput;

%. Initialize ouput storage arrays
   [nS,Hist] = TrussCreateHist(nOutFreq,nSteps,S);

%. Put applied loads in DOF order and initialize displacement vectors
   [Fo,u,uPre] = Force(f,id,S);

%. Compute stiffness, mass, damping matries and Q, D, EQ matrices
   [A,~,~,~,~] = TrussAssemble(x,u,ix,id,d,iv,S);
   [M,Q,P,D] = TrussMassMatrix(mass,x,d,ix,id,S);
   [C,Evec,freq,Damp] = DampingMatrix(M,A,Damp,DF);

%. Establish initial conditions
   [uo,vo] = TrussICs(Evec,S);

%. Get earthquake ground motion
   [EQaccel,~,EQdispl] = EQGroundMotion(h,nSteps,S.EQon,DF);

%. Initialize displacement, velocity. Compute initial acceleration.
   t = 0;
   uold = uo; vold = vo;
   u = [uold; uPre];
   F = TimeFunction(t,S.TimeType,1)*Fo + P;
   [~,g,~,iv,EF] = TrussAssemble(x,u,ix,id,d,iv,S);
   g = g + C*vo - F;
   aold = -M\g;

%. Compute motion by numerical integration over nsteps
   for n = 1:nSteps

%... Take care of the storage of the response
      Eq = EQdispl(n);  u = [uold; uPre];
      [nS,Hist] = TrussStoreHist(nS,Hist,t,u,iv,EF,Eq,F);

%.... Compute the values for the next time step
      t = t + h; err = 1; its = 0;
      bn = uold + h*vold + beta*h^2*aold;
      cn = vold + gamma*h*aold;
      anew = aold;
      F = TimeFunction(t,S.TimeType,0)*Fo + P;
      EQt = EQaccel(n)*Q;

%.... Newton's method to find displacement
      while (err > tol) && (its < itmax)
        unew = bn + zeta*anew;
        vnew = cn +  eta*anew;
        u = [unew; uPre];
        [A,g,~,~,~] = TrussAssemble(x,u,ix,id,d,iv,S);
        A = M + eta*C + zeta*A;
        g = M*anew + C*vnew + g - F + EQt;
        anew = anew - A\g;
```

```
      err = norm(g);
      its = its + 1;
   end

%.... Final update for velocity and displacement
      unew = bn + zeta*anew;
      vnew = cn + eta*anew;

%.... Update state for next iteration
      aold = anew;
      vold = vnew;
      uold = unew;

%... Update internal variables for new state
      u = [unew; uPre];
      [~,~,~,iv,EF] = TrussAssemble(x,u,ix,id,d,iv,S);

   end % loop over time steps

%. Output
   TrussEchoInput(S,d,uo,vo,Evec,freq,Fo,Q)
   TrussGraphs(x,ix,id,d,D,Hist,S,nS,Q,Evec,Fo,freq,DF)
```

Notice that the routines TimeFunction (Code 3.4) and DampingMatrix (Code 4.1) for specifying the temporal load function and the damping matrix, respectively, are exactly the same as previous codes. The periodic storage scheme built on the two functions TrussCreateHist and TrussStoreHist is nearly identical to that used in the NDOF code (see Codes 6.4 and 6.5). Retrieving the earthquake ground motion and setting up the state at $t = 0$ is identical to the NDOF code. There is almost no difference in the organization of the actual time-stepping part of the code.

The function TrussAssemble, Code 8.5, computes the residual and tangent in accord with Eqs. 8.44 and 8.46. It also computes the reaction forces, updates the internal variables, and passes back the internal axial force and deformation (to store for later plotting).

Code 8.5 MATLAB code to compute the residual and tangent matrices for the truss with nonlinear elements

```
function [T,g,reac,iv,ElmFD] = TrussAssemble(x,u,ix,id,d,iv,S)
%    Assemble tangent and residual from element contributions
%
%        x : Nodal coordinates
%        u : nodal displacements
%       ix : Element connectivities
%       id : Global DOF numbers
%        d : Material properties
%       iv : Internal variables
%        S : Struct with problem size parameters

%. Zero global stiffness matrix A, residual g, and element force/def
   A = zeros(S.nDOF,S.nDOF);
   g = zeros(S.nDOF,1);
   ElmFD = zeros(S.nElem,2);

%. Loop over all elements in the structure to assemble A
   for n = 1:S.nElem

%... Retrieve element local quantities for element n
```

```
      [xe,ue,de,ive,ii] = TrussLocalize(n,x,u,d,iv,ix,id);
%... Retrieve element stiffness and residual for element n
      [s,r,N,E,ive] = TrussElem(xe,ue,de,ive,S.nDim);
%... Assemble element quantities
         g(ii) = g(ii) + r;            % element residual
         A(ii,ii) = A(ii,ii) + s;      % element tangent
         ElmFD(n,:) = [N,E];           % force/def (for plotting)
         iv(n) = ive;                  % update internal variables

      end

%. Trim tangent, residual, and reactions
      reac = g(S.nFree+1:S.nDOF);
      T = A(1:S.nFree,1:S.nFree);
      g = g(1:S.nFree);

end
```

We use `TrussAssemble` in several locations in the main program. First, we compute the linear stiffness matrix by feeding it with zero displacements. With that stiffness and the mass matrix, we can compute the natural frequencies and mode shapes of the structure and use them to construct the damping matrix. One thing to be cautious of in a three-dimensional truss is the possibility of repeated eigenvalues, especially in the lower modes. This would happen, for example, with a symmetric structure. The problem that arises is in the specification of damping with the Rayleigh or Caughey models because they require the specification of the damping ratios in certain modes. That will not turn out well for the remaining modes if, for example, the two selected modes had the same frequency in Rayleigh damping.

We use the assembly function again to compute the residual for the initial acceleration. In that case we need to pass back the internal variables because it is possible that yielding can happen due to the initial conditions. We also pass back the element forces and displacements, `EF`, because they need to be stored in `Hist`. We use `TrussAssemble` a third time in the Newton loop where the residual and tangent are needed to iterate to the new acceleration for the time step. Finally, we use the function to update the internal variables after the Newton iteration has converged. In this call we also compute the element forces and deformations for output to `Hist`.

The `TrussAssemble` code has three main tasks that it accomplishes as it loops over the elements in the structure. First, the function `TrussLocalize`, Code E.2, extracts the quantities from the global arrays that are relevant to element e and sets up an index array of global DOF numbers associated with the element. In essence, this function does most of what the matrices $\mathbf{B}_e$ in the mathematical formulation do (just not through matrix multiplication). The localized element quantities are then fed to `TrussElem`, Code 8.1, to compute the element contribution to the tangent and residual. While in the element function, we also compute the element internal force, deformation, and the updated internal variable. Finally, the element quantities are assembled into the global tangent and residual arrays.

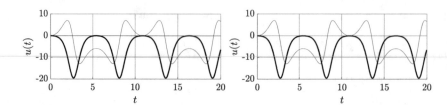

Fig. 8.11 *Truss code verification.* The pendulum problem solved by the truss code (on the left) and the pendulum code from Chap. 2 (on the right)

Aside from the functions that echo the input values and do the graphics, this section has outlined all of the major functions of the dynamic truss analysis code. Before moving on to the example in the next section, we provide a small code verification problem (the pendulum).

Code Verification We can verify the truss code by doing the Example 2.3 from Chap. 2. The truss has a single element of length $L = 10$, density $\rho = 0.1$ with one node at the origin and the other at $\mathbf{x} = L[\sin\theta, \cos\theta]$, where the initial angle is $\theta = 0.3$ rad. The result for both codes is shown in Fig. 8.11. Clearly, the two codes produce identical results.

The significance of this verification problem is that it proves that the code works for fully nonlinear motions and that the distributed element mass matrix is correct. The actual motions as the pendulum swings are on the same order as the dimensions of the structure itself. If there were issues with the description of the kinematics, they would appear here.

8.11 Example

There are two purposes for looking at an example. First, we need to see how the input file is put together. And second, it provides an opportunity to see what the program produces. Consider the two-dimensional truss shown in Fig. 8.12. The structure is pinned at Node 5 and sits on a roller at Node 6. There are applied downward loads at Nodes 4, 7, and 8 in the amounts of 1.0, 0.8, and 0.9, respectively. This represents the constant spatial pattern of loads. The time variation (including scaling of the magnitude) of the external loads is done in the function `TimeFunction`. All elements have the same material properties. Thus, there is only one material set with $EA = 30000$, $N_o = 4$, and $\rho A = 0.1$. The material type is elastoplastic so the nonlinear elasticity parameter C is not needed. There are no nodal masses and the self-weight of the elements is not included in the loading.

Before we get into the response to dynamic loads, let's look at some of the dynamic properties of the truss. Figure 8.13 shows the first three mode shapes of the structure, corresponding to the natural frequencies $\omega_1 = 4.65$, $\omega_2 = 15.41$, and

Fig. 8.12 *Example truss geometry.* The properties of this structure are defined in Code 8.3 in the previous section. The structure is pinned at Node 5 and sits on a roller at Node 6. Downward nodal loads are applied at Node 4, Node 7, and Node 8

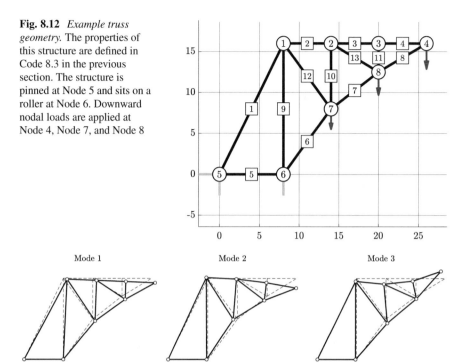

Fig. 8.13 *First three mode shapes for example structures.* The first three mode shapes of the example structure out of 13 possible are plotted with the undeformed structure in the background for reference

$\omega_3 = 26.48$. There are, of course, 13 modes for this structure because there are 13 free degrees of freedom. The highest frequency is $\omega_{13} = 217.5$. Unlike the shear building, the mode shapes of a truss do not organize into any intuitive pattern. The more complex the structure, the more difficult it is to ascribe particular geometric meaning to the mode shapes. However, the mode shapes play exactly the same role for the truss as for any structure in forming the damping matrix or in carrying out a modal analysis of the response of the structure to load.

For this example, we use Rayleigh damping to form the damping matrix. Figure 8.14 shows that we are specifying damping ratio of $\xi = 0.05$ in modes 1 and 3, leaving the damping ratio in the other modes to accrue as required by the Rayleigh damping model. Unlike the shear building, the spacing of the frequencies is not regular for the truss structure, making the variation of damping versus mode number (the left side of the figure) look a little irregular. As promised by the model (see Eq. 4.41), the damping values are inversely proportional to the mass and directly proportional to the stiffness, giving the curve on the right side of the figure.

Figure 8.14 gives an indication of the nature of the frequency spectrum of the example truss. Notice how some of the modes cluster at frequencies that are very close together. This feature is common in complex structures, especially in three

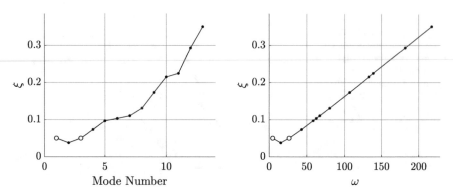

Fig. 8.14 *Rayleigh damping.* Rayleigh damping of the example structure with damping specified as $\xi = 0.05$ in modes 1 and 3

dimensions. In fact, it is possible to have two modes with identical frequencies. It is important to be aware of this when using a damping model like Rayleigh damping. This phenomenon was not possible for the shear building.

In what follows, we will look at the dynamic response of the example truss in three different scenarios. First, we will consider a sinusoidal variation of the loads shown in Fig. 8.12. Then, we will consider two earthquakes, or rather the same earthquake with two different directions. In each case we will examine the displacement and force responses.

Sinusoidal Load First consider an applied load $\mathbf{F}(t) = \mathbf{F}_o \sin(2t)$, where $\mathbf{F}_o$ is the spatial variation of loads shown in Fig. 8.12 with loads at Nodes 4, 7, and 8 in the amounts of 1.0, 0.8, and 0.9, respectively. The response of the structure to these load is shown in Fig. 8.15.

The top left plot shows the applied loads. The darker line is the load at Node 4. The two lighter lines are Nodes 7 and 8. The top right plot shows the force vs. deformation response of three of the 13 elements. We have selected Elements 1, 6, and 9 because they are the most stressed and give a good representation of what is happening in the elements. Element 1 shows the most significant yielding, while Element 6 remains elastic and Element 9 barely yields. The bottom right plot shows the element forces (for all elements) vs. time. This plot confirms the yielding and shows that Element 1 (the dark line) yields for considerable amounts of time. One can pick Element 9 out from the other lines because it is the only other element that shows any yielding. The remaining elements respond elastically, exhibiting some vibration at a higher frequency than the driving frequency.

The bottom left plot shows the displacement for all of the nodes in the structure, with the vertical displacement of Node 4 shown in the heavier line. It is evident that Node 4 exhibits the largest displacements. Note that there are 13 free degrees of freedom, so there should be 13 lines in the displacement versus time plot. The

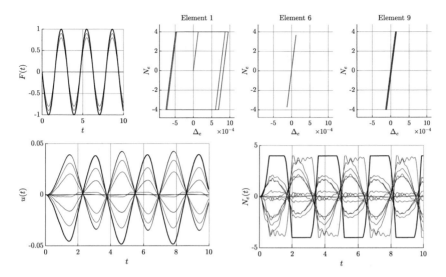

Fig. 8.15 *Example of truss subjected to sinusoidal load.* The upper left plot is the applied force vs. time. The upper right plots are the axial force vs. deformation of elements 1, 6, and 9. The lower right plot gives the axial force vs. time for all elements. The bottom left plot gives the nodal displacement vs. time for all free DOF. These results are for the truss shown in Fig. 8.12

reason that not all are visible is because some plot along the exact same line or cluster around zero (many of the vertical displacements are very small).

Earthquake Ground Motion Next, let us consider the response of the truss given in Fig. 8.12 to earthquake ground motion. We will use the artificial earthquake documented in Fig. 3.15. The analysis duration will be 50 s, containing the maximum response to the excitation. We will consider two cases. The first case will set the earthquake direction to be horizontal, i.e., $\mathbf{n} = [\,1, 0\,]$. The second will set the earthquake direction to be $\mathbf{n} = [\,0.6, 0.8\,]$. We will look at the differences in results that accrue due to the difference in direction of the earthquake.

One of the best indicators of the overall response is the earthquake participation factors ϑ_m similar to those defined for the shear building in Eq. 6.5. For the truss, the participation factors take the form

$$\vartheta_n = \frac{\varphi_n^T \, \mathbf{Q} \mathbf{n}}{\varphi_n^T \, \mathbf{M} \varphi_n}$$

where $\mathbf{Q}$ is given by Eq. 8.60. The earthquake participation factors for the two directions of earthquake motion are shown in Fig. 8.16. It is evident that the direction has a significant influence on the modal participation. Most importantly in this case, the participation of the fundamental mode for the direction $\mathbf{n} = [\,0.6, 0.8\,]$ is small. Ordinarily, the fundamental mode of vibration plays a large role in the response.

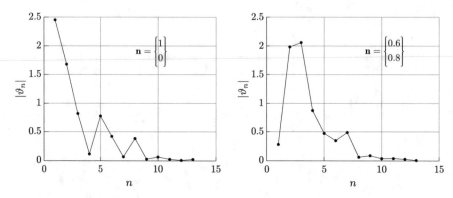

Fig. 8.16 *Participation factors.* Earthquake participation factors for two different earthquake directions. Plot on left has $\mathbf{n} = [\,1, 0\,]$. Plot on right has $\mathbf{n} = [\,0.6, 0.8\,]$

Therefore, we should expect this case to be substantially different from the other one.

One reason that the participation of the fundamental mode might be so small for this direction is the orientation of Elements 1 and 6, which roughly align with $\mathbf{n}$. This alignment could limit the amount of amplification of the motion due to sway of the structure.

The nodal displacement response for the two cases is shown in Fig. 8.17. The plot on top is for $\mathbf{n} = [\,1, 0\,]$ and the plot on the bottom is for $\mathbf{n} = [\,0.6, 0.8\,]$. Note that the scale of the ordinate is ten times greater on the top plot compared with the bottom plot. There is evidence of significant yielding in the structure in the first case as the displacements drift away from the axis and eventually settle on values associated with permanent deformation.

The yielding can be observed in Fig. 8.18, which shows the element axial force vs. axial deformation for Elements 1, 6, and 9. For $\mathbf{n} = [\,1, 0\,]$, Elements 1 and 9 experience significant yielding, while Element 6 remains elastic. For $\mathbf{n} = [\,0.6, 0.8\,]$, all three elements remain elastic, but notice that Element 6 now has the largest axial forces, coming close to the yield level of $N_o = 4$.

Finally, Fig. 8.19 shows the axial force versus time for all elements. It is obvious from this plot that all elements remain elastic for the case with $\mathbf{n} = [\,0.6, 0.8\,]$. The yielding events can be seen for the case with $\mathbf{n} = [\,1, 0\,]$. The yield events are relatively short in duration (especially when compared to those of the sinusoidal loading). It appears that Elements 6 and 9 are the main ones, if not the only ones, involved in yielding. It is also evident from Figs. 8.17 and 8.19 that all of the damage is done by 30 s.

Summary   Like the shear building example of the previous chapter, this truss example serves to show what is possible within this modeling context. Among the features we could explore are different temporal variations of load, different

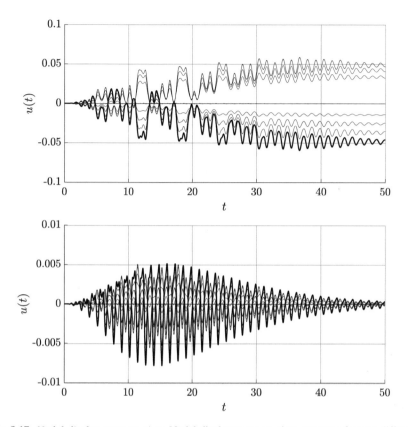

Fig. 8.17 *Nodal displacement vs. time.* Nodal displacement vs. time response for two different earthquake directions for all nodes in the example structure. Plot on top has **n** = [1, 0]. Plot on the bottom has **n** = [0.6, 0.8]

material properties (and different nonlinear models), earthquake ground motions as excitation, vibration in natural modes, etc.

The simple example explored in this section brings to light the importance of dynamic analysis. Sometimes what might seem to be subtle changes in the conditions have a dramatic impact on the results. This example suggests that there might be a worst direction for the earthquake ground motion. Identifying the worst-case scenario is one of the key aspects of the design of structures.

One of the features that is present in the code that is not possible to show in a static printed book is the animation. Producing a movie as output is valuable in dynamics because it is important to visualize the motion. The movie is to dynamics what the deflected shape is to static structural analysis. The animation of the motion allows visualization of the internal forces by using a color scheme to represent internal force levels. The value of the animation is even more acute for complex structures as the normal $x-y$ plots fall short of clearly elucidating the response of the structure.

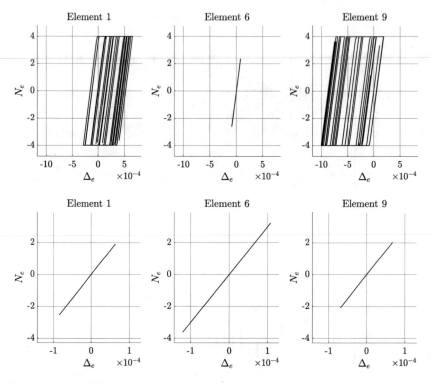

Fig. 8.18 *Force–deformation response.* Force–deformation response for two different earthquake directions for three of the elements in the example structure. The plots on top have $\mathbf{n} = [\,1, 0\,]$. The plots on the bottom have $\mathbf{n} = [\,0.6, 0.8\,]$

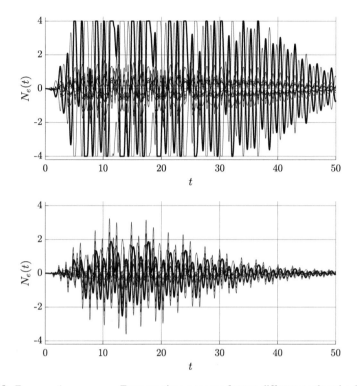

Fig. 8.19 *Force vs. time response.* Force vs. time response for two different earthquake directions for all elements in the example structure. Plot on top has $\mathbf{n} = [\,1, 0\,]$. Plot on the bottom has $\mathbf{n} = [\,0.6, 0.8\,]$

Chapter 9
Axial Wave Propagation

Up to this point we have considered only discrete dynamical systems. The primary assumption in a discrete system is that elements providing restoring forces have no associated mass and that the masses are rigid. Once we remove these restrictions, we are in the realm of *continuous systems*. In a continuous system, both the mass and the elasticity are distributed throughout the body. All materials have this feature, so it is a more realistic model than the discrete system idealization. The price of this additional fidelity is more mathematical complication. The remainder of the book concerns continuous systems. In this chapter we will explore the simplest version of a continuous system in solid mechanics—the *axial bar* problem.

Discrete systems are governed by ordinary differential equations while continuous systems are governed by partial differential equations. We have solved the ordinary differential equations of the discrete systems both classically (for linear systems) and numerically (e.g., using Newmark's method). Numerical integration works for both linear and nonlinear problems.

Continuous systems require a bit more mathematical effort for both classical and numerical solutions. To get classical solutions, we will make use of the technique of *separation of variables*. To get numerical solutions, we will first *discretize* the continuous spatial domain using a Ritz approximation to transform the problem into an ordinary differential equation in time. We will then solve the resulting ordinary differential equation either classically or numerically, as we have for the discrete systems of the previous chapters.

Both discrete and continuous systems exhibit wave propagation. Discrete structures must have multiple degrees of freedom for a disturbance to propagate. For

The original version of the chapter has been revised. A correction to this chapter can be found at https://doi.org/10.1007/978-3-030-89944-8_15.

Electronic Supplementary Material The online version of this chapter (https://doi.org/10.1007/978-3-030-89944-8_9) contains supplementary material, which is available to authorized users.

K. D. Hjelmstad, *Fundamentals of Structural Dynamics*, https://doi.org/10.1007/978-3-030-89944-8_9

continua, the propagation of waves is really a central feature of dynamic response. We will be able to see some interesting connections between vibration and wave propagation in continuous systems that were not directly evident in our analysis of discrete systems.

9.1   The Axial Bar Problem

Consider a bar with cross-sectional area A, Young's modulus E, mass density ρ, and length L, as shown in Fig. 9.1. The bar is subjected to a distributed axial force $p(x, t)$ and has certain boundary conditions. The bar shown in the figure is fixed at the left end and free at the right end with a concentrated load $P(t)$ applied at the free end. Of course, other end conditions are also possible (i.e., fixed–fixed, free–fixed, or free–free).

Motion   The motion of the bar is constrained to displace along its axis by an amount $u(x, t)$, as shown in Fig. 9.2. The kinematic hypothesis for the axial bar is that all points in a cross section displace in the axial direction by the same amount (plane sections remain plane). Cross sections do not rotate or displace transverse to the axis of the bar. The position vector for the typical particle $\mathcal{A}$, located a distance x from the left end in the undeformed configuration, is given by

$$\mathbf{x}(x, t) = x\,\mathbf{e}_1 + u(x, t)\,\mathbf{e}_1 \tag{9.1}$$

We can compute the velocity $\mathbf{v}$ and acceleration $\mathbf{a}$ from the position vector by differentiating the position vector with respect to time. To wit,

$$\mathbf{v}(x, t) = \frac{\partial u(x, t)}{\partial t}\mathbf{e}_1, \quad \mathbf{a}(x, t) = \frac{\partial^2 u(x, t)}{\partial t^2}\mathbf{e}_1 \tag{9.2}$$

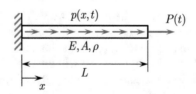

Fig. 9.1 *Axial bar.* Bar of area A, density ρ, Young's modulus E, and length L subjected to distributed axial force $p(x, t)$. The bar is fixed at $x = 0$ and subjected to a concentrated force $P(t)$ at $x = L$

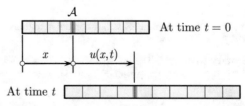

Fig. 9.2 *Motion.* Particle $\mathcal{A}$, initially located at x, moves by an amount $u(x, t)$ in the axial direction at time t

Fig. 9.3 *Deformation.* The small segment of length Δx deforms because the motion at its ends differ

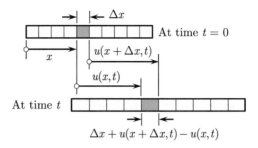

Because we are tracking the location of material particles in the body, this is a *Lagrangian* description of motion. Hence, the velocity and acceleration are simply partial derivatives of the displacement with respect to time. We will use the acceleration in the balance of linear momentum to get the equations of motion.

Deformation To analyze the deformation, consider the motion of a small piece of the bar that has length Δx at time $t = 0$, as shown in Fig. 9.3. The left end of the piece is originally located at x and moves by an amount $u(x, t)$. The right end starts at $x + \Delta x$ and moves by an amount $u(x + \Delta x, t)$. The deformed length of the piece (at time t) is

$$\ell(t) = \Delta x + u(x + \Delta x, t) - u(x, t)$$

The (engineering) strain[1] is the change in length, $\ell(t) - \Delta x$, divided by the original length, Δx. Computing this value and taking the limit as the size of the piece shrinks to zero gives

$$\varepsilon(x, t) = \lim_{\Delta x \to 0} \frac{u(x + \Delta x, t) - u(x, t)}{\Delta x} = \frac{\partial u(x, t)}{\partial x} \tag{9.3}$$

In words, the strain is the spatial rate of change of displacement. We refer to this equation as the *strain–displacement* relationship for the axial bar.

Stress Resultants If we take a cut at the cross section initially located at x, then we expose the traction field $\sigma(x, t)$. The net effect of those tractions is the sum

$$N(x, t) = \int_A \sigma(x, t)\, dA \tag{9.4}$$

which we call the resultant *axial force.* Since the cross section moves uniformly, the strain is constant over the cross section, and hence the stress is, too. Therefore, we

---

[1] The formulation in this chapter is linear. We use engineering strain for simplicity.

can write the net internal axial force as

$$N(x, t) = \sigma(x, t)A \tag{9.5}$$

where A is the area of the cross section. If the bar is *non-prismatic*, then the cross-sectional area varies with x.

Hooke's law for uniaxial tension is $\sigma(x, t) = E\varepsilon(x, t)$, where $\varepsilon(x, t)$ is the axial strain and E is Young's modulus. Substitute Hooke's law and the strain–displacement relationship, Eq. 9.3, into Eq. 9.5 to get

$$N(x, t) = EA\frac{\partial u(x, t)}{\partial x} \tag{9.6}$$

which shows that the axial force is proportional to the rate of change of the axial displacement with respect to x and the constant of proportionality is EA, which we call the *axial stiffness* of the bar.

Equations of Motion To establish the equations of motion, we take a free body diagram of the piece of the bar between stations x and $x + \Delta x$, as shown in Fig. 9.4. Setting the sum of the forces equal to mass times acceleration (and dotting the equation with $\mathbf{e}_1$) gives

$$N(x + \Delta x, t) - N(x, t) + p(\xi, t)\Delta x = \rho A \Delta x \frac{\partial^2 u}{\partial t^2}$$

where ξ is some value of x within the region (fundamental theorem of calculus). Note that the mass is the volume times the density. Dividing through by Δx and taking the limit as Δx approaches zero, we get

$$\frac{\partial N(x, t)}{\partial x} + p(x, t) = \rho A \frac{\partial^2 u(x, t)}{\partial t^2} \tag{9.7}$$

Finally, substituting the relationship between axial force and displacement from Eq. 9.6, we get

$$\frac{\partial}{\partial x}\left(EA\frac{\partial u}{\partial x}\right) + p = \rho A \frac{\partial^2 u}{\partial t^2} \tag{9.8}$$

Fig. 9.4 *Free body diagram.* We can find the equation of motion by taking a free body diagram of the small segment of length Δx

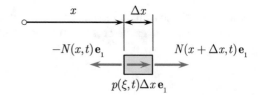

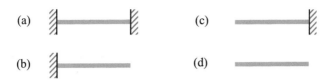

Fig. 9.5 *Boundary conditions.* Either the motion or the force must be specified at each end of the bar. The four possible cases are (**a**) fixed–fixed, (**b**) fixed–free, (**c**) free–fixed, and (**d**) free–free

This is the equation of motion for the dynamic axial bar. It holds for prismatic or non-prismatic bars and for any form of distributed axial loading. We will examine certain special cases of this general equation in the sequel.

Boundary and Initial Conditions For the problem to be well posed, we must specify both *boundary conditions*, which are conditions that are true for all t at the two ends of the bar, and *initial conditions*, which are conditions that are true for all x at time $t = 0$.

The two ends of the bar can be either fixed or free, as shown in Fig. 9.5. The two possible boundary conditions have the following characteristics:

$$u(\bullet, t) = 0 \qquad \text{fixed at end } \bullet$$

$$N(\bullet, t) = P(t) \qquad \text{free at end } \bullet$$

where $\bullet$ stands for either 0 or L and $P(t)$ is the applied load at that point.[2] Noting Eq. 9.6, we can write the condition at the free end in terms of the displacement as:

$$N(\bullet, t) = P(t) \quad \rightarrow \quad EA\frac{\partial u(\bullet, t)}{\partial x} = P(t) \tag{9.9}$$

The initial conditions prescribe the displacement and velocity at all points in the bar at $t = 0$ and can be expressed as

$$u(x, 0) = u_o(x), \qquad \frac{\partial u}{\partial t}(x, 0) = v_o(x) \tag{9.10}$$

where $u_o(x)$ and $v_o(x)$ are known functions of x alone. An initial displacement would be the result of applying an external agent to create displacement and then releasing that agent at $t = 0$. An initial velocity might be the result of an impact just prior to $t = 0$.

---

[2] The value of the applied load can be zero, of course. To find the appropriate boundary condition at a free end take a tiny free body diagram of a piece at that point to expose the internal force. Equilibrium then determines the boundary condition.

9.2 Motion Without Applied Loading

Let us first look at the special case where there is no applied load (i.e., the distributed load $p=0$ and the end load $P=0$). Further, assume that the bar is prismatic (i.e., EA and ρ are constant along the length of the bar). In this case the equation of motion reduces to

$$EA\frac{\partial^2 u}{\partial x^2} = \rho A\frac{\partial^2 u}{\partial t^2} \tag{9.11}$$

Let us define a constant $c = \sqrt{E/\rho}$ (we will soon identify this as the *wave speed*). Divide Eq. 9.11 through by ρA and write the equation of motion as

$$\frac{\partial^2 u}{\partial t^2} - c^2\frac{\partial^2 u}{\partial x^2} = 0 \tag{9.12}$$

This equation is often called the *wave equation*.

To solve this equation, consider any function $f(x+ct)$. The chain rule for differentiation gives

$$\frac{\partial^2}{\partial x^2}f(x+ct) = f''(x+ct), \qquad \frac{\partial^2}{\partial t^2}f(x+ct) = c^2 f''(x+ct)$$

where $(\bullet)'$ simply stands for the derivative of f with respect to its argument, i.e., the $\bullet$ in $f(\bullet)$. Express the axial displacement function in the particular form $u(x,t)=f(x+ct)$. Substituting this expression into Eq. 9.12 gives

$$\frac{\partial^2 u}{\partial t^2} - c^2\frac{\partial^2 u}{\partial x^2} = c^2 f'' - c^2 f'' = 0 \tag{9.13}$$

Therefore, this displacement function satisfies the equation of motion and hence is a general solution to the differential equation.[3] Following the same reasoning we can show that the function $u(x,t)=f(x-ct)$ also satisfies the equation.

Wave Propagation We have found a general solution to the equation of motion without forcing function:

$$u(x,t) = f_1(x+ct) + f_2(x-ct) \tag{9.14}$$

where we have left open the possibility that the two functions might be different. We can observe a few things about this solution. First, it implies that waves propagate in both directions at the speed c. Therefore, we refer to c as the wave speed. We can

---

[3] The term *general solution* means a function that satisfies the domain equation but does not yet consider the boundary or initial conditions.

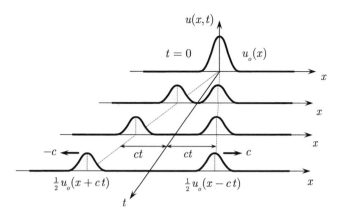

Fig. 9.6 *Axial wave propagation.* The displacement function starts with a value $u_o(x)$. The wave splits and propagates in opposite directions at the wave speed c. This sketch shows snapshots of the wave at different times

verify that it has units of speed. Young's modulus E has units of force per area, and density ρ has units of mass per volume. Hence, $\sqrt{E/\rho}$ has units of length per time, which are the appropriate units for velocity.

If we take $u(x,0) = u_o(x)$ and $\dot{u}(x,0) = 0,$[4] then it is evident that

$$u(x,0) = f_1(x) + f_2(x) = u_o(x)$$
$$\dot{u}(x,0) = cf_1'(x) - cf_2'(x) = 0$$

(9.15)

From the second of these two equations, we can determine that the difference between the two function must be a constant, i.e., $f_2(x) = f_1(x) + B$. We can then use the first equation to get

$$f_1(x) = \tfrac{1}{2}\big(u_o(x) - B\big), \quad f_2(x) = \tfrac{1}{2}\big(u_o(x) + B\big)$$

Hence, the solution must be

$$u(x,t) = \tfrac{1}{2}u_o(x + ct) + \tfrac{1}{2}u_o(x - ct)$$

(9.16)

As Fig. 9.6 illustrates, this solution implies that the initial displacement $u_o(x)$ splits and propagates in both directions. The two waves that emanate from the initial pulse are exactly one half of the original function. The wave does not disperse, but rather travels intact at the wave speed c, which is determined by the properties of the material. During the time that the waves overlap, they reinforce each other and sum to the full magnitude of the initial pulse.

---

[4] A dot over a function indicates partial derivative with respect to time.

Fig. 9.7 *Axial wave propagation with boundary conditions.* We can find the solution to a wave reflecting from a fixed boundary by imagining waves propagating in two directions that meet fortuitously to cancel each other out. The end result is to enforce the boundary condition

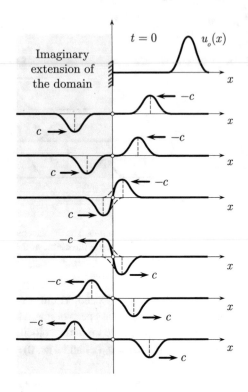

Boundary Conditions The solution we have derived works for an infinitely long bar. But what happens for a finite length bar with a boundary condition? Let us look at the half of the wave that is propagating to the left, heading for a fixed support, as shown in Fig. 9.7.

Imagine that there is no fixed support at $x = 0$ and that the domain of the bar is extended to the left. Now, we will treat the problem as if it did not have the support condition. Add a wave that is the inverted mirror image of the left-moving wave. This second wave is propagating to the right and is positioned such that the leading edge of the wave is exactly the same distance from $x = 0$ as the leading edge of the left-moving wave. Because the two waves are traveling at the same speed, they arrive at $x = 0$ at the same time. If we sum the two waves, the displacement is equal to zero at the point $x = 0$ at all times, thereby satisfying the fixed condition $u(0, t) = 0$.

Because the combined wave satisfies the boundary condition, this sum is the actual solution to the problem with the boundary condition. When the left-moving wave goes past the boundary, it moves outside of the actual domain of the bar (and ceases to be of interest), while the right-moving wave enters the domain. Hence, the right-moving wave appears to be the *reflection* of the left-moving wave at the fixed boundary. Therefore, we can observe that what happens to a displacement wave when it hits a fixed boundary is that it reflects with opposite sign and propagates in the opposite direction.

The internal axial force is proportional to the slope of the displacement function. For the initial wave shown in Fig. 9.7, the force is tensile on the left side of the left-moving wave and compressive on the right side. Since the right-moving wave is inverted, the right side is in tension and the left side is in compression. Hence, the stress wave does not change sign when reflected at a fixed boundary.

We can do a similar thought experiment for a free end at $x = 0$. Imagine that a wave is propagating to the left. Consider an infinite bar with a wave of the *same* sign propagating to the right at the same speed and equidistant from the location $x = 0$. When the two waves meet, they reinforce, but create a condition where the sum has zero slope at $x = 0$ at all times, which is the exact condition required at a free end. Hence, at a free end a displacement wave reflects with the same sign and propagates in the opposite direction. The stress wave would then change sign upon reflection at a free end.

9.3 Classical Solution by Separation of Variables

In this section we look at a second possibility to solve the axial wave propagation problem—a technique called *separation of variables*. There is really no good reason to think that the separation of variables assumption is going to lead to anything useful, particularly since the solution appears to have the form $f(x \pm ct)$. But by taking this route, we will discover a series solution to the problem.

Separation of Variables The idea of separation of variables (as a means of solving a partial differential equation) is to assume that the displacement function can be written in the form

$$u(x, t) = U(x)T(t) \tag{9.17}$$

where $U(x)$ and $T(t)$ are functions only of their respective single variables. With this assumption, we can compute the derivatives that we need for our equation of motion. In particular,

$$\frac{\partial^2 u}{\partial x^2} = U''T \qquad \frac{\partial^2 u}{\partial t^2} = U\ddot{T}$$

where a prime denotes a derivative with respect to x and a dot indicates a derivative with respect to t, i.e.,

$$U' = \frac{dU}{dx} \qquad \dot{T} = \frac{dT}{dt}$$

If we substitute these expressions into Eq. 9.12 (the wave equation), we get

$$c^2 U''T = U\ddot{T}$$

Divide both sides by $U(x)T(t)$ to get

$$c^2 \frac{U''}{U} = \frac{\ddot{T}}{T} = -\omega^2 \tag{9.18}$$

where ω is an as yet unknown constant. Why are both sides of this equation equal to a constant? Well, the left side is purely a function of x while the right side is purely a function of t. The only function that is both a function of x and a function of t is the constant function. We are calling the constant $-\omega^2$ for reasons that will soon become evident. Equation 9.18 actually gives us two separate ordinary different equations:

$$U''(x) + \lambda^2 U(x) = 0, \qquad \ddot{T}(t) + \omega^2 T(t) = 0 \tag{9.19}$$

where to simplify the notation, we have defined

$$\lambda = \frac{\omega}{c} \tag{9.20}$$

The solution to a linear ordinary differential equation with constant coefficients is an exponential function. In both of these equations, the exponent is purely imaginary. Hence, the solution can be written in terms of trigonometric functions. The solution to Eq. 9.19(b) is well known from our work on the SDOF system:

$$T(t) = A \cos \omega t + B \sin \omega t \tag{9.21}$$

where A and B are constants to be determined from the initial conditions. The spatial equation is in exactly the same form. Therefore, the general solution is

$$U(x) = D_1 \cos \lambda x + D_2 \sin \lambda x \tag{9.22}$$

where D_1 and D_2 are constants to be determined from the boundary conditions. It is easy to verify that these functions satisfy their respective differential equations. Keep in mind that we still do not know the value of λ (which, when found, also determines ω).

Boundary Conditions Let us first consider the boundary conditions. At this point we need to be specific, so let's consider the fixed–free case. Our boundary conditions are $u(0, t) = 0$ and $N(L, t) = 0$. Noting that the axial force is $N = EAu'$, we can write the second boundary condition in terms of displacement as $u'(L, t) = 0$. If these conditions are to hold for all values of t, then they imply that

$$U(0) = 0, \quad U'(L) = 0$$

Implementing both of these conditions into the general solution for $U(x)$ gives

$$U(0) = D_1 = 0$$

$$U'(L) = \lambda \left(-D_1 \sin \lambda L + D_2 \cos \lambda L\right) = 0$$

Since the first equation gives $D_1 = 0$, the second equation reduces to

$$D_2 \cos \lambda L = 0 \tag{9.23}$$

We call this equation the *characteristic equation*. This equation is satisfied if $D_2 = 0$ (the trivial solution) or if $\cos \lambda L = 0$, which happens for values of λL equal to

$$\lambda L = \tfrac{1}{2}\pi, \ \tfrac{3}{2}\pi, \ \tfrac{5}{2}\pi, \ \ldots \tag{9.24}$$

The values of λ that satisfy the characteristic equation are called the *eigenvalues*. In this case they are

$$\lambda_n = \left(\frac{2n-1}{2}\right)\frac{\pi}{L} \tag{9.25}$$

for $n = 1, 2, 3, \ldots$. There are an infinite number of eigenvalues, which is typical of continuous systems. For each of these values there is a function associated with the solution. Noting that $D_1 = 0$, Eq. 9.21(b) suggests that each of the functions

$$\varphi_n(x) = \sin \lambda_n x \tag{9.26}$$

satisfies the differential equation *and* boundary conditions. These functions are called *eigenfunctions*. There are an infinite number of eigenfunctions, each one associated with an eigenvalue. We do not include the constant D_2 because any value of D_2 will still satisfy the differential equation and boundary conditions. Thus, $\varphi_n(x)$ is a *mode shape*, analogous to how we interpreted mode shapes for discrete systems.

The first five mode shapes of the fixed–free bar are shown in Fig. 9.8. Notice that they satisfy the boundary conditions $\varphi_n(0) = 0$ and $\varphi'_n(L) = 0$. Also observe that the number of zeros (places where the function is equal to zero) of the function $\varphi_n(x)$ is equal to n, a feature typical of this type of problem.

It is worth noting that we get negative solutions to the characteristic equation, too. Thus,

$$\lambda L = -\tfrac{1}{2}\pi, \ -\tfrac{3}{2}\pi, \ -\tfrac{5}{2}\pi, \ldots$$

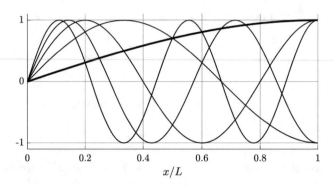

Fig. 9.8 *Mode shapes.* The first five natural mode shapes of the fixed–free axial bar. The first mode is the heavier line. All of the mode shapes satisfy the boundary conditions of the problem, $\varphi_n(0) = 0$ and $\varphi'_n(L) = 0$

are also solutions. Observe that $\lambda_{-n} = -\lambda_n$. These negative eigenvalues are associated with the eigenfunctions

$$\varphi_{-n}(x) = \sin \lambda_{-n} x = \sin (-\lambda_n x) = -\sin \lambda_n x = -\varphi_n(x)$$

which are redundant because they differ only by a constant multiplier. For that reason, we do not need to include the negative n values.

Natural Frequencies Since there are multiple characteristic values λ_n, there must also be multiple values of ω. From Eq. 9.20, we have

$$\omega_n = \lambda_n c \tag{9.27}$$

For the fixed–free bar, then, the natural frequencies are

$$\omega_n = \frac{(2n-1)\pi}{2L} \sqrt{\frac{E}{\rho}} \tag{9.28}$$

The natural frequencies have units of radians per second. The lowest frequency is associated with $n = 1$, often called the *fundamental frequency*, and the frequencies increase linearly from there. The ratio of adjacent frequencies is

$$\frac{\omega_{n+1}}{\omega_n} = \frac{2(n+1) - 1}{2n - 1} = \frac{2n + 1}{2n - 1} \tag{9.29}$$

The sequence is infinite, so there is no highest frequency (as there is in a discrete system). In practical computations we will truncate the higher modes. The eigenvalues are obviously distinct, but they get closer in value as n gets larger. There is no possibility of repeated eigenvalues.

Orthogonality of the Eigenfunctions Each eigenfunction satisfies the governing differential equation and boundary conditions (by construction)

$$\varphi_n'' + \lambda_n^2 \varphi_n = 0, \quad \varphi_n(0) = 0, \quad \varphi_n'(L) = 0 \tag{9.30}$$

Since this is true for all modes, we can write

$$\int_0^L \left[\left(\varphi_n'' + \lambda_n^2 \varphi_n \right) \varphi_m - \left(\varphi_m'' + \lambda_m^2 \varphi_m \right) \varphi_n \right] dx = 0$$

where n and m can represent any of the modes. This equation is true simply because the eigenfunctions satisfy the governing differential equation. Hence, the terms in parentheses are equal to zero. Multiplying them by a different eigenfunction and integrating over the domain will still give zero. To put this equation into a more useful form, integrate the first and third terms by parts to get

$$\int_0^L \left[-\varphi_n' \varphi_m' + \lambda_n^2 \varphi_n \varphi_m + \varphi_m' \varphi_n' - \lambda_m^2 \varphi_m \varphi_n \right] dx + \varphi_n' \varphi_m \Big|_0^L - \varphi_m' \varphi_n \Big|_0^L$$

The first and third terms in the integral cancel each other. All the boundary terms vanish because either $\varphi_n(0) = 0$ or $\varphi_n'(L) = 0$ (same for m).[5] Thus, this equation reduces to

$$\left(\lambda_n^2 - \lambda_m^2 \right) \int_0^L \varphi_n(x) \varphi_m(x) \, dx = 0$$

Because $\lambda_n \neq \lambda_m$ for $m \neq n$, the eigenfunctions satisfy

$$\int_0^L \varphi_n(x) \varphi_m(x) \, dx = 0, \quad n \neq m \tag{9.31}$$

For functions, this is what we mean by *orthogonality*. The product of two functions integrated over the domain is equal to zero. This sense of orthogonality is similar to the analogous notion of orthogonality of vectors as representing vectors that point in the most different directions possible (i.e., *perpendicular* for two and three dimensions). Orthogonal functions are "as different as possible" over the domain.

---

[5] It is easy to prove that these boundary terms vanish for *any* boundary conditions because either the displacement is prescribed or the axial force is prescribed at either end. Since the term that results from integration by parts has the product of an eigenfunction and the derivative of an eigenfunction, one of them must be zero at each end.

To demonstrate the orthogonality of the first derivative of the eigenfunctions, multiply the domain equation by φ_m and integrate from 0 to L:

$$\int_0^L \left(\varphi_n'' + \lambda_n^2 \varphi_n\right)\varphi_m \, dx = 0$$

which is zero because φ_n and λ_n satisfy the eigenvalue problem. Integrate the first term by parts to get

$$\int_0^L \left(-\varphi_n' \varphi_m' + \lambda_n^2 \varphi_n \varphi_m\right) dx + \varphi_n' \varphi_m \Big|_0^L = 0$$

The boundary term that results from integration by parts vanishes for all boundary conditions, so

$$\int_0^L \varphi_n' \varphi_m' \, dx = \lambda_n^2 \int_0^L \varphi_n \varphi_m \, dx = 0$$

Orthogonality of the eigenvectors thereby proves that the *first derivatives* of the eigenfunctions are also orthogonal for $n \neq m$. It is also worth noting that setting $m = n$ gives

$$\lambda_n^2 = \frac{\int_0^L \varphi_n'^2 \, dx}{\int_0^L \varphi_n^2 \, dx}$$

The General Solution By assuming that $u(x,t) = U(x)T(t)$, we found an infinite number of solutions to the differential equation governing $U(x)$. We also know that ω is determined by λ and that we have found the specific values of λ that work. In differential equations, if you find multiple functions that satisfy the homogeneous differential equation, then the linear combination of those functions comprise the *general solution* to the differential equation. Therefore, putting the pieces back together, we can write

$$u(x,t) = \sum_{n=1}^{\infty} \varphi_n(x)\left[A_n \cos \omega_n t + B_n \sin \omega_n t\right] \tag{9.32}$$

where the constants A_n and B_n must be determined by the initial conditions. It should be evident from this expression why we did not include the D_2 coefficient in the mode shape $\varphi_n(x)$. Because the constants A_n and B_n are unknown, multiplying them by another unknown constant would be redundant.

Implementing the Initial Conditions The initial conditions are

$$u(x,0) = u_o(x), \qquad \dot{u}(x,0) = v_o(x) \tag{9.33}$$

where $u_o(x)$ and $v_o(x)$ are given functions. From Eq. 9.32 at time zero, we can write the initial displacement as

$$u(x, 0) = \sum_{n=1}^{\infty} \varphi_n(x) A_n = u_o(x) \qquad (9.34)$$

To solve for A_n, multiply both sides of the equation by the eigenfunction φ_m and integrate from 0 to L. To wit,

$$\int_0^L \left(\sum_{n=1}^{\infty} \varphi_n(x) A_n \right) \varphi_m(x) \, dx = \int_0^L u_o(x) \varphi_m(x) \, dx$$

Noting that A_n is constant, and that we can exchange the order of summation and integration, we have

$$\sum_{n=1}^{\infty} A_n \int_0^L \varphi_n(x) \varphi_m(x) \, dx = \int_0^L u_o(x) \varphi_m(x) \, dx$$

Finally, recognizing the orthogonality of the eigenfunctions (Eq. 9.31), only the mth term survives. Thus,

$$A_m \int_0^L \varphi_m^2(x) \, dx = \int_0^L u_o(x) \varphi_m(x) \, dx$$

Now, solve for A_m to get

$$A_m = \frac{\int_0^L u_o(x) \varphi_m(x) \, dx}{\int_0^L \varphi_m^2(x) \, dx} \qquad (9.35)$$

We can use a similar process with the initial velocity to determine the constants B_n. At time $t = 0$ we have

$$\dot{u}(x, 0) = \sum_{n=1}^{\infty} \varphi_n(x) \, \omega_n B_n = v_o(x)$$

Multiply by the eigenfunction φ_m and integrate over the domain to get

$$\int_0^L \left(\sum_{n=1}^{\infty} \varphi_n(x) \, \omega_n B_n \right) \varphi_m(x) \, dx = \int_0^L v_o(x) \varphi_m(x) \, dx$$

Again, recognizing orthogonality, we find that

$$\omega_m B_m \int_0^L \varphi_m^2(x)\,dx = \int_0^L v_o(x)\varphi_m(x)\,dx$$

Therefore, the constant B_m is

$$B_m = \frac{\int_0^L v_o(x)\varphi_m(x)\,dx}{\omega_m \int_0^L \varphi_m^2(x)\,dx} \tag{9.36}$$

Orthogonality of the eigenfunctions is essential in this process because it enables the determination of the individual constants, even though it is an infinite series.

We can introduce a shorthand notation for the inner product of two functions:

$$\langle v, w \rangle = \int_0^L v(x)w(x)\,dx \tag{9.37}$$

With this notation, the complete solution for the axial wave propagation problem can be expressed as

$$u(x, t) = \sum_{n=1}^{\infty} \varphi_n(x)\left[\frac{\langle \varphi_n, u_o \rangle}{\langle \varphi_n, \varphi_n \rangle} \cos \omega_n t + \frac{1}{\omega_n}\frac{\langle \varphi_n, v_o \rangle}{\langle \varphi_n, \varphi_n \rangle} \sin \omega_n t\right] \tag{9.38}$$

The similarity with the solution to the NDOF system, using the mode shapes (i.e., Eq. 4.32) is worthy of note.

Vibration in Natural Modes The solution of the spatial part of the problem gives rise to the natural modes and natural frequencies of the system. The boundary conditions restrict the functions that satisfy the governing differential equation to a discrete (but infinite) set. Each mode shape is associated with a value λ_n which, in turn, dictates a natural frequency $\omega_n = \lambda_n c$.

Because the natural modes satisfy an orthogonality relationship, we can show that an initial displacement in the shape of a mode will vibrate in exactly that mode. Let us say that the initial displacement is equal to a multiple of the nth mode shape

$$u_o(x) = U_o \, \varphi_n(x) \tag{9.39}$$

where U_o is the amplitude of the displacement. We will also take the initial velocity to be zero (so all $B_m = 0$). We can compute the coefficients of the modal expansion to be

$$A_m = \frac{\int_0^L U_o \, \varphi_n(x)\varphi_m(x)\,dx}{\int_0^L \varphi_m^2(x)\,dx} = \begin{cases} U_o & m = n \\ 0 & m \neq n \end{cases} \tag{9.40}$$

Thus, the infinite series solution reduces to

$$u(x, t) = U_o \, \varphi_n(x) \cos \omega_n t \tag{9.41}$$

In other words, the displacement is in the shape of the nth mode at all times with a magnitude that oscillates between U_o and $-U_o$ at the nth natural frequency ω_n. This is a very interesting result because it connects wave propagation with modal vibration. What is actually happening is that the initial displacement $u_o(x)$ propagates in the manner discussed earlier—i.e., the wave splits in two with each respective part propagating to the left or right at the wave speed c. These waves reflect off the boundaries (in accord with the boundary conditions). What this wave propagation process looks like is a function $\varphi_n(x)$ with amplitude oscillating sinusoidally with frequency ω_n. We call this phenomenon *modal vibration*. Vibration does not look like wave propagation and it is impossible to observe the wave speed, but modal vibration is a special case of wave propagation.

Example Consider the initial displacement of the form shown in Fig. 9.9, which also shows the internal axial force implied by the displacement. The function has the explicit expression

$$u_o(x) = 1 - \cos\left(2\pi \frac{x - a}{b - a}\right) \qquad a < x < b$$

and is zero elsewhere. The closer we pick a and b, the narrower is the blip.

We can evaluate the constants A_m from Eq. 9.35. Note that the constants $B_m = 0$ because there is no initial velocity. Let $h = b - a$ be the span of the nonzero portion

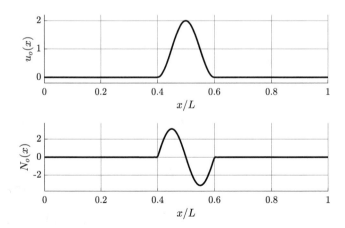

Fig. 9.9 *Example—Initial displacement.* The plot of the initial displacement and the associated axial force for the example. The bar has length $L = 10$, density $\rho = 1$, and modulus $E = 20$. The initial displacement $u_o(x)$ spans from $a = 4$ to $b = 6$. We have taken $EA = 1$ to compute the axial force

of the function $u_o(x)$ and let $\xi = (x - a)/h$. With this change of variable, we can write the integral of the eigenfunction times the initial displacement as

$$\langle \varphi_n, u_o \rangle = \int_0^1 \sin (\lambda_n a + \xi \lambda_n h)\big[1 - \cos (2\pi \xi)\big] h \, d\xi$$

which evaluates to

$$\langle \varphi_n, u_o \rangle = \frac{4\pi^2}{\lambda_n} \left(\frac{\cos (\lambda_n b) - \cos (\lambda_n a)}{\lambda_n^2 h^2 - 4\pi^2} \right)$$

Also, we can evaluate the integral of the square of the eigenfunction to be

$$\langle \varphi_n, \varphi_n \rangle = \int_0^L \sin^2 (\lambda_n x) \, dx = \frac{L}{2} \left(1 - \frac{\sin (2\lambda_n L)}{2\lambda_n L} \right) \tag{9.42}$$

The initial displacement $u_o(x)$ and the associated axial force $N_o(x)$ are shown in Fig. 9.9. This plot was created by computing the modal expansion of the functions using the coefficients $A_n = \langle \varphi_n, u_o \rangle / \langle \varphi_n, \varphi_n \rangle$, specifically,

$$u_o(x) = \sum_{n=1}^N \varphi_n(x) A_n, \qquad N_o(x) = EA \sum_{n=1}^N \varphi_n'(x) A_n$$

to verify that the A_n values are correct. The number of terms included in the truncated series was $N = 150$.

One can observe that the separation of variables solution is capable of representing wave propagation. The accuracy of the solution depends upon the number of terms included in the series. Figure 9.10 shows the propagation of the initial wave in Fig. 9.9 at four different times for a fixed–free bar with physical properties $L = 10$, $E = 20$, and $\rho = 1$. The number of terms required depends upon the smoothness of the initial wave function. This plot was generated using 150 modes. Very good results can be obtained with fewer modes, but at 30–40 modes some small wiggles in the axial force function at the edges of the wave were evident. The displacement approximation requires far fewer terms for accuracy, which is typical of this type of series approximation because the axial force is the derivative (with respect to x) of the displacement.

One way to understand the convergence properties of the series is to recognize that λ_n is proportional to n. The coefficients A_n are inversely proportional to λ_n^3 (and therefore n^3). The displacements are computed directly from A_n while the axial forces are proportional to $\lambda_n A_n$ and, hence, are inversely proportional to n^2. As a result, more terms are needed for convergence of the force approximation than

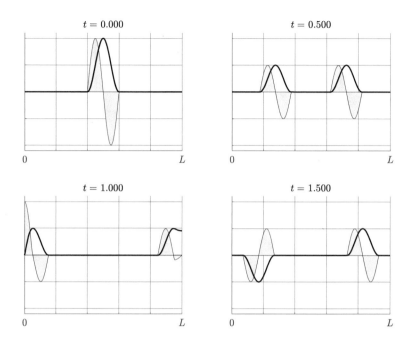

Fig. 9.10 *Example—wave propagation.* The plot of the displacement and the associated axial force (both scaled to fit on the same graph) for the example problem of Fig. 9.9 at four different times. The displacement is the solid line and the internal axial force is the thin line with gray shading

are needed for displacements. The coefficients for the displacement series (A_n) and the coefficients for the axial force series ($\lambda_n A_n$) are plotted for the first 40 modes in Fig. 9.11. The plot also shows the natural frequencies and the modal wave speeds $c_n = \lambda_n / \omega_n$ (which, of course, are all equal to the nominal wave speed c).

In this example you can observe that the displacement and stress waves reflect as expected at the fixed and free boundaries. Notice that the stress wave at the fixed end reflects with the same sign and, therefore, doubles its amplitude briefly during the reflection (e.g., at $t = 1.0$). The displacement wave at the fixed end reverses sign. At the free end, the stress wave reverses sign and the displacement wave reflects with the same sign. You can see that the slope of the displacement wave is zero at the free end at $t = 1.0$.

The program used to generate Figs. 9.9 and 9.10 is contained in Code 9.1. The graphical output functions are not included, but they produce the graphs shown in the figures, so it is evident what they do.

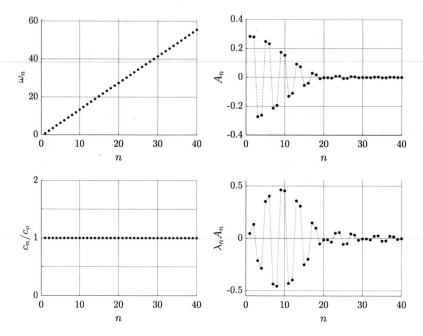

Fig. 9.11 *Example—mode coefficients.* The plots on the left show that frequency is proportional to mode number and modal wave speed is equal to $c_o = \sqrt{E/\rho}$. The plots on the right show the first 40 coefficients of the modal expansion of displacement and force for the example problem of Fig. 9.9

Code 9.1 MATLAB code to compute the response of the axial bar to an initial displacement using superposition of modes from the separation of variables solution

```
%  Axial Wave Modal Analysis - Fixed-free bar
   clear; clc;

%. Problem parameters
   nModes = 150;                        % Number of modes
   rho = 1.0;                           % Mass density
   E = 20.0;                            % Young's modulus
   EA = 1;                              % EA is axial stiffness
   c = sqrt(E/rho);                     % Wave speed
   L = 10.0;                            % Total length of bar
   WaveType = 2;                        % =1 sine, =2 cosine

% |Start time    |Stop time   |Time steps   |Spatial points
   T.to = 0;       T.tf = 1.5;   T.nT = 61;     T.nS = 250;

%. Compute response
   [Time,u,N,Freq,Phi,A] = ModalSum(T,L,c,EA,nModes,WaveType);

%. Plot results
   PlotModes(Phi,T);
   PlotInitialWave(u,N,T);
   PlotCoefficients(A,Freq,nModes);
   PlotSnapshots(u,N,Time,T);
   AnimateAxialBar(u,N,Time,T);

%- End of main program -------------------------------------------
```

```
function [Time,u,N,Freq,Phi,A] = ModalSum(T,L,c,EA,nModes,Type)
% Compute response of fixed-free axial bar with initial displacement

    x = linspace(0,L,T.nS);                  % Spatial grid points
    u = zeros(T.nT,T.nS);                    % Initialize u(x,t)
    N = zeros(T.nT,T.nS);                    % Initialize N(x,t)
    Time = linspace(T.to,T.tf,T.nT)';        % Time grid points
    Phi = zeros(nModes,T.nS);                % Mode shapes
    Freq = zeros(nModes,2);                  % Natural frequencies
    A = zeros(nModes,2);                     % Coefficients

    for i=1:T.nT
      t = Time(i);
      for n=1:nModes
        Ln = (2*n-1)*pi/(2*L);
        Freq(n,1) = Ln*c;
        Freq(n,2) = 1;
        A(n,1) = GetCoeff(Ln,L,Type);
        A(n,2) = A(n,1)*Ln;
        Phi(n,:) = sin(Ln.*x);
        dPhi_n = Ln*cos(Ln.*x);
        u(i,:) = u(i,:) + Phi(n,:)*A(n,1)*cos(Freq(n)*t);
        N(i,:) = N(i,:) + EA*dPhi_n*A(n,1)*cos(Freq(n)*t);
      end
    end

end

%-------------------------------------------------------------------------

function An = GetCoeff(Ln,L,Type)
% Compute the coefficients for the fixed-free axial bar
%     Type = 1: uo = 2*sin(pi*xi)    xi = (x-a)/(b-a)   a < x < b
%     Type = 2: uo = 1-cos(2*pi*xi)  xi = (x-a)/(b-a)   a < x < b

    a = L/2 - L/10;                          % Start of initial wave
    b = L/2 + L/10;                          % End of initial wave
    h = b-a;

    switch Type
      case 1   % uo = 2*sin(pi*xi)
        sLa = sin(Ln*a); sLb = sin(Ln*b);
        Term1 = h*pi*(sLa+sLb)/(pi^2-Ln^2*h^2);
        Term2 = (L/2)*(1 - sin(2*Ln*L)/(2*Ln*L));
        An = 2*Term1/Term2;
      case 2   % uo = 1-cos(2*pi*xi)
        cLa = cos(Ln*a); cLb = cos(Ln*b);
        Term1 = (4*pi^2/Ln)*(cLb-cLa)/(Ln^2*h^2-4*pi^2);
        Term2 = (L/2)*(1 - sin(2*Ln*L)/(2*Ln*L));
        An = Term1/Term2;
    end

end
```

9.4   Modal Analysis with Applied Loads

In the previous section we developed a solution method for the wave equation based upon separation of variables. That method led to an infinite series solution for the displacement function. Along the way we encountered the natural frequencies and mode shapes of the axial bar along with the proof of orthogonality of the mode shapes and their derivatives.

In this section we extend our analysis to applied forces, including the ones shown in Fig. 9.1. We will, again, consider the specific problem of the fixed–free bar. The equation of motion and boundary conditions for this problem are

$$EAu''(x, t) + p(x, t) = \rho A\ddot{u}(x, t)$$

$$u(0, t) = 0 \tag{9.43}$$

$$EAu'(L, t) = P(t)$$

where the prime stands for the partial derivative with respect to x and the dot stands for partial derivative with respect to t. It should be fairly obvious how to adjust these equations for other boundary conditions.

Distributed Load Let us first take a look at the case where we have a distributed load $p(x, t) = p_o \sin \Omega t$ but no end load (i.e., $P(t) = 0$). The spatial distribution of load is constant over the length of the bar and varies sinusoidally in time. The general solution can be expressed as

$$u(x, t) = u_h(x, t) + u_p(x, t) \tag{9.44}$$

where the homogeneous part $u_h(x, t)$ is given by Eq. 9.32. The challenge in solving driven problems is to find a particular solution. Lets's take

$$u_p(x, t) = \sum_{m=1}^{\infty} C_m \varphi_m(x) \sin \Omega t \tag{9.45}$$

where the C_m are, as yet, undetermined constants and the $\varphi_m(x)$ are the eigenfunctions. Observe that the particular solution satisfies the boundary conditions because the eigenfunctions do individually. We can determine C_m by plugging $u_p(x, t)$ into the equation of motion, Eq. 9.43(a). We know that $\varphi_m'' = -\lambda_m^2 \varphi_m$ and $\omega_m = \lambda_m \sqrt{E/\rho}$. Equating coefficients of $\sin \Omega t$ gives[6]

$$\rho A \sum_{m=1}^{\infty} \left(\omega_m^2 - \Omega^2 \right) C_m \varphi_m(x) = p_o$$

Multiply both sides of the equation by $\varphi_n(x)$ and integrate over the domain, noting orthogonality of the eigenfunctions, to get

$$C_n = \frac{p_o}{\rho A \left(\omega_n^2 - \Omega^2 \right)} \frac{\langle 1, \varphi_n \rangle}{\langle \varphi_n, \varphi_n \rangle}$$

---

[6] Note that A is the area of the cross section of the bar and should not be confused with A_n, which is the nth coefficient of the modal expansion.

where the notation $\langle \bullet, \bullet \rangle$ is defined in Eq. 9.37. Substituting these values into the general solution gives

$$u(x,t) = \sum_{n=1}^{\infty} \varphi_n(x)\left[A_n \cos \omega_n t + B_n \sin \omega_n t + C_n \sin \Omega t\right] \tag{9.46}$$

We can implement the initial conditions in exactly the same way as for the unforced problem. To wit,

$$u(x,0) = \sum_{n=1}^{\infty} \varphi_n(x)A_n = u_o(x)$$

$$\dot{u}(x,0) = \sum_{n=1}^{\infty} \varphi_n(x)\left(\omega_n B_n + \Omega C_n\right) = v_o(x) \tag{9.47}$$

To determine the constants A_n and B_n, we multiply both equations by an eigenfunction and integrate from 0 to L. Orthogonality of the eigenfunctions gives the solution to the constants as

$$A_n = \frac{\langle u_o, \varphi_n \rangle}{\langle \varphi_n, \varphi_n \rangle}, \qquad B_n = \frac{1}{\omega_n}\left[\frac{\langle v_o, \varphi_n \rangle}{\langle \varphi_n, \varphi_n \rangle} - \Omega C_n\right] \tag{9.48}$$

Obviously, the constants reduce to the values found for the unforced problem if we set $p_o = 0$. We will not provide any computational results for this case because the next section will take us through a similar solution.

Load at End of Bar Let us now take a look at the case where we have a sinusoidal load end load $P(t) = P_o \sin \Omega t$ applied at $x = L$ but no distributed load, i.e., $p(x,t) = 0$. Finding a particular solution for this case is more challenging. We know we need to add something to the homogeneous part because all of the eigenfunction derivatives are zero at the end where the load is applied. We need the internal axial force to be equal to the sinusoidally varying end force at that point while at the same time keeping the displacement equal to zero at $x = 0$. We can accomplish that objective with a particular solution of the form

$$u_p(x,t) = \sum_{n=1}^{\infty} \varphi_n(x)C_n \sin \Omega t + Dx \sin \Omega t \tag{9.49}$$

The last term gives us something that, when differentiated with respect to x, will match up with the applied load in the boundary condition at the right end. In particular, we have

$$EAu'_p(L,t) = EA(D \sin \Omega t) = P_o \sin \Omega t \tag{9.50}$$

Note that all of the other terms drop out because $\varphi_n'(L) = 0$. Solving for the constant D gives

$$D = \frac{P_o}{EA} \tag{9.51}$$

The particular solution must also satisfy the governing equation of motion in the domain, Eq. 9.43(a). Substituting $u_p(x, t)$, we get

$$EA \sum_{n=1}^{\infty} \varphi_n''(x) C_n \sin \Omega t = -\rho A \left[\sum_{n=1}^{\infty} \varphi_n(x) C_n + Dx \right] \Omega^2 \sin \Omega t$$

We know a few things that will help. First, the eigenfunctions satisfy the equation $\varphi_n'' = -\lambda_n^2 \varphi_n$. Second, $\omega_n = c\lambda_n$, where $c = \sqrt{E/\rho}$ is the wave speed. Equating the coefficients of the $\sin \Omega t$ terms gives

$$\sum_{n=1}^{\infty} \omega_n^2 \varphi_n(x) C_n = \sum_{n=1}^{\infty} \Omega^2 \varphi_n(x) C_n + \Omega^2 Dx$$

We can multiply both sides of this equation by an eigenfunction and integrate over the domain, noting orthogonality, to get

$$C_n = D\left(\frac{\Omega^2 \alpha_n}{\omega_n^2 - \Omega^2} \right) \tag{9.52}$$

where to consolidate the notation, we have made the definition

$$\alpha_n = \frac{\langle x, \varphi_n \rangle}{\langle \varphi_n, \varphi_n \rangle} \tag{9.53}$$

For the fixed–free bar, we have already computed $\langle \varphi_n, \varphi_n \rangle$ (see Eq. 9.42). The second integral can be evaluated as

$$\langle x, \varphi_n \rangle = \int_0^L x \sin(\lambda_n x)\, dx = \frac{1}{\lambda_n^2} \left[\sin(\lambda_n L) - \lambda_n L \cos(\lambda_n L) \right]$$

With that fairly lengthy preamble, we are now in the position to implement the boundary conditions. As we have done before, we can write

$$u(x, 0) = \sum_{n=1}^{\infty} \varphi_n(x) A_n = u_o(x)$$

$$\dot{u}(x, 0) = \sum_{n=1}^{\infty} \varphi_n(x) \left(\omega_n B_n + \Omega C_n \right) + \Omega Dx = v_o(x)$$

As usual, multiplying by an eigenfunction and integrating over the domain for both equations (noting orthogonality) gives the constants. To wit,

$$A_n = \frac{\langle u_o, \varphi_n \rangle}{\langle \varphi_n, \varphi_n \rangle}, \qquad B_n = \frac{1}{\omega_n} \left[\frac{\langle v_o, \varphi_n \rangle}{\langle \varphi_n, \varphi_n \rangle} - \Omega \Big(C_n + \alpha_n D \Big) \right] \qquad (9.54)$$

The modal analysis for this case has been implemented in Code 9.2, which assumes that there are no initial displacements or velocities. We will use this code to examine the behavior of the system in the next example.

Code 9.2   MATLAB code to compute the response of a fixed–free axial bar to a load $P_o \sin \Omega t$ applied at the right end. Solution by superposition of modes from the classical separation of variables solution

```
%  Axial Bar with sinusoidal end load
   clear; clc;

%. Problem parameters
   nModes = 50;                          % Number of modes
   rho = 1.0;                            % Mass density
   A = 1.0;                              % Area of bar
   E = 20.0;                             % Young's modulus
   EA = E*A;                             % EA is axial stiffness
   c = sqrt(E/rho);                      % Wave speed
   L = 10.0;                             % Total length of bar
   Po = -1;                              % End force magnitude
   OMEGA = 3.8;                          % End force frequency

%  |Start time   |Stop time   |Time steps    |Spatial points
   T.to = 0;      T.tf = 4;     T.nT = 61;     T.nS = 250;

%. Compute response
   [Time,u,N,Freq,Phi,A] = ModalSum(T,L,c,EA,Po,OMEGA,nModes);

%. Plot results
   PlotModes(Phi,T);
   PlotCoefficients(A,Freq,nModes);
   PlotSnapshots(u,N,Time,T);
   PlotTimeResults(Time,u,N,T)
   AnimateAxialBar(u,N,Time,T);

%- End of main program -------------------------------------------

function [Time,u,N,Freq,Phi,A] = ModalSum(T,L,c,EA,Po,OMEGA,nModes)
%  Compute n and N by modal sum for fixed-free bar driven by force
%  P=Po sin(OMEGA t) at end

   x = linspace(0,L,T.nS);               % Spatial grid points
   I = linspace(1,1,T.nS);               % Constant unit function
   u = zeros(T.nT,T.nS);                 % Initialize u(x,t)
   N = zeros(T.nT,T.nS);                 % Initialize N(x,t)
   Time = linspace(T.to,T.tf,T.nT)';     % Time grid points
   Phi = zeros(nModes,T.nS);             % Mode shapes
   Freq = zeros(nModes,2);               % Natural frequencies
   A = zeros(nModes,2);                  % Coefficients
   FD = @(t) sin(OMEGA*t);               % Forcing functions

   for i=1:T.nT
     t = Time(i);
     for n=1:nModes
       Ln = (2*n-1)*pi/(2*L);
       Wn = Ln*c;
```

```
           [An,Bn,Cn,D] = GetCoeff(Ln,Wn,L,Po,OMEGA,EA);
           Fn = An*cos(Wn*t) + Bn*sin(Wn*t) + Cn*FD(t);
           Phi(n,:) = sin(Ln.*x);
           dPhi_n = Ln*cos(Ln.*x);
           u(i,:) = u(i,:) + Phi(n,:)*Fn;
           N(i,:) = N(i,:) + EA*dPhi_n*Fn;
           A(n,:) = [Bn,Cn];
           Freq(n,:) = [Wn,1];
        end
        u(i,:) = u(i,:) + (D.*x)*FD(t);
        N(i,:) = N(i,:) + EA*(D.*I)*FD(t);
     end

  end

%-----------------------------------------------------------------------

function [An,Bn,Cn,D] = GetCoeff(Ln,Wn,L,Po,OMEGA,EA)
% Compute the coefficients An and Bn for case where the load
% is applied as x=L for the axial bar. P(t) = Po*sin(OMEGA*t)

  LnL = Ln*L;
  pnpn = (L/2)*(1 - sin(2*LnL)/(2*LnL));    % <phi_n,phi_n>
  pnx  = (sin(LnL) - LnL*cos(LnL))/(Ln^2);  % <x,phi_n>
  an = pnx/pnpn;

  An = 0;                                   % cos(Wn*t) coeff.
  D = Po/EA;                                % Po sin(OMEGA*t) coeff.
  Cn = -(D*an*OMEGA^2)/(OMEGA^2-Wn^2);      % Po sin(OMEGA*t) coeff.
  Bn = -OMEGA*(Cn + D*an)/Wn;               % sin(Wn*t) coeff.

end
```

The heart of the modal computation is in the function ModalSum, where the solution is computed on a spatial and temporal grid for a specified number of modes (nModes). The coefficients of the modal expansion are computed in the function GetCoeff.

The functions that generate the figures, i.e., PlotSnapshots (which produced Fig. 9.12) and PlotTimeResults (which produced Figs. 9.13 and 9.14) are not included here, as usual. What they do is evident from the figures themselves. The function AnimateAxialBar creates a movie that shows the propagation of the waves. The animation is helpful to understand the dynamic response. In the following example, the animation shows the transition from the initial waves to a steady vibration in the third mode (whose natural frequency is closest to the driving frequency).

Example Let us revisit the axial bar from the previous example, but this time with a sinusoidal load applied at the end. The bar has properties $E = 20$, $\rho = 1$, $A = 1$, and $L = 10$. The end load has magnitude one and driving frequency $\Omega = 3.8$. We use 50 modes in the analysis. The reason fewer modes are required for this example is because the loading generates a smoother response than the initial displacement of the previous example did.

Figure 9.12 shows four snapshots of the evolution of the state of the bar subjected to the end load. There is no need to show the state of the system at $t = 0$ because it starts from rest. At $t = 1$ you can see the wave propagating from the right end moving left. At this time, it has not yet reached the middle of the bar. All points on

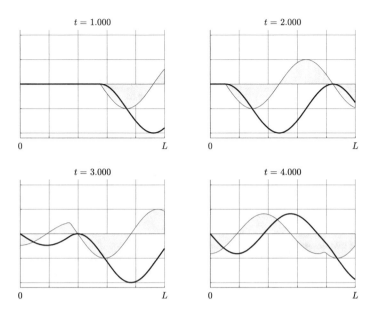

Fig. 9.12 *Example—wave propagation from forcing.* The plot of the displacement and the associated axial force (both scaled to fit on the same graph) for the example problem at four different times. The fixed–free bar is subject to a sinusoidal load at the right end and starts from rest. The displacement is the solid line and the internal axial force is the thin line with gray shading

the left half of the bar still have not felt the disturbance. At $t = 2$, the wave has nearly reached the left end. The wave speed is $c = 4.47$, so the arrival time of the wave at the left end should be $t = 2.24$. By $t = 3$ the wave has reflected and is moving to the right. Now, the right-moving wave is interacting with the train of left-moving waves that are being generated by the force at the right end. The process continues to $t = 4$.

Figure 9.13 gives a different view of those same 4 s for the same problem. In this figure we plot the displacement $u(x, t)$ and the internal axial force $N(x, t)$ for 25 values of x equally spaced along the bar. Each line represents one of those spatial points. The darker line in the displacement plot is $x = L$ and the darker line in the axial force plot is $x = 0$. The displacement of the right end begins immediately, but the other points initiate motion at later times. The left end of the bar does not feel any internal axial force until the first wave reaches it at $t = 2.24$ s.

The lowest frequency of the bar is $\omega_1 = 0.7025$. It is interesting to see what happens when we set the driving frequency equal to the fundamental frequency, i.e., $\Omega = \omega_1$. Figure 9.14 shows the result of 400 s of forcing. The motion and force at the two ends of the bar are shown as functions of time. The response clearly shows resonance. What happens is that the force at the right end propagates waves moving to the left continuously. Those waves reflect off the fixed boundary at the left end and propagate to the right. As they move right, they interact with the waves moving left. When the right-moving waves reach the right end, they reflect again.

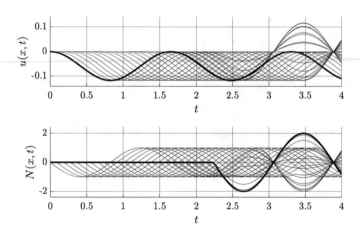

Fig. 9.13 *Example—displacement and force vs. time.* The plot of the displacement and the associated axial force vs. time for 25 equally spaced points along the bar. The fixed–free bar is subject to a sinusoidal load at the right end and starts from rest

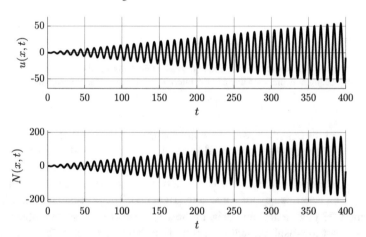

Fig. 9.14 *Example—resonance.* The plot of the displacement and the associated axial force vs. time at the end points. The fixed–free bar is subject to a sinusoidal load at the right end with a driving frequency $\Omega = \omega_1 = 0.7025$. The bar starts from rest

The reflected waves add to the waves that the force is newly creating. The reflected waves will either reinforce or cancel the other waves bouncing around in the system. When the frequency of the wave generator is synchronized with the natural wave propagation speed, the reflecting waves continually reinforce the new waves. With each reflection the magnitude of the overall wave increases. That is how resonance occurs in a system governed by wave propagation.

In addition to the interesting phenomena that we can observe through this analysis, this section has shown the complications associated with constructing a classical solution via separation of variables. When you can find a classical solution, it is a great tool. But there are significant limits on the types of loading systems we can find solutions for. In the next section we will take a different approach to the analysis of the axial bar problem that resolves those limitations.

9.5   The Ritz Method and Finite Element Analysis

It is reasonable to ask if we could make any progress without the eigenfunctions, or if we could solve more complex problems like the dynamics of non-prismatic bars. The answer is yes, and the key is to use *Ritz functions* in the framework of the principle of virtual work.

Let us imagine that we have selected a set of functions $\psi_n(x)$ intended to provide a basis for representing the spatial variation of the displacement function over the domain. These functions must be linearly independent, but they need not be orthogonal, and they do not have to individually satisfy the equation of motion, as the eigenfunctions $\varphi_n(x)$ do. They *do* have to satisfy the displacement boundary conditions (also known as *essential boundary conditions*), but they do not have to satisfy the force boundary conditions (also known as the *natural boundary conditions*) because those conditions will be satisfied naturally in our formulation.

We will build our solution with a scheme reminiscent of the separation of variables approach. To wit, we assume

$$u(x,t) = \sum_{n=1}^{N} \psi_n(x)\, u_n(t) \tag{9.55}$$

where $u_n(t)$ is an as yet unknown time-dependent function that represents the contribution of the base function $\psi_n(x)$ to the sum.[7] It should be evident that we could use the eigenfunctions as the Ritz functions. In a way, the eigenfunctions are sort of an ideal choice because of their orthogonality properties. But eigenfunctions are available only for those problems we are able to solve classically. We will also see that we can use a completely different approach using locally compact finite element shape functions to generate Ritz base functions.

9.5.1   Dynamic Principle of Virtual Work

To set the framework for using the Ritz approximation, we will derive a dynamic principle of virtual work. Let us take the original differential equation and natural

---

[7] The symbol u has two distinct uses here. The symbol without subscript refers to $u(x, t)$, whereas the symbol with subscript is a component of the Ritz expansion. The context should make the distinction clear.

boundary conditions and express them in residual form as

$$\rho A \ddot{u}(x, t) - E A u''(x, t) - p(x, t) = 0$$

$$E A u'(L, t) - P(t) = 0$$

(the left side of each equation is what we call the *residual*). Now, multiply the domain residual by an arbitrary function $\bar{u}(x)$, the *virtual displacement*, and integrate over the length of the bar. Add to that the boundary residual multiplied by $\bar{u}(L)$. The result is the *virtual work functional*

$$\mathcal{G} = \int_0^L \left(\rho A \ddot{u} - E A u'' - p \right) \bar{u} \, dx + \left[E A u'(L, t) - P \right] \bar{u}(L)$$

where $u(x, t)$ and $p(x, t)$ are both functions of space and time, but $\bar{u}(x)$ is a function only of the spatial variable x. By the fundamental theorem of the calculus of variations, if $\mathcal{G} = 0$ for all $\bar{u}(x)$, then the original classical governing equation and the natural boundary condition are satisfied. We can integrate the second term by parts (with respect to x) to get

$$\mathcal{G} = \int_0^L \left(\rho A \ddot{u} \bar{u} + E A u' \bar{u}' - p \bar{u} \right) dx - P \bar{u}(L) \tag{9.56}$$

In order for the boundary terms resulting from integration by parts to vanish, we note that we must assert that $\bar{u}(0) = 0$ to match the real displacement boundary condition $u(0, t) = 0$. Observe that the first term in square brackets for the boundary term vanished naturally when we integrated by parts. The condition $\mathcal{G} = 0$ for all $\bar{u}(x)$ is called the *dynamic principle of virtual work*, and is the launching point for numerical approximations using Ritz functions.

Discretization of the Equations of Motion Let us assume that the real and virtual displacements are approximated as

$$u(x, t) = \boldsymbol{\psi}^T(x) \, \mathbf{u}(t), \qquad \bar{u}(x) = \bar{\mathbf{u}}^T \boldsymbol{\psi}(x) \tag{9.57}$$

where we have introduced the matrix notation for $\boldsymbol{\psi}(x)$, $\mathbf{u}(t)$, and $\bar{\mathbf{u}}$ so that we can replace the summations with vector operations. Specifically, let

$$\boldsymbol{\psi}(x) = \left\{ \begin{array}{c} \psi_1(x) \\ \vdots \\ \psi_N(x) \end{array} \right\}, \quad \mathbf{u}(t) = \left\{ \begin{array}{c} u_1(t) \\ \vdots \\ u_N(t) \end{array} \right\}, \quad \bar{\mathbf{u}} = \left\{ \begin{array}{c} \bar{u}_1 \\ \vdots \\ \bar{u}_N \end{array} \right\} \tag{9.58}$$

The Ritz functions $\boldsymbol{\psi}(x)$ are fixed and known. We will say more about how to choose them later. Substituting the Ritz expansion into the virtual work functional gives

$$\mathcal{G} = \int_0^L \rho A (\bar{\mathbf{u}}^T \boldsymbol{\psi})(\boldsymbol{\psi}^T \ddot{\mathbf{u}}) dx + \int_0^L EA(\bar{\mathbf{u}}^T \boldsymbol{\psi}')(\boldsymbol{\psi}'^T \mathbf{u}) dx$$
$$- \int_0^L (\bar{\mathbf{u}}^T \boldsymbol{\psi}) p\, dx - \bar{\mathbf{u}}^T \boldsymbol{\psi}(L) P$$

which we can write simply as

$$\mathcal{G} = \bar{\mathbf{u}}^T \left[\mathbf{M}\ddot{\mathbf{u}}(t) + \mathbf{K}\mathbf{u}(t) - \mathbf{f}(t) \right] \tag{9.59}$$

if we define the system mass and stiffness matrices, respectively, as

$$\mathbf{M} = \int_0^L \rho A\, \boldsymbol{\psi}(x)\boldsymbol{\psi}^T(x)\, dx, \quad \mathbf{K} = \int_0^L EA\, \boldsymbol{\psi}'(x)\boldsymbol{\psi}'^T(x)\, dx \tag{9.60}$$

and the effective force vector as

$$\mathbf{f}(t) = \int_0^L p(x,t)\boldsymbol{\psi}(x)\, dx + P(t)\,\boldsymbol{\psi}(L) \tag{9.61}$$

There are a few things worth noting about these matrices. First, they can be computed before we know the solution to the problem. Given the Ritz functions, the physical properties of the bar, and the loading functions, we can execute the integrals. Second, $\mathbf{M}$ and $\mathbf{K}$ are $N \times N$ matrices and $\mathbf{f}$ is an $N \times 1$ matrix. If the Ritz functions are the eigenfunctions, then $\mathbf{M}$ and $\mathbf{K}$ are diagonal. If not, then they could potentially be full matrices.

The fundamental theorem of the calculus of variations says that if $\mathcal{G} = 0$ for all $\bar{\mathbf{u}}$,[8] then the term in square brackets in Eq. 9.59 must itself be zero. This result gives the discrete equations of motion:

$$\mathbf{M}\ddot{\mathbf{u}}(t) + \mathbf{K}\mathbf{u}(t) = \mathbf{f}(t) \tag{9.62}$$

which can be solved either classically or numerically using Newmark's method, as we have done previously for the shear building and the truss.

The process we have just followed is called *spatial discretization*. It reduces a partial differential equation to a discrete system of ordinary differential equations. The quality of the solution depends upon the quality and number of Ritz functions used in the approximation.

---

[8] When we invoke the Ritz approximation, we go from a continuous problem to a discrete one. The condition "for all $\bar{u}(x)$" then reduces to "for all $\bar{\mathbf{u}}$".

Example We will demonstrate the process of computing the mass and stiffness matrices with a very simple Ritz basis. Let

$$
\psi = \begin{Bmatrix} \xi \\ \xi^2 \\ \xi^3 \end{Bmatrix}, \qquad \psi' = \frac{1}{L} \begin{Bmatrix} 1 \\ 2\xi \\ 3\xi^2 \end{Bmatrix} \tag{9.63}
$$

where $\xi = x/L$ (note that $dx = L\,d\xi$). The mass matrix can be computed as

$$
\mathbf{M} = \rho A L \int_0^1 \begin{bmatrix} \xi^2 & \xi^3 & \xi^4 \\ \xi^3 & \xi^4 & \xi^5 \\ \xi^4 & \xi^5 & \xi^6 \end{bmatrix} d\xi = \rho A L \begin{bmatrix} \frac{1}{3} & \frac{1}{4} & \frac{1}{5} \\ \frac{1}{4} & \frac{1}{5} & \frac{1}{6} \\ \frac{1}{5} & \frac{1}{6} & \frac{1}{7} \end{bmatrix}
$$

The stiffness matrix can be computed as

$$
\mathbf{K} = \frac{EA}{L} \int_0^1 \begin{bmatrix} 1 & 2\xi & 3\xi^2 \\ 2\xi & 4\xi^2 & 6\xi^3 \\ 3\xi^2 & 6\xi^3 & 9\xi^4 \end{bmatrix} d\xi = \frac{EA}{L} \begin{bmatrix} 1 & 1 & 1 \\ 1 & \frac{4}{3} & \frac{6}{4} \\ 1 & \frac{6}{4} & \frac{9}{5} \end{bmatrix}
$$

The simple polynomial Ritz functions will not be very useful for the wave propagation problem, and three base functions is too few, but this example shows how to compute the matrices.

9.5.2 Finite Element Functions

We can choose a special set of Ritz functions that are locally compact called the *finite element functions*. The "hat" functions shown in Fig. 9.15 are examples. The domain is divided up into regions called *elements*, distinguished by *nodes* at their ends. A Ritz base function ramps up over one element, down over the adjacent one, and is zero everywhere else. In the example in Fig. 9.15, the bar is divided into four subregions of length $L/4$. The left end is fixed so the Ritz functions must all be zero at $x = 0$. The Ritz functions are built from two finite element *shape functions*

$$
\varphi_1(\xi) = 1 - \xi, \qquad \varphi_2(\xi) = \xi \tag{9.64}
$$

where

$$
\xi = \frac{x - x_{i(e)}}{L_e} \tag{9.65}
$$

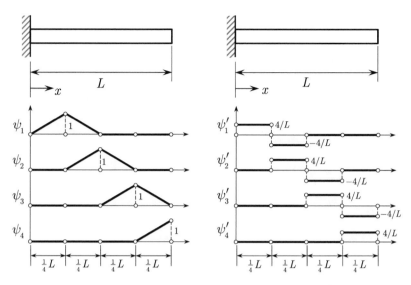

Fig. 9.15 *Finite element functions.* The simplest finite element Ritz functions are the so-called "hat" functions. Each Ritz function ramps up over the domain of one element and ramps down over the domain of the next one and is zero everywhere else

The index $i(e)$ is the node number at the left end of element e, $j(e)$ is the node number at the right end, and $L_e = x_{j(e)} - x_{i(e)}$ is the length of element. This change of variable maps the element domain onto $\xi \in [0, 1]$.

It is advantageous to break the integrals in Eqs. 9.60 and 9.61 into a sum of integrals over the element domains. To wit,

$$\int_0^L (\bullet)\, dx = \sum_{e=1}^M \int_{x_{i(e)}}^{x_{j(e)}} (\bullet)\, dx \tag{9.66}$$

where M is the number of elements. To see why this strategy is useful, consider the example

$$\int_0^L \psi_2(x)\psi_3(x)\, dx = \int_{L/2}^{3L/4} \varphi_1(x)\varphi_2(x)\, dx = \int_0^1 \varphi_1(\xi)\varphi_2(\xi)\, L_e\, d\xi$$

which is true because the product of ψ_2 and ψ_3 is nonzero only between $L/2$ and $3L/4$. In that region, $\psi_2 = \varphi_1$ (i.e., ramp down) and $\psi_3 = \varphi_2$ (i.e., ramp up). With the change of variable, we see that the result is

$$\int_0^1 \varphi_1(\xi)\varphi_2(\xi)\, L_e\, d\xi = \int_0^1 (1-\xi)\xi\, L_e\, d\xi = \tfrac{1}{6}L_e$$

The only integrals that appear in this entire process are

$$\int_0^1 \varphi_1 \varphi_1 \, L_e \, d\xi = \tfrac{1}{3} L_e, \qquad \int_0^1 \varphi_1 \varphi_2 \, L_e \, d\xi = \tfrac{1}{6} L_e, \qquad \int_0^1 \varphi_2 \varphi_2 \, L_e \, d\xi = \tfrac{1}{3} L_e$$

Notice that the terms that integrate $\psi_i \psi_i$ involve the integral of $\varphi_1 \varphi_1 + \varphi_2 \varphi_2$, which is $2L_e/3$. Therefore, if ρA is constant, the mass matrix is

$$\mathbf{M} = \frac{\rho A L_e}{6} \begin{bmatrix} 4 & 1 & 0 & 0 & \cdots & 0 \\ 1 & 4 & 1 & 0 & \cdots & 0 \\ 0 & 1 & 4 & 1 & \cdots & 0 \\ \vdots & & & \ddots & & \vdots \\ 0 & \cdots & 0 & 1 & 4 & 1 \\ 0 & \cdots & 0 & 0 & 1 & 2 \end{bmatrix}$$

The stiffness matrix can be dealt with in a similar manner. Note that the derivatives of the Ritz functions are piecewise constant. Therefore, the integrals are particularly simple. The result is

$$\mathbf{K} = \frac{EA}{L_e} \begin{bmatrix} 2 & -1 & 0 & 0 & \cdots & 0 \\ -1 & 2 & -1 & 0 & \cdots & 0 \\ 0 & -1 & 2 & -1 & \cdots & 0 \\ \vdots & & & \ddots & & \vdots \\ 0 & \cdots & 0 & -1 & 2 & -1 \\ 0 & \cdots & 0 & 0 & -1 & 1 \end{bmatrix} \tag{9.67}$$

Both the mass and stiffness matrices are tridiagonal. Recall that using the eigenfunctions as the Ritz basis would result in diagonal mass and stiffness matrices. So, we are not that far from orthogonality with finite element functions. The sparse structure of the matrices is due to the fact that the finite element Ritz functions are locally compact (i.e., they are zero everywhere except in the vicinity of the node in question). This also yields the benefit that these matrices are well conditioned.

9.5.3 A Slightly Different Formulation

We can set up the finite element method in a manner similar to how we formulated the shear building and the truss. Let us rewrite Eq. 9.56 in the form of a sum of integrals over the elements. To wit,

$$\mathcal{G} = \sum_{e=1}^{M} \int_{x_{i(e)}}^{x_{j(e)}} \left(\rho A \, \ddot{u}_e \bar{u}_e + E A \, u_e' \bar{u}_e' - p \, \bar{u}_e \right) dx - P \bar{u}(L) \tag{9.68}$$

where $u_e(x, t)$ and $\bar{u}_e(x)$ are the parts of $u(x, t)$ and $\bar{u}(x)$, respectively, that reside in element e. Again, $i(e)$ is the node number at the left end of element e and $j(e)$ is the node number at the right end. For the axial bar, we can number the nodes from 1 at $x = 0$ to N at $x = L$. We can number the elements from 1 at the left end to M at the right end. Note that $M = N - 1$.

Define a matrix $\mathbf{B}_e$ that picks the displacements at the two ends of element e from the global displacement array $\mathbf{u}(t)$, which is in DOF order. The two displacements associated with element e are, then, $\mathbf{B}_e\mathbf{u}(t)$. The $\mathbf{B}_e$ matrix takes the form

$$\mathbf{B}_e = \begin{bmatrix} 0 \cdots 0 & 1 & 0 & 0 \cdots 0 \\ 0 \cdots 0 & 0 & 1 & 0 \cdots 0 \end{bmatrix}$$

where the one in the first row is in the column with DOF number associated with node $i(e)$ and the one in the second row is in the column with DOF number associated with node $j(e)$. We can interpolate the displacement within the element using the finite element shape functions as

$$u_e(\xi, t) = \boldsymbol{\varphi}^T(\xi)\mathbf{B}_e\mathbf{u}(t), \qquad \bar{u}_e(\xi) = \boldsymbol{\varphi}^T(\xi)\mathbf{B}_e\bar{\mathbf{u}} \tag{9.69}$$

where $\bar{\mathbf{u}}$ is the global array of virtual displacements (in the same DOF order as $\mathbf{u}$) and the 2×1 finite element shape function array is

$$\boldsymbol{\varphi}(\xi) = \begin{Bmatrix} \varphi_1(\xi) \\ \varphi_2(\xi) \end{Bmatrix} = \begin{Bmatrix} 1 - \xi \\ \xi \end{Bmatrix} \tag{9.70}$$

It is simple to compute the derivatives of the element displacement as

$$u'_e(\xi, t) = \frac{1}{L_e}\boldsymbol{\varphi}'^T(\xi)\mathbf{B}_e\mathbf{u}(t), \qquad \ddot{u}_e(\xi) = \boldsymbol{\varphi}^T(\xi)\mathbf{B}_e\ddot{\mathbf{u}}(t) \tag{9.71}$$

The reason the $1/L_e$ shows up in the first equation is because the derivative u_e is with respect to x and the derivative of $\boldsymbol{\varphi}$ is with respect to ξ (note the change of variable). The virtual element displacements can be computed similarly. Now, we can evaluate the integrals that appear in Eq. 9.68. The first term gives

$$\sum_{e=1}^{M} \int_{x_{i(e)}}^{x_{j(e)}} \rho A \ddot{u}_e \bar{u}_e \, dx = \bar{\mathbf{u}}^T \sum_{e=1}^{M} \mathbf{B}_e^T \mathbf{m}_e \mathbf{B}_e \ddot{\mathbf{u}} \tag{9.72}$$

where the *element mass matrix* is

$$\mathbf{m}_e = \int_0^1 \rho A \boldsymbol{\varphi}\boldsymbol{\varphi}^T L_e \, d\xi = \frac{\rho A L_e}{6}\begin{bmatrix} 2 & 1 \\ 1 & 2 \end{bmatrix}$$

The second term in Eq. 9.68 gives

$$\sum_{e=1}^{M} \int_{x_{i(e)}}^{x_{j(e)}} E A\, u'_e \bar{u}'_e\, dx = \bar{\mathbf{u}}^T \sum_{e=1}^{M} \mathbf{B}_e^T \mathbf{k}_e \mathbf{B}_e \mathbf{u} \tag{9.73}$$

where the *element stiffness matrix* is

$$\mathbf{k}_e = \int_0^1 \frac{EA}{L_e} \boldsymbol{\varphi}' \boldsymbol{\varphi}'^T d\xi = \frac{EA}{L_e} \begin{bmatrix} 1 & -1 \\ -1 & 1 \end{bmatrix}$$

The third term gives

$$\sum_{e=1}^{M} \int_{x_{i(e)}}^{x_{j(e)}} p\, \bar{u}_e\, dx = \bar{\mathbf{u}}^T \sum_{e=1}^{M} \mathbf{B}_e^T \mathbf{f}_e \tag{9.74}$$

The loading terms that lead to $\mathbf{f}_e$ depend upon the specific load function. As an example, consider the loading $p(x,t) = p_o$. The *element load vector* is then

$$\mathbf{f}_e = \int_0^1 p_o L_e \boldsymbol{\varphi}\, d\xi = \frac{p_o L_e}{2} \begin{Bmatrix} 1 \\ 1 \end{Bmatrix}$$

Finding the global load vector for a load P_o at the end is very simple. Since only one of the Ritz functions is nonzero at the end, and its value is 1 there, we get

$$\mathbf{f}_{nodal} = P_o [\, 0 \;\cdots\; 0 \; 1\,]^T$$

i.e., there is only one entry equal to P_o in the last slot of $\mathbf{f}_{nodal}$. The total load vector is the sum of the nodal and element loads. To wit,

$$\mathbf{f}(t) = \sum_{e=1}^{M} \mathbf{B}_e^T \mathbf{f}_e + \mathbf{f}_{nodal} \tag{9.75}$$

As was the case for the shear building and the truss, $\mathbf{B}_e$ gives a means of assembling the element matrices into the global matrices

$$\mathbf{M} = \sum_{e=1}^{M} \mathbf{B}_e^T \mathbf{m}_e \mathbf{B}_e, \qquad \mathbf{K} = \sum_{e=1}^{M} \mathbf{B}_e^T \mathbf{k}_e \mathbf{B}_e$$

If we assemble these matrices, we get

$$
\mathbf{M} = \frac{\rho A L_e}{6}
\begin{bmatrix}
2 & 1 & 0 & 0 & \cdots & 0 \\
1 & 4 & 1 & 0 & \cdots & 0 \\
0 & 1 & 4 & 1 & \cdots & 0 \\
\vdots & & & \ddots & & \vdots \\
0 & \cdots & 0 & 1 & 4 & 1 \\
0 & \cdots & 0 & 0 & 1 & 2
\end{bmatrix}
$$

and

$$
\mathbf{K} = \frac{E A}{L_e}
\begin{bmatrix}
1 & -1 & 0 & 0 & \cdots & 0 \\
-1 & 2 & -1 & 0 & \cdots & 0 \\
0 & -1 & 2 & -1 & \cdots & 0 \\
\vdots & & & \ddots & & \vdots \\
0 & \cdots & 0 & -1 & 2 & -1 \\
0 & \cdots & 0 & 0 & -1 & 1
\end{bmatrix}
$$

Notice that the only difference with the matrices we got earlier is the top row and first column. These matrices are $N \times N$ whereas the previous ones were $(N-1) \times (N-1)$. What is the difference? When we assemble the matrices from element matrices, we do not account for the boundary conditions. In the fixed–free case the first node (degree of freedom number 1) is restrained. Hence, $u_1 = 0$. One way to implement that is to eliminate the first row and column of $\mathbf{M}$ and $\mathbf{K}$ and the first row of $\mathbf{f}$. Then the resulting equation of motion is

$$
\mathbf{M}\ddot{\mathbf{u}}(t) + \mathbf{K}\mathbf{u}(t) = \mathbf{f}(t) \tag{9.76}
$$

which is identical to the Eq. 9.62 obtained previously.

9.5.4   Boundary Conditions

A better way to deal with the boundary conditions is to identify the restrained degrees of freedom in the id array upon input and to convert the id array to global equation numbers, as we described for the shear building and truss. In that case, we assign the fixed nodes to have the largest global equation numbers. Then, all we need to do is trim off the rows and columns of $\mathbf{M}$, $\mathbf{K}$, and $\mathbf{f}$ associated with the DOF numbers larger than nFree, the number of free degrees of freedom in the system.

The details of the process for numbering the global DOF in a way that puts the restrained DOF at the end is described in Appendix D. In the appendix we present the MATLAB function GlobalDOF, which is a general approach to numbering the global degrees of freedom and creating the array eDOF which contains the association between element local DOF and global DOF—essentially the ingredients of $\mathbf{B}_e$.

9.5.5 Higher Order Interpolation

To see one of the main issues associated with low order finite element interpolation, consider the static response of an axial bar of length $L = 10$ and stiffness $EA = 500$ that is fixed at $x = 0$, free at $x = L$, and subjected to a constant load of magnitude $p_o = 2$.

The results are shown in Fig. 9.16 using ten elements. As expected, the displacements are piecewise linear. One ramification is that the axial force $N(x)$ is piecewise constant. The exact solution to this problem for $N(x)$ is a linear ramp down from $N = 20$ at the left end to zero at the right end. If we take more elements, the stair step solution gets closer and closer to the exact solution, but it takes many elements to get an accurate estimate of the maximum axial force (which is likely one of the most important response quantities for design).

One might imagine that the gradients of displacement will be much higher in a dynamic wave propagation problem than in this simple static problem. Hence, the refinement necessary to get an accurate solution make the "hat" functions less than ideal for interpolation. We can increase the fidelity of the finite element approximation by using bubble functions, as outlined in Appendix D. In fact, if you generalize the interpolation of the element displacement

$$u_e(\xi, t) = \boldsymbol{\varphi}^T(\xi)\mathbf{B}_e\mathbf{u}(t) \tag{9.77}$$

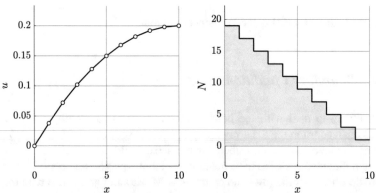

Fig. 9.16 *Static example of linear interpolation.* The static response of an axial bar fixed at the left end and free at the right end with $L = 10$, $EA = 500$, and a uniform load of $p_o = 2$ using the "hat" functions with ten elements

to have $\boldsymbol{\varphi}(\xi) = [\,\varphi_1(\xi), \ldots, \varphi_n(\xi)\,]^T$, where $n > 2$ are the bubble functions, and $\mathbf{B}_e$ is reinterpreted to add rows for the parameters that multiply the bubble functions (i.e., it has n rows). With this change, the derivation follows exactly as above. For the interpolation in the following example, we use

$$\boldsymbol{\varphi}(\xi) = \left[1 - \xi,\ \xi,\ 4b(\xi),\ \tfrac{27}{4}\xi b(\xi),\ \tfrac{256}{27}\xi^2 b(\xi),\ \tfrac{3125}{256}\xi^3 b(\xi)\right]^T$$

where $b(\xi) = \xi(1-\xi)$ is the primary bubble function. The coefficients are set so that the maximum value of the shape function is one. The derivation can be found in Appendix D. Higher order interpolation is better able to capture the response of the bar as it propagates waves, particularly for the axial force, which is the derivative of the axial displacement.

To illustrate the point, consider the example shown in Figs. 9.17 and 9.18. The fixed–free bar of length $L = 10$, modulus $E = 500$, and density $\rho = 1$ is subjected to an initial displacement

$$u_o(x) = U_o e^{-a(\xi-b)^2} \tag{9.78}$$

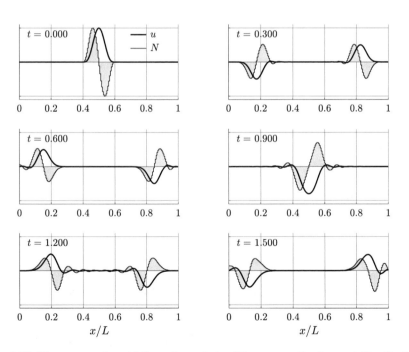

Fig. 9.17 *Wave propagation with linear interpolation.* Wave propagation in an axial bar fixed at the left end and free at the right end with $L = 10$, $E = 500$, $\rho = 1$, and an initial displacement given by Eq. 9.78. The progress of the wave is shown at six different times for a linear interpolation using 180 elements

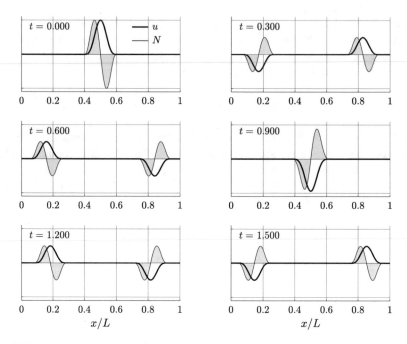

Fig. 9.18 *Wave propagation with bubble functions.* Wave propagation in an axial bar fixed at the left end and free at the right end with $L = 10$, $E = 500$, $\rho = 1$, and an initial displacement given by Eq. 9.78. The progress of the wave is shown at six different times for an interpolation using 20 elements with five bubble functions

where $\xi = x/L$, $a = 800$, and $b = 0.5$. The plots of u and N are normalized so that they can be shown on the same graph. Hence, the magnitude of the initial displacement is not relevant.

The problem is discretized in two ways: one with 180 linear elements (Fig. 9.17) and one with 20 elements having five bubble functions (Fig. 9.18). The total number of degrees of freedom is actually greater for the former (180) than the latter (120). At time $t = 0$ you can see the approximation of the initial wave form. The bubble functions do a good job of representing the function. The linear functions represents $u(x, t)$ well, but $N(x, t)$ is not smooth because the linear interpolation of displacement implies a piecewise constant axial force. It appears that the discretization does not affect the wave propagation speed very much. However, in the linear discretization, the response is prone to spurious artifacts.

It is possible to compute the wave speed from the simulation. The initial wave travels to the right, reflects and traverses the entire bar, reflects and traverses the entire bar again, then reflects and gets to $x/L = 0.85$ (this sequence is most evident from watching the animation). The total distance traveled by the peak of the wave is 33.5 units in a time of 1.5 s. The ratio of distance to time gives $33.5/1.5 = 22.33$, which is almost identical to the theoretical value of $\sqrt{500/1} = 22.36$.

Suffice it to say that wave propagation is a challenging problem for numerical methods, but it can be successful with an appropriate discretization. This example reinforces the importance of having the higher order element interpolation available in the code. The "hat" functions provide a nice introduction to finite element analysis, but do not generally have the performance attributes needed to do wave propagation studies. In fact, this example showed that the improved results from the higher-order interpolation actually reduced the number of degrees of freedom required to get a good solution.

9.5.6   Initial Conditions

The example shown in Fig. 9.18 is driven by the initial wave described by Eq. 9.78. The solution of the equations in that example use discrete modal analysis of Eq. 9.76 with $\mathbf{f}(t) = \mathbf{0}$. As was true for the NDOF problem, we require discrete initial displacement and velocity vectors to solve the problem. Let

$$\mathbf{u}_o = \mathbf{u}(0), \qquad \mathbf{v}_o = \dot{\mathbf{u}}(0)$$

These discrete quantities must be consistent with the initial displacement and velocity functions $u_o(x)$ and $v_o(x)$, respectively. In this section, we show how to compute the discrete initial conditions.

First note that if we knew $\mathbf{u}_o$, then we could compute the value of the displacement in element e using Eq. 9.77 as

$$\tilde{u}_e^o(\xi) = \boldsymbol{\varphi}^T(\xi) \mathbf{B}_e \mathbf{u}_o \tag{9.79}$$

where $\tilde{u}_e^o$ is simply the interpolated value of $\mathbf{u}_o$ within the element. Our problem is to compute $\mathbf{u}_o$, which is an overdetermined problem because there are many elements that claim Eq. 9.79, all sharing the same $\mathbf{u}_o$. We can solve the problem by minimizing the square of the difference between the specified value $u_e^o(\xi)$, which is the part of $u_o(x)$ in element e, and the value computed from Eq. 9.79. This is the *method of least squares*. Let us define the function

$$J(\mathbf{u}_o) = \sum_{e=1}^{M} \int_0^1 \frac{1}{2} \rho A \big(\tilde{u}_e^o(\mathbf{u}_o) - u_e^o\big)^2 L_e \, d\xi \tag{9.80}$$

Note that for the purposes of this calculation, we think of $\tilde{u}_e^o$ as being a function of the discrete initial diplacement parameters $\mathbf{u}_o$. The value of $\mathbf{u}_o$ that minimizes the function J can be found by setting the directional derivative equal to zero, i.e., $\mathcal{D}J(\mathbf{u}_o) \cdot \bar{\mathbf{u}} = 0$, where $\bar{\mathbf{u}}$ is an arbitrary variation of $\mathbf{u}_o$. The directional derivative is covered in Appendix B. Using the chain rule, the equation $\mathcal{D}J(\mathbf{u}_o) \cdot \bar{\mathbf{u}} = 0$ can be written as

$$\sum_{e=1}^{M} \int_0^1 \rho A \big(\tilde{u}_e^o(\mathbf{u}_o) - u_e^o\big)\big[\mathcal{D}\tilde{u}_e^o(\mathbf{u}_o) \cdot \bar{\mathbf{u}}\big] L_e \, d\xi = 0 \qquad (9.81)$$

The directional derivative of $\tilde{u}_e^o$ can be computed from Eq. 9.79 as

$$\mathcal{D}\tilde{u}_e^o(\mathbf{u}_o) \cdot \bar{\mathbf{u}} = \frac{d}{d\varepsilon}\Big[\boldsymbol{\varphi}^T(\xi)\mathbf{B}_e\big(\mathbf{u}_o + \varepsilon\bar{\mathbf{u}}\big)\Big]_{\varepsilon=0} = \boldsymbol{\varphi}^T(\xi)\mathbf{B}_e\bar{\mathbf{u}}$$

Substituting this result and the expression for $\tilde{u}_e^o$ from Eq. 9.79 into Eq. 9.81 gives

$$\bar{\mathbf{u}}^T \sum_{e=1}^{M} \mathbf{B}_e^T \int_0^1 \big(\rho A \boldsymbol{\varphi}\boldsymbol{\varphi}^T \mathbf{B}_e \mathbf{u}_o - \rho A \boldsymbol{\varphi} u_e^o\big) L_e \, d\xi = 0$$

Since $\bar{\mathbf{u}}$ is arbitrary, the sum itself must be equal to zero. Hence,

$$\sum_{e=1}^{M} \mathbf{B}_e^T \int_0^1 \rho A \boldsymbol{\varphi}\boldsymbol{\varphi}^T L_e \, d\xi \, \mathbf{B}_e \mathbf{u}_o = \sum_{e=1}^{M} \mathbf{B}_e^T \int_0^1 \rho A \boldsymbol{\varphi} u_e^o L_e \, d\xi$$

The sum on the left side of the equation is simply the global mass matrix $\mathbf{M}$, making the left side equal to $\mathbf{M}\mathbf{u}_o$. The reason for including ρA in the definition of J should now be evident. The right side represents the assembly of element integrals of the initial displacement function, which is very similar to the assembly of distributed loads. We can now solve for the initial displacement as

$$\mathbf{u}_o = \mathbf{M}^{-1} \sum_{e=1}^{M} \mathbf{B}_e^T \int_0^1 \rho A \boldsymbol{\varphi} u_e^o L_e \, d\xi \qquad (9.82)$$

Computation of initial velocities follows a similar path, resulting in

$$\mathbf{v}_o = \mathbf{M}^{-1} \sum_{e=1}^{M} \mathbf{B}_e^T \int_0^1 \rho A \boldsymbol{\varphi} v_e^o(\xi) L_e \, d\xi \qquad (9.83)$$

where v_e^o is the part of the initial velocity $v_o(x)$ in element e.

In the basic "hat" function interpolation, the nodal values $\mathbf{u}(t)$ are the approximate values of $u(x, t)$ at the nodal locations. It might seem like all we need to do is evaluate the initial displacement and velocity functions at the nodal locations and assign those values to $\mathbf{u}_o$. However, when we enhance the interpolation with bubble functions, we need to evaluate the coefficients of the bubble functions, which are also contained in $\mathbf{u}_o$. The above approach gives the route to assigning those values. The MATLAB code that does this computation is given in Code 9.3.

Code 9.3   MATLAB code to compute the initial displacements and velocity for the finite element interpolation

```
function [uo,vo] = AxialICs(V,M,x,d,ix,eDOF,S,DF)
%  Establish the initial displacement and velocity
%
%        V : Matrix of eigenvectors
%        M : Mass matrix
%        x : Nodal coordinates (to get length L)
%        d : Physical properties [Type,E,A,rho]
%       ix : Element connections (needed in AxialLocalize)
%     eDOF : Element global DOF numbers (needed in AxialLocalize)
%        S : Struct containing problem size parameters, etc.
%              S.Mode = 0 uo and vo set from function WaveForm
%                     > 0 uo and vo set as mode shape 'S.Mode' from V
%              S.Uo magnitude of initial displacement
%              S.Vo magnitude of initial velocity
%       DF : Struct with output control items

%. Set uo from function WaveForm
   if S.Mode==0

%... Plot the initial wave
     PlotWaveForm(S,DF)

%... Establish length of bar
     L = max(x(:,1))-min(x(:,1));

%... Loop over elements, form element ue, and assemble to uo
     uo = zeros(S.nDOF,1);
     for e=1:S.nElms
        [~,xe,jj] = AxialLocalize(e,uo,x,ix,eDOF,S);
        [ue] = AxialElementIC(xe,L,d,S);
        uo(jj) = uo(jj) + ue;
     end

%... Trim to unconstrained DOF and multiply by M inverse
     uo = uo(1:S.nFree);
     uo = M\uo;

%. Set initial displacements to mode if Mode > 0
   else
     uo = V(:,S.Mode);

   end

%. Set initial velocities to be same as uo and scale
   vo = uo*S.Vo;
   uo = uo*S.Uo;

end

%-----------------------------------------------------------------------

function [ue] = AxialElementIC(x,L,d,S)
%  Compute and assemble initial conditions for initial displacement.
%
%        x : Nodal coordinates for element
%        L : Length of beam
%        d : Physical properties [Type,E,A,rho]
%        S : Struct containing problem size parameters, etc.

%. Total DOF per element
   nDOFe = 2+sum(S.nExt);

%. Compute element length and location of left end
   Le = L/S.nElms;
```

```
    x0 = x(1,1);
%. Numerically integrate the wave form
    ue = zeros(nDOFe,1);
    for i=1:S.nQpts
      xi = S.st(i);                        % Local coordinate
      xx = x0 + xi*Le;                     % Global coordinate
      [uo,~] = WaveFunction(xx,L,S);       % Wave function value
      [h,~] = BubbleFcns(xi,S.nExt(1));    % Shape functions
      ue = ue + d(3)*d(4)*uo*h*Le*S.wt(i); % Element contribution
    end

end
```

We call this function from the main dynamics program when the initial displacement and velocity arrays are needed. Note that the function creates uo and vo to be the same. The scaling parameters S.Uo and S.Vo provide an easy mechanism to turn them on and off. When S.Mode is greater than zero, the initial displacements and velocities are set to mode shape number equal to S.Mode. Hence, Code 9.3 requires the eigenvectors, which are passed in through the array V (whose columns are the mode shapes). When S.Mode is zero, the waveform is computed from Code 9.4.

Code 9.4 MATLAB code to generate waveforms for the axial bar problem. Note that additional waveforms can easily be added to this function

```
function [w,dw] = WaveFunction(z,L,S)
%  Compute the value of wave function and its derivative
%
%          z : Coordinate of point to evalute
%          L : Length of bar

%. Convert to range [0,1]
   z = z/L;

   switch S.WaveType

      case 1  % Wave = 1-cos(2*pi*zi)   zi=(z-a)/(b-a)   a<z<b
         Width = 0.1; Center = 0.5;
         a = max(Center - Width,0);
         b = min(Center + Width,1);
         h = b - a;
         w = 0; dw = 0;
         if (z>=a) && (z<=b)
            zi = (z-a)/h;
            w  = (1 - cos(2*pi*zi))/2;
            dw = sin(2*pi*zi)/h;
         end

      case 2  % Wave = exp{-a*(z-b)^2]
         a = 800;  b = 0.5;
         zi = z - b;
         w  = exp(-a*zi^2);
         dw = -2*a*zi*exp(-a*zi^2);

      case 3  % Wave = (4*zi*(1-zi))^n,   zi=(z-a)/(b-a)   a<z<b
         n = 4;
         Width = 0.1; Center = 0.5;
         a = max(Center - Width,0);
         b = min(Center + Width,1);
         h = b - a;
         w = 0; dw = 0;
```

```
            if (z>=a) && (z<=b)
              zi = (z-a)/h;   wo = 4*zi*(1-zi);
              w  = wo^n;
              dw = n*wo^(n-1)*4*(1-2*zi)/h;
            end

      end

%. Convert to derivative with respect to original coordinates
   dw = dw/L;

end
```

The initial waveforms are specified completely in this function. The location and width of the wave can be adjusted and additional waveforms can be added.

9.6   Axial Bar Dynamics Code

The code that generated Figs. 9.17 and 9.18 is worth looking at because it shows the general organization of the calculation. This code uses modal analysis (not numerical integration in time). It is set up only to do the wave propagation from initial displacement or velocity. Code 9.5 is the main program.

Code 9.5 MATLAB code to do modal analysis of waves propagating in an axial bar subjected to an initial displacement or velocity

```
%  Axial Bar Dynamics (Modal Analysis)
   clear; clc;

%. Get physical properties and set up data structures
   [x,ix,id,eDOF,d,Load,F,~,S,DF] = AxialInput;

%. Put forces into DOF order
   [f,u,uPre] = AxialForce(F,id,S,DF);

%. Establish time and displ. arrays
   tf = 1.5;                        % Final time
   nSteps = 151;                    % Number of time steps
   Time = linspace(0,tf,nSteps)';   % Time
   Disp = zeros(nSteps,S.nDOF);     % Displacements

%. Form mass, stiffness and load matrices
   uTot = [u; uPre];
   [M,K,~] = AxialAssemble(uTot,x,d,eDOF,ix,Load,S);

%. Get eigenvectors and natural frequencies
   [Evec,Eval] = eig(K,M);
   freq = diag(sqrt(Eval));

%. Establish initial conditions
   [uo,vo] = AxialICs(Evec,M,x,d,ix,eDOF,S,DF);

%. Compute coefficients for classical modal solution
   An = Evec'*M*uo;
   Bn = (Evec'*M*vo)./freq;

%. Compute motion by modal decomposition
   for i=1:nSteps
```

```
      u = Evec*(An.*cos(freq.*Time(i)) + Bn.*sin(freq.*Time(i)));
      Disp(i,:) = [u; uPre]';
   end

%. Generate graphical output of response
   AxialGraphs(DF,x,ix,Time,Disp,id,eDOF,S,freq);
```

The problem inputs are created in the function `AxialInput` (see Code 9.8). This function describes the model, establishing the nodal coordinates x, the element connections ix, the global DOF numbers by node id, the global DOF numbers by element eDOF, the physical parameters d (i.e., E, A, and ρ), the time-dependent load parameters Load (not used in this code, but will be in others), the spatial pattern of nodal loads F, and the problem size parameters S (which is a 'struct' object in MATLAB). The DF 'struct' is just a convenient way to manage the various graphical outputs generated by the program in the function `AxialGraphics`.

The function `AxialForce` takes the forces and prescribed displacements in the input array F, which are in nodal order, and puts them in global DOF order in the array f. Note that the convention we use is that if a DOF is restrained, then the value input in F is interpreted as a prescribed displacement. The outcome of this function is the global force vector, the initialized displacement vector u (with size nFree), and the prescribed displacement vector uPre.

Of course, we need to give the final time tf of the analysis and the time increment dt. Since we are doing modal analysis, the time increment is not needed for accuracy of the results, so the time increment is selected primarily for the animation of the propagating wave.

The function `AxialAssemble`, Code 9.6, is where the element formation and global assembly takes place.

Code 9.6　MATLAB code to assemble mass and stiffness matrices and load vector for axial bar

```
function [M,K,g] = AxialAssemble(u,x,d,eDOF,ix,Load,S)
%       Form mass, M, tangent K, and residual g for axial bar
%
%              x : Nodal coordinates
%              u : Nodal displacements
%              d : Material properties
%           eDOF : Global DOF for each element
%             ix : Element nodal connections
%           Load : Distributed load parameters
%         S.Free : Number of free DOF in structure
%         S.nDOF : Number of total DOF in structure

%. Establish length of bar
   L = max(x(:,1));

%. Initialize global arrays
   M = zeros(S.nDOF,S.nDOF);
   K = zeros(S.nDOF,S.nDOF);
   g = zeros(S.nDOF,1);

%. Loop over elements
   for e=1:S.nElms

%... Find local coordinates and assembly pointers
       [~,xe,jj] = AxialLocalize(e,u,x,ix,eDOF,S);
```

```
%... Form element matrices
     [me,ke,fe] = AxialElement(xe,d,L,Load,S);

%... Assemble element matrices into global arrays
     M(jj,jj) = M(jj,jj) + me;
     K(jj,jj) = K(jj,jj) + ke;
     g(jj) = g(jj) + fe;

  end

%. Trim to unconstrained DOF
  M = M(1:S.nFree,1:S.nFree);
  K = K(1:S.nFree,1:S.nFree);
  g = g(1:S.nFree);

end
```

This is a very simple function to which we pass all of the necessary information to build the element matrices and carry out the assembly by looping over the elements and adding the element contributions to the global matrices. This function relies on two other functions. The function AxialLocalize, Code E.7 in Appendix E, localizes the global arrays and extracts the indices jj needed for assembly. With the relevant arrays localized, we are ready to perform the computation of the element arrays $\mathbf{m}_e$, $\mathbf{k}_e$, and $\mathbf{f}_e$. These computations are done in Code 9.7, which is the function AxialElement.

Code 9.7 MATLAB code to compute element mass, stiffness, and load arrays ($\mathbf{m}_e$, $\mathbf{k}_e$, and $\mathbf{f}_e$) for the axial bar

```
function [me,ke,fe] = AxialElement(x,d,L,Load,S)
%  Element mass, stiffness, and force matrices, integrated numerically.
%
%        x : Nodal coordinates
%        d : Material properties [Type,E,A,rho]
%        L : Length of beam
%     Load : Distributed load information
%        S : Struct with problem size parameters

%. Establish total number of DOF for element
   nDOFe = 2+sum(S.nExt);

%. Extract load type and load magnitudes from Load
   LoadType = Load(1);
   po = Load(2);

%. Compute location of left end of element e
   x0 = x(1,1);
   Le = L/S.nElms;

%. Initialize arrays
   me = zeros(nDOFe,nDOFe);           % Element mass matrix
   ke = zeros(nDOFe,nDOFe);           % Element stiffness matrix
   fe = zeros(nDOFe,1);               % Element force vector

%. Numerically integrate arrays
     for i=1:S.nQpts

%... Set element coordinate for integration point, get load
       xi = S.st(i);    xx = x0 + xi*Le;
       p = po*LoadFunction(xx,L,LoadType);

%... Compute shape functions
```

```
        [h,dh] = BubbleFcns(xi,S.nExt(1));

%... Compute cross sectional moduli and mass
        EA   = d(2)*d(3);
        rhoA = d(3)*d(4);

%... Compute the contributions mass, stiffnessm force
        me = me + (rhoA*Le)*(h*h')*S.wt(i);
        ke = ke + (EA/Le)*(dh*dh')*S.wt(i);
        fe = fe +  p*h*Le*S.wt(i);

    end
end
```

In this function we do the integration numerically, looping over the integration stations S.st(i) and accumulating the contribution to each of the three quantities being integrated (i.e., me, ke, and fe). The integration weights and stations are extracted from the function NumInt (see Code F.1 in Appendix F), which is called in the input function AxialInput. The shape functions and their derivatives are obtained from the function BubbleFcns (see Code D.2 in Appendix D). The loading function $p(x)$ is the spatial distribution of load, and is obtained from the function LoadFunction. Note that the loading is defined in terms of the real x coordinate and not the local element coordinate. The graphical output is created by the function AxialGraphics, which takes the response (stored in Time and Disp) and produces the graphs like Fig. 9.17 and the animation of the motion.

Finally, we present the input function for the axial bar in Code 9.8. This function takes care of setting up all of the problem parameters, including the physical properties of the bar (d), the basic data structure arrays (e.g, x, ix, id, eDOF), the loading, and the boundary conditions.

Code 9.8 MATLAB code to establish all of the features of the problem, including the physical parameters and the basic database arrays, (e.g., x, ix, id, eDOF, d)

```
function [x,ix,id,eDOF,d,Load,f,Damp,S,DF] = AxialInput
%    Problem input for Axial Bar codes
%
%      x : Coordinates of nodes
%     ix : I-node, J-node for each element
%     id : Global DOF numbers for each node
%   eDOF : Global DOF numbers for each element
%      d : Material properties [ConstModel, E, A, rho]
%   Load : Distributed loading [LoadType, po]
%      f : Nodal applied forces in node order
%   Damp : Struct with damping properties (not used)
%      S : Struct with problem size parameters, etc.
%     DF : Struct to manage graphic output

%. Numerical discretization
    S.nElms = 180;                % Number of elements
    S.nExt = 0;                   % Number of extra DOF for U
    S.nNodes = S.nElms + 1;       % Number of FE nodes
    S.nDOFpn = 1;                 % Number of DOF per node for U

%. Physical parameters
    L = 10;                       % Bar length
    rho = 1;                      % Density
    A = 1;                        % Cross sectional area
    E = 500;                      % Young's modulus
```

```
%. Distributed loading
   LoadType = 1;                  % =1 1, =2 x/L, =3 1-x/L, =4 sin(pi*x/L)
   po = 0;                        % Load magnitude
   Load = [LoadType,po];

%. Select wave type and time function type
   S.WaveType = 2;                % =1 cosine, =2 normal, =3 poly.
   S.TimeType = 0;                % =0 none, =1 const, =2 sine, etc.

%. Initial displacement, velocity, and waveform information
   S.Mode = 0;                    % Mode =0 wave, >0 mode shape Mode
   S.Uo = 0.1;                    % Magnitude of initial displacement
   S.Vo = 0.0;                    % Magnitude of initial velocity

%. Boundary conditions
   BCL = 2;                       % BC at x=0: 1 = free, 2 = fixed
   BCR = 1;                       % BC at x=L: 1 = free, 2 = fixed

%. Total number of DOF
   S.nDOF = S.nNodes + S.nElms*sum(S.nExt);

%. Physical parameters (ConstModel =1 Elastic is available)
   ConstModel = 1;
   d = [ConstModel,E,A,rho];

%. Establish the nodal coordinates
   x = zeros(S.nNodes,2);
   for n=1:S.nNodes
      x(n,:) = L*[n-1, 0]/S.nElms;
   end

%. Compute element connection array
   ix = zeros(S.nElms,2);
   for e=1:S.nElms
      ix(e,:) = e:e+1;
   end

%. Establish BCs: =0 DOF free, =1 DOF restrained
   BC = [0; 1];
   id = zeros(S.nNodes,1);
   id(1,:) = BC(BCL,:);
   id(S.nNodes,:) = BC(BCR,:);

%. Compute local-to-global DOF mappings eDOF and id
   [eDOF,id,S] = GlobalDOF(id,ix,S);

%. Nodal force, each row is (x-force) in node order
   f = zeros(S.nNodes,1);

%. Establish quadrature rule
   S.QType = 1;                   % =1 Gauss-Legendre
   S.nQpts = S.nExt(1)+1;         % Number of integration points
   [S.wt,S.st,S.nQpts] = NumInt(S.QType,S.nQpts);

%. Damping (not used in the AxialBar code)
   Damp = [];

%. Set which figures to do and establish properties
   [DF] = AxialManageFigures;
   AxialEchoInput(L,A,E,rho,BCL,BCR,po,LoadType,S);

end
```

In addition to establishing the inputs, this function does some processing of those inputs. First, it computes the local-to-global DOF mapping eDOF that is used in the localization and assembly process. This array is created by the function `GlobalDOF`, which is described in Appendix E. It also establishes the numerical integration weights and stations that will be used to form the element matrices. Many of the problem parameters are stored in the struct `S` for convenience in passing information to various functions.

Code for Newmark Integration It is quite straightforward to create a code that solves the problem by numerical integration. In addition to the parts already mentioned for the modal analysis code, we must also include the numerical analysis parameters and organize the code accordingly. Code 9.9 is the implementation of the axial bar problem using Newmark's method for time integration.

Code 9.9 MATLAB code to solve the axial bar problem with numerical integration by Newmark's method

```
%   Axial Bar Dynamics (Newmark)
    clear; clc;

%. Get physical properties and set up data structures
    [x,ix,id,eDOF,d,Load,F,damp,S,DF] = AxialInput;

%. Put nodal forces into DOF order
    [fn,u,uPre] = AxialForce(F,id,S,DF);

%. Time parameters
    tf = 1.5;                       % final time
    h = 0.001;                      % analysis time step

%. Retrieve parameters for Newmark integration
    [beta,gamma,eta,zeta,nSteps,tol,itmax,tf] = NewmarkParams(tf,h);

%. Form mass, stiffness matrices and distributed load vector
    uTot = [u; uPre];
    [M,K,fd] = AxialAssemble(uTot,x,d,eDOF,ix,Load,S);
    f = fn + fd;

%. Get eigenvectors and natural frequencies
    [Evec,Eval] = eig(K,M);
    freq = diag(sqrt(Eval));

%. Set up arrays to save results
    Time = zeros(nSteps,1);         % History of time
    Disp = zeros(nSteps,S.nDOF);    % History of displacements

%. Establish initial conditions
    [uo,vo] = AxialICs(Evec,M,x,d,ix,eDOF,S,DF);

%. Initialize time, position, velocity, and acceleration
    t = 0;
    uold = uo;
    vold = vo;
    Ft = TimeFunction(t,S.TimeType,0);
    aold = M\(Ft*f - K*uo);

%. Compute motion by modal decomposition
    for i=1:nSteps

%... Store results for plotting later
        Time(i,:) = t;
        Disp(i,:) = [uold; uPre]';
```

```
%... Compute the values for the next time step
       t = t + h; its = 0; err = 1;
       Ft = TimeFunction(t,S.TimeType,0);
       bn = uold + h*vold + beta*h^2*aold;
       cn = vold + gamma*h*aold;
       anew = aold;

%... Newton iteration to determine the new acceleration
       while (err > tol) && (its < itmax)
          its = its + 1;
          unew = bn + zeta*anew;
          g = M*anew + K*unew - Ft*f;
          A = M + zeta*K;
          anew = anew - A\g;
          err = norm(g);
       end

%... Compute velocity and displacement by numerical integration
       unew = bn + zeta*anew;
       vnew = cn + eta*anew;

%... Update state for next time step
       aold = anew;
       vold = vnew;
       uold = unew;

    end

%. Generate graphical output of response
    AxialGraphs(DF,x,ix,Time,Disp,id,eDOF,S,freq);
```

This code does the propagation of initial waves, as the previous one did, but it also has the capability to include time-dependent forcing functions through `TimeFunction` (Code 3.4). While this code solves only the linear problem, it is set up with a Newton iteration to find the new acceleration. This structure would easily allow the inclusion of nonlinear problems if a new residual `g` and tangent `A` is evaluated at every iteration. With `maxit` set to 1, the iteration is avoided.

Example The wave propagation from Figs. 9.17 and 9.18 can be done with the Newmark code (using only the higher order interpolation). The results are shown in Figs. 9.19 and 9.20. If the time step is taken as $h = 0.001$ (Fig. 9.20) then the results are almost exactly the same as the modal analysis solution (which is exact with respect to time integration). However, if the time step is $h = 0.01$ the results are as shown in Fig. 9.19. The time discretization introduces spatial numerical artifacts. The upshot of this example is that care must be taken to assure that the solution is converged, whatever the nature of the discretization is.

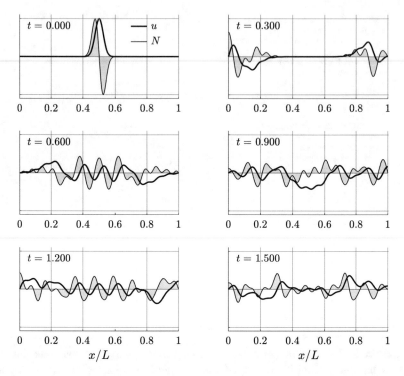

Fig. 9.19 *Wave propagation with Newmark's method* ($h = 0.01$). Wave propagation in an axial bar fixed at the left end and free at the right end with $L = 10$, $E = 500$, $\rho = 1$, and an initial displacement given by Eq. 9.78. The progress of the wave is shown at six different times for an interpolation with 20 elements using five bubble functions. The Newmark parameters are $\beta = 0.25$, $\gamma = 0.5$

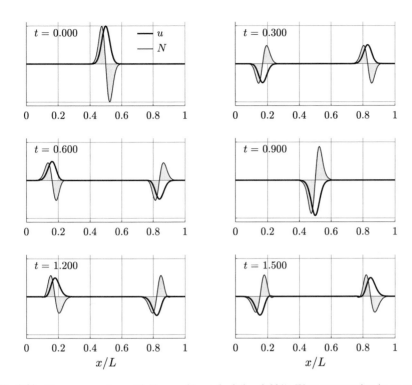

Fig. 9.20 *Wave propagation with Newmark's method* ($h = 0.001$). Wave propagation in an axial bar fixed at the left end and free at the right end with $L = 10$, $E = 500$, $\rho = 1$, and an initial displacement given by Eq. 9.78. The progress of the wave is shown at six different times for an interpolation with 20 elements using five bubble functions. The Newmark parameters are $\beta = 0.25$, $\gamma = 0.5$

Chapter 10
Dynamics of Planar Beams: Theory

In this chapter we establish the theory of beam dynamics, confining our attention to planar motion of beams. These developments will provide the theoretical foundation needed for the remainder of the book. In the following two chapters we explore methods of solution of the linearized equations of motion, both classical solutions based on separation of variables and finite element solutions. In Chap. 13 we extend the finite element solution method to nonlinear beams and in Chap. 14 we tackle the problem of analyzing planar frames.

There are some similarities between the axial bar problem and the beam bending problem, but there are also some major differences. Figure 10.1 illustrates the two cases. For the axial bar, the applied load $p(x, t)$ is oriented along the axis of the bar. As a result, the bar displaces only in the longitudinal direction. For beams, loads can be applied transverse to the beam axis, e.g., $q(x, t)$ shown in the sketch. We also embrace the possibility of applied couple loads. As a result, the beam displaces transverse to the longitudinal axis of the beam and cross sections of the beam rotate. Besides length L, the cross-sectional area A is the primary geometric feature of the axial bar. For the beam, the moment of inertia I of the cross section is also important. Young's modulus E is the only relevant material property for the axial bar. For certain beam theories (specifically Timoshenko beam theory), the shear modulus G is also important. Of course, the density ρ is important to both the axial bar and the beam. For both, we take the line connecting the centroids of the cross sections to be the x axis.

The axial bar is a second order differential equation in space so there are only two boundary conditions (one at each end). The beam is a fourth order differential equation in space. Hence, there are four boundary conditions (two at each end). Both the axial bar and beam are second order differential equations in time, so the initial conditions operate similarly.

© The Author(s), under exclusive license to Springer Nature Switzerland AG 2022 285
K. D. Hjelmstad, *Fundamentals of Structural Dynamics*,
https://doi.org/10.1007/978-3-030-89944-8_10

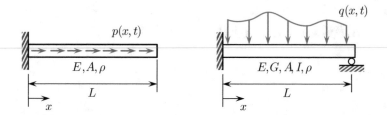

Fig. 10.1 *Comparison between axial bar and beam.* The axial bar is subjected to purely axial forces and responds axially. The beam is subjected to transverse forces and responds by deflecting transverse to the original axis

10.1 Beam Kinematics

Beam theory is based upon the kinematic hypothesis that plane sections remain plane after deformation. As such, a cross section acts like a rigid body, but adjacent cross sections can rotate and displace relative to each other. This hypothesis reduces the three-dimensional continuum equations to differential equations that depend upon only one spatial variable—the axial coordinate x. From the perspective of dynamics, the theory connects with rigid body dynamics through the treatment of the cross section, which is assumed to be rigid in its own plane. Hence, the dynamics of a cross section *is* rigid body dynamics.

In this section we derive the equations of motion for the planar beam, framing the problem in the context of finite deformations. Hence, the resulting equations of kinematics will be *geometrically exact*. In the next chapter, we will linearize the equations of motion to examine how wave propagation goes in beams. It is always possible to get linear equations from nonlinear ones, but not vice versa, so this is a natural starting point. Starting with the nonlinear formulation provides the opportunity to see what the linear theory neglects. Also, linear beam theories are not capable of modeling structural instabilities, which is an important physical phenomenon.

Consider a beam with cross-sectional area A, moment of inertia I, Young's modulus E, shear modulus G, mass density ρ, and length L as shown in Fig. 10.1 (on the right). The beam is subjected to a distributed transverse force $q(x, t)$, as shown in the figure, but it can also be subjected to a distributed axial force $p(x, t)$ and a distributed moment $m(x, t)$. The beam has certain boundary conditions at each end. For example, the beam in Fig. 10.1 is fixed at the left end and simply supported at the right end.

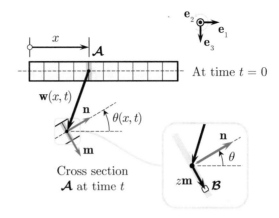

Fig. 10.2 *Beam kinematics.*
The primary assumption of
beam theory is that plane
cross section remain plane
after deformation. The
magnified view of the cross
section at $\mathcal{A}$ shows the typical
particle $\mathcal{B}$ a distance z from
the centroidal axis

10.1.1 Motion of a Beam Cross Section

If we assume that plane cross sections remain plane after deformation, then the
motion of the beam can be completely characterized by displacement of the centroid
$\mathbf{w}(x, t)$ and rotation of the cross section $\theta(x, t)$, as illustrated in Fig. 10.2. The
position vector for the typical particle $\mathcal{B}$ (in cross section $\mathcal{A}$, which is originally
located a distance x from the left end) is given by

$$\mathbf{x}(x, z, t) = x\,\mathbf{e}_1 + \mathbf{w}(x, t) + z\,\mathbf{m}(x, t) \tag{10.1}$$

where x is the distance along the axis of the undeformed beam and z is the distance
from the centroid of the cross section to the typical particle. The unit vector $\mathbf{m}(x, t)$
is oriented in the direction of the rotated cross section and is, therefore, a function
of $\theta(x, t)$. The unit vector $\mathbf{n}(x, t)$ is perpendicular to $\mathbf{m}(x, t)$ and, therefore, normal
to the cross section. Since the motion is restricted to the x–z plane, the motion does
not depend upon y.

The displacement vector can be expressed in terms of components relative to the
standard base vectors as

$$\mathbf{w}(x, t) = u(x, t)\mathbf{e}_1 + w(x, t)\mathbf{e}_3 \tag{10.2}$$

where $u(x, t)$ is the displacement along, and $w(x, t)$ is the displacement transverse
to, the *original* axis of the beam. The deformation map in Eq. 10.1 tracks the
motion of each particle in the beam and respects the kinematic hypothesis that plane
sections remain plane after deformation.

The unit vectors $\mathbf{n}$ and $\mathbf{m}$ track the rotation of the cross section at x, and can be
expressed in terms of components relative to the standard base vectors as

$$\mathbf{n}(x, t) = \cos\theta(x, t)\,\mathbf{e}_1 - \sin\theta(x, t)\,\mathbf{e}_3$$
$$\mathbf{m}(x, t) = \sin\theta(x, t)\,\mathbf{e}_1 + \cos\theta(x, t)\,\mathbf{e}_3 \tag{10.3}$$

Therefore, both vectors depend upon x and t implicitly through the function $\theta(x, t)$, the angle of rotation of the cross section measured relative to the orientation of the undeformed axis of the beam, as shown in Fig. 10.2.

It is straightforward to show that the time derivatives of these vectors are

$$\dot{\mathbf{n}} = -\dot{\theta}\,\mathbf{m}, \qquad \dot{\mathbf{m}} = \dot{\theta}\,\mathbf{n}$$

Just do the indicated time derivative, factor out $\dot{\theta}$, and note that the remaining terms can be recognized as the expression (or negative of) the other unit vector. Similarly, the spatial derivatives (with respect to x) of these vectors are

$$\mathbf{n}' = -\theta'\mathbf{m}, \qquad \mathbf{m}' = \theta'\mathbf{n}$$

These relationships will come in handy in some of the later derivations. Note that we are using a superposed dot to represent derivatives with respect to t and a prime to represent derivatives with respect to x. For example,

$$\dot{\mathbf{w}} = \frac{\partial\mathbf{w}}{\partial t}, \quad \ddot{\mathbf{w}} = \frac{\partial^2\mathbf{w}}{\partial t^2}, \quad \mathbf{w}' = \frac{\partial\mathbf{w}}{\partial x}, \quad \mathbf{w}'' = \frac{\partial^2\mathbf{w}}{\partial x^2}$$

This is possible because there is only one spatial variable, x, and (of course) only one time variable, t. To keep the notation compact, we will often suppress the dependence on x and t (e.g., write $\mathbf{w}$ rather than $\mathbf{w}(x, t)$) in most of the equations, unless doing so creates ambiguity.

We can compute the velocity $\dot{\mathbf{x}}$ and acceleration $\ddot{\mathbf{x}}$ for particle $\mathcal{B}$ from the position vector by differentiation with respect to time. To wit,

$$\dot{\mathbf{x}}(x, z, t) = \dot{\mathbf{w}} + z\dot{\theta}\mathbf{n}$$
$$\ddot{\mathbf{x}}(x, z, t) = \ddot{\mathbf{w}} + z(\ddot{\theta}\mathbf{n} - \dot{\theta}^2\mathbf{m}) \tag{10.4}$$

Because we are tracking the location of material particles in the body, this is a *Lagrangian description of motion*, which is why the velocity and acceleration are simply partial derivatives of the displacement with respect to time. We will use the acceleration in balance of linear and angular momentum to get the equations of motion of the beam.

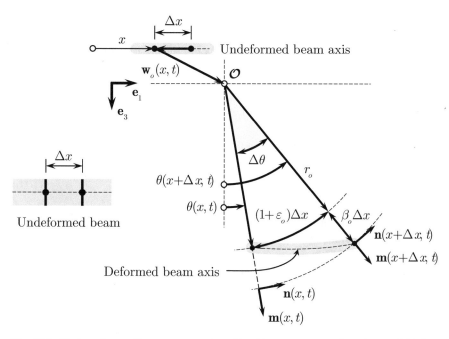

Fig. 10.3 *Motion of two adjacent cross sections.* Consider two cross sections that are initially a distance Δx apart. The centroid of each cross section is at the black dot. The motion includes translation and rotation of the cross section and the deformation includes stretching of the axis, $1 + \varepsilon_o$, and shearing of the beam segment, β_o. The cross sections translate and rotate

10.1.2 Strain–Displacement Relationships

To see what happens as a result of the plane sections hypothesis, consider the sketch in Fig. 10.3. We have two neighboring cross sections, one initially located a distance x and the other a distance $x + \Delta x$ from the left end of the beam. Both cross sections translate and rotate relative to their initial positions. We track the translations of the centroidal axis of each cross section. The translation of the cross section at x is $\mathbf{w}(x, t)$—a vector going from the black dot in the undeformed configuration to the associated black dot in the deformed configuration. Similarly, the cross section on the right translates by $\mathbf{w}(x + \Delta x, t)$. The two cross sections rotate by $\theta(x, t)$ and $\theta(x + \Delta x, t)$, respectively. Note that the unit vectors $\mathbf{n}$ and $\mathbf{m}$ also have different values at x and $x + \Delta x$.

The "plane sections" kinematic hypothesis allows neighboring cross sections to rotate relative to each other. If they do, then the lines extending along the two cross sections must intersect at some point $\mathcal{O}$. The deformed axis of the beam is the line that connects the two black dots—i.e., the line of centroids. Let us scribe a circular arc through the deformed centroid at x. Call the arc length between the cross sections $(1 + \varepsilon_o)\Delta x$. The right cross section displaces in the $\mathbf{m}$ direction relative to the left cross section by $\beta_o \Delta x$. We identify the quantities ε_o and β_o as *strain resultants*, the former being the axial strain and the latter the shear strain.

Let us designate the radial distance from $\mathcal{O}$ to the circular arc as r_o, the *radius of curvature* of the beam, which we can relate to the relative rotations of the cross sections. The difference in angles $\theta(x + \Delta x) - \theta(x)$ subtends the arc. Hence,

$$r_o\big(\theta(x + \Delta x, t) - \theta(x, t)\big) = (1 + \varepsilon_o)\Delta x \tag{10.5}$$

Dividing through by $r_o \Delta x$ and taking the limit as Δx goes to zero, we get

$$\frac{\partial \theta}{\partial x} = \frac{1 + \varepsilon_o}{r_o} \tag{10.6}$$

This is the first *strain–displacement* relationship of beam theory. Strain–displacement relationships connect the motion variables (like θ) with the strain variables (like ε_o). Let us define the *curvature*[1] of the beam to be

$$\kappa_o = \frac{1 + \varepsilon_o}{r_o} \tag{10.7}$$

The curvature is a strain-like quantity that measures the bending of the beam. Using Eqs. 10.6 and 10.7, we can express the *rotation–curvature* relationship as

$$\kappa_o = \theta' \ . \tag{10.8}$$

The displacement from the centroidal black dot in the undeformed configuration to the centroidal black dot in the deformed configuration for the cross section originally located at x is, by vector addition,

$$\mathbf{w}(x, t) = \mathbf{w}_o + r_o\,\mathbf{m}(x)$$

where $\mathbf{w}_o$ is the vector from the centroid of the cross section in the undeformed configuration at x to the point $\mathcal{O}$, as shown in Fig. 10.3. Similarly, the displacement from the undeformed to deformed centroid of the cross section originally located at $x + \Delta x$ is

$$\mathbf{w}(x + \Delta x, t) = -\Delta x\,\mathbf{e}_1 + \mathbf{w}_o + \big(r_o + \beta_o \Delta x\big)\mathbf{m}(x + \Delta x, t)$$

The partial derivative of $\mathbf{w}(x, t)$ with respect to x is, by definition,

---

[1] Note that this definition of beam curvature is not the conventional inverse of the radius of curvature. One could just as easily adopt the latter measure. The difference will manifest when we establish constitutive equations.

$$\frac{\partial \mathbf{w}}{\partial x} = \lim_{\Delta x \to 0} \frac{\mathbf{w}(x + \Delta x, t) - \mathbf{w}(x, t)}{\Delta x}$$

Substituting the expressions for the displacements of the two cross sections and taking the limit as $\Delta x \to 0$ gives

$$\frac{\partial \mathbf{w}}{\partial x} = \lim_{\Delta x \to 0} \left\{ r_o \frac{\mathbf{m}(x + \Delta x, t) - \mathbf{m}(x, t)}{\Delta x} + \beta_o\, \mathbf{m}(x + \Delta x, t) - \mathbf{e}_1 \right\}$$

Noting that $\mathbf{m}' = \theta' \mathbf{n}$, and substituting Eq. 10.6, we find that

$$\mathbf{w}' = (1 + \varepsilon_o)\mathbf{n} + \beta_o\, \mathbf{m} - \mathbf{e}_1 \tag{10.9}$$

This is the second *strain–displacement* relationship for beam theory.

We can write component expressions for Eq. 10.9 either as strain in terms of displacement or as displacement in terms of strain. First, dot Eq. 10.9 with $\mathbf{n}$ and $\mathbf{m}$, respectively, to get

$$\begin{aligned}
\varepsilon_o &= (1 + u') \cos\theta - w' \sin\theta - 1 \\
\beta_o &= (1 + u') \sin\theta + w' \cos\theta
\end{aligned} \tag{10.10}$$

To get the displacement derivatives in terms of the strains, dot Eq. 10.9 with $\mathbf{e}_1$ and $\mathbf{e}_3$, respectively, to get

$$\begin{aligned}
u' &= (1 + \varepsilon_o) \cos\theta + \beta_o \sin\theta - 1 \\
w' &= -(1 + \varepsilon_o) \sin\theta + \beta_o \cos\theta
\end{aligned} \tag{10.11}$$

Equations 10.9, 10.10, and 10.11 are equivalent.

Equations 10.8 and 10.9 constitute the complete *strain–displacement* equations for the beam. The generalized strain resultants ε_o, β_o, and κ_o are adequate to completely characterize the deformation of the beam within the context of the kinematic hypothesis.

10.1.3 Normal and Shear Strain

Figure 10.4 shows the deformed positions of the two cross section originally located at x and $x + \Delta x$. This picture will aid us in deriving expressions for the normal and shear strains at other points in the beam cross section. Recall that z measures position in the cross section relative to the centroidal axis. The length of the circular arc at z is

Fig. 10.4 *Strain and curvature.* Longitudinal strain comes from bending of the beam. Some fibers get longer, some get shorter. The component of strain ε_o measures the change in length of the centroidal fiber. The component of strain β_o accrues when the two cross sections move parallel to each other without rotating

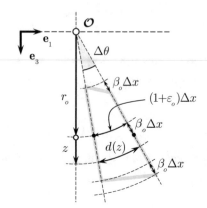

$$d(z) = (r_o + z)\,\Delta\theta = (r_o + z)\left(\frac{1 + \varepsilon_o}{r_o}\right)\Delta x$$

where $\Delta\theta = \theta(x + \Delta x) - \theta(x)$, and we have substituted Eq. 10.5. The normal strain of a fiber in the beam is the change in length divided by the original length (all fibers start out with length Δx). Thus,

$$\varepsilon(x, z, t) = \frac{d(x, z) - \Delta x}{\Delta x} = \varepsilon_o + z\left(\frac{1 + \varepsilon_o}{r_o}\right)$$

Noting the definition of curvature from Eq. 10.7, we can simplify this expression for the normal strain at a cross section to

$$\varepsilon(x, z, t) = \varepsilon_o(x, t) + z\,\kappa_o(x, t) \tag{10.12}$$

which implies that normal strain varies linearly over the cross section.

If the curvature is zero, then the deformation map simply has the two cross sections moving parallel to each other—i.e., a shearing deformation. Therefore, we can identify the shear strain in the beam as

$$\gamma(x, z, t) = \beta_o(x, t) \tag{10.13}$$

Observe that the shear strain is constant over the cross section. This artifact of the plane-sections hypothesis is slightly at odds with the need for the shear stress to vanish on the boundaries of the cross section. There are various ways to amend this shortcoming, the most common being the introduction of a *shear coefficient* for the cross sectional area in the constitutive equations.

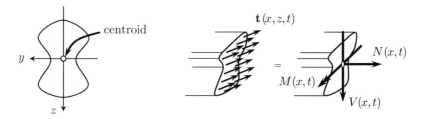

Fig. 10.5 *Planar beam stress resultants.* A cut through a cross section at x exposes a traction field $\mathbf{t}$ that can be characterized by the stress resultants N, V, and M

10.2 Beam Kinetics

Let us now turn our attention to the force side of the picture. The body is in a state of stress, as illustrated in Fig. 10.5. When we take a cut through a cross section, we expose the traction field

$$\mathbf{t} = \sigma\mathbf{n} + \tau\mathbf{m}$$

where σ is the normal component and τ is the shear component of the traction. The *stress resultants* represent the net effect of those tractions, which we compute by integrating the tractions over the cross section. To wit,

$$N(x,t) = \int_A \sigma(x,z,t)\,dA$$

$$V(x,t) = \int_A \tau(x,z,t)\,dA \qquad (10.14)$$

$$M(x,t) = \int_A z\,\sigma(x,z,t)\,dA$$

We call N the axial force, V the shear force, and M the bending moment. Note that the vector direction of the moment is the double-headed arrow, which is defined by the right-hand rule. The positive sign convention on the stress resultants, displacements, and rotation are shown in Fig. 10.6.

10.3 Constitutive Equations

Hooke's law for the beam can be expressed as $\sigma = E\varepsilon$ for the normal stresses, where E is Young's modulus, and as $\tau = G\gamma$ for the shear stresses, where G is the shear modulus. The reason the normal stress constitutive equation manifests in this simple

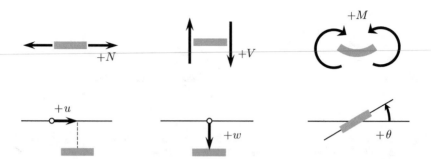

Fig. 10.6 *Sign convention for stress resultants and motions.* The positive sense for the stress resultants, displacements, and rotation. The strains ε_o, β_o, and κ_o have the same positive sense as their associated stress resultants N, V, and M

form is because we assume that there are no confining normal stresses in the cross section (i.e., the stress components $\sigma_{yy} = 0$ and $\sigma_{zz} = 0$).

We can find a relationship between the stress resultants and the strain resultants as follows. For the axial force, we substitute Hooke's law and Eq. 10.12 into the definition of resultant axial force from Eq. 10.14 to get

$$N = \int_A \sigma \, dA = \int_A E\varepsilon \, dA = \int_A E(\varepsilon_o + z\kappa_o)dA = EA\,\varepsilon_o$$

Note that the term with curvature vanishes due to the definition of the centroid. If z is the centroidal axis, then

$$\int_A z \, dA = 0$$

Similarly, for the shear we substitute Hooke's law and Eq. 10.13 into the definition of resultant shear from Eq. 10.14 to get

$$V = \int_A \tau \, dA = \int_A G\gamma \, dA = \int_A G\beta_o \, dA = GA\beta_o$$

where A is the area of the cross section. Due to the fact that the shear stresses τ cannot vanish at the top and bottom fibers of the cross section (because of the kinematic hypothesis), GA is an overestimate of the shear stiffness of the cross section. Often, this value is replaced by GA^*, where $A^* = \nu A$ is called the *shear area* and ν the *shear coefficient*. For example, one can show that the shear coefficient for a rectangular cross section is about 5/6 and for a solid circular section it is 9/10. The shear coefficient for an I-beam is approximately equal to the ratio of the area of the web to total area. For simplicity, we will take the shear coefficient as one throughout the remainder of the book.

Finally, for the moment, we substitute Hooke's law and Eq. 10.12 into the definition of resultant moment from Eq. 10.14 to get

$$M = \int_A z\,\sigma\,dA = \int_A z(E\varepsilon)dA = \int_A zE(\varepsilon_o + z\,\kappa_o)dA = EI\kappa_o$$

where I is the second moment of the area relative to the centroid, i.e.,

$$I = \int_A z^2\,dA$$

The elastic constitutive equations can be summarized as

$$N = EA\,\varepsilon_o, \qquad V = GA\beta_o, \qquad M = EI\kappa_o \qquad (10.15)$$

This derivation is often viewed as a starting point for more general constitutive equations relating stress and strain resultants. While it is possible to derive the linear constitutive equations starting from three-dimensional continuum constitutive theory, as we have done, there are some limitations to extending these equations to nonlinear problems. It is complicated, for example, to derive equations where inelasticity is involved because yielding progresses through the cross section when flexure is involved. The way we have defined strain will affect the constitutive response functions as well.

Another widely accepted approach is to simply start the modeling exercise at the stress resultant level. What this means is that we could write the constitutive relationships in the form

$$N = \hat{N}(\varepsilon_o), \qquad V = \hat{V}(\beta_o), \qquad M = \hat{M}(\kappa_o) \qquad (10.16)$$

where the symbol with a hat simply stands for a function. For example, we compute the axial force N using the function $\hat{N}$, whose argument is the strain ε_o. What these equations imply is that there is no coupling between the resultant forces and the resultant strains—an assumption that could certainly be revised. The functions can take any form. For example, one might postulate nonlinear elastic relationships like

$$N = \frac{EA\,\varepsilon_o}{\sqrt{1 + a\,\varepsilon_o^2}}, \qquad V = \frac{GA\,\beta_o}{\sqrt{1 + b\beta_o^2}}, \qquad M = \frac{EI\kappa_o}{\sqrt{1 + c\,\kappa_o^2}}$$

where a, b, and c are additional constitutive parameters, which could be determined for a particular material by physical testing.

For inelastic constitutive models it is common to include coupling between the components because the presence of one stress resultant may reduce the capacity of the other due to interaction. For example, the presence of axial force might induce yielding at a lower bending moment than it would in absence of axial force. This model is treated in more detail in Chap. 14 in Sect. 14.3.

10.4   Equations of Motion

To establish the equations of motion, we take a free body diagram of a segment of beam between x and $x+\Delta x$. The forces and moments acting on the deformed free body diagram are shown in Fig. 10.7. The externally applied loads are $\mathbf{q}(x,t)$ and the applied moment is $m(x,t)$ (about the $\mathbf{e}_2$ axis). Sometimes it is convenient to break the applied load into components as

$$\mathbf{q}(x,t) = p(x,t)\,\mathbf{e}_1 + q(x,t)\,\mathbf{e}_3$$

where $p(x,t)$ is the axial load and $q(x,t)$ is the transverse load.

Assume that the applied force does not change with elongation, i.e., the total force on the beam segment is $\mathbf{q}\,\Delta x$, even though the beam axis stretches to $(1+\varepsilon_o)\Delta x$ because whatever caused the loading is fixed in the initial configuration. For example, the load might be caused by gravity, but stretching does not increase the number of particles contributing to the weight. Thus, the force associated with weight does not change, but just spreads out a little.

The net force acting on the cross section is designated by the vector $\mathbf{R}(x)$. This force can be resolved into components as

$$\mathbf{R} = N\mathbf{n} + V\mathbf{m} = H\mathbf{e}_1 + Q\mathbf{e}_3 \tag{10.17}$$

where N and V are the axial force and shear force, respectively, and H and Q are the components of force relative to the fixed basis $\{\mathbf{e}_1, \mathbf{e}_2, \mathbf{e}_3\}$. The vectors $\mathbf{n}$ and $\mathbf{m}$ are defined in Eq. 10.3. It is straightforward to show that the relationships among the force components are

$$
\begin{aligned}
H &= N\cos\theta + V\sin\theta & N &= H\cos\theta - Q\sin\theta \\
Q &= -N\sin\theta + V\cos\theta & V &= H\sin\theta + Q\cos\theta
\end{aligned}
\tag{10.18}
$$

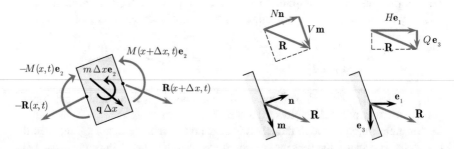

Fig. 10.7 *Free body diagram of beam segment of length Δx. The free body diagram of a piece of beam between x and $x + \Delta x$. The applied loads are $\mathbf{q}(x,t)$ and the applied moment is $m(x,t)$*

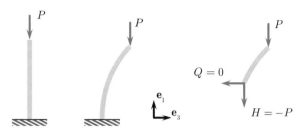

Fig. 10.8 *Example—Euler's elastica.* A cantilever column subjected to the end force P. This problem usually goes by the name *Euler's elastica* in honor of mathematician Leonard Euler, who was the first to study it in detail

There are a few reasons to distinguish between the two sets of components that represent the internal resultant force $\mathbf{R}$. First, when we do constitutive equations, it is natural to refer to the resultant axial and shear forces. Equations 10.15 show that these two components respond in accord with their associated strains. For some problems (e.g., *Euler's elastica* shown in Fig. 10.8) it is sometimes more convenient to refer to the components relative to the standard basis. For example, a cantilever column with an axial force P acting at the end has internal forces $H = -P$ and $Q = 0$. The corresponding relationships for N and V are not so simple for this case.

10.4.1 Balance of Linear Momentum

Using the free body diagram in Fig. 10.7, establish balance of linear momentum by setting the sum of forces equal to mass times acceleration. To wit,

$$\mathbf{R}(x + \Delta x) - \mathbf{R}(x) + \mathbf{q}\,\Delta x = \int_A \rho\,\ddot{\mathbf{x}}\,\Delta x\,dA$$

Divide this equation through by Δx, take the limit as $\Delta x \to 0$, and note that the integral of density times acceleration over the cross section is

$$\int_A \rho\,\ddot{\mathbf{x}}\,dA = \int_A \rho\left(\ddot{\mathbf{w}} + z\left(\ddot{\theta}\mathbf{n} - \dot{\theta}^2\mathbf{m}\right)\right) dA = \rho A \ddot{\mathbf{w}}$$

to show that balance of linear momentum holds if

$$\mathbf{R}' + \mathbf{q} = \rho A\,\ddot{\mathbf{w}} \tag{10.19}$$

It is interesting to note that we do not need to specify the components of $\mathbf{R}$ for balance of linear momentum. This equation encompasses both the axial part and the transverse part.

10.4.2   Balance of Angular Momentum

Balance of angular momentum can be done by taking moments about the centroid on the left side (i.e., at x) of the free body diagram in Fig. 10.7. The moment arm for the stress resultant at $x + \Delta x$ is the vector from the centroid at x to the centroid at $x + \Delta x$, which is most easily defined as the difference between the two radial vectors meeting at $\mathcal{O}$ (see Fig. 10.3). To wit,

$$\Big((r_o + \beta_o \Delta x)\mathbf{m}(x + \Delta x) - r_o \mathbf{m}(x) \Big) \times \mathbf{R}(x + \Delta x)$$

$$+ \Big(M(x + \Delta x) - M(x) + m(\xi, t)\Delta x \Big) \mathbf{e}_2 = \int_A z\mathbf{m} \times \rho \ddot{\mathbf{x}} \, \Delta x \, dA$$

where ξ is a location between x and Δx. Again, divide the equation through by Δx, take the limit as $\Delta x \to 0$, and note that

$$\int_A z\mathbf{m} \times \rho \ddot{\mathbf{x}} \, dA = \int_A z\mathbf{m} \times \rho \left(\ddot{\mathbf{w}} + z(\ddot{\theta}\mathbf{n} - \dot{\theta}^2 \mathbf{m}) \right) dA = \rho I \ddot{\theta} \, \mathbf{e}_2$$

to show that balance of angular momentum holds if

$$\big(M' + m \big)\mathbf{e}_2 + \big(r_o \, \mathbf{m}' + \beta_o \, \mathbf{m} \big) \times \mathbf{R} = \rho I \ddot{\theta} \, \mathbf{e}_2$$

This is a vector equation, but it should be evident that all terms point in the $\mathbf{e}_2$ direction. We can get a scalar equation by dotting the equation with $\mathbf{e}_2$. For the two terms that include the cross product, recall the cyclic nature of the triple scalar product $(\mathbf{u} \times \mathbf{v}) \cdot \mathbf{w} = (\mathbf{w} \times \mathbf{u}) \cdot \mathbf{v}$. Specifically, note the following relationships:

$$(\mathbf{n} \times \mathbf{R}) \cdot \mathbf{e}_2 = -\mathbf{m} \cdot \mathbf{R} \qquad (\mathbf{m} \times \mathbf{R}) \cdot \mathbf{e}_2 = \mathbf{n} \cdot \mathbf{R}$$

It is simple to verify that $\mathbf{e}_2 \times \mathbf{n} = -\mathbf{m}$ and $\mathbf{e}_2 \times \mathbf{m} = \mathbf{n}$. From the geometry of deformation we know that $r_o \mathbf{m}' = (1 + \varepsilon_o)\mathbf{n}$. Putting all of these facts together, we can express the balance of angular momentum for the beam as

$$M' + \big(\beta_o \mathbf{n} - (1 + \varepsilon_o) \, \mathbf{m} \big) \cdot \mathbf{R} + m = \rho I \ddot{\theta} \qquad (10.20)$$

This is the most basic form of the equation, as it does not commit to any set of components of the internal force vector $\mathbf{R}$. However, recognizing that $\mathbf{R} = N\mathbf{n} + V\mathbf{m}$, we can express Eq. 10.20 in the form

$$M' + \beta_o N - (1 + \varepsilon_o) V + m = \rho I \ddot{\theta} \qquad (10.21)$$

Alternatively, substituting the expressions for N and V from Eq. 10.18 and using Eqs. 10.10, we can express Eq. 10.20 as

$$M' + Hw' - Q\left(1+u'\right) + m = \rho I \ddot{\theta} \tag{10.22}$$

We can use any of these forms of balance of angular momentum (i.e., Eqs. 10.20, 10.21, and 10.22) interchangeably.

10.5 Summary of Beam Equations

The equations that govern the planar beam are summarized in the box below.

Planar Beam Equations

- *Strain–displacement equations*

$$\mathbf{w}' = \left(1+\varepsilon_o\right)\mathbf{n} + \beta_o\mathbf{m} - \mathbf{e}_1, \qquad \theta' = \kappa_o$$

- *Balance of linear and angular momentum*

$$\mathbf{R}' + \mathbf{q} = \rho A\,\ddot{\mathbf{w}}, \qquad M' + \left(\beta_o\mathbf{n} - (1+\varepsilon_o)\,\mathbf{m}\right) \cdot \mathbf{R} + m = \rho I \ddot{\theta}$$

- *Constitutive equations*

$$N = \hat{N}(\varepsilon_o), \qquad V = \hat{V}(\beta_o), \qquad M = \hat{M}(\kappa_o)$$

10.6 Linear Beam Theory

The equations derived in the previous section are highly nonlinear. We will develop a numerical approach to these equations in Chap. 13, but it is instructive to consider linearized versions of the equations—especially if we want to investigate wave propagation, which is the subject of Chap. 11 or the general linear response of beams, which is the subject of Chap. 12. The linear equations will lend some insight into the dynamic behavior of beams and various approximations that have been used to represent that behavior (specifically, Bernoulli–Euler, Rayleigh, and Timoshenko beam theories).

10.6.1 Linearized Kinematics

If we assume that the rotations are small, i.e., $\theta \ll 1$, then $\cos\theta \approx 1$, $\sin\theta \approx \theta$, and the strain–displacement equations (Eqs. 10.10 and 10.8) reduce to the linear form

$$\varepsilon_o = u', \qquad \beta_o = w' + \theta, \qquad \kappa_o = \theta' \qquad (10.23)$$

These assumptions lead to what is commonly called *Timoshenko beam theory*. If we make the assumption that shear strain is negligible, i.e., $\beta_o = 0$, then $\theta = -w'$. Differentiating this relationship once and substituting it into the curvature equation, we get

$$\kappa_o = -w'' \qquad (10.24)$$

This assumption leads to what is commonly called *Bernoulli–Euler beam theory*. In this approximation, the curvature is directly related to the transverse displacement.

10.6.2   Linearized Kinetics

If we assume that the rotations and strains are small, the equations of balance of linear and angular momentum, Eqs. 10.19 and 10.21, reduce to

$$N' + p = \rho A \ddot{u}$$
$$V' + q = \rho A \ddot{w} \qquad (10.25)$$
$$M' - V + m = \rho I \ddot{\theta}$$

The first of these three equations is the axial bar problem which we considered in great detail in the Chap. 9. Note that it is not coupled with the other two equations in the linear theory. The second and third equations are coupled through the shear force. In these two equations, there are two internal force unknowns (V and M) and two motion unknowns (w and θ).

10.6.3   Linear Equations of Motion

Using the linear strain–displacement equations in conjunction with the linear constitutive equations, we can express Eqs. 10.25(b,c) in terms of the motion variables as

$$GA(w'' + \theta') + q = \rho A \ddot{w}$$
$$EI\theta'' - GA(w' + \theta) + m = \rho I \ddot{\theta} \qquad (10.26)$$

This set of equations often goes by the name *Timoshenko beam theory*, and is the most general linear theory that conforms to the plane sections kinematic hypothesis. We can eliminate the shear force from the equations by differentiating Eq. 10.25(c) with respect to x and substituting V' from Eq. 10.25(b). If we assume that shear deformations are negligible, then $\theta = -w'$, and the equation of motion reduces to

$$\left(EIw''\right)'' + \rho A\ddot{w} - \rho I\ddot{w}'' = q + m' \tag{10.27}$$

This equation is generally called *Rayleigh beam theory*. If we also neglect the rotary inertia term $\rho I\ddot{w}''$, then the equation becomes

$$\left(EIw''\right)'' + \rho A\ddot{w} = q + m' \tag{10.28}$$

This equation is generally called *Bernoulli–Euler beam theory*. We will consider classical solution methods for these partial differential equations in the next chapter. It should be evident that the beam bending equations of motion are quite different from the axial wave equations.

10.6.4   Boundary Conditions

For the dynamic beam, we must specify both boundary and initial conditions. Boundary conditions are conditions that are true for all time at the two ends of the beam. We can have any one of four conditions: fixed, slide, simple, or free at either end, as illustrated in Fig. 10.9, with the characteristics listed in Table 10.1.

For the Rayleigh and Bernoulli–Euler beams, the kinematic constraint means that $\theta(\bullet, t) = w'(\bullet, t) = 0$ where rotation is prevented. Noting the constitutive

Fig. 10.9 *Beam boundary conditions.* There are two boundary conditions at each end of the beam. The boundary conditions are the four possibilities shown here. Note that a pin has the same conditions as a roller (i.e., $w = 0$ and $M = 0$) but does not allow motion in the axial direction

Table 10.1 *Beam boundary conditions.* There must be two boundary conditions at each end of the beam. The various combinations are illustrated in Fig. 10.9. Note that $\bullet$ stands for either $x = 0$ or $x = L$

BC 1	BC 2	Description of support
$w(\bullet, t) = 0$	$\theta(\bullet, t) = 0$	Fixed at end $\bullet$
$\theta(\bullet, t) = 0$	$V(\bullet, t) = 0$	Slide at end $\bullet$
$w(\bullet, t) = 0$	$M(\bullet, t) = 0$	Simple at end $\bullet$
$M(\bullet, t) = 0$	$V(\bullet, t) = 0$	Free at end $\bullet$

relationships and the definition of the stress resultant M, we can write the moment conditions (simple and free end) in terms of motion variables as

$$M(\bullet, t) = EI\theta'(\bullet, t)$$

for Timoshenko beam theory or as

$$M(\bullet, t) = -EIw''(\bullet, t)$$

for Rayleigh and Bernoulli–Euler beam theories. For Timoshenko beam theory, we can use the constitutive equation to write the shear boundary condition in terms of motion variables as

$$V(\bullet, t) = GA\big(w'(\bullet, t) + \theta(\bullet, t)\big)$$

For the Rayleigh and Bernoulli–Euler beam, the shear is actually a reaction force because of the kinematic constraint that forces the shear deformation to be zero. Hence, the shear boundary condition in these theories must be obtained from balance of angular momentum as

$$V(\bullet, t) = M'(\bullet, t) + m(\bullet, t) - \rho I \ddot{\theta}(\bullet, t)$$

Substituting the constitutive equation for moment and the displacement derivative for the rotation, the shear boundary conditions can be written in terms of motion variables as

$$V(\bullet, t) = -EIw'''(\bullet, t) + m(\bullet, t) + \rho I \ddot{w}'(\bullet, t) \tag{10.29}$$

The last term is neglected in Bernoulli–Euler beam theory. We will generally write all boundary conditions in terms of the displacement variables in the solution approaches that follow.

It is possible to have a concentrated force or moment applied at the end of a beam. If a concentrated transverse force P is applied at a free (or slide) boundary condition, then the shear boundary condition would be

$$V(\bullet, t) = P(t)$$

A similar modification would be done for a concentrated moment applied at a free or simple boundary condition.

10.6.5 *Initial Conditions*

For the dynamic beam we must specify initial conditions (conditions that are true for all x at time $t = 0$). The initial conditions apply to the displacement and velocity at all points in the beam

$$w(x, 0) = w_o(x), \qquad \dot{w}(x, 0) = v_o(x) \qquad (10.30)$$

where $w_o(x)$ and $v_o(x)$ are known functions of x alone. For the Timoshenko beam we must also specify initial conditions on $\theta(x, t)$. To wit,

$$\theta(x, 0) = \theta_o(x), \qquad \dot{\theta}(x, 0) = \omega_o(x) \qquad (10.31)$$

where $\theta_o(x)$ and $\omega_o(x)$ are known functions of x and are specified as problem inputs.

Chapter 11
Wave Propagation in Linear Planar Beams

In Chap. 10 we derived the equations governing the dynamic response of planar beams. A simpler set of equations emanated from the linearization of the equations of motion. Further, we introduced assumptions to yield *Timoshenko, Rayleigh*, and *Bernoulli–Euler* beam theories. Timoshenko beam theory is the most complete, including rotary inertia effects and shear deformation. Rayleigh beam theory neglects shear deformation but includes rotary inertia. Bernoulli–Euler beam theory neglects both shear deformations and rotary inertia. While these differences are usually not very important in the context of static response, they have significant consequences in dynamics. In this chapter we explore the phenomenon of *wave propagation* in beams through classical solutions of the linearized equations of motion of these three theories.

We first consider the propagation of an infinite train of sinusoidal waves in an infinite beam. This idealized problem gives insight into wave speeds in beams. From this simple analysis we will see how the beam problem differs substantially from the axial bar problem and it gives some clues about phenomena that show up in our subsequent analysis. To deal with the more practical problem of finite beams with end supports, we will consider the classical *separation of variables* approach to solve the differential equations of motion. As it did for the axial bar problem, this approach will show how modal vibration and wave propagation are related. Each of the three beam theories have unique aspects, so all are examined in detail.

The original version of the chapter has been revised. A correction to this chapter can be found at https://doi.org/10.1007/978-3-030-89944-8_15.

Electronic Supplementary Material   The online version of this chapter (https://doi.org/10.1007/978-3-030-89944-8_11) contains supplementary material, which is available to authorized users.

K. D. Hjelmstad, *Fundamentals of Structural Dynamics*, https://doi.org/10.1007/978-3-030-89944-8_11

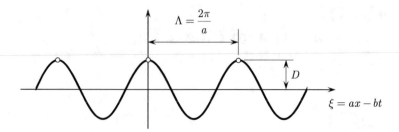

Fig. 11.1 *Propagating wave train.* Consider a beam of infinite extent on which a sinusoidal wave is propagating. The *wave length* is Λ and the amplitude is D

11.1 Propagation of a Train of Sinusoidal Waves

Because of the nature of the differential equations of beam theory, there are no general solutions of the type $w(x, t) = f(x \pm ct)$, as was the case for the axial bar problem. To verify that this is true, just substitute that function back into any of the equations of motion for our three beam theories and observe that there is no way to satisfy them. We can, however, get a sense of wave propagation by considering the motion of a train of *sinusoidal* waves in an infinite beam. To wit, consider a transverse displacement in the form

$$w(x, t) = D \cos(ax - bt) \tag{11.1}$$

where D, a, and b are constants. To get an idea of what those constants mean, consider Fig. 11.1, where $\xi = ax - bt$. The point at the origin (i.e., $\xi = 0$) has the property that $ax - bt = 0$. The actual location x of the point that occupies the origin $\xi = 0$ is

$$x = \frac{b}{a} t \tag{11.2}$$

Therefore, the cosine wave travels to the right with constant speed $c = b/a$. Further, note that at $t = 0$ the second peak is located at $\xi = 2\pi$. Therefore, the *wavelength* is

$$\Lambda = \frac{2\pi}{a} \quad \rightarrow \quad a = \frac{2\pi}{\Lambda} \tag{11.3}$$

The infinite beam is an idealization that requires some suspension of intuitions based on physical reality. A long beam would sag under its own weight. These though experiments, then, must be considered in the absence of gravity forces (which is essentially what $q = 0$ presumes). It is only in such an idealized situation that one can contemplate a beam of infinite extent. Let us examine what this candidate solution implies about our three beam theories.

11.1.1 Bernoulli–Euler Beam

The equation of motion of the Bernoulli–Euler beam with constant area A, bending modulus EI, and no external loads is

$$EIw'''' + \rho A\ddot{w} = 0 \tag{11.4}$$

Let's see if we can find values of the constants D, a, and b for which the equations of motion are satisfied. First, compute the derivatives that appear in the equation. Specifically,

$$w'''' = Da^4 \cos(ax - bt), \qquad \ddot{w} = -Db^2 \cos(ax - bt) \tag{11.5}$$

Substituting these values into Eq. 11.4 gives

$$\left(EIa^4 - \rho Ab^2\right)D\cos\left(ax - bt\right) = 0$$

To satisfy this equation, the leading term in parentheses must be zero. Noting that $b = ca$ and $a = 2\pi/\Lambda$, we can solve for the wave speed c as

$$\frac{c}{c_o} = \frac{1}{\lambda} \tag{11.6}$$

where the reference wave speed c_o, also known as the *longitudinal wave speed*, and the normalized wavelength λ are, respectively,

$$c_o = \sqrt{\frac{E}{\rho}}, \qquad \lambda = \frac{\Lambda}{2\pi k} \tag{11.7}$$

and $k = \sqrt{I/A}$ is the *radius of gyration* of the cross section. Because k and Λ both have units of length, λ is unitless. This result implies that the wave speed is inversely proportional to the wavelength for the Bernoulli–Euler beam. A shorter wave will move faster than a longer wave. Further, in the limit as $\lambda \to 0$, the wave speed approaches infinity. We will see why this is a problem later in the chapter.

11.1.2 Rayleigh Beam

As we did for the Bernoulli–Euler beam, we can examine the propagation of an infinite train of sinusoidal waves in a Rayleigh beam. The equation of motion of the Rayleigh beam (see Eq. 10.27), assuming that there are no transverse loads or moments and that EI is constant, is

$$EIw'''' + \rho A\ddot{w} - \rho I\ddot{w}'' = 0 \tag{11.8}$$

Again, consider a solution in the form of Eq. 11.1. The additional derivative we need beyond those already computed for the Bernoulli–Euler beam (i.e., Eq. 11.5) is

$$\ddot{w}'' = Da^2b^2 \cos(ax - bt)$$

Substituting the derivatives into the equation of motion gives

$$\left(EIa^4 - \rho Ab^2 - \rho Ia^2b^2\right) D \cos(ax - bt) = 0$$

Again, the leading term in parentheses must be zero. Substituting $b = ca$ and $a = 2\pi/\Lambda$ and solving for c gives

$$\frac{c}{c_o} = \frac{1}{\sqrt{1 + \lambda^2}} \tag{11.9}$$

where c_o and λ are defined in Eq. 11.7. Observe that the wave speed for the Rayleigh beam does not go to infinity as the wavelength approaches zero. Rather, it approaches the longitudinal wave speed $c_o = \sqrt{E/\rho}$. As λ gets large, the wave speed for the Rayleigh beam is asymptotic to the wave speed for the Bernoulli–Euler beam.

Effects of Axial Force The presence of axial force in a beam changes the wave propagation and vibration properties. The linear theory does not include this effect, but we can revisit the nonlinear equations to include the term that is responsible for the most prevalent effect of axial force. We will go into much more detail on nonlinear beam problems later in the book, but we will find that we can get some insight about the effects of axial force from the linearized problem.

Starting with Eq. 10.22 and assuming that shear deformation is negligible (i.e., $\theta = -w'$) and that axial strains are small compared to one, balance of angular and linear momentum then take the form

$$M' + Hw' - Q + \rho I \ddot{w}' = 0, \qquad Q' + q = \rho A \ddot{w} \tag{11.10}$$

where H is the force acting in the direction of the undeformed axis of the beam (positive in tension). Eliminate Q by substitution and consider the case where H is constant and $q = 0$. Substitute $M = -EIw''$ to get

$$EIw'''' - Hw'' + \rho A\ddot{w} - \rho I\ddot{w}'' = 0$$

which is the equation of motion including axial force.

Now, reconsider the propagation of the wave train with the axial force term included. Following the same process, noting that $w'' = -Da^2 \cos(ax - bt)$, we find that

$$\frac{c}{c_o} = \sqrt{\frac{1 + \mu}{1 + \lambda^2}}$$

where c_o and λ are defined in Eq. 11.7 and

$$\mu = \frac{H\Lambda^2}{4\pi^2 EI}$$

Note that if H is positive (tensile), then the wave speed is increased over a beam with no axial force. However, if H is negative (compressive), then the wave speed is reduced. In fact, as $\mu \to -1$, the wave speed approaches zero. That value of the axial force is what we would call the *buckling load* in a static beam. It is clear that the longer the wavelength, the smaller the compressive axial force required to make the wave speed zero. A wave speed of zero simply means that waves cannot propagate.

11.1.3 Timoshenko Beam

Consider the propagation of the train of sinusoidal waves for the Timoshenko beam. In order to do that, we must first put the equations of motion into a single equation involving only $w(x, t)$. Assuming constant EI and GA and no applied loads, the equations of motion can be written as

$$GA\left(w'' + \theta'\right) = \rho A \ddot{w}$$
$$EI\theta''' - \rho A \ddot{w} = \rho I \ddot{\theta}' \tag{11.11}$$

We can solve the first of these equations to get

$$\theta' = \frac{\rho}{G}\ddot{w} - w''$$

With this equation, we can compute the derivatives of θ that appear in the second equation

$$\theta''' = \frac{\rho}{G}\ddot{w}'' - w'''', \qquad \ddot{\theta}' = \frac{\rho}{G}\ddddot{w} - \ddot{w}''$$

Substituting these into the second of Eqs. 11.11 gives

$$EI\left(\frac{\rho}{G}\ddot{w}'' - w''''\right) - \rho A \ddot{w} - \rho I\left(\frac{\rho}{G}\ddddot{w} - \ddot{w}''\right) = 0$$

Noting that $\ddddot{w} = Db^4 \cos(ax - bt)$, using the derivatives of w already computed for the other two theories, and dividing through by ρA, this equation can be put into the form

$$\left[\frac{E}{\rho}a^2 k^2\left(\frac{\rho}{G}b^2 - a^2\right) + b^2 - k^2\left(\frac{\rho}{G}b^4 - a^2 b^2\right)\right] D\cos(ax - bt) = 0$$

As before, the term in square brackets must be zero. Substituting the expressions $b = ca$ and $a = 2\pi/\Lambda$, dividing through by $a^4 k^2$, where k is the radius of gyration of the cross section, and simplifying, we get an equation for the wave speed c

$$\left(\frac{\rho}{E}\frac{E}{G}\right) c^4 - \left(1 + \frac{E}{G} + \lambda^2\right) c^2 + \frac{E}{\rho} = 0$$

where λ is defined in Eq. 11.7. Finally, defining $\gamma = c/c_o$, where $c_o = \sqrt{E/\rho}$, and dividing through by E/ρ gives the quartic equation

$$\left(\frac{E}{G}\right) \gamma^4 - \left(1 + \frac{E}{G} + \lambda^2\right) \gamma^2 + 1 = 0 \qquad (11.12)$$

This equation is actually just a quadratic equation for γ^2 and is easily solved. One way to think about the constraint of zero shear deformation is to consider the limit as $G \to \infty$. In that limit, the ratio E/G goes to zero and Eq. 11.12 reverts to the wave-speed equation for the Rayleigh beam.

The wave speeds as a function of wavelength for the Bernoulli–Euler, Rayleigh, and Timoshenko beams are shown in Fig. 11.2. It is evident that the three models give the same result for large wavelengths but depart substantially for small wavelengths. Observe that the Timoshenko beam has two wave speeds corresponding to the possibility of propagating both flexural and shear waves.

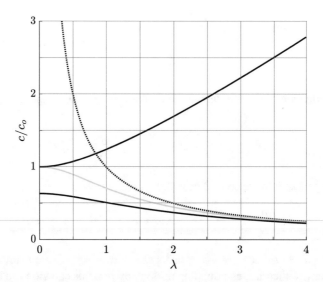

Fig. 11.2 *Wave speeds.* Wave speeds for Bernoulli–Euler (dotted line), Rayleigh (gray line), and Timoshenko (black lines) beams for the propagation of a train of sinusoidal waves on an infinite beam. Note that $c_o = \sqrt{E/\rho}$ and the dimensionless wavelength is $\lambda = \Lambda/2\pi k$, where k is the radius of gyration of the cross section. For the Timoshenko beam this chart has $E/G = 2.5$

11.2 Solution by Separation of Variables

In solving the axial bar problem in Chap. 9, we encountered the strategy of separation of variables. This approach allowed us to break the partial differential equation in x and t into two ordinary differential equations, for which we have a well-developed solution approach. Along the way we encountered eigenvalues and eigenfunctions with orthogonality properties that aided the completion of the solution. The separation of variables technique seeks a solution $w(x, t)$ to the equations of motion in the form

$$w(x, t) = W(x)T(t) \tag{11.13}$$

where $W(x)$ is a pure function of the spatial variable x and $T(t)$ is a pure function of time t. As with the axial bar problem, partial derivatives of w will be ordinary derivatives of W or T.

11.3 The Bernoulli–Euler Beam

The simplest version of beam theory is the Bernoulli–Euler beam. Beyond linearization, this theory assumes that the shearing deformation is zero, thereby introducing a constraint that relates the transverse displacement to the rotation (specifically, $\theta = -w'$). The equation of motion for the Bernoulli–Euler beam (see Eq. 10.28), assuming no applied transverse loads or moments and constant EI, is

$$EI w'''' + \rho A \ddot{w} = 0 \tag{11.14}$$

This equation represents a partial differential equation in the single variable $w(x, t)$. If we substitute Eq. 11.13 into Eq. 11.14, and divide through by EI, we get

$$W''''(x)T(t) + \left(\frac{\rho A}{EI}\right) W(x)\ddot{T}(t) = 0$$

Divide this equation through by $W(x)T(t)$ to get

$$\frac{W''''(x)}{W(x)} = -\left(\frac{\rho A}{EI}\right)\frac{\ddot{T}(t)}{T(t)} = \lambda^4 \tag{11.15}$$

where λ is an as yet unknown constant.[1] The only function that is both a pure function of x and a pure function of t is a constant. This gives rise to two equations:

---

[1] This use of λ here is not to be confused with the normalized wavelength for the analysis of a train of waves in an infinite beam. The context should make the usage clear.

$$W''''(x) - \lambda^4 W(x) = 0, \qquad \ddot{T}(t) + \omega^2 T(t) = 0 \tag{11.16}$$

where, for convenience, we have introduced the constant ω, defined through the relationship

$$\omega^2 = \left(\frac{EI}{\rho A}\right)\lambda^4 \tag{11.17}$$

As we did with the axial bar, we will consider the spatial differential equation first. That will determine the values of λ that work. Once we have λ, we can compute the frequency ω. The reason we call ω the *frequency* is that the solution to the time differential equation is (recall the SDOF system)

$$T(t) = A \cos \omega t + B \sin \omega t \tag{11.18}$$

Thus, the time function oscillates with frequency ω.

For the ordinary differential equation for $W(x)$, we will assume that the solution is in the form of an exponential, i.e., $W(x) = Ce^{sx}$. The derivatives of this function are simple to compute. We only need the fourth derivative, which is $W'''' = s^4 Ce^{sx}$. Substituting back into the governing equation gives

$$\left(s^4 - \lambda^4\right) Ce^{sx} = 0$$

The exponential can never be zero and $C = 0$ is a trivial solution (i.e., no motion). Hence, we must satisfy the characteristic equation $s^4 = \lambda^4$, which has four solutions:

$$s_1 = \lambda, \quad s_2 = -\lambda, \quad s_3 = i\lambda, \quad s_4 = -i\lambda$$

Because we have four roots, we have four independent solutions. The general solution is the sum of those. To wit,

$$W(x) = C_1 e^{\lambda x} + C_2 e^{-\lambda x} + C_3 e^{i\lambda x} + C_4 e^{-i\lambda x}$$

The complex exponential form is not the most convenient one for our purposes. We can change the form of the equation by first noting the Euler identity $e^{ix} = \cos x + i \sin x$, from which we can write

$$\cos \lambda x = \tfrac{1}{2}\left(e^{i\lambda x} + e^{-i\lambda x}\right), \qquad i \sin \lambda x = \tfrac{1}{2}\left(e^{i\lambda x} - e^{-i\lambda x}\right)$$

Rearranging the equation and introducing new constants, we can write an equivalent form of the general solution as

$$W(x) = D_1 e^{-\lambda(L-x)} + D_2 e^{-\lambda x} + D_3 \cos \lambda x + D_4 \sin \lambda x \tag{11.19}$$

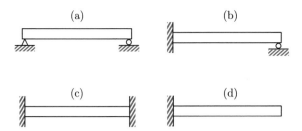

Fig. 11.3 *Boundary conditions for the beam.* A beam of length L, density ρ, modulus E, cross-sectional area A, and moment of inertia I. The variable x is measured from the left end. The boundary conditions cases are (**a**) simple–simple, (**b**) fixed–simple, (**c**) fixed–fixed, and (**d**) fixed–free

The constants D_1, D_2, D_3, and D_4 will be determined for specific cases using the boundary conditions. There are always exactly four boundary conditions, and therefore we have four equations at our disposal. However, there are five unknowns: D_1, D_2, D_3, D_4, and λ. It is straightforward to show that the equations that emanate from the boundary conditions constitute a linear homogeneous system of equations for the constants D_i. The coefficient matrix of that linear system depends upon λ, generally in a nonlinear fashion. The conditions for a nontrivial solution to this system of equations will lead to an equation to determine λ, which we call the *characteristic equation*. The following examples will show how this process works. We will consider the simple–simple, fixed–simple, fixed–fixed, and fixed–free support conditions. The boundary conditions for these cases are illustrated in Fig. 11.3

11.3.1   Implementing Boundary Conditions

As a first example, let us examine the boundary conditions for the *simple–simple* case. Since $M = 0$ implies $w'' = 0$, the four boundary conditions are

$$w(0, t) = 0, \quad w''(0, t) = 0, \quad w(L, t) = 0, \quad w''(L, t) = 0$$

Since $w(x, t) = W(x)T(t)$, these boundary conditions can only be satisfied if

$$W(0) = 0, \quad W''(0) = 0, \quad W(L) = 0, \quad W''(L) = 0$$

From our general expression for $W(x)$ in Eq. 11.19 we can compute the second derivative as

$$W''(x) = \lambda^2 \left(D_1 e^{-\lambda(L-x)} + D_2 e^{-\lambda x} - D_3 \cos \lambda x - D_4 \sin \lambda x \right) \qquad (11.20)$$

To simplify the ensuing computations, we will introduce a shorthand notation. Let

$$\bar{e} = e^{-\lambda L}, \qquad \hat{c} = \cos \lambda L, \qquad \hat{s} = \sin \lambda L$$

The boundary conditions at $x, = 0$ give

$$W(0) = D_1 \bar{e} + D_2 + D_3 = 0$$
$$W''(0) = \lambda^2 (D_1 \bar{e} + D_2 - D_3) = 0$$

If we multiply the first of these equations by λ^2 and subtract the second, we find that $D_3 = 0$. The boundary conditions at $x = L$ give

$$W(L) = D_1 + D_2 \bar{e} + D_3 \hat{c} + D_4 \hat{s} = 0$$
$$W''(L) = \lambda^2 (D_1 + D_2 \bar{e} - D_3 \hat{c} - D_4 \hat{s}) = 0$$

If we multiply the first of these equations by λ^2 and subtract the second, we find that $D_4 \hat{s} = 0$. Substituting this result back into the original equations gives

$$D_1 \bar{e} + D_2 = 0$$
$$D_1 + D_2 \bar{e} = 0$$

It is straightforward to solve these two equations to find $D_1 = D_2 = 0$. Now, three of the four constants have been proven to be zero. The fourth condition, i.e., $D_4 \hat{s} = 0$, written out is

$$D_4 \sin \lambda L = 0 \tag{11.21}$$

There are several ways to satisfy this equation. First, if $D_4 = 0$ then all four of the constants are zero and we have $W(x) = 0$. While this is a valid solution, it implies no motion and is therefore not very interesting. We call such a case a *trivial solution*. Therefore, we are interested in finding solutions for which $D_4 \neq 0$. That is possible only if

$$\sin \lambda L = 0$$

which we call the *characteristic equation*. This equation has an infinite number of solutions $\lambda L = \pi, 2\pi, 3\pi, \ldots$[2] These specific values of λL are called the *eigenvalues* of the boundary value problem. Let

---

[2] Note that $\lambda L = 0$ is a solution, but it also a trivial solution. Also, the negative multiples of π are solutions, but these are redundant because $\sin(-\lambda L) = -\sin(\lambda L)$, which can be absorbed into D_4.

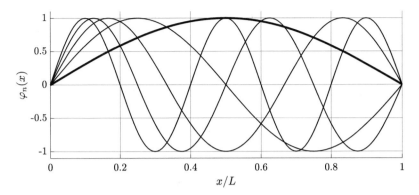

Fig. 11.4 *Mode shapes for the simple–simple Bernoulli–Euler beam.* The first five mode shapes for the simple–simple beam. The first mode is shown with a heavier line. Note that each of these functions satisfy all four boundary conditions

$$\lambda_n = \frac{n\pi}{L} \tag{11.22}$$

for $n = 1, 2, 3, \ldots$. Note that we cannot solve for D_4 (we have run out of boundary conditions), but we can identify the *eigenfunctions* associated with the eigenvalues

$$\varphi_n(x) = \sin \lambda_n x \tag{11.23}$$

The eigenfunctions are also called *mode shapes* for the same reasons we use that term in discrete eigenvalue problems—they capture the shape of the mode, but not the magnitude.

As with the discrete eigenvalue problem, each eigenfunction is associated with an eigenvalue. Unlike the discrete problem, there are an infinite number of eigenpairs. As we have seen in other problems, the modes shapes have orthogonality properties, and that opens up the possibility that we might be able to use the mode shapes to construct a complete solution to the problem, as we did for the axial bar problem.

The first five eigenfunctions (mode shapes) for the simple–simple beam are shown in Fig. 11.4. The first mode is shown with a heavier line. Each subsequent mode has more tortuosity than its predecessors. It should be evident that every mode shape satisfies all four boundary conditions by definition (because we enforced the boundary conditions to get them). Of course, the mode shapes also satisfy the governing differential equation for $W(x)$.

11.3.2 Natural Frequencies

The natural frequency is defined in terms of λ through Eq. 11.17. Now that we have values for λ, we can compute the corresponding natural frequencies. There are an infinite number of natural frequencies defined through the relationship

$$\omega_n^2 = \left(\frac{EI}{\rho A}\right)\lambda_n^4$$

For the simple–simple beam, the values of the frequencies are

$$\omega_n = \left(\frac{n\pi}{L}\right)^2 \sqrt{\frac{EI}{\rho A}} = c_o k \left(\frac{n\pi}{L}\right)^2 \tag{11.24}$$

where $c_o = \sqrt{E/\rho}$ is the longitudinal wave speed and $k = \sqrt{I/A}$ is the radius of gyration of the cross section.

We are now ready to put the complete solution together from the separation of variables assumption $w(x, t) = W(x)T(t)$. Because we got multiple solutions to the problem, the general solution is the linear combination of those solutions. The solution to the temporal part of the problem is

$$T(t) = A \cos \omega t + B \sin \omega t$$

Hence, the general solution to the problem can be written as the infinite series

$$w(x, t) = \sum_{n=1}^{\infty} \varphi_n(x)\left[A_n \cos \omega_n t + B_n \sin \omega_n t\right] \tag{11.25}$$

Each of the spatial modes gives rise to a frequency ω_n and those frequencies are associated with the mode shapes $\varphi_n(x)$. The values of the constants A_n and B_n will be determined from the initial conditions.

11.3.3   Orthogonality of the Eigenfunctions

The eigenfunctions have the property that they are orthogonal to each other. The proof goes as follows. The eigenfunctions satisfy the differential equation

$$\varphi_n''''(x) - \lambda_n^4 \varphi_n(x) = 0$$

and all boundary conditions. Multiply this equation by $\varphi_m(x)$ and integrate over the length

$$\int_0^L \left(\varphi_n'''' - \lambda_n^4 \varphi_n\right) \varphi_m \, dx = 0$$

Integrate the first term by parts *twice* to get

$$\int_0^L \left(\varphi_n'' \varphi_m'' - \lambda_n^4 \, \varphi_n \varphi_m \right) dx + \varphi_n''' \varphi_m \Big|_0^L - \varphi_n'' \varphi_m' \Big|_0^L = 0$$

The boundary terms vanish because the eigenfunctions satisfy all of the boundary conditions. For example, the simple boundary conditions have the conditions $\varphi(\bullet) = 0$ and $\varphi''(\bullet) = 0$ at both ends. In fact, the boundary terms that appear due to integration by parts vanish for *any* boundary conditions. Hence, we can write

$$\int_0^L \left(\varphi_n'' \varphi_m'' - \lambda_n^4 \, \varphi_n \varphi_m \right) dx = 0 \tag{11.26}$$

We can follow the same steps starting with the mth eigenvalue problem and multiplying by the nth eigenfunction, with the result

$$\int_0^L \left(\varphi_m'' \varphi_n'' - \lambda_m^4 \, \varphi_m \varphi_n \right) dx = 0 \tag{11.27}$$

Subtracting Eq. 11.27 from Eq. 11.26 gives

$$\left(\lambda_m^4 - \lambda_n^4 \right) \int_0^L \varphi_m(x) \varphi_n(x) \, dx = 0$$

Since $\lambda_n \neq \lambda_m$, the eigenfunctions must satisfy

$$\int_0^L \varphi_m(x) \varphi_n(x) \, dx = 0, \qquad n \neq m \tag{11.28}$$

i.e., they are orthogonal. It is also straightforward to prove that the second derivatives of the eigenfunctions are also orthogonal. Return to Eq. 11.26 or 11.27 and note that the eigenfunctions are orthogonal (and that $\lambda_n \neq 0$). The equation reduces to

$$\int_0^L \varphi_m''(x) \varphi_n''(x) \, dx = 0, \qquad n \neq m \tag{11.29}$$

That completes the proof of orthogonality of the eigenfunctions. Orthogonality of the eigenfunctions plays a key role in developing the complete solution of the differential equation.

11.3.4   Implementing the Initial Conditions

Implementing the initial conditions to find the constants in Eq. 11.25 depends crucially on the orthogonality of the eigenfunctions because the solution is in the form of an infinite series. The general solution to the differential equation of motion for the Bernoulli–Euler beam is (see Eq. 11.25)

$$w(x, t) = \sum_{n=1}^{\infty} \varphi_n(x) \left[A_n \cos \omega_n t + B_n \sin \omega_n t \right]$$

and the first derivative with respect to time is

$$\dot{w}(x, t) = \sum_{n=1}^{\infty} \varphi_n(x) \, \omega_n \left[-A_n \sin \omega_n t + B_n \cos \omega_n t \right]$$

The initial conditions require that

$$w(x, 0) = \sum_{n=1}^{\infty} \varphi_n(x) A_n = w_o(x)$$

$$\dot{w}(x, 0) = \sum_{n=1}^{\infty} \varphi_n(x) \, \omega_n B_n = v_o(x)$$

where $w_o(x)$ is the initial displacement and $v_o(x)$ is the initial velocity. Multiplying both sides of the equations by $\varphi_m(x)$ and integrating from 0 to L gives, in the first case,

$$\int_0^L \varphi_m(x) \sum_{n=1}^{\infty} \varphi_n(x) A_n \, dx = \int_0^L \varphi_m(x) w_o(x) \, dx$$

and in the second case,

$$\int_0^L \varphi_m(x) \sum_{n=1}^{\infty} \varphi_n(x) \, \omega_n B_n \, dx = \int_0^L \varphi_m(x) v_o(x) \, dx$$

On the left side of both of these equations, we can move the integral inside the sum and use orthogonality to find that only the mth term is nonzero. The constants can then be computed as (relabeling the index m to n in the final result for future reference)

$$A_n = \frac{\int_0^L \varphi_n(x) w_o(x) \, dx}{\int_0^L \varphi_n^2(x) \, dx}, \qquad B_n = \frac{\int_0^L \varphi_n(x) v_o(x) \, dx}{\omega_n \int_0^L \varphi_n^2(x) \, dx} \qquad (11.30)$$

which holds for all values of n. With the constants A_n and B_n determined, we have a complete solution to the problem of wave propagation in the beam subjected to initial displacements or velocities. Even though there are an infinite number of terms, we can compute each one independent of the others. If the values of A_n and B_n are small for large values of n, then the infinite series can be reasonably approximated by a finite number of terms.

11.3.5 Modal Vibration

Consider the very special case where the initial velocity is zero and the initial displacement is proportional to the mth eigenfunction, $w_o(x) = W_o \varphi_m(x)$. From Eq. 11.30(b) we find that all $B_n = 0$ and from Eq. 11.30(a)

$$A_n = \begin{cases} W_o & n = m \\ 0 & n \neq m \end{cases}$$

Therefore, the complete solution to the problem is

$$w(x, t) = W_o \varphi_m(x) \cos \omega_m t \tag{11.31}$$

From this equation it is evident that the beam simply vibrates in the mth mode, with amplitude W_o.

As was the case for the axial bar, modal vibration can be interpreted as waves propagating back and forth across the beam, with the net result of reflections from the boundaries being an oscillating wave in the form of the mth mode shape. To see this in the case of the simple–simple beam, note that we can write the modal vibration solution as

$$w(x, t) = W_o \sin \lambda_m x \cos \omega_m t$$

Noting the trigonometric identity $2 \sin a \cos b = \sin(a + b) + \sin(a - b)$, we can write

$$w(x, t) = \tfrac{1}{2} W_o \big(\sin(\lambda_m x + \omega_m t) + \sin(\lambda_m x - \omega_m t) \big)$$

which shows that the initial wave splits in two, with half propagating to the right and half to the left at a speed of

$$c_m = \frac{\omega_m}{\lambda_m} = c_o k \frac{m \pi}{L}$$

Observe that the wave speed is proportional to m, showing that the higher modes propagate with very high speeds, which is not very reasonable from a physical point of view. We will be able to see how this feature manifests when we look at wave propagation using the series solution of Eq. 11.25.

11.3.6   Other Boundary Conditions

The simple–simple beam results in a deceptively simple form for the eigenvalues
and eigenfunctions. In this section we examine some other cases where the
outcome is a little more complex. Specifically, we will compute the eigenvalues
and eigenfunctions for the fixed–simple, fixed–fixed, and fixed–free beams. The
boundary conditions for these cases are shown in Fig. 11.3

Fixed–Simple Beam   Consider the fixed–simple beam shown in Fig. 11.3b. The
boundary conditions are

$$W(0) = 0, \quad W'(0) = 0, \quad W(L) = 0, \quad W''(L) = 0$$

The first and second derivatives of $W(x)$ are

$$W'(x) = \left(D_1 e^{-\lambda(L-x)} - D_2 e^{-\lambda x} - D_3 \sin \lambda x + D_4 \cos \lambda x \right) \lambda$$

$$W''(x) = \left(D_1 e^{-\lambda(L-x)} + D_2 e^{-\lambda x} - D_3 \cos \lambda x - D_4 \sin \lambda x \right) \lambda^2$$

The boundary conditions at $x = 0$ give

$$W(0) = D_1 \bar{e} + D_2 + D_3 = 0$$

$$W'(0) = \lambda \left(D_1 \bar{e} - D_2 + D_4 \right) = 0$$

where, for short, $\bar{e} = e^{-\lambda L}$. It is straightforward to solve these two equations
to find $D_3 = -(D_2 + D_1\bar{e})$ and $D_4 = (D_2 - D_1\bar{e})$. The conditions $W(L) = 0$ and
$W''(L) = 0$ give

$$W(L) = D_1 + D_2 \bar{e} + D_3 \hat{c} + D_4 \hat{s} = 0$$

$$W''(L) = \lambda^2 \left(D_1 + D_2 \bar{e} - D_3 \hat{c} - D_4 \hat{s} \right) = 0$$

where, for short, $\hat{c} = \cos \lambda L$ and $\hat{s} = \sin \lambda L$. Dividing the second of these equations
through by λ^2 and adding the two equations gives $D_1 + D_2 \bar{e} = 0$. Substituting this
result and the expressions for D_3 and D_4 into the first equation gives

$$D_2 \left[(1 + \bar{e}^2) \hat{s} - (1 - \bar{e}^2) \hat{c} \right] = 0$$

If $D_2 = 0$ then all four of the constants are zero. If $D_2 \neq 0$ then, we get the
characteristic equation $(1 + \bar{e}^2) \hat{s} - (1 - \bar{e}^2) \hat{c} = 0$, which can be rewritten as

$$\tan \lambda L = \tanh \lambda L \tag{11.32}$$

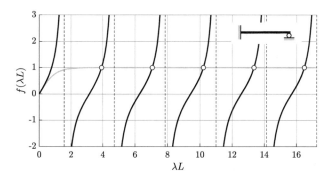

Fig. 11.5 *Characteristic equation for the fixed–simple Bernoulli–Euler beam.* The values of λ for the fixed–simple beam are the solutions to the equation $\tan \lambda L = \tanh \lambda L$

by noting that $\tanh \lambda L = (1 - \bar{e}^2)/(1 + \bar{e}^2)$. Figure 11.5 has a plot of the functions $\tan \lambda L$ and $\tanh \lambda L$. The point of intersection of these curves solves Eq. 11.32 (shown as open circles in the plot). This equation can be solved with Newton's method with residual and tangent

$$g(y) = \tan y - \tanh y, \qquad g' = \frac{1}{\cos^2 y} - \frac{1}{\cosh^2 y}$$

where $y = \lambda L$. The starting point for the nth root of $y = (4n + 1)\pi/4$ converges for all n. It is evident that the function $\tanh y$ levels off at one very quickly as y grows. Therefore, the roots are almost exactly equal to $y = (4n + 1)\pi/4$ for n greater than two. The only root that departs from this value is the first one $y_1 = 3.92660$, albeit only slightly.

Knowing the eigenvalues λ_n, we can find the eigenfunctions $\varphi_n(x)$. The magnitude of the eigenfunction is arbitrary. Thus, we can assign the value of $D_2 = 1$, which then determines all of the others (from the boundary condition equations). To wit,

$$D_1^{(n)} = -\bar{e}_n, \quad D_2^{(n)} = 1, \quad D_3^{(n)} = -(1 - \bar{e}_n^2), \quad D_4^{(n)} = 1 + \bar{e}_n^2$$

where $\bar{e}_n = e^{-\lambda_n L}$ and the notation $D_i^{(n)}$ simply means the value of D_i evaluated using λ_n. Hence, the constants have different values for each mode n. Substituting these constants back into the expression for $W(x)$ gives the eigenfunction

$$\varphi_n(x) = D_1^{(n)} e^{-\lambda_n(L-x)} + D_2^{(n)} e^{-\lambda_n x} + D_3^{(n)} \cos \lambda_n x + D_4^{(n)} \sin \lambda_n x$$

The first five mode shapes are shown in Fig. 11.6. It is evident that each function satisfies the boundary conditions (verify this analytically!).

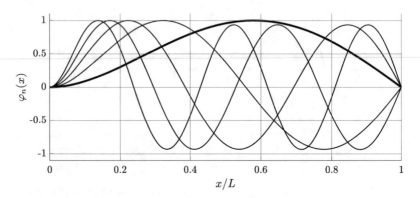

Fig. 11.6 *Mode shapes for the fixed–simple beam.* The first five eigenfunctions are plotted. The functions are normalized to have maximum value of one. The heavier line is the first mode

Fixed–Fixed Beam Consider the fixed–fixed beam shown in Fig. 11.3c. The boundary conditions are

$$W(0) = 0, \quad W'(0) = 0, \quad W(L) = 0, \quad W'(L) = 0$$

From our general expression for $W(x)$, we find that the conditions at $x = 0$ imply that

$$W(0) = D_1\bar{e} + D_2 + D_3 = 0$$
$$W'(0) = \lambda\left(D_1\bar{e} - D_2 + D_4\right) = 0$$

as before. Again, $D_3 = -(D_2 + D_1\bar{e})$ and $D_4 = (D_2 - D_1\bar{e})$. The conditions $W(L) = 0$ and $W'(L) = 0$ give

$$W(L) = D_1 + D_2\bar{e} + D_3\hat{c} + D_4\hat{s} = 0$$
$$W'(L) = \lambda\left(D_1 - D_2\bar{e} - D_3\hat{s} + D_4\hat{c}\right) = 0$$

where, again, $\bar{e} = e^{-\lambda L}$, $\hat{c} = \cos \lambda L$ and $\hat{s} = \sin \lambda L$. Divide the second of these equations through by λ and substitute the expressions for D_3 and D_4 to get

$$\begin{bmatrix} 1 - \bar{e}(\hat{c} + \hat{s}) & \bar{e} - (\hat{c} - \hat{s}) \\ 1 - \bar{e}(\hat{c} - \hat{s}) & -\bar{e} + (\hat{c} + \hat{s}) \end{bmatrix} \begin{Bmatrix} D_1 \\ D_2 \end{Bmatrix} = \begin{Bmatrix} 0 \\ 0 \end{Bmatrix}$$

If $D_1 = 0$ and $D_2 = 0$ then all four of the constants are zero. A nonzero solution exists only if the determinant of the coefficient matrix is zero. Setting the determinant to zero gives *characteristic equation* $(1 + \bar{e}^2)\hat{c} - 2\bar{e} = 0$, which can be written out as

$$\left(1 + e^{-2\lambda L}\right) \cos \lambda L = 2e^{-\lambda L} \tag{11.33}$$

This equation can be solved by Newton's method with residual and tangent

$$g(y) = \left(1 + \bar{e}^2\right) \cos y - 2\bar{e}, \quad g'(y) = -2\bar{e}^2 \cos y - \left(1 + \bar{e}^2\right) \sin y + 2\bar{e}$$

where $y = \lambda L$ and $\bar{e}(y) = e^{-y}$. A good starting point for finding the nth root is $y = (2n + 1)\pi/2$. The solution of the characteristic equation yields values 4.7300, 7.8532, and 10.996 for the first three roots. The values are very close to $y = (2n + 1)\pi/2$ for $n > 3$.

Knowing the eigenvalues λ_n, we can find the eigenfunctions $\varphi_n(x)$. We can assign the value of $D_2 = 1 - \bar{e}(\hat{c} + \hat{s})$, which then determines all the others as

$$D_1^{(n)} = -\bar{e}_n + \hat{c}_n - \hat{s}_n \qquad D_2^{(n)} = 1 - \bar{e}_n(\hat{c}_n + \hat{s}_n)$$
$$D_3^{(n)} = -1 + \bar{e}_n^2 + 2\bar{e}_n\hat{s}_n \qquad D_4^{(n)} = 1 + \bar{e}_n^2 - 2\bar{e}_n\hat{c}_n$$

where $\bar{e}_n = e^{-\lambda_n L}$, $\hat{c}_n = \cos \lambda_n L$, and $\hat{s}_n = \sin \lambda_n L$. Substituting these constants back into the expression for $W(x)$ gives the eigenfunction, as before. The first five mode shapes are shown in Fig. 11.7.

Fixed–Free Beam As a last example, consider the fixed–free beam shown in Fig. 11.3d. The boundary conditions are

$$W(0) = 0, \quad W'(0) = 0, \quad W''(L) = 0, \quad W'''(L) = 0$$

because $M = -EIw''$ and $V = M'$. The third derivative of $W(x)$ is

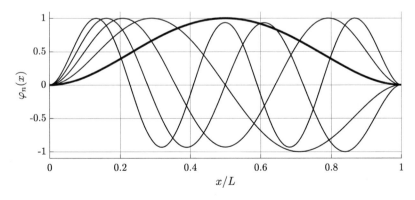

Fig. 11.7 *Mode shapes for the fixed–fixed Bernoulli–Euler beam. The first five eigenfunctions are plotted. The functions are normalized to have maximum value of one. The heavier line is the first mode*

$$W'''(x) = \left(D_1 e^{-\lambda(L-x)} - D_2 e^{-\lambda x} + D_3 \sin \lambda x - D_4 \cos \lambda x\right)\lambda^3$$

The equations emanating from the boundary conditions are

$$D_1 \bar{e} + D_2 + D_3 = 0$$

$$D_1 \bar{e} - D_2 + D_4 = 0$$

$$D_1 + D_2 \bar{e} - D_3 \hat{c} - D_4 \hat{s} = 0$$

$$D_1 - D_2 \bar{e} + D_3 \hat{s} - D_4 \hat{c} = 0$$

The process of solving these equations is similar to the previous cases, yielding the *characteristic equation* $(1 + \bar{e}^2)\hat{c} + 2\bar{e} = 0$, which can be written out as

$$\left(1 + e^{-2\lambda L}\right) \cos \lambda L + 2e^{-\lambda L} = 0 \tag{11.34}$$

This equation is very similar to the fixed–fixed beam, and can be solved by Newton's method with residual and tangent

$$g(y) = \left(1 + \bar{e}^2\right) \cos y + 2\bar{e}, \quad g'(y) = -2\bar{e}^2 \cos y - \left(1 + \bar{e}^2\right) \sin y - 2\bar{e}$$

where, as before, $y = \lambda L$ and $\bar{e}(y) = e^{-y}$. A good starting point for finding the nth root is $y = (2n - 1)\pi/2$. The solution of the characteristic equation yields values 1.8751, 4.6941, and 7.8548 for the first three modes. The values are very close to $y = (2n - 1)\pi/2$ for $n > 3$. Knowing the eigenvalues λ_n, we can find the eigenfunctions $\varphi_n(x)$. The magnitude of the eigenfunction is arbitrary. Thus, we can assign the value of $D_2 = 1 + \bar{e}(\hat{c} + \hat{s})$, which then determines all of the others as

$$D_1^{(n)} = -\bar{e}_n - \hat{c}_n + \hat{s}_n \quad D_2^{(n)} = 1 + \bar{e}_n(\hat{c}_n + \hat{s}_n)$$

$$D_3^{(n)} = -1 + \bar{e}_n^2 - 2\bar{e}_n \hat{s}_n \quad D_4^{(n)} = 1 + \bar{e}_n^2 + 2\bar{e}_n \hat{c}_n$$

where $\bar{e}_n = e^{-\lambda_n L}$, $\hat{c}_n = \cos \lambda_n L$, and $\hat{s}_n = \sin \lambda_n L$. Substituting these constants back into the expression for $W(x)$ gives the eigenfunction. The first five mode shapes are shown in Fig. 11.8.

Implementation The computations associated with finding the eigenvalues and eigenfunctions of the Bernoulli–Euler beam for these four boundary conditions are contained in Code 11.1. For each different value of BC, the code solves the characteristic equation by Newton's method, using the starting values mentioned in the previous sections, and establishes the values of the eigenfunction constants D_1, D_2, D_3, and D_4 for each mode. The eigenvalues are passed back in the array Ln and the eigenfunction constants in the array D.

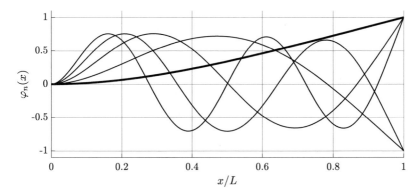

Fig. 11.8 *Mode shapes for the fixed–free Bernoulli–Euler beam.* The first five eigenfunctions are plotted. The functions are normalized to have maximum value of one. The heavier line is the first mode

Code 11.1   MATLAB code to find the eigenvalues and the coefficients of the eigenfunction $\varphi_n(x)$ for a Bernoulli–Euler beam with different boundary conditions

```
function [Ln,D] = BEEigenproblem(nModes,BC)
%  Eigenvalues and eigenfunctions for the Bernoulli-Euler beam
%     nModes : Number of modes
%         BC : Boundary condition flag

%. Initialize arrays
   Ln = zeros(nModes,1);      % Eigenvalues (Lambda*L)
   D = zeros(nModes,4);       % Coefficients for Eigenfunctions

%. Set tolerance for Newton's method
   tol = 1.e-8;

   switch BC

     case 1   % Simple-Simple
        for n=1:nModes
          Ln(n) = n*pi;
          D(n,:) = [0,0,0,1];
        end

     case 2   % Fixed-Simple
        for n=1:nModes
          y = (4*n-1)*pi/4;   err = 1; its = 0;
          while err>tol
            g = tan(y) - tanh(y);
            Dg = 1/cos(y)^2 - 1/cosh(y)^2;
            y = y - g/Dg;
            err = abs(g);   its = its + 1;
          end
          ebar = exp(-y);
          Ln(n) = y;
          D(n,:)=[-ebar, 1, -(1-ebar^2), (1+ebar^2)];
        end

     case 3   % Fixed-Fixed
        for n=1:nModes
          y = (2*n+1)*pi/2;   err = 1;   its = 0;
          while err>tol
            ebar = exp(-y);
```

```
        g = (1+ebar^2)*cos(y)  - 2*ebar;
        Dg = -2*ebar^2*cos(y)  - (1+ebar^2)*sin(y) + 2*ebar;
        y = y - g/Dg;
        err = abs(g); its = its + 1;
      end
      ebar = exp(-y); chat = cos(y); shat = sin(y);
      Ln(n) = y;
      D(n,:) = [-ebar + chat  - shat,...
                1 - ebar*(chat + shat),...
                -1 + ebar^2 + 2*shat*ebar,...
                1 + ebar^2 - 2*chat*ebar];
    end

  case 4    % Fixed-Free
    for n=1:nModes
      y = (2*n-1)*pi/2;   err = 1;   its = 0;
      while err>tol
        ebar = exp(-y);
        g = (1+ebar^2)*cos(y) + 2*ebar;
        Dg = -2*ebar^2*cos(y)  - (1+ebar^2)*sin(y) - 2*ebar;
        y = y - g/Dg;
        err = abs(g); its = its + 1;
      end
      ebar = exp(-y); chat = cos(y); shat = sin(y);
      Ln(n) = y;
      D(n,:) = [-ebar - chat + shat,...
                1 + ebar*(chat + shat),...
                -1 + ebar^2 - 2*ebar*shat,...
                1 + ebar^2 + 2*ebar*chat];
    end

  end
end
```

11.3.7 Wave Propagation

While the eigenvalues (and, by association, natural frequencies) and eigenfunctions are important in their own right for the beam, the main reason for computing them is to build the solution to the wave propagation problem. The first step in this process is to compute the eigenfunctions $\varphi_n(x)$, based upon $\lambda_n L$ and the coefficients D_1, D_2, D_3, and D_4 associated with the boundary conditions. The second step is to compute the coefficients A_n of the initial wave $w_o(x)$ (and B_n if there are initial velocities). We will consider the specific initial displacement function

$$
w_o(\xi) = \begin{cases} 0 & \xi < 0 \\ \left(4\xi(1-\xi)\right)^4 & 0 < \xi < 1 \\ 0 & \xi > 1 \end{cases} \tag{11.35}
$$

where $\xi = (x-a)/(b-a)$ with $x = a$ being the location of the start of the nonzero portion of the wave and $x = b$ being the end. The function has a maximum value of one. The values could be easily scaled, but since the codes normalize the displacements for plotting, we have not included the scale factor. Code 11.2 does the computations associated with computing the eigenfunctions and the coefficients.

Code 11.2 MATLAB code to find the eigenfunction and coefficients A_n associated with an initial wave $w_o(x)$ for a Bernoulli–Euler beam. The boundary conditions are implicit in the values of Ln and D passed in

```
function [Phi,ddPhi,A] = BEModeShapes(Ln,D,nModes,y)
%   Compute eigenvectors and coefficients of initial wave, BE beam
%        Ln = Lambda_n*L
%         D = Coefficients of eigenfunctions
%    nModes = Number of modes
%         y = Spatial grid of points

%. Initialize arrays
    nPts = size(y,2);               % Number of grid points
    Phi = zeros(nModes,nPts);       % Mode shapes
    ddPhi = zeros(nModes,nPts);     % 2nd derivative of mode shapes
    A = zeros(nModes,2);            % Coefficients of wave function

%. Initial wave function
    a = 0.4;                        % Start of initial wave
    b = 0.6;                        % End of initial wave
    N = 4;                          % Exponent for waveform
    w =@(x) (4.*x.*(1-x)).^N;       % Wave function
    z =@(x) (b-a)*x + a;            % z(x) maps [0,1] -> [a,b]

    for n=1:nModes

%... Form functions to make mode shapes
        s1 =@(xi) D(n,1)*exp(-Ln(n)*(1-xi));
        s2 =@(xi) D(n,2)*exp(-Ln(n)*xi);
        s3 =@(xi) D(n,3)*cos(Ln(n)*xi);
        s4 =@(xi) D(n,4)*sin(Ln(n)*xi);

%... Create the mode shape functions
        phi   =@(xi)  s1(xi) + s2(xi) + s3(xi) + s4(xi);
        ddphi =@(xi) (s1(xi) + s2(xi) - s3(xi) - s4(xi))*Ln(n)^2;

%... Put in numerical values for Phi and ddPhi
        Phi(n,:) = phi(y);
        ddPhi(n,:) = ddphi(y);

%... Compute the initial displacement coefficient
        NormFcn =@(x) phi(x).*phi(x);
        WaveFcn =@(x) w(x).*phi(z(x))*(b-a);
        aa = integral(NormFcn,0,1);
        bb = integral(WaveFcn,0,1);
        A(n,1) = bb/aa;
        A(n,2) = Ln(n);

    end

end
```

Finally, the complete solution to the problem is accomplished by modal summation, i.e., implementing the calculation of displacement and moment (see Eq. 11.25 with zero initial velocity). To wit,

$$w(x,t) = \sum_{n=1}^{N} \varphi_n(x) A_n \cos \omega_n t$$

$$M(x, t) = -EI \sum_{n=1}^{N} \varphi_n''(x) A_n \cos \omega_n t$$

where N is the number of modes included in the sum (nModes in the code). Code 11.3 executes this computation. This code produced Figs. 11.9 and 11.10. The computational core of this program in the function BEModalSum, which computes the response at various times, calling BEEigenproblem and BEModeShapes in the process. There are several graphical output functions that are not included here, but they generate the figures shown in the examples (e.g., PlotModes generated Figs. 11.4, 11.6, 11.7, and 11.8).

Code 11.3 MATLAB code to generate the modal solution of the Bernoulli–Euler beam with an initial displacement $w_o(x)$. The boundary conditions are implicit in the eigenfunctions

```matlab
%   Bernoulli-Euler Beam Wave Modal Analysis
    clear; clc;

%.  Problem parameters
    nModes = 150;                     % Number of modes
    L = 10.0;                         % Length of beam
    Density = 0.6;                    % Mass density per unit volume
    E = 80;                           % Young's modulus
    Area = 0.05;                      % Cross-sectional area
    Inertia = 0.2;                    % Moment of inertia
    k = sqrt(Inertia/Area);          % Radius of gyration
    co = sqrt(E/Density);            % Nominal wave speed
    EI = 1;                           % Bending stiffness (only for M)
    BC = 1;                           % =1 S-S, =2 Fx-S, =3 Fx-Fx, =4 Fx-Fr

%.  Compute and plot results
%   Start time      Stop time      Time steps       Spatial points
    T.to = 0;       T.tf = 0.1;    T.nT = 61;       T.nS = 250;
    [Time,x,w,M,An,Freq,Phi] = BEModalSum(T,L,co,k,EI,nModes,BC);

    PlotModes(Phi,5);
    PlotInitialWave(x,w,M);
    PlotCoefficients(An,Freq,k,L,nModes);
    PlotSnapshots(x,w,M,Time,T);
    AnimateBeam(T,x,w,M,Time)

%- End of main program -------------------------------------------------

function [Time,x,w,M,A,Freq,Phi] = BEModalSum(T,L,co,k,EI,nModes,BC)
%   Compute w and M by modal sum for Bernoulli-Euler beam
%           T = Struct with time parameters
%           L = Length of beam
%          co = Nominal wave speed
%           k = Radius of gyration
%          EI = Bending stiffness
%      nModes = Number of modes
%          BC = Boundary condition code

%.  Initialize storage
    x = linspace(0,1,T.nS);                      % Spatial grid points
    w = zeros(T.nT,T.nS);                        % Initialize w(x,t)
    M = zeros(T.nT,T.nS);                        % Initialize M(x,t)
    Time = linspace(T.to,T.tf,T.nT)';            % Time grid points

%.  Compute eigenvalues, mode shape coefficients, and frequencies
    [Ln,D] = BEEigenproblem(nModes,BC);
```

```
    Freq = co*k*(Ln/L).^2;
%. Compute mode shapes and coefficients for initial wave
    [Phi,ddPhi,A] = BEModeShapes(Ln,D,nModes,x);
%. Compute response time history
    for i=1:T.nT
      t = Time(i);
      for n=1:nModes
        CosWt = cos(Freq(n)*t);
        w(i,:) = w(i,:) + Phi(n,:)*A(n,1)*CosWt;
        M(i,:) = M(i,:) - ddPhi(n,:)*A(n,1)*CosWt*EI;
      end
    end
end
```

11.3.8 Example: Simple–Simple Beam

Let us examine the response of a simple–simple beam of length $L = 10$, $E = 80$, $\rho = 0.6$, $A = 0.05$, and $I = 0.2$, subjected to an initial displacement $w_o(x)$ defined by Eq. 11.35 with $a = 4$ and $b = 6$. We use 150 modes in the analysis, an analysis time of $t = 0.006$ s, and a spatial grid with 250 points.

The propagation of the initial wave is shown in Fig. 11.9. Note that we are looking at the very early response of the beam—the first 0.006 s of motion. The

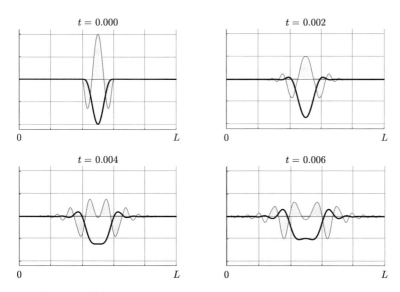

Fig. 11.9 *Bernoulli–Euler beam wave evolution.* Simple–simple beam of length $L = 10$, $E = 80$, $\rho = 0.6$, $A = 0.05$, and $I = 0.2$. The four plots show the displacement (dark line) and moment (thin line with shading) at four different times

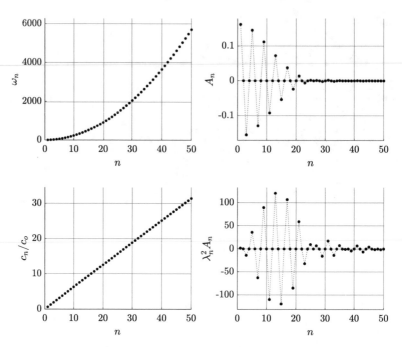

Fig. 11.10 *Bernoulli–Euler beam wave propagation—modal components.* Simple–simple beam of Fig. 11.11. Top left is the natural frequencies, bottom left is the wave speed of the nth component, top right is the coefficients for w, and the bottom right is the coefficients for M

reason for taking this early segment is to try to capture the dispersion of the moment wave caused by the nature of the modal wave speeds of Bernoulli–Euler theory that we first observed in the analysis of the train of sinusoidal waves at the beginning of the chapter. The shape of the initial displacement $w_o(x)$ is evident in the plot for $t = 0$. The associated moment is also shown. As time progresses, dispersion of the initial wave manifests very quickly. This figure clearly shows the higher frequency modal waves racing out in front of the main wave, indicating that they are associated with higher wave speeds. Running the analysis for a longer time period shows that the wave propagation problem quickly deteriorates into chaotic motion as the dispersed waves bounce off the boundaries and interact with each other. It is difficult to discern a wave speed for the Bernoulli–Euler beam.

Figure 11.10 plots the natural frequencies ω_n, the relative modal wave speeds c_n/c_o, the modal displacement coefficients A_n, and the modal moment coefficients $\lambda_n^2 A_n$ against the mode number n. This figure gives some hints as to why the Bernoulli–Euler beam behaves the way it does. First, note the quadratic growth of the natural frequencies. In the next two sections we will contrast that with the approximately linear growth in frequencies for the Rayleigh and Timoshenko beams. Second, observe that the modal wave speeds grow linearly with mode number n. They very quickly exceed the nominal wave speed by an extraordinary amount (we are only showing 50 of the 150 modes used in the analysis). The plot of the modal coefficients A_n indicate that the main contributions come from the first 20 modes. Even in those first 20 modes, the wave speeds get very large. That explains why dispersion happens so fast.

This example shows why Bernoulli–Euler beam theory is not useful for wave propagation problems. It is the simplest of the three theories, and it has long been a staple of static structural analysis. We include this deficient theory here because it is the stopping point of many books on structural dynamics—the *only* theory covered (usually only up to the computation of modes and frequencies). It is a cautionary tale that makes the pursuit of the other two theories more compelling.

11.4   The Rayleigh Beam

The difference between the Bernoulli–Euler beam and the Rayleigh beam is that the former neglects the rotary inertia of the cross section while latter does not. We will proceed along the same lines that we did for the Bernoulli–Euler beam to see how these two theories compare. For the case with EI constant and no external loads, i.e., $q = 0$ and $m = 0$, the equation of motion of the Rayleigh beam reduces to

$$EIw'''' + \rho A\ddot{w} - \rho I\ddot{w}'' = 0$$

Further, consider the possibility of expressing the solution to this equation in the form $w(x, t) = W(x)T(t)$, i.e., separation of variables. Substituting this expression into the equation gives

$$EI\, W''''T + \rho A\left(W - k^2 W''\right)\ddot{T} = 0$$

where $k = \sqrt{I/A}$ is the radius of gyration of the cross section. Now we can move the $\ddot{T}$ terms to the right side and divide through by $EI\left(W - k^2 W''\right)T$ to get

$$\frac{W''''}{W - k^2 W''} = -\left(\frac{\rho A}{EI}\right)\frac{\ddot{T}}{T} = \lambda^4$$

where λ is a constant yet to be determined. As was the case for the Bernoulli–Euler beam, both sides of this equation must equal a constant because that is the only way a pure function of x can be equal to a pure function of t. The consequence of this observation is that we have two equations

$$W'''' + \lambda^4 k^2 W'' - \lambda^4 W = 0, \qquad \ddot{T} + \omega^2 T = 0 \qquad (11.36)$$

where, as before, the square of the natural frequencies in the temporal equation is defined as

$$\omega^2 = \left(\frac{EI}{\rho A}\right)\lambda^4 \tag{11.37}$$

The value of λ is determined by the boundary conditions. Those values then determine the frequencies ω.

Equation 11.36(a) is an ordinary differential equation with constant coefficients. Therefore, the solution should be in the form $W(x) = Ce^{sx}$, where C and s are as yet undetermined constants. Substituting the candidate solution into the equation of motion results in

$$\left(s^4 + \lambda^4 k^2 s^2 - \lambda^4\right)C\,e^{st} = 0$$

The only possible nontrivial solution comes from the *characteristic equation* for s

$$s^4 + \lambda^4 k^2 s^2 - \lambda^4 = 0$$

This equation is quadratic in s^2 and can be solved with the quadratic formula. Let us define

$$\alpha = \tfrac{1}{2}k^2\lambda^2 \tag{11.38}$$

The solution to the quadratic is $s^2 = \lambda^2(-\alpha \pm \sqrt{1+\alpha^2})$. Observe that one root is strictly positive because $\sqrt{1+\alpha^2} > \alpha$ and the other strictly negative. Make the definitions

$$\hat{\lambda} = \lambda\sqrt{\sqrt{1+\alpha^2}+\alpha}, \qquad \bar{\lambda} = \lambda\sqrt{\sqrt{1+\alpha^2}-\alpha} \tag{11.39}$$

Note that if $\alpha = 0$, which eliminates the rotary inertia term from the equation of motion, then $\hat{\lambda} = \bar{\lambda} = \lambda$, and the solution degenerates to Bernoulli–Euler beam theory. With the definitions of Eq. 11.39, the four solutions to the characteristic equation are

$$s_1 = \bar{\lambda}, \qquad s_2 = -\bar{\lambda}, \qquad s_3 = i\hat{\lambda}, \qquad s_4 = -i\hat{\lambda}$$

which means that the general solution is

$$W(x) = C_1 e^{\bar{\lambda}x} + C_2 e^{-\bar{\lambda}x} + C_3 e^{i\hat{\lambda}x} + C_4 e^{-i\hat{\lambda}x}$$

As we did for the Bernoulli–Euler beam, we will write the general solution in the more convenient form

$$W(x) = D_1 e^{-\bar{\lambda}(L-x)} + D_2 e^{-\bar{\lambda}x} + D_3 \cos\hat{\lambda}x + D_4 \sin\hat{\lambda}x \tag{11.40}$$

where the constants D_i have been suitably reinterpreted. There is a subtle difference between Eqs. 11.40 and 11.19 from the Bernoulli–Euler theory. We explore that difference in this section.

The process of solution is similar to what we did for the Bernoulli–Euler beam. In particular, we first find the values of λ that work by implementing the boundary conditions in the spatial part of the problem. It should come as no surprise that only discrete values will work, and those *eigenvalues* will give rise to *eigenfunctions*. What we will find is that it is more convenient to establish the discrete values of $\hat{\lambda}$, from which we can always compute λ and $\bar{\lambda}$. Hence, it is worthwhile to express λ in terms of $\hat{\lambda}$.

Start with the definition of $\hat{\lambda}$ in Eq. 11.39. Square both sides of the equation to get

$$\lambda^2\left(\sqrt{1+\alpha^2}+\alpha\right) = \hat{\lambda}^2$$

Multiply both sides of the equation by $\sqrt{1+\alpha^2} - \alpha$ and simplify to

$$\lambda^2 = \hat{\lambda}^2\left(\sqrt{1+\alpha^2} - \alpha\right)$$

Now, substitute $\alpha = k^2\lambda^2/2$, move the second term on the right side to the left side, and regroup to get

$$\left(1 + \tfrac{1}{2}k^2\hat{\lambda}^2\right)\lambda^2 = \hat{\lambda}^2\sqrt{1 + \tfrac{1}{4}k^4\lambda^4}$$

Finally, square both sides of the equation and solve for λ^4 to get

$$\lambda^4 = \frac{\hat{\lambda}^4}{1 + k^2\hat{\lambda}^2} \tag{11.41}$$

This relationship will allow us to go back and forth between λ and $\hat{\lambda}$. It will also be useful to define the ratio of $\bar{\lambda}$ to $\hat{\lambda}$ as

$$\beta = \frac{\bar{\lambda}}{\hat{\lambda}} = \frac{\sqrt{\sqrt{1+\alpha^2} - \alpha}}{\sqrt{\sqrt{1+\alpha^2} + \alpha}} = \frac{1}{\sqrt{1+\alpha^2} + \alpha} \tag{11.42}$$

To see how the last equality comes about, square both sides of the equation, multiply top and bottom by $\sqrt{1+\alpha^2} + \alpha$, and take the square root of the result. We will use α and β for notational convenience. It is important to keep in mind that, ultimately, they are both functions of λ, which is what we seek when we implement the boundary conditions. Note also that we can express α in terms of $\hat{\lambda}$ as

$$\alpha = \tfrac{1}{2}k^2\lambda^2 = \frac{1}{2}\frac{k^2\hat{\lambda}^2}{\sqrt{1 + k^2\hat{\lambda}^2}} \tag{11.43}$$

Also, the frequency ω from Eq. 11.37 can be expressed in terms of $\hat{\lambda}$ as

$$\omega = c_o k \lambda^2 = \frac{c_o k \hat{\lambda}^2}{\sqrt{1 + k^2 \hat{\lambda}^2}} \tag{11.44}$$

where $c_o = \sqrt{E/\rho}$ is the nominal longitudinal wave speed.

11.4.1 Simple–Simple Rayleigh Beam

To compare the solution by separation of variables with what we got for the Bernoulli–Euler beam, we will consider a beam simply supported at both ends. The boundary conditions imply that

$$W(0) = 0, \qquad W''(0) = 0, \qquad W(L) = 0, \qquad W''(L) = 0$$

The general solution $W(x)$ (see Eq. 11.40) and its second derivative are

$$W(x) = D_1 e^{-\bar{\lambda}(L-x)} + D_2 e^{-\bar{\lambda}x} + D_3 \cos \hat{\lambda}x + D_4 \sin \hat{\lambda}x$$
$$W''(x) = \bar{\lambda}^2 \left(D_1 e^{-\bar{\lambda}(L-x)} + D_2 e^{-\bar{\lambda}x} \right) - \hat{\lambda}^2 \left(D_3 \cos \hat{\lambda}x + D_4 \sin \hat{\lambda}x \right)$$

From the conditions at $x = 0$ we get

$$D_1 \bar{e} + D_2 + D_3 = 0$$
$$\beta^2 \left(D_1 \bar{e} + D_2 \right) - D_3 = 0$$

where $\bar{e} = e^{-\bar{\lambda}L}$, which is a slight reinterpretation of the notation used for the Bernoulli–Euler beam. Adding these equations proves that $D_1 \bar{e} + D_2 = 0$, and therefore that $D_3 = 0$. The conditions $W(L) = 0$ and $W''(L) = 0$ give, respectively,

$$D_1 + D_2 \bar{e} + D_3 \hat{c} + D_4 \hat{s} = 0$$
$$\beta^2 \left(D_1 + D_2 \bar{e} \right) - D_3 \hat{c} - D_4 \hat{s} = 0$$

where $\hat{c} = \cos \hat{\lambda}L$ and $\hat{s} = \sin \hat{\lambda}L$. Adding the equations gives $D_1 + D_2 \bar{e} = 0$. Substituting $D_2 = -D_1 \bar{e}$ proves that $D_1 = 0$ and $D_2 = 0$. That leaves, from either of the two equations, $D_4 \hat{s} = 0$. If $D_4 = 0$ then all of the constants are zero and there is no motion. Therefore, the *characteristic equation* is

$$\sin \hat{\lambda}L = 0 \tag{11.45}$$

There are an infinite number of solutions to this equation

$$\hat{\lambda}_n = \frac{n\pi}{L}$$

each with an associated eigenfunction

$$\varphi_n(x) = \sin \hat{\lambda}_n x \tag{11.46}$$

It is interesting to note that the eigenfunctions for the Rayleigh beam are exactly the same as the eigenfunctions for the Bernoulli–Euler beam for the simple–simple boundary conditions. Recall that five of these functions are shown in Fig. 11.4. The general solution to the Rayleigh beam problem for free vibration is given by Eq. 11.25, also identical to the Bernoulli–Euler beam. However, the evaluation of the constants A_n and B_n are quite different because the orthogonality properties are not the same.

11.4.2   Orthogonality Relationships

As was the case for the axial bar and the Bernoulli–Euler beam, the eigenfunctions (mode shapes) of the Rayleigh beam have orthogonality properties. However, they are not quite as simple as those other two theories. Let us derive the orthogonality properties for the Rayleigh beam. First note that the eigenfunctions satisfy the differential equation (see Eq. 11.36)

$$\varphi_n''''(x) + \lambda_n^4 k^2 \varphi_n''(x) - \lambda_n^4 \varphi_n(x) = 0$$

along with the boundary conditions. Next, multiply this equation by $\varphi_m(x)$ and integrate over the length

$$\int_0^L \left[\varphi_n'''' + \lambda_n^4 k^2 \varphi_n'' - \lambda_n^4 \varphi_n \right] \varphi_m \, dx = 0 \tag{11.47}$$

Now, integrate the first term by parts *twice* and the second term *once* to get

$$\int_0^L \left[\varphi_n'' \varphi_m'' - \lambda_n^4 \left(k^2 \varphi_n' \varphi_m' + \varphi_n \varphi_m \right) \right] dx$$

$$-\varphi_n'' \varphi_m' \Big|_0^L + \left(\varphi_n''' + k^2 \lambda_n^4 \varphi_n' \right) \varphi_m \Big|_0^L = 0$$

The boundary terms that result from integration by parts vanish because the eigenfunctions satisfy the boundary conditions. In fact, these terms vanish for *all* boundary conditions. For example, a simple boundary condition has the conditions

$\varphi(\bullet) = 0$ and $\varphi''(\bullet) = 0$. A fixed support has $\varphi(\bullet) = 0$ and $\varphi'(\bullet) = 0$. A free end has $\varphi''(\bullet) = 0$ and $V(\bullet) = 0$ (vanishing shear). The shear is defined by the equation of motion (see Eq. 10.29) to be

$$V = -EIw''' + \rho I \ddot{w}'$$

Noting that $w(x, t) = W(x)T(t)$, that $\ddot{T} = -\omega^2 T$, and that ω can be expressed in terms of $\hat{\lambda}$ through Eq. 11.44, we can show that the vanishing shear conditions can be written as $W''' + k^2 \lambda_n^4 W' = 0$, which means that the eigenfunctions for a problem with a vanishing shear boundary condition satisfy

$$\varphi_n'''(\bullet) + k^2 \lambda_n^4 \varphi_n'(\bullet) = 0$$

at the place where the shear is zero.

Since all boundary terms that resulted from integration by parts vanish, we can write Eq. 11.47 as

$$\int_0^L \left[\varphi_n'' \varphi_m'' - \lambda_n^4 \left(k^2 \varphi_n' \varphi_m' + \varphi_n \varphi_m \right) \right] dx = 0 \qquad (11.48)$$

We can follow the same steps starting with the mth eigenvalue problem and multiplying by the nth eigenfunction, with the result

$$\int_0^L \left[\varphi_m'' \varphi_n'' - \lambda_m^4 \left(k^2 \varphi_m' \varphi_n' + \varphi_m \varphi_n \right) \right] dx = 0 \qquad (11.49)$$

Subtracting Eq. 11.49 from Eq. 11.48 gives

$$\left(\lambda_m^4 - \lambda_n^4 \right) \int_0^L \left(k^2 \varphi_m' \varphi_n' + \varphi_m \varphi_n \right) dx = 0$$

Since $\lambda_n \neq \lambda_m$, we have proven that the eigenfunctions satisfy the orthogonality relationship

$$\int_0^L \left(k^2 \varphi_m' \varphi_n' + \varphi_m \varphi_n \right) dx = 0, \qquad n \neq m \qquad (11.50)$$

It is also straightforward to prove that the second derivatives of the eigenfunctions are also orthogonal. Return to Eq. 11.48 or 11.49 and note Eq. 11.50 (and that $\lambda_n \neq 0$). These equations then reduce to

$$\int_0^L \varphi_m'' \varphi_n'' \, dx = 0, \qquad n \neq m \qquad (11.51)$$

Orthogonality of the eigenfunctions plays a key role in developing the complete solution of the differential equation. Recall that we need to use it to determine the coefficients A_n and B_n in the expression

$$w(x, t) = \sum_{n=1}^{N} \varphi_n(x)\left[A_n \cos \omega_n t + B_n \sin \omega_n t \right] \tag{11.52}$$

The form of the orthogonality relationship expressed in Eq. 11.50 is not quite suited to this task. Let us do one more step to get a result that we can use to find the modal coefficients. Integrate the first term in Eq. 11.50 by parts to get

$$\int_0^L \left(\varphi_n - k^2 \varphi_n''\right)\varphi_m \, dx + \varphi_n' \varphi_m \Big|_0^L = 0 \tag{11.53}$$

Observe that the boundary term that appears due to integration by parts vanishes only for simple and fixed conditions but does not vanish at a free end. To see how this version of the orthogonality relationship is useful for modal analysis, consider the case with initial displacement $w(x, 0) = w_o(x)$ and no initial velocity. Equation 11.52 at $t = 0$ takes the form

$$w_o(x) = \sum_{n=1}^{N} \varphi_n(x) A_n \tag{11.54}$$

We need to find A_n. To prepare, let us consolidate the notation as follows

$$\langle u, v \rangle = \int_0^L u(x)v(x)\,dx, \qquad \left[uv\right]_0^L = u(L)v(L) - u(0)v(0)$$

for any functions $u(x)$ and $v(x)$. We will do two operations on Eq. 11.54. First, multiply both sides of the equation by $\varphi_m - k^2 \varphi_m''$ and integrate from 0 to L to get

$$\langle w_o, \varphi_m - k^2 \varphi_m'' \rangle = \sum_{n=1}^{N} A_n \langle \varphi_n, \varphi_m - k^2 \varphi_m'' \rangle$$

Second, multiply both sides of that same equation by φ_m' and evaluate the difference between 0 and L to get

$$\left[w_o \varphi_m' \right]_0^L = \sum_{n=1}^{N} A_n \left[\varphi_n \varphi_m' \right]_0^L$$

Adding these two results together, noting that the range of the sum is the same for both, gives

$$\langle w_o, \varphi_m - k^2 \varphi_m'' \rangle + \left[w_o \varphi_m' \right]_0^L = \sum_{n=1}^{N} A_n \left[\langle \varphi_n, \varphi_m - k^2 \varphi_m'' \rangle + \left[\varphi_n \varphi_m' \right]_0^L \right]$$

Now, due to the orthogonality relationship in Eq. 11.53, the only term in the sum on the right side of the equation that is not equal to zero is when $n = m$. Therefore, we can solve for A_n as

$$A_n = \frac{\langle w_o, \varphi_n - k^2 \varphi_n'' \rangle + \left[w_o \varphi_n' \right]_0^L}{\langle \varphi_n, \varphi_n - k^2 \varphi_n'' \rangle + \left[\varphi_n \varphi_n' \right]_0^L}$$

The process for computing the modal coefficients B_n from an initial velocity $v_o(x)$ is quite similar. We leave the calculation as an exercise for the reader.

11.4.3   Wave Propagation: Simple–Simple Beam

We can now complete the modal analysis of the propagation of an initial displacement. Let us return to the example of a simple–simple beam of length $L = 10$, $E = 80$, $\rho = 0.6$, $A = 0.05$, and $I = 0.2$, subjected to an initial displacement $w_o(x)$ defined by Eq. 11.35 with $a = 4$ and $b = 6$. We use 150 modes in the analysis, an analysis time of $t = 0.3$ s, and a spatial grid with 250 points.

The progress of the wave propagation is shown in Fig. 11.11 for four different times. The heavy line is the displacement w and the thin line with shading is the moment M (scaled so that the two can be plotted on the same graph). Taking the position of the peak and dividing by the time gives the apparent wave speed. In this case the wave speed is approximately 11.40. The nominal wave speed of the material is $c_o = 11.547$. Some dispersion of the wave is evident.

We can get a better understanding of what is happening to the wave by considering the modal expansion

$$w(x, t) = \sum_{n=1}^{N} A_n \sin \hat{\lambda}_n x \, \cos \omega_n t \tag{11.55}$$

where N is the number of terms included in the sum. The coefficients A_n tell how important the nth mode is to the overall response. Using the trigonometric identity $2 \sin a \cos b = \sin(a + b) + \sin(a - b)$ we can rewrite the Eq. 11.55 in the form

$$w(x, t) = \sum_{n=1}^{N} \tfrac{1}{2} A_n \left[\sin(\hat{\lambda}_n x + \omega_n t) + \sin(\hat{\lambda}_n x - \omega_n t) \right]$$

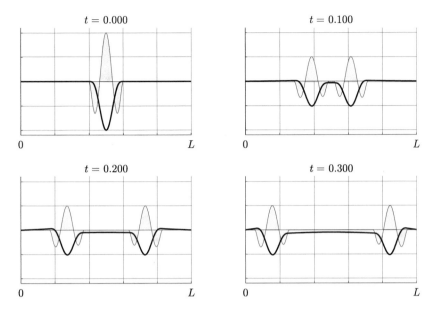

Fig. 11.11 *Rayleigh beam wave propagation.* Simple–simple beam of length $L = 10$, $E = 80$, $\rho = 0.6$, $A = 0.05$, and $I = 0.2$, subjected to an initial displacement of Eq. 11.35 with $a = 4$ and $b = 6$. The four plots show the displacement (dark line) and moment (thin line with shading) at four different times

What this equation indicates is that each modal component is a pure sine wave that splits in half, one part propagating right and one left, both at the modal wave speed of

$$c_n = \frac{\omega_n}{\hat{\lambda}_n} = \frac{c_o\, k\hat{\lambda}_n}{\sqrt{1 + k^2\hat{\lambda}_n^2}}$$

The values of the natural frequencies ω_n and the modal wave speeds c_n/c_o are shown in Fig. 11.12. It is clear that the modal wave speeds are asymptotic to the nominal wave speed c_o and that most of the modes share nearly the same wave speed. The snapshots of the wave in Fig. 11.11 includes $N = 150$ modes. Only the first 50 are shown in Fig. 11.12.

If all the modal wave speeds were the same, then the wave would propagate with coherence like waves do in the axial bar problem. However, because some of the modal wave speeds associated with the lower frequencies are slower, the wave tends to spread out. This is the source of the dispersion that one can observe in the propagation of waves in beams. However, most of the modal wave speeds are the same for this set of physical parameters, which explains why the observed wave speed is nearly equal to the nominal wave speed and why there is good coherence of the wave, at least in the short term. Decreasing the moment of inertia I (and thereby

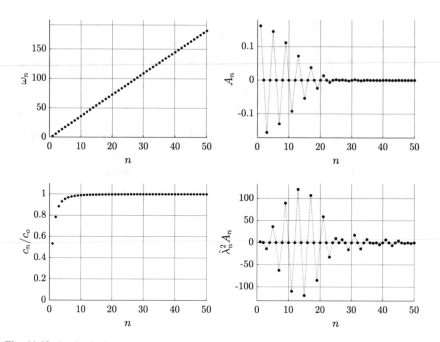

Fig. 11.12 *Rayleigh beam wave propagation. Modal components.* Simple–simple beam of Fig. 11.11. Top left shows the natural frequencies, bottom left shows the modal wave speeds, top right shows the modal coefficients for w, and bottom right shows the modal coefficients for M

the radius of gyration k) tends to cause more dispersion, showing attributes more like the Bernoulli–Euler beam.

Figure 11.12 shows the modal coefficients A_n, which govern the expansion of $w(x, t)$. Several things are clear from this plot. First, every other coefficient is zero for this wave form. Second, the coefficients themselves decay very rapidly, with little contribution of modes past $n = 25$. Because the modal coefficients are multiplied by $\hat{\lambda}_n^2$, the coefficients for the moment sum continue to be significant for larger values of n. In other words, to capture the moment waves accurately, more modes are required in the sum. However, because the first few components of the moment are small, the moment wave tends to be more coherent than the displacement wave.

11.4.4   Other Boundary Conditions

As was the case for the Bernoulli–Euler beam, the simple–simple boundary conditions lead to deceptively simple expressions for the eigenvalues and eigenfunctions. In this section we will examine the same boundary conditions we did for the Bernoulli–Euler beam: Fixed–simple, fixed–fixed, and fixed–free. Each case follows a process identical to the Bernoulli–Euler beam, but it is instructive to follow through the details because each case brings in some new feature.

Fixed–Simple Beam Consider the fixed–simple beam shown in Fig. 11.3b. The boundary conditions are

$$W(0) = 0, \quad W'(0) = 0, \quad W(L) = 0, \quad W''(L) = 0$$

Using our general expression for $W(x)$ (Eq. 11.40), we find that the conditions at $x = 0$ imply that

$$W(0) = D_1 \bar{e} + D_2 + D_3 = 0$$
$$W'(0) = \bar{\lambda}(D_1 \bar{e} - D_2) + \hat{\lambda} D_4 = 0$$

where, $\bar{e} = e^{-\bar{\lambda}L}$. These two equations can be solved to give

$$D_3 = -(D_2 + D_1 \bar{e}), \qquad D_4 = \beta(D_2 - D_1 \bar{e})$$

where $\beta = \bar{\lambda}/\hat{\lambda}$. Next, the conditions $W(L) = 0$ and $W''(L) = 0$ give

$$W(L) = D_1 + D_2 \bar{e} + D_3 \hat{c} + D_4 \hat{s} = 0$$
$$W''(L) = \bar{\lambda}^2 (D_1 + D_2 \bar{e}) - \hat{\lambda}^2 (D_3 \hat{c} + D_4 \hat{s}) = 0$$

where $\hat{c} = \cos \hat{\lambda} L$ and $\hat{s} = \sin \hat{\lambda} L$. Dividing the second of these equations through by $\hat{\lambda}^2$ and adding the two equations gives $D_1 + D_2 \bar{e} = 0$. Substituting this result and the expressions for D_3 and D_4 into the first equation gives

$$D_2 \left[\beta(1 + \bar{e}^2)\hat{s} - (1 - \bar{e}^2)\hat{c} \right] = 0$$

If $D_2 = 0$ then all four of the constants are zero. If $D_2 \neq 0$ then, we get the *characteristic equation* $\beta(1 + \bar{e}^2)\hat{s} - (1 - \bar{e}^2)\hat{c} = 0$, which can be rewritten as

$$\beta \tan \hat{\lambda} L = \tanh(\beta \hat{\lambda} L) \tag{11.56}$$

This equation can be solved with Newton's method. Letting $y = \hat{\lambda} L$, the residual and tangent take the form

$$g(y) = \beta \tan y - \tanh(\beta y), \qquad g' = \beta' \tan y + \frac{\beta}{\cos^2 y} - \frac{\beta' y + \beta}{\cosh^2 y}$$

Note that β is a function of α (Eq. 11.42), which is a function of y in accord with Eq. 11.43. Letting $z = (k/L)y$, α and its derivative can be expressed as

$$\alpha = \frac{1}{2}\frac{z^2}{\sqrt{1+z^2}}, \qquad \alpha' = \frac{d\alpha}{dz}\frac{dz}{dy} = \frac{1}{2}\frac{z(2+z^2)}{(1+z^2)^{3/2}}\frac{k}{L}$$

Likewise, β and its derivative can be expressed as

$$\beta = \frac{1}{\sqrt{1+\alpha^2}+\alpha}, \qquad \beta' = \frac{d\beta}{d\alpha}\frac{d\alpha}{dy} = \frac{-\beta\alpha'}{\sqrt{1+\alpha^2}}$$

It is evident that the characteristic equation for the Rayleigh beam is more complicated than the Bernoulli–Euler beam. But since we are already using Newton's method to solve the latter, it is not much additional effort to solve the former. All of the characteristic equations for the Rayleigh beam will involve β, so it makes sense to create a function that takes care of computing the ingredients required for the Newton iteration. Code 11.4 does just that. The code also computes gamma and its derivative, which is needed for the fixed–free boundary condition.

Code 11.4 MATLAB code to compute quantities related to α and β for the Rayleigh beam. Note that δ and γ are needed for the fixed–free case

```
function [beta,dbeta,dby,gamma,dgamma] = GetParams(koL,y)
%   Establish the common parameters used in the characteristic
%   equations and their derivatives (to be used in Newton's method)
%   Incoming:     koL = k/L   and   y = lambda_hat*L
    z = koL*y;

    alpha = 0.5*z^2/sqrt(1+z^2);
    da = 0.5*z*koL*(2+z^2)/sqrt(1+z^2)^3;

    beta = 1/(sqrt(1+alpha^2) + alpha);
    dbeta = -da*beta/sqrt(1+alpha^2);
    dby = dbeta*y + beta;

    delta = z^2/(1+z^2);
    ddelta = 2*koL*z/(1+z^2)^2;

    gamma = (beta^2+delta)/(1-delta);
    dgamma = 2*beta*dbeta/(1-delta) + ddelta*(1+beta^2)/(1-delta)^2;

end
```

The starting point for the nth root of $y^{(n)} = n\pi - \tan^{-1}\beta^{(n-1)}$ converges for all n, starting with a value $\beta^{(0)} = 1$ (as long as the physical parameters are reasonable). Knowing $\hat{\lambda}_n$, we can find the coefficients of the eigenfunctions $\varphi_n(x)$. The magnitude of the eigenfunction is arbitrary. Thus, we can assign the value of $D_2 = 1$, which then determines all of the others as

$$D_1^{(n)} = -\bar{e}_n, \quad D_2^{(n)} = 1, \quad D_3^{(n)} = -(1-\bar{e}_n^2), \quad D_4^{(n)} = \beta(1-\bar{e}_n^2)$$

where $\bar{e}_n = e^{-\hat{\lambda}_n L}$. It is worth taking note of the similarities with the corresponding case for the Bernoulli–Euler theory. Substituting these constants back into the expression for $W(x)$ gives the eigenfunction

$$\varphi_n(x) = D_1^{(n)} e^{-\bar{\lambda}_n(L-x)} + D_2^{(n)} e^{-\bar{\lambda}_n x} + D_3^{(n)} \cos \hat{\lambda}_n x + D_4^{(n)} \sin \hat{\lambda}_n x$$

The mode shapes are similar to those of the Bernoulli–Euler beam shown in Fig. 11.6, with deviations that depend upon the value of β.

Fixed–Fixed Beam Consider the fixed–fixed beam shown in Fig. 11.3c. The boundary conditions are

$$W(0) = 0, \quad W'(0) = 0, \quad W(L) = 0, \quad W'(L) = 0$$

The conditions $W(0) = 0$ and $W'(0) = 0$ are the same as the fixed–simple beam and imply that

$$D_3 = -(D_2 + D_1 \bar{e}), \qquad D_4 = \beta(D_2 - D_1 \bar{e})$$

The conditions $W(L) = 0$ and $W'(L) = 0$ give

$$D_1 + D_2 \bar{e} + D_3 \hat{c} + D_4 \hat{s} = 0$$

$$\bar{\lambda}(D_1 - D_2 \bar{e}) - \hat{\lambda}(D_3 \hat{s} - D_4 \hat{c}) = 0$$

where $\hat{c} = \cos \hat{\lambda} L$ and $\hat{s} = \sin \hat{\lambda} L$. Divide the second of these equations through by $\hat{\lambda}$ and substitute the expressions for D_3 and D_4 in both to get

$$\begin{bmatrix} 1 - \bar{e}(\hat{c} + \beta \hat{s}) & \bar{e} - \hat{c} + \beta \hat{s} \\ \beta + \bar{e}(\hat{s} - \beta \hat{c}) & -\beta \bar{e} + \hat{s} + \beta \hat{c} \end{bmatrix} \begin{Bmatrix} D_1 \\ D_2 \end{Bmatrix} = \begin{Bmatrix} 0 \\ 0 \end{Bmatrix}$$

If $D_1 = 0$ and $D_2 = 0$ then all four of the constants are zero. A nonzero solution exists only if the determinant of the coefficient matrix is zero. Setting the determinant to zero gives *characteristic equation*

$$2\beta(1 + \bar{e}^2)\hat{c} + (1 - \beta^2)(1 - \bar{e}^2)\hat{s} - 4\beta \bar{e} = 0 \tag{11.57}$$

This equation can also be solved by Newton's method (see Code 11.5). The constants for the eigenfunctions $\varphi_n(x)$ for the fixed–fixed case are

$$D_1^{(n)} = -\bar{e}_n + \hat{c}_n - \beta_n \hat{s}_n \qquad D_2^{(n)} = 1 - \bar{e}_n(\hat{c}_n + \beta_n \hat{s}_n)$$

$$D_3^{(n)} = -1 + \bar{e}_n^2 + 2\beta_n \bar{e}_n \hat{s}_n \qquad D_4^{(n)} = (1 + \bar{e}_n^2 + 2\bar{e}_n \hat{c}_n)\beta_n$$

where $\bar{e}_n = e^{-\bar{\lambda}_n L}$, $\hat{c}_n = \cos \hat{\lambda}_n L$, $\hat{s}_n = \sin \hat{\lambda}_n L$, and $\beta_n = \hat{\lambda}_n / \bar{\lambda}_n$. The mode shapes are similar to those associated with the Bernoulli–Euler beam with fixed–fixed boundary conditions, shown in Fig. 11.7, but now dependent on β.

Fixed–Free Beam Finally, consider the fixed–free beam shown in Fig. 11.3d. The boundary conditions are

$$W(0) = 0, \quad W'(0) = 0, \quad W''(L) = 0, \quad V(L) = 0$$

the third being true because $M(L) = 0$ and $M = -EIW''$. The first three boundary conditions are the same as the fixed–simple case. Thus, the equations emanating from those boundary conditions are

$$D_1 \bar{e} + D_2 + D_3 = 0$$

$$D_1 \bar{e} - D_2 + D_4 = 0$$

$$\beta^2 (D_1 + D_2 \bar{e}) - D_3 \hat{c} - D_4 \hat{s} = 0$$

The vanishing shear boundary condition can be expressed in terms of the displacement W through the relationship $W'''(L) + k^2 \lambda^4 W'(L) = 0$. Substituting Eq. 11.42, we can write this condition as

$$W'''(L) + \delta \hat{\lambda}^2 W'(L) = 0 \tag{11.58}$$

where we have defined

$$\delta = \frac{k^2 \hat{\lambda}^2}{1 + k^2 \hat{\lambda}^2}$$

This is the δ that appears in Code 11.4 (along with its derivative, which is straightforward to compute). From the expression for $W(x)$, we can evaluate the third and first derivatives of W at $x = L$ as

$$W'''(L) = \bar{\lambda}^3 \left(D_1 - D_2 \bar{e} \right) + \hat{\lambda}^3 \left(D_3 \hat{s} - D_4 \hat{c} \right)$$

$$W'(L) = \bar{\lambda} \left(D_1 - D_2 \bar{e} \right) + \hat{\lambda} \left(-D_3 \hat{s} + D_4 \hat{c} \right)$$

Now, the vanishing shear boundary condition can be written as

$$\beta (\beta^2 + \delta)(D_1 - D_2 \bar{e}) + (1 - \delta)(D_3 \hat{s} - D_4 \hat{c}) = 0$$

As a final economization of notation, let

$$\gamma = \frac{\beta^2 + \delta}{1 - \delta} \tag{11.59}$$

which is the γ that appears in Code 11.4 (along with its derivative, which is straightforward to compute). Eliminating D_3 and D_4 from the boundary conditions at $x = 0$ in favor of D_1 and D_2 leaves the result

$$\begin{bmatrix} \beta^2 + \bar{e}(\hat{c} + \beta\hat{s}) & \beta^2 \bar{e} + \hat{c} - \beta\hat{s} \\ \beta\gamma - \bar{e}(\hat{s} - \beta\hat{c}) & -\beta\gamma\bar{e} - \hat{s} - \beta\hat{c} \end{bmatrix} \begin{Bmatrix} D_1 \\ D_2 \end{Bmatrix} = \begin{Bmatrix} 0 \\ 0 \end{Bmatrix}$$

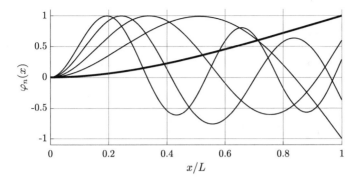

Fig. 11.13 *Mode shapes for the fixed–free Rayleigh beam.* The first five eigenfunctions are plotted for the value $k/L = 0.2$. The functions are normalized to have maximum value of one. The heavier line is the first mode

The process of solving these equations is similar to the previous cases, yielding the *characteristic equation*

$$2(1+\beta^2\gamma)\bar{e} + \beta(1-\gamma)\hat{s}\,(1 - \bar{e}^2) + (\beta^2+\gamma)\hat{c}\,(1 + \bar{e}^2) = 0$$

This equation can be solved by Newton's method (see Code 11.5). Knowing the eigenvalues λ_n, we can find the eigenfunctions $\varphi_n(x)$. The magnitude of the eigenfunction is arbitrary. We can assign the value of $D_2 = \beta^2 + \bar{e}(\hat{c} + \beta\hat{s})$, which then determines the others as

$$D_1^{(n)} = -\beta_n^2\bar{e}_n - \hat{c}_n + \beta_n\hat{s}_n \qquad D_2^{(n)} = \beta_n^2 + \bar{e}_n(\hat{c}_n + \beta_n\hat{s}_n)$$
$$D_3^{(n)} = -\beta_n^2(1-\bar{e}_n^2) - 2\beta_n\bar{e}_n\hat{s}_n \qquad D_4^{(n)} = \beta_n^3(1+\bar{e}_n^2) + 2\beta_n\bar{e}_n\hat{c}_n$$

The first five mode shapes are shown in Fig. 11.13 for a value of $k/L = 0.2$. It is interesting to compare the corresponding modes shapes for the Bernoulli–Euler beam in Fig. 11.8.

It should be evident that all of the expressions for the coefficients of the mode shapes of the Rayleigh beam reduce to those of the Bernoulli–Euler beam with $\beta = 1$. However, the eigenvalues do not unless we also set $\delta = 0$, which is tantamount to ignoring the rotary inertia term in the equation of motion. If $\delta = 0$, then $\gamma = 1$.

11.4.5   Implementation

The computations associated with finding the eigenvalues and eigenfunctions of the Rayleigh beam for these four boundary conditions are contained in Code 11.5. For each different value of BC, the code solves the characteristic equation by Newton's method, using the starting values mentioned in the previous sections, and establishes the values of the eigenfunction constants D_1, D_2, D_3, and D_4 for each mode. The eigenvalues are passed back in the array Ln and the eigenfunction constants in the array D. Clearly, the biggest difference between the Bernoulli–Euler beam code and the Rayleigh beam code is the additional complexity of computing the eigenvalues and eigenfunctions.

Code 11.5 MATLAB code to find the eigenvalues and the coefficients of the eigenfunction $\varphi_n(x)$ for a Rayleigh beam with different boundary conditions

```
function [Ln,D] = RayEigenproblem(nModes,koL,BC)
%  Eigenvalues and eigenfunctions for the Rayleigh beam
%     nModes : Number of modes
%        koL : k/L ratio of radius of gyration to length
%         BC : Boundary condition flag

%. Initialize arrays
   Ln = zeros(nModes,2);      % Eigenvalues Lambda_hat L
   D = zeros(nModes,4);       % Coefficients for Eigenfunctions
   beta = 1;

%. Set tolerance for Newton's method
   tol = 1.e-8;

   switch BC

     case 1    % Simple-Simple
       for n=1:nModes
         y = n*pi;
         [beta,~,~,~,~] = GetParams(koL,y);
         Ln(n,:) = [y,beta*y];
         D(n,:) = [0,0,0,1];
       end

     case 2    % Fixed-Simple
       for n=1:nModes
         its = 0; err = 1;
         y = n*pi + atan(1/beta);
         while err>tol
           [beta,db,dby,~,~] = GetParams(koL,y);
           g = beta*tan(y) - tanh(beta*y);
           Dg = db*tan(y) + beta/cos(y)^2 - dby/cosh(beta*y)^2;
           y = y - g/Dg;
           err = abs(g);   its = its+1;
         end
         [beta,~,~,~,~] = GetParams(koL,y);
         ebar = exp(-beta*y);
         Ln(n,:) = [y,beta*y];
         D(n,:)=[-ebar, 1, -(1-ebar^2), (1+ebar^2)*beta];
       end

     case 3    % Fixed-Fixed
       for n=1:nModes
         its = 0; err = 1;
         y = (n+1-beta/2)*pi;
         while err>tol
           [beta,db,dby,~,~] = GetParams(koL,y);
           chat = cos(y); shat = sin(y);
           ebar = exp(-beta*y);   Ec = 1+ebar^2;        Es = 1-ebar^2;
           debar = -dby*ebar;   dEc = 2*ebar*debar;    dEs = -dEc;
           g = 2*beta*Ec*chat + (1-beta^2)*Es*shat - 4*beta*ebar;
           Dg = 2*(db*Ec*chat + beta*dEc*chat - beta*Ec*shat) ...
                - 2*beta*db*Es*shat + (1-beta^2)*(dEs*shat + Es*chat) ...
                - 4*db*ebar - 4*beta*debar;
           y = y - g/Dg;
```

```
          err = abs(g);   its = its+1;
        end
        [beta,~,~,~,~] = GetParams(koL,y);
        Ln(n,:) = [y,beta*y];
        ebar = exp(-beta*y); chat = cos(y); shat = sin(y);
        D(n,:) = [-ebar + chat - beta*shat,...
                  1 - ebar*(chat + beta*shat),...
                  -1 + ebar^2 + 2*beta*shat*ebar,...
                  (1 + ebar^2 - 2*chat*ebar)*beta];
      end

  case 4    % Fixed-Free
    for n=1:nModes
      its = 0; err = 1;
      y = (2*n-1)*pi/2;
      while err>tol
        [beta,db,dby,gam,dgam] = GetParams(koL,y);
        C1 = 1 + beta^2*gam;    dC1 = 2*beta*db*gam + beta^2*dgam;
        C2 = beta*(1-gam);      dC2 = db*(1-gam) - beta*dgam;
        C3 = beta^2 + gam;      dC3 = 2*beta*db + dgam;
        chat = cos(y); shat = sin(y);
        ebar = exp(-beta*y); debar = -dby*ebar;
        Ec = 1+ebar^2;  dEc = 2*ebar*debar;
        Es = 1-ebar^2;  dEs =-2*ebar*debar;
        g = 2*C1*ebar + C2*Es*shat + C3*Ec*chat;
        Dg =   2*(dC1*ebar + C1*debar)...
             + dC2*Es*shat + C2*dEs*shat + C2*Es*chat ...
             + dC3*Ec*chat + C3*dEc*chat - C3*Ec*shat;
        y = y - g/Dg;
        err = abs(g);   its = its+1;
      end
      [beta,~,~,~,~] = GetParams(koL,y);
      ebar = exp(-beta*y); chat = cos(y); shat = sin(y);
      Ln(n,:) = [y,beta*y];
      D(n,:) = [-beta^2*ebar - chat + beta*shat,...
                beta^2 + ebar*(chat + beta*shat),...
                -beta^2*(1-ebar^2) - 2*beta*ebar*shat,...
                beta^3*(1+ebar^2) + 2*beta*ebar*chat];
    end

  end
end
```

The computation of the eigenfunctions is similar to the Bernoulli–Euler beam, except that we need to account for the orthogonality conditions of the Rayleigh beam. Code 11.6 carries out the computation of the modal coefficients, based upon the eigenvalues and eigenfunction coefficients computed in `RayEigenproblem`. The variable names in the codes correspond closely with the equations the come from. For example, the code uses `chat` to represent $\hat{c} = \cos\hat{\lambda}L$, `shat` for $\hat{s} = \sin\hat{\lambda}L$, and `ebar` for $\bar{e} = e^{-\hat{\lambda}L}$.

Code 11.6   MATLAB code to find the modal coefficient for an initial wave for a Rayleigh beam. The boundary conditions are implicit in the eigenvalues and eigenfunction coefficients passed in from `RayEigenproblem`

```
function [Phi,ddPhi,A] = RayModeShapes(Ln,D,koL,nModes,y)
%   Compute eigenvectors and coefficients of initial wave, Rayleigh beam
%       Ln = Lambda_n*L
%        D = Coefficients of eigenfunctions
%      koL = k/L Radius of gyration divided by length
%   nModes = Number of modes
%        y = Spatial grid of points
```

```
%. Initialize arrays
   nPts = size(y,2);                   % Number of grid points
   Phi = zeros(nModes,nPts);           % Mode shapes
   ddPhi = zeros(nModes,nPts);         % 2nd derivative of mode shapes
   A = zeros(nModes,2);                % Coefficients of wave function

%. Initial wave function
   a = 0.4;                            % Start of initial wave
   b = 0.6;                            % End of initial wave
   N = 4;                              % Exponent for waveform
   w =@(x) (4.*x.*(1-x)).^N;           % Wave function
   z =@(x) (b-a)*x + a;                % z(x) maps [0,1] -> [a,b]

   for n=1:nModes

%... Form functions to make mode shapes
   LnH = Ln(n,1);   LnB = Ln(n,2);
   s1 =@(xi) D(n,1)*exp(-LnB*(1-xi));
   s2 =@(xi) D(n,2)*exp(-LnB*xi);
   s3 =@(xi) D(n,3)*cos(LnH*xi);
   s4 =@(xi) D(n,4)*sin(LnH*xi);

%... Create the mode shape functions
   phi   =@(xi) (s1(xi) + s2(xi))      + (s3(xi) + s4(xi));
   dphi  =@(xi) (s1(xi) - s2(xi))*LnB  - (s3(xi) - s4(xi))*LnH;
   ddphi =@(xi) (s1(xi) + s2(xi))*LnB^2 - (s3(xi) + s4(xi))*LnH^2;

%... Put in numerical values for Phi and ddPhi
   Phi(n,:) = phi(y);
   ddPhi(n,:) = ddphi(y);

%... Compute the initial displacement coefficient
   pdp   =@(xi) phi(xi) - koL^2*ddphi(xi);
   phi0 = phi(0);     dphi0 = dphi(0);
   phiL = phi(1);     dphiL = dphi(1);
   NormFcn =@(x) phi(x).*pdp(x);
   WaveFcn =@(x) w(x).*pdp(z(x))*(b-a);
   aa = integral(NormFcn,0,1) + koL^2*(phiL*dphiL - phi0*dphi0);
   bb = integral(WaveFcn,0,1);
   A(n,1) = bb/aa;
   A(n,2) = Ln(n);

   end

end
```

The structure of the computations for wave propagation is very similar to the associated code for the Bernoulli–Euler beam. The other functions, which are not included here are responsible for creating the graphical output. For example, the function PlotCoefficients produced Fig. 11.12 and the function PlotSnapshots produced Fig. 11.11.

Code 11.7 MATLAB code to generate the modal solution of the Rayleigh beam with an initial displacement $w_o(x)$

```
%  Modal Analysis of Rayleigh Beam Wave Propagation
   clear; clc;

%. Problem parameters
   nModes = 150;                       % Number of modes
   L = 10.0;                           % Length of beam
   Density = 0.6;                      % Mass density per unit volume
   E = 80;                             % Young's modulus
```

```
   Area = 0.05;                     % Cross-sectional area
   Inertia = 0.2;                   % Moment of inertia
   k = sqrt(Inertia/Area);          % Radius of gyration
   co = sqrt(E/Density);            % Nominal wave speed
   EI = E*Inertia;                  % Bending stiffness
   koL = k/L;                       % Ratio of radius of gyration to length
   BC = 1;                          % =1 S-S, =2 Fx-S, =3 Fx-Fx, =4 Fx-Fr

%. Compute and plot results
%  Start time      Stop time       Time steps      Spatial points
   T.to = 0;       T.tf = 0.3;     T.nT = 61;      T.nS = 250;
   [Time,x,w,M,A,Freq,Phi] = RayModalSum(T,L,co,koL,EI,nModes,BC);

   PlotModes(Phi,5);
   PlotInitialWave(x,w,M);
   PlotCoefficients(A,Freq,k,L,nModes);
   PlotSnapshots(x,w,M,Time,T);
   AnimateBeam(T,x,w,M,Time)

%- End of main program -----------------------------------------------

function [Time,x,w,M,A,Freq,Phi] = RayModalSum(T,L,co,koL,EI,nModes,BC)
%  Compute w and M by modal sum for Bernoulli-Euler beam
%          T = Struct with time parameters
%          L = Length of beam
%         co = Nominal wave speed
%        koL = Radius of gyration divided by length
%         EI = Bending stiffness
%     nModes = Number of modes
%         BC = Boundary condition code

%. Initialize storage
   x = linspace(0,1,T.nS);          % Spatial grid points
   w = zeros(T.nT,T.nS);            % Initialize w(x,t)
   M = zeros(T.nT,T.nS);            % Initialize M(x,t)
   Time = linspace(T.to,T.tf,T.nT)';  % Time grid points

%. Compute eigenvalues, mode shape coefficients, and frequencies
   [Ln,D] = RayEigenproblem(nModes,koL,BC);
   LLn = koL*Ln(:,1);
   Freq = (co/L)*LLn.*Ln(:,1)./sqrt(1 + LLn.^2);

%. Compute mode shapes and coefficients for initial wave
   [Phi,ddPhi,A] = RayModeShapes(Ln,D,koL,nModes,x);

%. Compute response time history
   for i=1:T.nT
     t = Time(i);
     for n=1:nModes
       CosWt = cos(Freq(n)*t);
       w(i,:) = w(i,:) + Phi(n,:)*A(n,1)*CosWt;
       M(i,:) = M(i,:) - ddPhi(n,:)*A(n,1)*CosWt*EI;
     end
   end

end
```

11.5   The Timoshenko Beam

The Timoshenko beam can be analyzed following the same process as the Bernoulli–Euler beam and the Rayleigh beam. For the case with EI and GA

constant and no external loads, i.e., $q = 0$ and $m = 0$, the equations of motion of the Timoshenko beam reduce to

$$GA(w'' + \theta') = \rho A \ddot{w}$$
$$EI\theta''' = \rho A(\ddot{w} + k^2 \ddot{\theta}')$$

(11.60)

where $k = \sqrt{I/A}$ is the radius of gyration of the cross section. Unlike the previous two theories, this one has two independent variables. The trick to making progress with the separation of variables strategy is to use different functions for the spatial variation but the same function for the temporal variation. Hence, we will consider the decomposition

$$w(x, t) = W(x)T(t), \qquad \theta(x, t) = \Theta(x)T(t)$$

Substituting these expressions into the equations of motion gives

$$GA(W'' + \Theta')T = \rho AW\ddot{T}, \qquad EI\Theta'''T = \rho A(W + k^2\Theta')\ddot{T}$$

As before, we posit that the temporal part of the solution can be written as

$$\ddot{T} + \omega^2 T = 0$$

This framework anticipates that the expected result will be oscillatory. With this assumption, we can substitute $\ddot{T} = -\omega^2 T$ into the previous equations, divide both equations through by T, and divide the first equation by GA and the second by EI to get

$$W'' + \Theta' + \frac{\omega^2 \rho}{G} W = 0, \qquad \Theta''' + \frac{\omega^2 \rho}{k^2 E}(W + k^2\Theta') = 0$$

(11.61)

We can eliminate Θ by solving the first equation for Θ'. To wit,

$$\Theta' = -\left(W'' + \frac{\omega^2 \rho}{G} W\right)$$

(11.62)

Now, differentiate this expression twice and substitute in the second equation to get

$$W'''' + \frac{\omega^2 \rho}{E}\left(1 + \frac{E}{G}\right)W'' + \frac{\omega^2 \rho}{E}\left(\frac{\omega^2 \rho}{E}\frac{E}{G} - \frac{1}{k^2}\right)W = 0$$

(11.63)

To simplify the notation, let

$$\lambda^4 = \frac{\omega^2 \rho}{k^2 E}, \qquad \alpha = \tfrac{1}{2}k^2\lambda^2, \qquad \mu = \frac{E}{G}$$

(11.64)

As we have done before, let us try solutions of the form $W(x) = Ce^{sx}$. Substituting all this into Eq. 11.63, we find that the associated characteristic equation is

$$s^4 + 2\alpha\lambda^2(1+\mu)s^2 - \lambda^4(1 - 4\mu\alpha^2) = 0 \qquad (11.65)$$

which is a quadratic equation in s^2. Note that for $\mu = 0$ this equation reduces to the characteristic equation for the Rayleigh beam. The solution to the quadratic equation is

$$s^2 = \lambda^2\left(-\alpha(1+\mu) \pm \sqrt{\alpha^2(1-\mu)^2 + 1}\right) \qquad (11.66)$$

It is obvious that the quantity under the radical is positive. Hence, the solutions for s^2 are both real.

The negative radical leads to the oscillatory sine and cosine terms. For the Rayleigh beam it was the case that the positive radical gave rise to purely positive roots and that led to the real exponential terms in the solution. For the Timoshenko beam that is not always the case. In order for the positive radical to produce a positive root, we must have

$$\sqrt{\alpha^2(1-\mu)^2 + 1} > \alpha(1+\mu)$$

Squaring both side and simplifying, we find that this condition reduces to $4\mu\alpha^2 < 1$. Thus, we get a positive root if

$$\frac{\omega^2\rho}{E} < \frac{GA}{EI} \qquad (11.67)$$

We will need to consider two cases in our subsequent analysis—one for which Eq. 11.67 holds, in which case the quartic characteristic equation will have two real roots and two imaginary roots, and one for which Eq. 11.67 does not hold, in which case all four of the roots will be imaginary. Note that the Rayleigh beam is the limit as $\mu \to 0$, in which case Eq. 11.67 always holds. We do not determine ω until we implement boundary conditions for a specific beam, so we cannot know *a priori* which case will govern.

Consider case 1 (i.e., Eq. 11.67 holds) first. To simplify the discussion, make the definitions

$$\hat{\lambda} = \lambda\sqrt{\sqrt{\alpha^2(1-\mu)^2 + 1} + \alpha(1+\mu)}$$
$$\bar{\lambda} = \lambda\sqrt{\sqrt{\alpha^2(1-\mu)^2 + 1} - \alpha(1+\mu)} \qquad (11.68)$$

Note that these two quantities are analogous to the ones defined for the Rayleigh beam. In fact, they reduce to those of the Rayleigh beam as $\mu \to 0$. Observe that both values are positive. The roots of the characteristic equation, Eq. 11.65, are $s_1 = \bar{\lambda}$, $s_2 = -\bar{\lambda}$, $s_3 = i\hat{\lambda}$, $s_4 = -i\hat{\lambda}$ and we can express the general solution to the differential equation as

$$W(x) = D_1 e^{-\bar{\lambda}(L-x)} + D_2 e^{-\bar{\lambda}x} + D_3 \cos \hat{\lambda}x + D_4 \sin \hat{\lambda}x \qquad (11.69)$$

This result is exactly the same as the solution we found for the Rayleigh beam, albeit with different values of $\bar{\lambda}$ and $\hat{\lambda}$. Hence, many of the subsequent analysis steps are similar.

Now consider case 2 (i.e., Eq. 11.67 does *not* hold). In this case the quantity under the top-level radical in the definition of $\bar{\lambda}$ is negative. Therefore, we define

$$\tilde{\lambda} = \lambda \sqrt{\alpha(1+\mu) - \sqrt{\alpha^2(1-\mu)^2 + 1}} \qquad (11.70)$$

which is positive. With this definition, the roots are $s_1 = i\tilde{\lambda}$, $s_2 = -i\tilde{\lambda}$, $s_3 = i\hat{\lambda}$, and $s_4 = -i\hat{\lambda}$, and the general solution takes the form

$$W(x) = D_1 \cos \tilde{\lambda}x + D_2 \sin \tilde{\lambda}x + D_3 \cos \hat{\lambda}x + D_4 \sin \hat{\lambda}x \qquad (11.71)$$

The only difference between the two cases is that one has exponentially decaying terms while the other has only pure oscillatory terms. In general, we find from the boundary conditions that there are an infinite number of natural frequencies. So, the possibility exists that Eq. 11.67 does not hold for some of those frequencies but does for others. Let's take another look at the solution to the simply supported beam.

11.5.1   Simple–Simple Beam

To compare the solution by separation of variables for the Timoshenko beam with what we got for the Bernoulli–Euler beam and the Rayleigh beam, we will consider the simply supported beam again. The boundary conditions are

$$W(0) = 0, \qquad \Theta'(0) = 0, \qquad W(L) = 0, \qquad \Theta'(L) = 0$$

We know from Eq. 11.62 the relationship between Θ and W. Hence, we can write the Θ boundary conditions as

$$\Theta'(\bullet) = -\left(W''(\bullet) + \frac{\omega^2 \rho}{G} W(\bullet) \right)$$

Therefore, we can restate the boundary conditions for the simple–simple beam as

$$W(0) = 0, \qquad W''(0) = 0, \qquad W(L) = 0, \qquad W''(L) = 0$$

From the conditions at $x = 0$, using Eq. 11.69, for Case 1 we get

$$D_1\bar{e} + D_2 + D_3 = 0$$

$$\bar{\lambda}^2(D_1\bar{e} + D_2) - \hat{\lambda}^2 D_3 = 0$$

where $\bar{e} = e^{-\bar{\lambda}L}$. Multiplying the first equation by $\bar{\lambda}^2$ and adding the two equations together proves that $D_3 = 0$ and, thus, $D_1\bar{e} + D_2 = 0$. The conditions at the right end give, for case 1,

$$D_1 + D_2\bar{e} + D_3\hat{c} + D_4\hat{s} = 0$$

$$\bar{\lambda}^2(D_1 + D_2\bar{e}) - \hat{\lambda}^2(D_3\hat{c} + D_4\hat{s}) = 0$$

where $\hat{c} = \cos\hat{\lambda}L$ and $\hat{s} = \sin\hat{\lambda}L$. Multiplying the first equation by $\hat{\lambda}^2$ and adding the equations together proves that $D_1 + D_2\bar{e} = 0$. Since we previously found that $D_1\bar{e} + D_2 = 0$, we can conclude that $D_1 = D_2 = 0$. Now, either equation gives $D_4 \sin\hat{\lambda}L = 0$. If $D_4 = 0$, then the entire solution is zero. Therefore, a nontrivial solution is possible only if

$$\sin\hat{\lambda}L = 0 \tag{11.72}$$

which is the *characteristic equation* for the Timoshenko beam. There are an infinite number of solutions

$$\hat{\lambda}_n = \frac{n\pi}{L}$$

for $n = 1, 2, \ldots$. We can solve for the corresponding λ_n (which will give us the frequency ω_n) from Eq. 11.68(b). Square both sides of Eq. 11.68(a) to get,

$$\hat{\lambda}^2 = \lambda^2\sqrt{\alpha^2(1-\mu)^2 + 1} + \lambda^2\alpha(1+\mu)$$

Move the second term on the right side to the left side, square both sides, and rearrange terms. After some algebra, we get

$$\hat{\lambda}^4 - 2\alpha(1+\mu)\hat{\lambda}^2\lambda^2 + (4\mu\alpha^2 - 1)\lambda^4 = 0$$

Substitute the definition for α from Eq. 11.64 to get

$$\hat{\lambda}^4 - k^2(1+\mu)\hat{\lambda}^2\lambda^4 + (\mu k^4\lambda^4 - 1)\lambda^4 = 0$$

Finally, let $z = k^2\lambda^4$ and rearrange to get a quadratic equation for z

$$\mu z^2 - \left(\hat{\lambda}^2(1+\mu) + \frac{1}{k^2}\right)z + \hat{\lambda}^4 = 0$$

This equation can be solved by the quadratic formula for each $\hat{\lambda}_n$ to yield two solutions: $\hat{z}_n$ and $\bar{z}_n$. Once these values are obtained, we can compute two associated natural frequencies from Eq. 11.64(a) as

$$\hat{\omega}_n = c_o\sqrt{\hat{z}_n}, \qquad \bar{\omega}_n = c_o\sqrt{\bar{z}_n} \qquad (11.73)$$

where $c_o = \sqrt{E/\rho}$ is the nominal wave speed of the material. Observe that we get two natural frequencies for each n because there are two roots to the quadratic equation for z_n. With the frequencies known, we can check to see if Eq. 11.67 is satisfied. In general, at the low end of the spectrum, some of the frequencies will be above the critical value. If that is the case, then we need to consider case 2.

It is interesting to note that the eigenfunction for the Timoshenko beam is exactly the same as the eigenfunction for the Bernoulli–Euler beam and the Rayleigh beam for the simple–simple boundary conditions, i.e.,

$$\varphi_n(x) = \sin \hat{\lambda}_n x \qquad (11.74)$$

This means that the remainder of the solution is the same as in those cases, except for the fact that we got two frequencies from the solution, not one.

What About Case 2? It is possible that some of the frequencies fall above the cutoff in Eq. 11.67. So, let us take a look at case 2. For this case, the conditions at the right end give

$$D_2 \sin \tilde{\lambda}L + D_4 \sin \hat{\lambda}L = 0, \qquad -D_2\tilde{\lambda}^2 \sin \tilde{\lambda}L - D_4\hat{\lambda}^2 \sin \hat{\lambda}L = 0,$$

It is not possible to prove that either of the coefficients is zero, but both conditions

$$\sin \tilde{\lambda}L = 0 \quad \text{or} \quad \sin \hat{\lambda}L = 0$$

are true, and both equations have the same solution, i.e.,

$$\tilde{\lambda}L = \hat{\lambda}L = n\pi$$

Therefore, the associated natural frequencies are exactly the same and they share the same eigenfunction (i.e., the one given by Eq. 11.74). In this case, there is no need to distinguish between Case 1 and Case 2. That may not be true for other boundary conditions, though.

Orthogonality of the Eigenfunctions To solve the wave propagation problem we need orthogonality relationships for the functions $\varphi_n(x)$ and $\psi_n(x)$, where $\psi'_n(x) = \varphi_n(x)$. We do not have a proof of orthogonality like the ones presented for the Bernoulli–Euler and Rayleigh beams, but we already know that $\varphi_n(x)$ and $\psi_n(x)$ are orthogonal for the simple–simple beam because the functions $\sin(n\pi x/L)$ and $\cos(n\pi x/L)$ have shown up in previous theories and have been proven orthogonal. Hence, we will use the orthogonality of those functions with the understanding that the eigenfunctions that would appear for other boundary conditions are not yet demonstrated.

11.5.2 Wave Propagation

Because we have two frequencies, the general solution to the problem can be expressed as

$$w(x, t) = \sum_{n=1}^{\infty} \varphi_n(x) \Big[A_n \cos \hat{\omega}_n t + B_n \sin \hat{\omega}_n t \tag{11.75}$$
$$+ C_n \cos \bar{\omega}_n t + D_n \sin \bar{\omega}_n t \Big]$$

where $\varphi(x)$ is given by Eq. 11.74 and $\hat{\omega}_n$ and $\bar{\omega}_n$ are given by Eq. 11.73. The coefficients A_n, B_n, C_n, and D_n must be determined by the initial conditions. Unlike the Bernoulli–Euler and Rayleigh beams, we have four constants rather than two. But we also have two fields that require specification of initial conditions. Let us assume that the initial velocities are zero, but we have initial displacements and rotation $w_o(x)$ and $\theta_o(x)$, respectively. The computation of the coefficients from the initial displacement is similar to the previous two theories. To wit,

$$w(x, 0) = \sum_{n=1}^{\infty} \varphi_n(x) \Big[A_n + C_n \Big] = w_o(x)$$
$$\dot{w}(x, 0) = \sum_{n=1}^{\infty} \varphi_n(x) \Big[\hat{\omega}_n B_n + \bar{\omega}_n D_n \Big] = 0 \tag{11.76}$$

To implement the initial conditions on θ, note that we have an expression for the first derivative

$$\theta' = \frac{\rho}{G} \ddot{w} - w'' \tag{11.77}$$

Differentiating the expression for w in Eq. 11.75 and substituting the results into Eq. 11.77 gives

$$\theta'(x,t) = \sum_{n=1}^{\infty} \varphi_n(x) \Big[\hat{\gamma}_n A_n \cos \hat{\omega}_n t + \hat{\gamma}_n B_n \sin \hat{\omega}_n t$$

$$+ \bar{\gamma}_n C_n \cos \bar{\omega}_n t + \bar{\gamma}_n D_n \sin \bar{\omega}_n t \Big] \tag{11.78}$$

where

$$\hat{\gamma}_n = \hat{\lambda}_n^2 - \frac{\rho}{G}\hat{\omega}_n^2 = \hat{\lambda}_n^2 - \mu \hat{z}_n, \qquad \bar{\gamma}_n = \hat{\lambda}_n^2 - \frac{\rho}{G}\bar{\omega}_n^2 = \hat{\lambda}_n^2 - \mu \bar{z}_n$$

To get θ, we integrate Eq. 11.78 with respect to x to get

$$\theta(x,t) = \sum_{n=1}^{\infty} \psi_n(x) \Big[\hat{\gamma}_n A_n \cos \hat{\omega}_n t + \hat{\gamma}_n B_n \sin \hat{\omega}_n t$$

$$+ \bar{\gamma}_n C_n \cos \bar{\omega}_n t + \bar{\gamma}_n D_n \sin \bar{\omega}_n t \Big] \tag{11.79}$$

where $\psi_n(x)$ is the antiderivative of the eigenfunction $\varphi_n(x)$. In the context of the present example, we have $\varphi_n(x) = \sin(\hat{\lambda}_n x)$. Therefore, we can find $\psi_n(x) = -\cos(\hat{\lambda}_n x)/\hat{\lambda}_n$. Now we can assert the initial conditions on the rotation field as

$$\theta(x,0) = \sum_{n=1}^{\infty} \psi_n(x) \Big[\hat{\gamma}_n A_n + \bar{\gamma}_n C_n \Big] = \theta_o(x)$$

$$\dot{\theta}(x,0) = \sum_{n=1}^{\infty} \psi_n(x) \Big[\hat{\gamma}_n \hat{\omega}_n B_n + \bar{\gamma}_n \bar{\omega}_n D_n \Big] = 0 \tag{11.80}$$

We can use the orthogonality properties of $\varphi_n(x)$ and $\psi_n(x)$ to solve for the constants in the sum. Denote

$$W_n^o = \frac{\langle \varphi_n, w_o \rangle}{\langle \varphi_n, \varphi_n \rangle}, \qquad \Theta_n^o = \frac{\langle \psi_n, \theta_o \rangle}{\langle \psi_n, \psi_n \rangle} \tag{11.81}$$

where the bracket notation $\langle \bullet, \bullet \rangle$ means the integral of the product of the two functions over the domain. Because we are considering initial conditions with zero velocity, it is straightforward to show that $B_n = 0$ and $D_n = 0$ from the second equation in each of Eqs. 11.76 and 11.80. From the first of each of these equations, we can solve for the other constants as

$$A_n = \frac{\bar{\gamma}_n W_n^o - \Theta_n^o}{\bar{\gamma}_n - \hat{\gamma}_n}, \qquad C_n = \frac{\hat{\gamma}_n W_n^o - \Theta_n^o}{\hat{\gamma}_n - \bar{\gamma}_n} \tag{11.82}$$

To compute the moment $M(x,t)$ for the Timoshenko beam, recall that moment is proportional to the derivative of rotation, i.e., $M = EI\theta'$. We can compute the derivative of the rotation function from Eq. 11.78 and substitute it into the moment equation to get

$$M(x,t) = EI \sum_{n=1}^{\infty} \varphi_n(x) \big[\hat{\gamma}_n A_n \cos \hat{\omega}_n t + \bar{\gamma}_n C_n \cos \bar{\omega}_n t \big] \qquad (11.83)$$

The shear can be computed similarly.

11.5.3   Numerical Example

The modal components of the solution for the simple–simple beam of length $L = 10$, $E = 80$, $G = 10$, $\rho = 0.6$, $A = 0.05$, and $I = 0.2$ are shown in Fig. 11.14. The analysis uses 150 modes, but only the first 30 are shown in the figure. These plots show that for the Timoshenko beam there are two frequencies for each mode number, each growing linearly with mode number. There are also two wave speeds associated with each mode. The larger wave speeds converge to the *longitudinal wave speed* wave speed $c_o = \sqrt{E/\rho}$. This wave speed is associated with flexural waves. The smaller wave speeds converge to the *shear wave speed* $c_s = \sqrt{G/\rho}$. The ratio of shear wave speed to longitudinal wave speed for this problem is $c_s/c_o = \sqrt{E/G} = 0.35$.

The low value of the shear modulus is selected to accentuate the different propagation speeds of the flexural and shear waves. Figure 11.15 shows the displacement, rotation, and moment at four different times. The initial wave is given by Eq. 11.35. In this case $a = 4$ and $b = 6$. The initial rotation is taken to be zero (which should excite shear waves). Observe that there is no moment initially because there is no rotation (and therefore, no first derivative of rotation). Because $w' + \theta$ is large, there are initial shears that quickly evolve moments (as one can see already at $t = 0.100$).

As expected, the initial wave splits in two and propagates in both directions. At the two later times, it is clear that there are two waves propagating and different speeds. Notice in Fig. 11.14 that the modal wave propagation speeds converge to a constant value very quickly (within the first few modes). Therefore, one would expect the propagating waves to maintain coherence. That appears to be true and can be confirmed by running the analysis for a longer period of time.

The structure of the computations is very similar to the classical Rayleigh wave programs for the separation of variables solution. Code 11.8 produced Figs. 11.14 and 11.15.

Code 11.8   MATLAB code to generate the modal solution of the simple–simple Timoshenko beam with an initial displacement $w_o(x)$

```
%. Timoshenko beam wave propagation
   clear; clc;

%. Problem parameters
```

```
    nModes = 150;                    % Number of modes
    L = 10.0;                        % Length of beam
    Density = 0.6;                   % Mass density (mass per unit volume)
    E = 80;                          % Young's modulus
    G = 10;                          % Shear modulus
    Area = 0.05;                     % Cross-sectional area
    Inertia = 0.2;                   % Moment of inertia
    k = sqrt(Inertia/Area);          % Radius of gyration
    co = sqrt(E/Density);            % Nominal wave speed
    mu = E/G;                        % Dimensionless ratio of E/G
    EI = E*Inertia;                  % Bending stiffness

%. Compute and plot results
%  Start time      Stop time        Time steps      Spatial points
    T.to = 0;       T.tf = 0.30;     T.nT = 61;      T.nS = 250;
    [Time,x,w,theta,M,A,Freq] = ModalSum(T,L,k,co,mu,EI,nModes);

    PlotInitialWave(x/L,w,theta,M);
    PlotCoefficients(A,Freq,nModes);
    PlotSnapshots(x/L,w,theta,M,Time,T);
    AnimateBeam(T,x/L,w,theta,M,Time)

%- End of main program ------------------------------------------------

function [Time,x,w,r,M,A,Freq] = ModalSum(T,L,k,co,mu,EI,nModes)
%   Compute w and M by modal sum for simple-simple Timoshenko beam

    x = linspace(0,L,T.nS);          % Spatial grid points
    w = zeros(T.nT,T.nS);            % Initialize w(x,t)
    r = zeros(T.nT,T.nS);            % Initialize Theta(x,t)
    M = zeros(T.nT,T.nS);            % Initialize M(x,t)
    Time = linspace(T.to,T.tf,T.nT)';% Time grid points
    Phi = zeros(nModes,T.nS);        % First mode shape
    Psi = zeros(nModes,T.nS);        % Second mode shape
    Freq = zeros(nModes,4);          % Natural frequencies
    A = zeros(nModes,4);             % Coefficients

%. Compute coefficients
    for n=1:nModes
      Ln = n*pi/L;
      Phi(n,:) = sin(Ln.*x);
      Psi(n,:) = -cos(Ln.*x)./Ln;
      aa = mu; bb = Ln^2*(1+mu)+1/k^2; cc = Ln^4;
      zHat = (bb+sqrt(bb^2-4*aa*cc))/(2*aa);
      zBar = (bb-sqrt(bb^2-4*aa*cc))/(2*aa);
      Freq(n,1) = co*sqrt(zHat);
      Freq(n,2) = co*sqrt(zBar);
      Wo = WoCoeff(Ln,L);
      gHat = Ln^2 - mu*zHat;
      gBar = Ln^2 - mu*zBar;
      A(n,1) = gBar*Wo/(gBar-gHat);
      A(n,2) = gHat*Wo/(gHat-gBar);
      A(n,3) = gHat*A(n,1);
      A(n,4) = gBar*A(n,2);
      Freq(n,3:4) = Freq(n,1:2)/(Ln*co);
    end

%. Compute response time history
    for i=1:T.nT
      t = Time(i);
      for n=1:nModes
        CosHat = cos(Freq(n,1)*t);
        CosBar = cos(Freq(n,2)*t);
        Wn = A(n,1)*CosHat + A(n,2)*CosBar;
        Rn = A(n,3)*CosHat + A(n,4)*CosBar;
        w(i,:) = w(i,:) + Phi(n,:)*Wn;
        r(i,:) = r(i,:) + Psi(n,:)*Rn;
```

```
        M(i,:) = M(i,:) + Phi(n,:)*Rn*EI;
      end
    end

end

%-------------------------------------------------------------------
function Wo = WoCoeff(Ln,L)
%  Compute the coefficients for initial displacement, Timoshenko beam
%    Wo = (4*xi*(1-xi))^N,    xi = (x-a)/(b-a)   a < x < b

    a = L/2 - L/10;                        % Start of initial wave
    b = L/2 + L/10;                        % End of initial wave
    N = 4;                                 % Exponent for waveform
    w =@(x) (4.*x.*(1-x)).^N;              % Wave function
    z =@(x) (b-a)*x + a;                   % z(x) maps [0,1] -> [a,b]

    phi =@(xi) sin(Ln*xi);
    NormFcn =@(x) phi(x).*phi(x);
    WaveFcn =@(x) w(x).*phi(z(x))*(b-a);
    aa = integral(NormFcn,0,L);
    bb = integral(WaveFcn,0,1);
    Wo = bb/aa;

end
```

The functions `PlotCoefficients` and `PlotSnapshots` produce the figures shown in the book, but it is worth noting that wave propagation is best understood through animation. The function `AnimateBeam` does essentially what the snapshot function does, except at more times and they are gathered together in a movie. The results provide an animation of the response. In the animation the presence of two wave speeds is much more obvious.

11.6 Summary

The classical solutions to the beam wave propagation problem found in this chapter using the technique of separation of variables are important for a few reasons. First, they demonstrate that classical solutions can be found and how to do it. Second, they show some key differences among the three beam theories that give insight into how these equations behave. Finally, they motivate the search for alternative methods to solve this problem. The subsequent chapters will explore the finite element method where it will be simpler to add features like applied loads and nonlinearity.

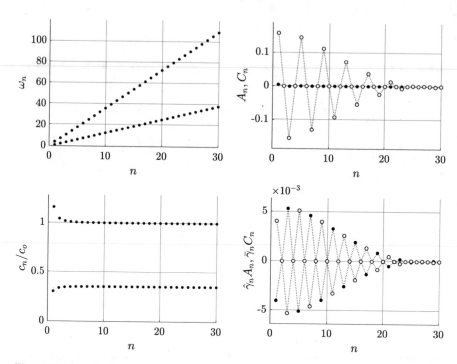

Fig. 11.14 *Timoshenko beam wave propagation. Modal components.* Simple–simple beam of length $L = 10$, $E = 80$, $G = 10$, $\rho = 0.6$, $A = 0.05$, and $I = 0.2$. The plots of frequency and wave speed show the two values that come from Timoshenko beam theory. The coefficients for displacement w and moment M are at right on top and bottom, respectively. Solid dots are A_n and $\hat{\gamma}_n A_n$ and the open circles are C_n and $\bar{\gamma}_n C_n$

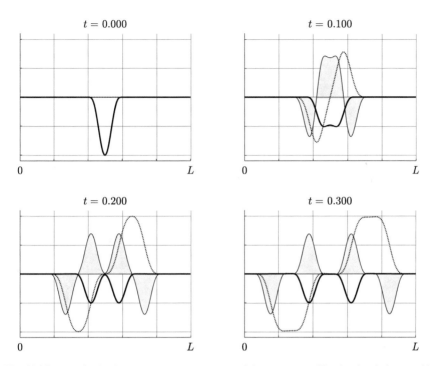

Fig. 11.15 *Timoshenko beam wave propagation. Modal components.* Simple–simple beam with the same parameters as those in Fig. 11.14. The four plots show the displacement (dark line), rotation (dotted line), and moment (thin line, shaded area) at four different times

Chapter 12
Finite Element Analysis of Linear Planar Beams

In the previous chapter we were able to find solutions to the beam dynamics problem using the separation of variables technique. The process went very much like it did for the axial bar problem. But the beam differential equation involves fourth order derivatives with respect to x, and consequently the spatial part of the solution was more complicated. While the classical techniques provide a good route to solving wave propagation problems in beams, we want to solve more general problems (e.g., problems with time-dependent forcing functions and possibly non-prismatic cross sections). In this chapter we will approach the problem in a different way. We will start by introducing the principle of virtual work and develop an approximate numerical approach using the Ritz method with finite element base functions.

The road map is as follows. We will start with the Bernoulli–Euler beam because it has the simplest mathematical description. As we develop the approach in that context, we will gain an appreciation for the method as it is applied to beam theory. With that background, we will move on to the Rayleigh beam and then the Timoshenko beam. This chapter concerns only the linear theories. We will extend the approach to the fully nonlinear case in Chaps. 13 and 14. Through some of the examples, we will connect results from this chapter with results obtained with the classical separation of variables approach from Chap. 11.

12.1   The Dynamic Principle of Virtual Work

Let us start with the Bernoulli–Euler beam model. Even though we know from the classical solution that this model is not a good one for capturing wave propagation in beams, it is mathematically the simplest and is the one most commonly included in textbooks on structural dynamics. The Bernoulli–Euler model assumes that the

Electronic Supplementary Material   The online version of this chapter (https://doi.org/10.1007/978-3-030-89944-8_12) contains supplementary material, which is available to authorized users.

K. D. Hjelmstad, *Fundamentals of Structural Dynamics*,
https://doi.org/10.1007/978-3-030-89944-8_12

rotary inertia associated with the cross section is negligible and shear deformations are zero. The equation of motion is

$$\rho A \ddot{w} + \left(E I w'' \right)'' - q = 0 \tag{12.1}$$

where ρ is the density, A is the cross-sectional area, EI is the bending stiffness, $q(x, t)$ is the applied transverse load, and $w(x, t)$ is the transverse displacement. It is possible that EI can depend upon x, in which case the beam would be *non-prismatic*.

One way to think about the solution to a differential equation is that any function $w(x, t)$ can be put into Eq. 12.1 as a candidate *configuration* of the system. The value of the left side of the equation for a candidate configuration is called the *residual*—the amount by which that configuration fails to satisfy the equation if it turns out not to be zero. In general, the residual varies with x and t. The *classical* solution to the differential equation is the configuration with zero residual for all x and t.

The *method of weighted residuals*, often referred to as *the principle of virtual work*, provides an alternative framework for solving the differential equation. Multiply the left side of Eq. 12.1 by an arbitrary function $\bar{w}(x)$ and integrate the result from 0 to L to get the virtual work functional

$$\mathcal{G} = \int_0^L \left(\rho A \ddot{w} + \left(E I w'' \right)'' - q \right) \bar{w} \, dx \tag{12.2}$$

If $w(x, t)$ is a solution to the differential equation, then $\mathcal{G}$ will be zero, no matter what $\bar{w}(x)$ is. The *principle of virtual work* says that if $\mathcal{G} = 0$ for all choices of the virtual displacement $\bar{w}(x)$, then $w(x, t)$ is the solution to the original equation of motion. The fundamental theorem of the calculus of variations proves this assertion.

Integrating the second term by parts (with respect to x) twice gives

$$\mathcal{G} = \int_0^L \left(\rho A \ddot{w} \bar{w} + E I w'' \bar{w}'' - q \bar{w} \right) dx + \left(E I w'' \right)' \bar{w} \Big|_0^L - E I w'' \bar{w}' \Big|_0^L$$

The virtual displacement $\bar{w}(x)$ is not a real configuration of the system, but rather an arbitrary function (of x, not t) with the same character as the real displacement $w(x, t)$. We generally require the virtual displacement to satisfy homogeneous essential boundary conditions associated with the real displacement and rotation. If we do so, then the boundary terms that appeared when we integrated by parts vanish, in which case we can write the virtual work functional as

$$\mathcal{G} = \int_0^L \left(\rho A \ddot{w} \bar{w} + E I w'' \bar{w}'' - q \bar{w} \right) dx \tag{12.3}$$

The statement $\mathcal{G} = 0$ for all $\bar{w}$ is sometimes called the *weak form* of the differential equation. We will relax the requirement "for all $\bar{w}$" to come up with an approximate numerical method to solve the problem. Notice that we have not assumed a prismatic cross section in the derivation of the equation of motion, either in the original form or in the virtual work form, and the integration by parts did not rely on that. Hence, these formulations are not restricted to prismatic beams.

A Specific Case of Boundary Conditions To get a feel for what happens with the boundary terms that came from integration by parts, consider the specific case of the fixed–simple beam. For this case we have the boundary conditions

$$w(0, t) = 0, \quad w'(0, t) = 0, \quad w(L, t) = 0, \quad M(L, t) = 0$$

We can convert the last boundary condition to one involving the displacement function by noting that $M = -EIw''$. Hence, the last condition can be rewritten as $w''(L, t) = 0$. The first three boundary conditions are called *essential boundary conditions*. The last one is called a *natural boundary condition*. For natural boundary conditions, the terms that appear when we integrate by parts vanish naturally. That is not the case for the essential boundary conditions. Hence, we will insist that the virtual displacements satisfy the boundary conditions:

$$\bar{w}(0) = 0, \quad \bar{w}'(0) = 0, \quad \bar{w}(L) = 0$$

These are called *homogeneous* essential boundary conditions.[1] With this restriction, all of the boundary terms that appeared when we integrated G by parts vanish—three of them because the virtual displacement or slope is zero at that point and the fourth because the real moment is zero at that point (a *natural boundary condition*).

12.1.1 The Ritz Approximation

The virtual work form of the equations of motion are ideal for using a Ritz approximation of the unknown function $w(x, t)$. Let the displacement field be written as

$$w(x, t) = \sum_{n=1}^{N} \psi_n(x) u_n(t) \tag{12.4}$$

where $\psi_n(x)$ is a known function of x and $u_n(t)$ is an as yet unknown function of t. The function $\psi_n(x)$ is called a *Ritz function*. These functions form a basis

---

[1] It is possible for the real displacement to be prescribed but not equal to zero, e.g., $w(0, t) = w_o$. That is a non-homogeneous essential boundary condition. The reason the essential boundary conditions must be homogeneous for the virtual displacements is precisely because we want the terms that result from integration by parts to vanish.

for our approximation space, so we often call them *base functions*. Taking the displacement in this form is reminiscent of the separation of variables strategy, but there is no requirement that the Ritz functions be eigenfunctions and the time-dependent functions are not necessarily sinusoidal. The most basic requirement is that the Ritz functions be linearly independent and complete.[2] We can write the sum in Eq. 12.4 in matrix form as

$$w(x, t) = \boldsymbol{\psi}^T(x) \, \mathbf{u}(t) \tag{12.5}$$

where the arrays of Ritz functions and Ritz parameters are defined as

$$\boldsymbol{\psi}(x) = \left\{ \begin{array}{c} \psi_1(x) \\ \vdots \\ \psi_N(x) \end{array} \right\}, \quad \mathbf{u}(t) = \left\{ \begin{array}{c} u_1(t) \\ \vdots \\ u_N(t) \end{array} \right\} \tag{12.6}$$

The virtual displacement function can also be written as a Ritz expansion

$$\bar{w}(x) = \bar{\mathbf{u}}^T \boldsymbol{\psi}(x) \tag{12.7}$$

where the constant array $\bar{\mathbf{u}}$ is organized in the same way as $\mathbf{u}(t)$. Using the same Ritz functions $\boldsymbol{\psi}(x)$ for the virtual displacement is called a *Galerkin approximation*. It is, of course, possible to interpolate the virtual displacements with functions different from the ones used for the real displacement, but there will be an advantage to using the same functions for the problems we are solving.

We can convert the equation of motion $\mathcal{G} = 0$ into a system of *ordinary* differential equations by substituting the Ritz approximation into the virtual work functional. To wit,

$$\mathcal{G} = \int_0^L \left[\rho A (\bar{\mathbf{u}}^T \boldsymbol{\psi})(\boldsymbol{\psi}^T \ddot{\mathbf{u}}) + EI (\bar{\mathbf{u}}^T \boldsymbol{\psi}'')(\boldsymbol{\psi}''^T \mathbf{u}) - q(\bar{\mathbf{u}}^T \boldsymbol{\psi}) \right] dx$$

The virtual constants $\bar{\mathbf{u}}$ do not depend upon x, nor do the real functions $\mathbf{u}(t)$, so they can be pulled out of the integrals. If we define the matrices

$$\mathbf{M} = \int_0^L \rho A \, \boldsymbol{\psi} \boldsymbol{\psi}^T \, dx, \quad \mathbf{K} = \int_0^L EI \, \boldsymbol{\psi}'' \boldsymbol{\psi}''^T \, dx, \quad \mathbf{f} = \int_0^L q \, \boldsymbol{\psi} \, dx \tag{12.8}$$

then the virtual work functional reduces to

---

[2] The notion of *completeness* is essentially a measure of the ability of the Ritz base functions to represent certain basic functions. For example, if the Ritz functions are based on polynomials, you would not leave out the x^2 term because that would hamper the ability to represent low-order polynomials. One commonly-used strategy is the *patch test*, which tests the ability of the base functions to represent a constant.

$$\mathcal{G} = \bar{\mathbf{u}}^T \big[\mathbf{M}\ddot{\mathbf{u}}(t) + \mathbf{K}\mathbf{u}(t) - \mathbf{f}(t) \big] \tag{12.9}$$

Due to the discretization, the statement "for all $\bar{w}(x)$" in the continuous version of the virtual work equation reduces to "for all $\bar{\mathbf{u}}$" in the discrete version. Note that the matrices $\mathbf{M}$, $\mathbf{K}$, and $\mathbf{f}$ can be computed without knowing the solution $\mathbf{u}(t)$ to the problem.

If $\mathcal{G} = 0$ is to hold for all $\bar{\mathbf{u}}$, then the quantity in square brackets in Eq. 12.9 must be zero. The proof is simple. Just pick a $\bar{\mathbf{u}}$ exactly equal to the term in square brackets (at any particular time). The inner product of any vector with itself is the length of that vector. The only way the length of a vector can be zero is if it is the zero vector. Thus, the virtual work equation implies the discrete equation of motion

$$\mathbf{M}\ddot{\mathbf{u}}(t) + \mathbf{K}\mathbf{u}(t) - \mathbf{f}(t) = \mathbf{0} \tag{12.10}$$

This process is often called *semi-discretization* because it only eliminates one of the two independent variables and the result is still a differential equation, albeit an *ordinary* differential equation. The beauty of this result is that the discrete equations of motion are identical in mathematical form to what we got for the shear building, truss, and discretized axial bar problems. Therefore, all of the methods to solve those problems are available to us. In particular, we could integrate those equations in time either with modal analysis or numerically with Newmark's method.

12.1.2 Initial Conditions

The initial conditions for the dynamic beam are restrictions specified for all x at time $t = 0$. Since the primary variable is the displacement $w(x, t)$, the initial conditions apply to the displacement and velocity. Specifically,

$$w(x, 0) = w_o(x), \qquad \dot{w}(x, 0) = v_o(x) \tag{12.11}$$

where $w_o(x)$ and $v_o(x)$ are known functions of x alone. To implement the initial conditions in the context of the Ritz approximation, we must specify initial values of the time-dependent functions $\mathbf{u}(t)$. Let $\mathbf{u}_o$ and $\mathbf{v}_o$ be discrete specified values such that

$$\mathbf{u}(0) = \mathbf{u}_o, \qquad \dot{\mathbf{u}}(0) = \mathbf{v}_o \tag{12.12}$$

These discrete initial values must be consistent with the functions $w_o(x)$, $v_o(x)$, and the Ritz interpolation functions $\boldsymbol{\psi}(x)$. From the Ritz expansion, Eq. 12.5, we can write

$$\tilde{w}_o(x) = \boldsymbol{\psi}^T(x)\,\mathbf{u}_o, \qquad \tilde{v}_o(x) = \boldsymbol{\psi}^T(x)\,\mathbf{v}_o \tag{12.13}$$

where $\tilde{w}_o(x)$ and $\tilde{v}_o(x)$ are simply the interpolated values of $\mathbf{u}_o$ and $\mathbf{v}_o$, respectively. We can find the discrete initial values by minimizing the square of the difference between the specified functions and the interpolated functions. For the initial displacement, for example, define the function

$$J(\mathbf{u}_o) = \int_0^L \tfrac{1}{2}\rho A(\tilde{w}_o(\mathbf{u}_o) - w_o)^2 dx \qquad (12.14)$$

We can find the minimum of this function by setting its directional derivative equal to zero, i.e., $DJ(\mathbf{u}_o) \cdot \bar{\mathbf{u}} = 0$, where $\bar{\mathbf{u}}$ is an arbitrary variation of $\mathbf{u}_o$. Using the chain rule, this equation can be written as

$$\int_0^L \rho A(\tilde{w}_o(\mathbf{u}_o) - w_o)\big[D\tilde{w}_o(\mathbf{u}_o) \cdot \bar{\mathbf{u}}\big] dx = 0 \qquad (12.15)$$

The directional derivative of $\tilde{w}^o$ can be computed from Eq. 12.13(a) as

$$D\tilde{w}_o(\mathbf{u}_o) \cdot \bar{\mathbf{u}} = \frac{d}{d\varepsilon}\Big[\boldsymbol{\psi}^T(x)\big(\mathbf{u}_o + \varepsilon\bar{\mathbf{u}}\big)\Big]_{\varepsilon=0} = \boldsymbol{\psi}^T(x)\bar{\mathbf{u}}$$

Substituting this result and the expression for $\tilde{w}_o$ from Eq. 12.13(a) into Eq. 12.15, noting that $\bar{\mathbf{u}}$ is constant, we get

$$\bar{\mathbf{u}}^T \int_0^L \big(\rho A\boldsymbol{\psi}\boldsymbol{\psi}^T\mathbf{u}_o - \rho A\boldsymbol{\psi}w_o\big) dx = 0$$

In order for this to be true for any choice of $\bar{\mathbf{u}}$, we must have

$$\int_0^L \rho A\,\boldsymbol{\psi}\boldsymbol{\psi}^T dx\,\mathbf{u}_o = \int_0^L \rho A\,\boldsymbol{\psi}\,w_o\,dx$$

Following the same process, we can derive an analogous condition for determining the initial velocity as

$$\int_0^L \rho A\,\boldsymbol{\psi}\boldsymbol{\psi}^T dx\,\mathbf{v}_o = \int_0^L \rho A\,\boldsymbol{\psi}\,v_o\,dx$$

Recognizing that the coefficient matrix on the left side of both equations is simply the mass matrix $\mathbf{M}$, we can solve these equations to get the initial values of the Ritz displacement and velocity parameters as

$$\mathbf{u}_o = \mathbf{M}^{-1}\int_0^L \rho A\,\boldsymbol{\psi}\,w_o\,dx, \qquad \mathbf{v}_o = \mathbf{M}^{-1}\int_0^L \rho A\,\boldsymbol{\psi}\,v_o\,dx$$

This is essentially a projection of the functions $w_o(x)$ and $v_o(x)$ onto the subspace spanned by the Ritz functions $\psi(x)$. The quality of the projection will depend upon the quality of the approximating subspace.

12.1.3 Selection of Ritz Functions

The virtual work approach requires the selection of functions $\psi(x)$ to form the Ritz basis. The functions must make sense from the point of view of establishing and improving the approximation of the original problem. The approximation has a finite number of terms. Contrast that with the infinite series that characterized the separation of variables approach. Orthogonality of the eigenfunctions played an important role in implementing the infinite series solution that emanated from separation of variables. The Ritz functions could be orthogonal, but in general are not. For a non-orthogonal basis, the evaluation of the coefficients generally involves the solution of a system of equations. Hence, having a finite number of terms is essential to the success of the method.

We could use the eigenfunctions from separation of variables to form the Ritz basis, but they can be a little cumbersome to work with. We could use a set of functions like the polynomials, but without orthogonalization they quickly become ill-conditioned. To capture a dynamic motion usually requires a fairly high order approximation, so polynomials defined over the entire length of the beam are not workable.

The approach that we will take to establish the Ritz functions is the *finite element method*. In this approach we will use simple polynomial functions defined over subregions of the beam (called elements). The functions will, therefore, be defined *piecewise*. The finite element functions that we used for the axial bar problem (see Fig. 9.15) were C^0 continuous (even with the addition of bubble functions). The functional $\mathcal{G}$ for the Bernoulli–Euler beam contains second derivatives of w with respect to x. The second derivatives of C^0 functions do not exist.[3] Thus, for beam theory, we will require Ritz functions that are continuous and have continuous first derivatives. Functions that satisfy this level of continuity are called C^1 continuous functions and will assure that all terms in $\mathcal{G}$ can be computed.

---

[3] More precisely, the second derivatives are zero *almost everywhere* with infinite jumps at the locations where the first derivative is discontinuous. These functions fail to be square integrable, which is the mathematical requirement for functions to be admissible in the virtual work setting.

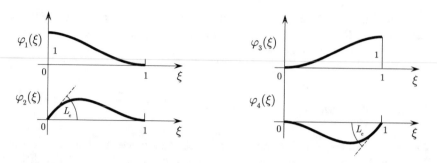

Fig. 12.1 *Cubic Hermitian polynomials.* The cubic Hermitian polynomials allow us to specify the displacement and rotation at each end of the segment

12.1.4   Beam Finite Element Functions

We can build suitable beam finite element functions from the cubic Hermitian polynomials shown in Fig. 12.1. These functions are familiar from elementary structural analysis as the solution to the static prismatic Bernoulli–Euler beam problem subjected only to end loads. These functions specify the values of the function and its first derivative, thereby providing the tools to enforce C^1 continuity of the approximate solution built from them.

Assume that the range of x in element e is from $x_{i(e)}$ to $x_{j(e)}$. Introduce the change of variable $\xi = (x - x_{i(e)})/L_e$, where $L_e = x_{j(e)} - x_{i(e)}$ is the length of the element. With this definition, the cubic Hermitian functions have the form

$$
\begin{aligned}
\varphi_1(\xi) &= 1 - 3\xi^2 + 2\xi^3 &\qquad \varphi_3(\xi) &= 3\xi^2 - 2\xi^3 \\
\varphi_2(\xi) &= \left(\xi - 2\xi^2 + \xi^3\right)L_e &\qquad \varphi_4(\xi) &= \left(-\xi^2 + \xi^3\right)L_e
\end{aligned}
\tag{12.16}
$$

These functions are called *shape functions* because they establish the shape but not magnitude of the solution, which will come from the $\mathbf{u}(t)$ values once we solve the discrete equations of motion. We will use the shape functions to build the Ritz functions $\psi(x)$.

Notice that the functions $\varphi_1(\xi)$ and $\varphi_3(\xi)$ are dimensionless, but the functions $\varphi_2(\xi)$ and $\varphi_4(\xi)$ have dimensions of length due to the presence of the L_e multiplier. The reason for this difference is the change of variable implicit in ξ. In order to have unit slope, we must make sure that

$$
\frac{d}{dx}\left(\varphi_2(\xi)\right)\Big|_{\xi=0} = 1
$$

When we differentiate with respect to x, our change of variable implies, by the chain rule for differentiation, that

$$\frac{d}{dx}\big(f(\xi)\big) = \frac{df}{d\xi}\frac{d\xi}{dx} = \frac{1}{L_e}f'(\xi)$$

where the prime means differentiation with respect to the argument of the function (which is ξ). In the case of $\varphi_2(\xi)$, that means

$$\frac{d}{dx}\big(\varphi_2(\xi)\big) = \frac{1}{L_e}\frac{d}{d\xi}\Big(\big(\xi - 2\xi^2 + \xi^3\big)L_e\Big) = 1 - 4\xi + 3\xi^2$$

which evaluates to 1 at $\xi = 0$, as it should. With this strategy, the rotations that we take as degrees of freedom for the beam problem will directly relate to the first derivative of the displacement function with respect to x and thereby enforce continuity of the first derivative of w from one element to the next.

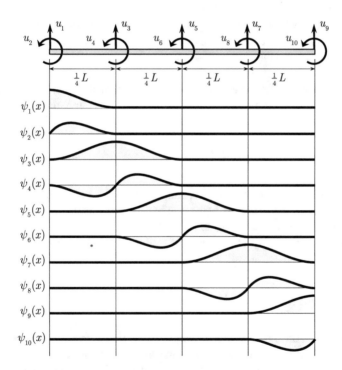

Fig. 12.2 *Degrees of freedom and associated Ritz functions.* The nodal displacements and rotations are labeled $u_1, \ldots, u_{10}$. The functions $\psi_1(x), \ldots, \psi_{10}(x)$ are shown

12.1.5   Ritz Functions and Degrees of Freedom

The beam is a collection of *nodes* connected by *elements*. For simplicity, we will take the nodes to be equally spaced and number them sequentially from left to right. Number the components of the displacement variables $\mathbf{u}(t)$ in nodal order. The translational degree of freedom is numbered first, followed by the rotational degree of freedom as shown in Fig. 12.2.

The displacement variables correspond with the Ritz functions shown in Fig. 12.2. The number of Ritz functions equals the number of degrees of freedom in the system. Here we have not shown any restrained degrees of freedom, so boundary conditions are not yet enforced. There are several things to notice about the Ritz functions sketched in Fig. 12.2. The beam has been broken up into four equal segments of length $L_e = L/4$. There are five nodes. Each Ritz function is built from the shape functions $\varphi_i(\xi)$. We pair φ_1 with φ_3 and φ_2 with φ_4 to accomplish the task of assuring continuity. The Ritz functions are continuous and smooth (i.e., their first derivatives are also continuous) and they are compact in the sense that after they have ramped up and back down, they are zero outside of that range.

We can use these Ritz functions directly in Eq. 12.8. Because the Ritz functions are defined over the four regions (the *elements*), it makes sense to imagine the integrals from 0 to L being done as the sum of integrals over elements. To wit,

$$\int_0^L (\bullet)\,dx = \sum_{e=1}^M \int_{x_{i(e)}}^{x_{j(e)}} (\bullet)\,dx$$

where $x_{i(e)}$ is the x location of the left end of element e, $x_{j(e)}$ is the location of the right end of the element, and M is the number of elements. Since an integral is the limit of a sum, it is always possible to break it up into a sum of integrals over the subdomains. We need a systematic way of setting up this calculation.

12.1.6   Local to Global Mapping

We can set up a convenient way to assemble the global stiffness matrix $\mathbf{K}$, the mass matrix $\mathbf{M}$, and global force vector $\mathbf{f}$ by introducing element degrees of freedom as shown in Fig. 12.3. Define an array that holds the four displacements and rotations

Fig. 12.3 *Degrees of freedom associated with element e*. We can localize the degrees of freedom that are associated with the element

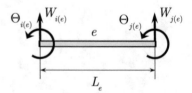

that are relevant to element e:

$$\mathbf{W}_e = \begin{Bmatrix} W_{i(e)} \\ \Theta_{i(e)} \\ W_{j(e)} \\ \Theta_{j(e)} \end{Bmatrix} \tag{12.17}$$

where $i(e)$ is the node number at the left end of element e and $j(e)$ is the node number at the right end of element e. As we have done previously, we will define an $M \times 2$ array ix, for which row e contains the $i(e)$ and $j(e)$ node numbers. In addition, row i of the array id contains the global DOF numbers associated with node i (translation followed by rotation). These arrays will help us to localize the portions of the array $\mathbf{u}(t)$ (stored in global DOF order) relevant to each element.

Define a matrix $\mathbf{B}_e$ that selects the displacements associated with element e from the global array $\mathbf{u}(t)$ as

$$\mathbf{W}_e(t) = \mathbf{B}_e \mathbf{u}(t) \tag{12.18}$$

where $\mathbf{B}_e$ is a $4 \times \mathcal{N}$ matrix with the form

$$\mathbf{B}_e = \begin{bmatrix} 0 & 1 & 0 & 0 & \cdots & 0 & 0 & 0 & 0 & \cdots \\ 0 & 0 & 1 & 0 & \cdots & 0 & 0 & 0 & 0 & \cdots \\ 0 & 0 & 0 & 0 & \cdots & 0 & 1 & 0 & 0 & \cdots \\ 0 & 0 & 0 & 0 & \cdots & 0 & 0 & 1 & 0 & \cdots \end{bmatrix} \tag{12.19}$$

where $\mathcal{N}$ is the number of degrees of freedom in the structure. The first row of $\mathbf{B}_e$ has a 1 in column id(ix(e, 1), 1); the second row has a 1 in column id(ix(e, 1), 2); the third row has a 1 in column id(ix(e, 2), 1); and the fourth row has a 1 in column id(ix(e, 2), 2). Now, the displacement field within element e can be written as

$$w_e(\xi, t) = \boldsymbol{\varphi}^T(\xi) \mathbf{W}_e(t) = \boldsymbol{\varphi}^T(\xi) \mathbf{B}_e \, \mathbf{u}(t) \tag{12.20}$$

The matrix of shape functions is

$$\boldsymbol{\varphi}^T(\xi) = \begin{bmatrix} \varphi_1(\xi), & \varphi_2(\xi), & \varphi_3(\xi), & \varphi_4(\xi) \end{bmatrix}$$

where the $\varphi_i(\xi)$ are the cubic Hermitian functions defined in Eq. 12.16.

12.1.7   Element Matrices and Assembly

The virtual work functional, Eq. 12.3, involves an integral over the domain of the beam, which can be expressed as the sum of integrals over the element subdomains.

To wit,

$$G = \sum_{e=1}^{M} \int_{x_{i(e)}}^{x_{j(e)}} \left(\rho A \bar{w}_e \ddot{w}_e + E I \bar{w}_e'' w_e'' - q \bar{w}_e \right) dx \qquad (12.21)$$

Within each element we can use the change of variable

$$\xi = \frac{x - x_{i(e)}}{L_e}, \qquad L_e = x_{j(e)} - x_{i(e)}$$

to give

$$G = \sum_{e=1}^{M} \int_0^1 \left(\rho A \bar{w}_e \ddot{w}_e + E I \bar{w}_e'' w_e'' - q \bar{w}_e \right) L_e \, d\xi \qquad (12.22)$$

Keep in mind that w_e and $\bar{w}_e$ are functions of x and the derivatives (shown as primes) in the equation are with respect to x. In the change of variable, we have only adjusted the limits of integration and noted $dx = L_e \, d\xi$. Since

$$w_e(x, t) = \boldsymbol{\varphi}^T(\xi) \mathbf{B}_e \mathbf{u}(t), \quad \bar{w}_e(x) = \bar{\mathbf{u}}^T \mathbf{B}_e^T \boldsymbol{\varphi}(\xi)$$

expresses the element displacement in terms of $\xi(x)$, we can compute the derivatives by the chain rule as

$$w_e'' = \frac{1}{L_e^2} \boldsymbol{\varphi}''^T \mathbf{B}_e \mathbf{u}, \quad \bar{w}_e'' = \frac{1}{L_e^2} \bar{\mathbf{u}}^T \mathbf{B}_e^T \boldsymbol{\varphi}''$$

where the primes on w_e and $\bar{w}_e$ mean derivatives with respect to x and the primes on the shape functions $\boldsymbol{\varphi}$ mean derivatives with respect to ξ.

With these definitions, the virtual work function takes the form

$$G = \sum_{e=1}^{M} \int_0^1 \left(\rho A \, \bar{\mathbf{u}}^T \mathbf{B}_e^T \boldsymbol{\varphi} \boldsymbol{\varphi}^T \mathbf{B}_e \ddot{\mathbf{u}} + \frac{E I}{L_e^4} \bar{\mathbf{u}}^T \mathbf{B}_e^T \boldsymbol{\varphi}'' \boldsymbol{\varphi}''^T \mathbf{B}_e \mathbf{u} \right) L_e \, d\xi$$

$$+ \sum_{e=1}^{M} \int_0^1 \left(q \, \bar{\mathbf{u}}^T \mathbf{B}_e^T \boldsymbol{\varphi} \right) L_e \, d\xi$$

Note that some of the terms in this equation depend upon the variable of integration ξ and others do not (e.g., $\mathbf{B}_e$, $\bar{\mathbf{u}}$, and $\mathbf{u}(t)$ do not). Those terms can be pulled out of the integral. Further, some of the terms do not depend upon the summation index e (e.g., $\mathbf{u}(t)$ and $\bar{\mathbf{u}}$). Those terms can be pulled out of the sum. Let us first define element mass, stiffness, and load matrices as

$$\mathbf{m}_e = \int_0^1 \rho A L_e \boldsymbol{\varphi} \boldsymbol{\varphi}^T d\xi$$

$$\mathbf{k}_e = \int_0^1 \frac{EI}{L_e^3} \boldsymbol{\varphi}'' \boldsymbol{\varphi}''^T d\xi \tag{12.23}$$

$$\mathbf{f}_e = \int_0^1 q\, L_e \boldsymbol{\varphi}\, d\xi$$

With these matrices defined, the virtual work functional can be written as

$$\mathcal{G} = \bar{\mathbf{u}}^T \left(\sum_{e=1}^M \mathbf{B}_e^T \mathbf{m}_e \mathbf{B}_e \ddot{\mathbf{u}}(t) + \sum_{e=1}^M \mathbf{B}_e^T \mathbf{k}_e \mathbf{B}_e \mathbf{u}(t) - \sum_{e=1}^M \mathbf{B}_e^T \mathbf{f}_e(t) \right) \tag{12.24}$$

Now, define *global* mass, stiffness, and applied force as

$$\mathbf{M} = \sum_{e=1}^M \mathbf{B}_e^T \mathbf{m}_e \mathbf{B}_e, \qquad \mathbf{K} = \sum_{e=1}^M \mathbf{B}_e^T \mathbf{k}_e \mathbf{B}_e, \qquad \mathbf{f} = \sum_{e=1}^M \mathbf{B}_e^T \mathbf{f}_e$$

It should be evident that these matrices are *assembled* in the standard way using the $\mathbf{B}_e$ arrays. Of course, the information needed for assembly is contained in the ix, id, and eDOF arrays in the computer implementation, so we will not directly form those arrays or do the matrix multiplications indicated when we implement those equations in a computer code.

With the definitions of the global matrices, Eq. 12.24 can be written as

$$\mathcal{G} = \bar{\mathbf{u}}^T \left[\mathbf{M}\ddot{\mathbf{u}}(t) + \mathbf{K}\mathbf{u}(t) - \mathbf{f}(t) \right] \tag{12.25}$$

Because the virtual displacement parameters $\bar{\mathbf{u}}$ are arbitrary, $\mathcal{G} = 0$ for all $\bar{\mathbf{u}}$ implies, by the fundamental theorem of the calculus of variations, that

$$\mathbf{M}\ddot{\mathbf{u}}(t) + \mathbf{K}\mathbf{u}(t) - \mathbf{f}(t) = \mathbf{0}, \tag{12.26}$$

which is exactly what we got from the basic Ritz approach. This formulation shows how the finite element shape functions factor in and what that implies about the assembly of the equations. The reason for going down this path is that it is important not only to define the Ritz functions, but also to formulate the problem in a way that we can enhance it and implement in a computer code.

Example (Closed form Solution for Prismatic Beams) We can compute the element mass and stiffness matrices for a prismatic beam. Using the cubic Hermitian shape functions, the element mass matrix for the prismatic beam is

$$\mathbf{m}_e = \frac{\rho A L_e}{420} \begin{bmatrix} 156 & 22L_e & 54 & -13L_e \\ 22L_e & 4L_e^2 & 13L_e & -3L_e^2 \\ 54 & 13L_e & 156 & -22L_e \\ -13L_e & -3L_e^2 & -22L_e & 4L_e^2 \end{bmatrix} \tag{12.27}$$

and the element stiffness matrix is

$$\mathbf{k}_e = \frac{EI}{L_e^3} \begin{bmatrix} 12 & 6L_e & -12 & 6L_e \\ 6L_e & 4L_e^2 & -6L_e & 2L_e^2 \\ -12 & -6L_e & 12 & -6L_e \\ 6L_e & 2L_e^2 & -6L_e & 4L_e^2 \end{bmatrix} \tag{12.28}$$

The details of the calculation are left as an exercise for the reader. As a sample from the calculation for the mass matrix, we can compute the (3,4) component

$$\begin{aligned} (\mathbf{m}_e)_{34} &= \rho A L_e^2 \int_0^1 \varphi_3(\xi) \varphi_4(\xi) \, d\xi \\ &= \rho A L_e^2 \int_0^1 \left(3\xi^2 - 2\xi^3 \right) \left(-\xi^2 + \xi^3 \right) d\xi \\ &= \rho A L_e^2 \int_0^1 \left(-3\xi^4 + 5\xi^5 - 2\xi^6 \right) d\xi \\ &= \rho A L_e^2 \left(-\tfrac{3}{5} + \tfrac{5}{6} - \tfrac{2}{7} \right) = -\tfrac{22}{420} \rho A L_e^2 \end{aligned}$$

which matches the term in the matrix above. The computation of the element force vector depends upon the nature of the distributed load $q(x, t)$. Often, this function will be expressed as a spatial distribution multiplied by a function of time. We will not pursue the implementation of the Bernoulli–Euler beam further because we will be able to get it as a special case of the Rayleigh beam.

12.2   The Rayleigh Beam

The difference between the Bernoulli–Euler beam and the Rayleigh beam is that the latter does not neglect the rotary inertia of the cross section while the former does. The equation of motion of the Rayleigh beam, Eq. 10.27, is (assuming that the distributed applied moment $m = 0$)

$$\left(EIw'' \right)'' - \left(Hw' \right)' - q + \rho A \ddot{w} - \rho I \ddot{w}'' = 0 \tag{12.29}$$

The axial force H, technically a second-order effect, is included in the formulation because it is simple to do so. We will proceed along the same lines that we did for the Bernoulli–Euler beam to see how these two theories compare.

12.2.1   Virtual Work for the Rayleigh Beam

To derive the virtual work form of the equations of motion, start with the weighted residual form of the virtual work functional

$$\mathcal{G} = \int_0^L \left((EIw'')'' - (Hw')' - q + \rho A\ddot{w} - \rho I\ddot{w}'' \right) \bar{w}\, dx \qquad (12.30)$$

Of course, $\mathcal{G}=0$ for all $\bar{w}$ implies Eq. 12.29. As usual, we will balance the spatial derivatives using integration by parts. In this case, integrate the first term by parts twice, the second term once, and the fifth term once to get the equivalent form

$$\mathcal{G} = \int_0^L \left(EIw''\bar{w}'' + Hw'\bar{w}' - q\bar{w} + \rho A\ddot{w}\bar{w} + \rho I\ddot{w}'\bar{w}' \right) dx$$
$$+ \left(EIw''' - Hw' - \rho I\ddot{w}' \right)\bar{w}\,\Big|_0^L - EIw''\bar{w}'\,\Big|_0^L \qquad (12.31)$$

From Eq. 11.10 we can write $V = -EIw''' + Hw' + \rho I\ddot{w}'$ and from the constitutive equation for moment we have $M = -EIw''$. When we implement the boundary conditions for a particular problem, we will have either $V(\bullet, t) = 0$ or $w(\bullet, t) = 0$ at $x = \bullet$. If the shear force V is zero, it is either a free or slide boundary condition. If the displacement w is zero, it is either a simple or fixed condition. In any event, shear and displacement are mutually exclusive boundary conditions—you can specify one or the other, but not both. By the same token, we will have either $M(\bullet, t) = 0$ or $w'(\bullet, t) = 0$ at $x = \bullet$. Moment is zero at a free or a simple end condition. Rotation is zero at a fixed or a slide end condition. Therefore, if we make $\bar{w}$ satisfy the homogeneous essential boundary conditions associated with w, the boundary terms in the virtual work equation will vanish, and the virtual work functional reduces to

$$\mathcal{G} = \int_0^L \left(EIw''\bar{w}'' + Hw'\bar{w}' - q\bar{w} + \rho A\ddot{w}\bar{w} + \rho I\ddot{w}'\bar{w}' \right) dx \qquad (12.32)$$

This expression is identical to what we had previously with the Bernoulli–Euler beam, but there are two additional terms, one for the effect of axial force and one from the effect of rotary inertia. This equation must hold for all virtual displacements. The principle of virtual work assures us that if $\mathcal{G}=0$ holds for all $\bar{w}(x)$, then the system satisfies the equations of motion.

12.2.2   Finite Element Discretization

The nodal degrees of freedom for the Rayleigh beam are exactly the same as the ones defined for the Bernoulli–Euler beam, illustrated in Fig. 12.2. Divide the beam into M finite elements of length L_e and use the same change of variable for element e, i.e., $\xi = (x - x_{i(e)})/L_e$ that we did previously. The displacements associated with element e are also the same as the ones defined for the Bernoulli–Euler beam, as illustrated in Fig. 12.3. Interpolate the element displacement and virtual displacement as

$$w_e(\xi, t) = \boldsymbol{\varphi}^T(\xi)\mathbf{B}_e\mathbf{u}(t), \quad \bar{w}_e(\xi) = \bar{\mathbf{u}}^T\mathbf{B}_e^T\boldsymbol{\varphi}(\xi) \tag{12.33}$$

A little experimentation with this formulation shows that the cubic Hermitian shape functions are not enough to accurately solve wave propagation problems. In particular, they imply that moments are linear within an element and shears are constant. To remedy this problem, we can introduce bubble functions to the C^1 interpolation, as shown in Sect. D.6 of Appendix D. All we need to do is add bubble functions to $\boldsymbol{\varphi}(\xi)$ and add appropriate rows to $\mathbf{B}_e$ to associate those functions with the additional global degrees of freedom. All of what we did for the Bernoulli–Euler beam previously can be modified in the same way.

Substitute the Ritz approximation into the virtual work form of the equations to generate the discrete version of the virtual work functional. To wit,

$$\mathcal{G} = \bar{\mathbf{u}}^T \left(\sum_{e=1}^{M} \mathbf{B}_e^T \mathbf{m}_e \mathbf{B}_e \ddot{\mathbf{u}}(t) + \sum_{e=1}^{M} \mathbf{B}_e^T \mathbf{k}_e \mathbf{B}_e \mathbf{u}(t) - \sum_{e=1}^{M} \mathbf{B}_e^T \mathbf{f}_e(t) \right)$$

where the element matrices are defined as

$$\mathbf{m}_e = \int_0^1 \left(\rho A L_e \boldsymbol{\varphi}\boldsymbol{\varphi}^T + \frac{\rho I}{L_e}\boldsymbol{\varphi}'\boldsymbol{\varphi}'^T \right) d\xi$$

$$\mathbf{k}_e = \int_0^1 \left(\frac{EI}{L_e^3}\boldsymbol{\varphi}''\boldsymbol{\varphi}''^T + \frac{H}{L_e}\boldsymbol{\varphi}'\boldsymbol{\varphi}'^T \right) d\xi \tag{12.34}$$

$$\mathbf{f}_e = \int_0^1 q\, L_e \boldsymbol{\varphi}\, d\xi$$

Note that H is the axial force and I is the moment of inertia in element e. The additional terms for the mass and stiffness matrices due to rotary inertia and geometric effects are

$$\mathbf{m}_e^R = \frac{\rho I}{L_e} \int_0^1 \boldsymbol{\varphi}'\boldsymbol{\varphi}'^T d\xi, \qquad \mathbf{k}_e^G = \frac{H}{L_e} \int_0^1 \boldsymbol{\varphi}'\boldsymbol{\varphi}'^T d\xi$$

Both of these element matrices have the common factor

$$\mathbf{g}_e = \frac{1}{L_e} \int_0^1 \boldsymbol{\varphi}' \boldsymbol{\varphi}'^T d\xi$$

such that $\mathbf{m}_e^R = \rho I \mathbf{g}_e$ and $\mathbf{k}_e^G = H \mathbf{g}_e$. Using the cubic Hermitian shape functions (without additional bubble modes), the matrix $\mathbf{g}_e$ can be computed explicitly as

$$\mathbf{g}_e = \frac{1}{30L_e} \begin{bmatrix} 36 & 3L_e & -36 & 3L_e \\ 3L_e & 4L_e^2 & -3L_e & -L_e^2 \\ -36 & -3L_e & 36 & -3L_e \\ 3L_e & -L_e^2 & -3L_e & 4L_e^2 \end{bmatrix} \tag{12.35}$$

Tensile axial force increases the stiffness while compressive axial force decreases it. In fact, if the compressive axial force is large enough, it will cause the beam to buckle. In a dynamic context, as the compressive force approaches the static buckling load, the lowest natural frequency of the system approaches zero. The implication of a zero frequency is that the period of vibration is infinite—the system displaces outward and does not return (in finite time).

Code 12.1 shows the organization of the computations to form the element mass, stiffness, and force arrays for the Rayleigh beam.

Code 12.1 MATLAB code to compute the element mass, stiffness, and force matrices for the Rayleigh beam

```
function [me,ke,fe] = RayleighElement(x,d,Load,S)
%   Form element mass, stiffness, and force for Rayleigh beam.
%
%         x : Nodal coordinates
%         d : Constitutive properties [L,E,A,I,rho,H,RotOn]
%         L : Length of beam
%      Load : Distributed load information [qo, mo]
%         S : Problem size parameters (nExt,nQpts,wt,st)

%. Extract load magnitudes and load type from Load
    LoadType = Load(1);                    % spatial load form
    qo = Load(2);                          % qo
    L = d(1);                              % Length of beam
    x0 = x(1,1);                           % Coordinate of i-node
    Le = x(2,1)-x(1,1);                    % Length of element

%. Initialize arrays
    nDOFe = 4+sum(S.nExt);                 % Number of DOF for element
    me = zeros(nDOFe,nDOFe);               % Element mass matrix
    ke = zeros(nDOFe,nDOFe);               % Element stiffness matrix
    fe = zeros(nDOFe,1);                   % Element force vector

%. Compute element tangent and residual by numerical integration
    for i=1:S.nQpts

%... Set element coordinate for integration point, get load
      xi = S.st(i);   xx = (x0 + xi*Le)/L;
      q = qo*LoadFunction(xx,1,LoadType);

%... Compute shape functions and kernel matrices
```

```
    [h,dh,ddh,~] = BubbleFcnsC1(xi,S.nExt,Le);
    hh = (h*h')*Le;
    DhDh = (dh*dh')/Le;
    DDhDDh = (ddh*ddh')/Le^3;

%... Compute cross sectional moduli and mass, d = [L,E,A,I,rho,H,RotOn]
    EI = d(2)*d(4);        H = d(6);
    rhoA = d(5)*d(3);      rhoI = d(5)*d(4)*d(7);

%... Compute the contributions to the element matrices
    ke = ke + (EI*DDhDDh + H*DhDh)*S.wt(i);
    me = me + (rhoA*hh + rhoI*DhDh)*S.wt(i);
    fe = fe + q*h*Le*S.wt(i);

    end

end
```

Observe that the element matrices are computed by numerical quadrature with the weights and stations set by the function NumInt called from the input routine and resident in the 'struct' S. The shape functions are found using the function BubbleFcnsC1 (Code D.3 in Appendix D). The physical properties of the beam are contained in the array d. Note that if the value of RotOn stored as d(7) is set to zero, then the code does the Bernoulli–Euler beam by ignoring the additional contributions to the element mass matrix from rotary inertia. The form of the distributed load is set in LoadFunction, with the first entry being the type of load function and the second being the magnitude. In this code, the axial force H is input as an element property (like E and ρ) in the array d. This allows some simple exploration of the effect of axial force on the dynamic properties of a beam. In the broader context of structures, the internal axial force develops as needed through the equations of motion.

Of course, the element matrices assemble to the global mass $\mathbf{M}$, stiffness $\mathbf{K}$, and load $\mathbf{f}$ matrices in the usual manner. Noting the revised definitions of the element and global mass, stiffness, and load matrices, the virtual work functional reduces to

$$\mathcal{G} = \bar{\mathbf{u}}^T \left[\mathbf{M}\ddot{\mathbf{u}}(t) + \mathbf{K}\mathbf{u}(t) - \mathbf{f}(t) \right]$$

Again, if $\mathcal{G} = 0$ for all $\bar{\mathbf{u}}$, the discrete equations of motion are, as before,

$$\mathbf{M}\ddot{\mathbf{u}}(t) + \mathbf{K}\mathbf{u}(t) - \mathbf{f}(t) = \mathbf{0}$$

12.2.3 Initial Conditions for Wave Propagation

We can examine the propagation of an initial displacement wave for the Rayleigh beam. Assume that the initial transverse displacement is $w_o(x)$. We need to establish the value of $\mathbf{u}_o = \mathbf{u}(0)$ as the initial condition for the discretized system. Observe that if we knew $\mathbf{u}_o$, then we could compute the displacement in element e using Eq. 12.33 as

$$\tilde{w}_e^o(\xi) = \boldsymbol{\varphi}^T(\xi)\mathbf{B}_e\mathbf{u}_o \tag{12.36}$$

where $\tilde{w}_e^o$ is simply the interpolated value of $\mathbf{u}_o$ within the element. Our problem is to compute $\mathbf{u}_o$, which is overdetermined because there are many elements that claim Eq. 12.36. We can solve the problem by minimizing the square of the difference between the specified value $w_e^o(\xi)$, which is the part of $w_o(x)$ in element e, and the value computed from Eq. 12.36. Define the function

$$J(\mathbf{u}_o) = \sum_{e=1}^{M} \int_0^1 \tfrac{1}{2}\rho A\big(\tilde{w}_e^o(\mathbf{u}_o) - w_e^o\big)^2 L_e \, d\xi \tag{12.37}$$

Note that for the purposes of this calculation we are considering $\tilde{w}_e^o$ to be a function of $\mathbf{u}_o$. The function J is minimized by setting the directional derivative equal to zero, i.e., $\mathcal{D}J(\mathbf{u}_o) \cdot \bar{\mathbf{u}} = 0$, where $\bar{\mathbf{u}}$ is an arbitrary variation of $\mathbf{u}_o$. Using the chain rule, this equation can be written as

$$\sum_{e=1}^{M} \int_0^1 \rho A\big(\tilde{w}_e^o(\mathbf{u}_o) - w_e^o\big)\big[\mathcal{D}\tilde{w}_e^o(\mathbf{u}_o) \cdot \bar{\mathbf{u}}\big] L_e \, d\xi = 0 \tag{12.38}$$

The directional derivative of $\tilde{w}_e^o$ can be computed from Eq. 12.36 as

$$\mathcal{D}\tilde{w}_e^o(\mathbf{u}_o) \cdot \bar{\mathbf{u}} = \frac{d}{d\varepsilon}\Big[\boldsymbol{\varphi}^T(\xi)\mathbf{B}_e\big(\mathbf{u}_o + \varepsilon\bar{\mathbf{u}}\big)\Big]_{\varepsilon=0} = \boldsymbol{\varphi}^T(\xi)\mathbf{B}_e\bar{\mathbf{u}}$$

Substituting this result and the expression for $\tilde{w}_e^o$ from Eq. 12.36 into Eq. 12.38 gives

$$\bar{\mathbf{u}}^T \sum_{e=1}^{M} \mathbf{B}_e^T \int_0^1 \big(\rho A\boldsymbol{\varphi}\boldsymbol{\varphi}^T\mathbf{B}_e\mathbf{u}_o - \rho A_e\boldsymbol{\varphi}w_e^o\big) L_e \, d\xi = 0$$

Since $\bar{\mathbf{u}}$ is arbitrary, the sum itself must be equal to zero. Hence,

$$\sum_{e=1}^{M} \mathbf{B}_e^T \int_0^1 \rho A\boldsymbol{\varphi}\boldsymbol{\varphi}^T L_e \, d\xi \, \mathbf{B}_e\mathbf{u}_o = \sum_{e=1}^{M} \mathbf{B}_e^T \int_0^1 \rho A\boldsymbol{\varphi}w_e^o L_e \, d\xi$$

Define the element *partial* mass matrix and the discrete element initial displacements as

$$\mathbf{m}_e^o = \int_0^1 \rho A\boldsymbol{\varphi}\boldsymbol{\varphi}^T L_e \, d\xi, \qquad \mathbf{w}_e^o = \int_0^1 \rho A\boldsymbol{\varphi}w_e^o L_e \, d\xi \tag{12.39}$$

Observe that this element mass matrix does not include the term that comes from rotary inertia. Now, we can assemble these element quantities to get

$$\mathbf{w}_o = \sum_{e=1}^{M} \mathbf{B}_e^T \mathbf{w}_e^o, \qquad \mathbf{M}_o = \sum_{e=1}^{M} \mathbf{B}_e^T \mathbf{m}_e^o \mathbf{B}_e \qquad (12.40)$$

and solve for the discrete initial displacement as

$$\mathbf{u}_o = \mathbf{M}_o^{-1} \mathbf{w}_o \qquad (12.41)$$

Computation of initial velocities follows a similar path.

Code 12.2 shows how the initial displacement and velocity vectors are established for the Rayleigh beam, in accord with Eq. 12.41. This routine calls `RayleighElementIC`, which refers to `WaveFunction` (Code 9.4), which generates the value of the initial wave at the coordinate location passed into the function. There are three different waveforms available in this function. The global partial mass matrix is assembled at the same time.

Code 12.2   MATLAB code to compute the initial displacement and velocity for the Rayleigh beam

```
function [wo,vo] = RayleighICs(V,x,d,ix,eDOF,S,DF)
%   Establish the initial conditions for Rayleigh beam
%
%        V : Matrix of eigenvectors
%        x : Nodal coordinates
%        d : Physical properties [L,E,A,I,rho,H,RotOn]
%       ix : Element connections
%     eDOF : Element global DOF numbers
%        S : Problem size parameters (nDOF,nFree)
%            S.Mode = 0 wo and vo set from WaveFunction
%                   > 0 wo and vo set as mode shape 'S.Mode' from V

    if S.Mode==0

%... Plot the wave
        RayleighPlotWaveForm(S,DF);

%... Form initial displacement and mass matrix
        wo = zeros(S.nDOF,1);
        M = zeros(S.nDOF,S.nDOF);
        for e=1:S.nElms
            [~,xe,jj] = RayleighLocalize(e,wo,x,ix,eDOF,S);
            [me,ue] = RayleighElementIC(xe,d,S);
            wo(jj) = wo(jj) + ue;
            M(jj,jj) = M(jj,jj) + me;
        end

%... Trim to unconstrained DOF
        wo = wo(1:S.nFree);
        M = M(1:S.nFree,1:S.nFree);
        wo = M\wo;

%. Set initial displacements to a mode if mode > 0
    else
        wo = V(:,S.Mode);
    end

%. Multiply by amplitudes to give uo and vo
    vo = S.Vo*wo;
    wo = S.Uo*wo;
```

```
end

function [me,we] = RayleighElementIC(x,d,S)
%   Compute and assemble initial conditions for initial displacement.
%
%         x : Nodal coordinates for element
%         d : Physical properties [L,E,A,I,rho,H,RotOn];
%         S : Problem size parameters (nExt,nQpts,wt,st)

%. Geometric properties
    L = d(1);    x0 = x(1,1);    Le = x(2,1)-x(1,1);

%. Initialize matrices
    nDOFe = 4+sum(S.nExt);                  % Total DOF per element
    we = zeros(nDOFe,1);                    % Element initial displacement
    me = zeros(nDOFe,nDOFe);                % Element mass matrix

%. Compute element tangent and residual by numerical integration
    for i=1:S.nQpts

%... Set element coordinate for integration point, get wave value
       xi = S.st(i);    z = x0 + xi*Le;
       [weo,~] = WaveFunction(z,L,S);

%... Compute element initial displacement and mass
       [h,~,~,~] = BubbleFcnsC1(xi,S.nExt,Le);
       rhoA = d(5)*d(3);
       we = we + rhoA*weo*h*Le*S.wt(i);
       me = me + rhoA*(h*h')*Le*S.wt(i);

    end

end
```

12.2.4   The Rayleigh Beam Code

The implementation of the Rayleigh beam code is very similar to the axial bar (Code 9.5 for the modal version and Code 9.9 for the Newmark version). Code 12.3 is the main program for solving the Rayleigh beam with the *modal analysis* approach. This version is limited to wave propagation by specifying either an initial displacement or velocity. The primary purpose of this code is to provide comparisons with the previous chapter for wave propagation. The initial displacement and velocity can be set to a mode shape to give modal vibration. More general loading is possible with Code 12.4, the Newmark version of the code. This code allows the applied forces to be time dependent through the function TimeFunction, Code 3.4. A typical input file, which works with both the modal and Newmark main programs, is shown in Code 12.5.

Code 12.3 Main program for the Rayleigh beam with modal analysis approach

```
%   Rayleigh Dynamic Beam (modal)
    clear; clc;

%. Get physical properties and set up mesh
```

```
      [x,ix,id,eDOF,d,Load,F,Damp,S,DF] = RayleighInput;

%. Establish final time and time step for plotting
   tf = 0.3;    dt = 0.1;

%. Create nodal force vector, initialize u and uPre
   [f,u,uPre] = RayleighForce(F,id,S,DF);

%. Form mass, stiffness and load matrices
   [M,K,~] = RayleighAssemble(u,x,d,eDOF,ix,Load,S);

%. Compute natural frequencies and mode shapes
   [Evec,Eval] = eig(K,M);
   freq = diag(sqrt(Eval));

%. Set up arrays to save results
   nSteps = ceil(tf/dt)+1;              % Number of time steps
   tf = (nSteps-1)*dt;                  % Reset final time
   Time = linspace(0,tf,nSteps)';       % Storage for time
   Disp = zeros(nSteps,S.nDOF);         % Storage for displacements

%. Establish initial conditions
   [uo,vo] = RayleighICs(Evec,x,d,ix,eDOF,S,DF);

%. Compute coefficients for classical solution
   BB = Evec'*M*uo;
   CC = (Evec'*M*vo)./freq;

%. Compute motion by modal decomposition
   for n=1:nSteps
      u = Evec*(BB.*cos(freq.*Time(n) + CC.*sin(freq.*Time(n))));
      Disp(n,:) = [u; uPre]';
   end

%. Plot response
   RayleighGraphs(DF,x,ix,Time,Disp,id,eDOF,S,freq);
```

Code 12.4 Main program for the Rayleigh beam with Newmark time integration approach

```
%   Rayleigh Beam Dynamics (Newmark)
    clear; clc;

%. Get physical properties and set up mesh
   [x,ix,id,eDOF,d,Load,F,damp,S,DF] = RayleighInput;

%. Put nodal forces into DOF order
   [fn,u,uPre] = RayleighForce(F,id,S,DF);

%. Final time and analysis time step
   tf = 0.30;  h = 0.001;

%. Retrieve parameters for Newmark integration
   [beta,gamma,eta,zeta,nSteps,tol,itmax,tf] = NewmarkParams(tf,h);

%. Form mass, stiffness matrices and distributed load vector
   [M,K,fd] = RayleighAssemble(u,x,d,eDOF,ix,Load,S);
   f = fn + fd;

%. Get eigenvectors and natural frequencies
   [Evec,Eval] = eig(K,M);
   freq = diag(sqrt(Eval));

%. Initialize arrays to store results
   Time = zeros(nSteps,1);
   Disp = zeros(nSteps,S.nDOF);
```

```
%. Establish initial conditions
   [uo,vo] = RayleighICs(Evec,x,d,ix,eDOF,S,DF);

%. Initialize position, velocity, and acceleration
   t = 0;
   Ft = TimeFunction(t,S.TimeType,0);
   uold = uo;
   vold = vo;
   aold = M\(Ft*f - K*uo);

%. Compute motion by modal decomposition
   for i=1:nSteps

%... Store results for plotting later
      Time(i,:) = t;
      Disp(i,:) = [uold; uPre]';

%... Compute the values for the new time step
      t = t + h; its = 0; err = 1;
      Ft = TimeFunction(t,S.TimeType,0);
      bn = uold + h*vold + beta*h^2*aold;
      cn = vold + gamma*h*aold;
      anew = aold;

%... Newton iteration to determine the new acceleration
      while (err > tol) && (its < itmax)
        its = its + 1;
        unew = bn + zeta*anew;
        g = M*anew + K*unew - Ft*f;
        A = M + zeta*K;
        anew = anew - A\g;
        err = norm(g);
      end

%... Compute final velocity and displacement for time step
      unew = bn + zeta*anew;
      vnew = cn + eta*anew;

%... Update state for next time step
      aold = anew;  vold = vnew;  uold = unew;

   end

%. Generate graphical output
   PlotTimeFunction(S,DF,tf)
   RayleighGraphs(DF,x,ix,Time,Disp,id,eDOF,S,freq);
```

Code 12.5 Input function for the Rayleigh beam

```
function [x,ix,id,eDOF,d,Load,f,Damp,S,DF] = RayleighInput
%   Problem description for Rayleigh beam codes
%
%       x : (x,y) coordinates of nodes
%      ix : i-node, j-node for each element
%      id : Global DOF numbers for each node
%    eDOF : Global DOF numbers for each element
%       d : Material properties [L,E,A,I,rho,H,Ray]
%    Load : Distributed loading [LoadType, qo, mo]
%       f : Nodal applied forces in node order
%    Damp : Struct containing damping information
%       S : Struct for problem size, quadrature, etc.
%      DF : Struct for output management

%. Numerical discretization
```

```
      S.nElms = 30;                    % Number of elements
      S.nNodes = S.nElms + 1;          % Number of FE nodes
      S.nExt = 5;                      % Number of bubble DOF for W
      S.nDOFpn = 2;                    % Number of DOF per node for W

%. Total number of DOF
      S.nDOF = 2*S.nNodes + S.nElms*sum(S.nExt);

%. Physical parameters (input)
      L = 10;                          % Beam length
      rho = 0.6;                       % Density
      A = 0.05;                        % Cross sectional area
      I = 0.2;                         % Moment of inertia
      E = 80;                          % Young's modulus
      H = 0;                           % Prescribed axial force
      RotOn = 1;                       % =1 Rayleigh, =0 Bernoulli-Euler
      d = [L,E,A,I,rho,H,RotOn];

%. Distributed loading
      qo = [2,0];                      % Load magnitude [qo,mo]
      LoadType = 1;                    % =1 1, =2 x/L, =3 1-x/L, =4 sin(pi*x/L)
      Load = [LoadType,qo];

%. Boundary conditions
      BCL = 2;        % x=0 : 1=free, 2=fixed, 3=simple, 4=slide
      BCR = 2;        % x=L : 1=free, 2=fixed, 3=simple, 4=slide

%. Establish the nodal coordinates
      x = zeros(S.nNodes,2);
      x(:,1) = linspace(0,L,S.nNodes);

%. Compute element connection array
      ix = zeros(S.nElms,2);
      for e=1:S.nElms; ix(e,:) = e:e+1; end

%. Establish BCs: =0 DOF free, =1 DOF restrained [W,Theta]
      BC = [0,0; 1,1; 1,0; 0,1];
      id = zeros(S.nNodes,2);
      id(1,:) = BC(BCL,:); id(S.nNodes,:) = BC(BCR,:);

%. Compute global DOF numbers and store back in id array
      [eDOF,id,S] = GlobalDOF(id,ix,S);

%. Nodal force, each row is (x-force, y-force, moment) in node order
      f = zeros(S.nNodes,2);

%. Damping (not included)
      Damp = [];

%. Initial condition, Time function
      S.Mode = 0;                      % Mode for initial condition
      S.Uo = 0.1;                      % Magnitude of initial displacement
      S.Vo = 0;                        % Magnitude of initial velocity
      S.WaveType = 3;                  % =1 cosine, =2 normal, =3 poly.
      S.TimeType = 0;                  % Type of time function

%. Establish quadrature rule
      S.Type = 1;                      % = 1 Gauss-Legendre
      S.nQpts = S.nExt(1)+2;           % Number of integration points
      [S.wt,S.st,S.nQpts] = NumInt(S.Type,S.nQpts);

%. Control graphical output
      [DF] = RayleighManageFigures;
      RayleighEchoInput(d,BCL,BCR,Load,S)

end
```

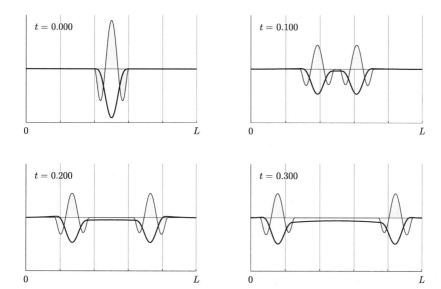

Fig. 12.4 *Rayleigh beam wave propagation. Finite element solution.* Simple–simple beam of length $L = 10$, $E = 80$, $\rho = 0.6$, $A = 0.05$, and $I = 0.2$, subjected to an initial displacement of Eq. 11.35 with $a = 4$ and $b = 6$. The four plots show the displacement (dark line) and moment (thin line with shading) at four different times. The solution was done using 30 elements, each with four bubble functions

12.2.5 Example

With this formulation we can solve the wave propagation problem for the example done in the previous chapter by the separation of variables approach. Consider the simple–simple beam with the properties given in Fig. 11.11. The response from the finite element analysis (modal version) is shown at four snapshot times in Fig. 12.4. The solution used 30 elements, each having four bubble functions in addition to the four cubic Hermitian shape functions. The polynomial order of the shape functions, therefore, is seven. The difference between the finite element solution and the classical solution by separation of variables is almost imperceptible. If you plot the shear forces, this level of discretization shows some small glitches at the interelement boundaries.

12.3 The Timoshenko Beam

Bernoulli–Euler and Rayleigh beam theory both neglect the effects of shear deformation. If we include the effects of shear deformation we have *Timoshenko beam theory*. Recall from Eq. 10.23 that the shear strain and curvature are, respectively, related to the transverse displacement w and the rotation θ through the strain–

displacement relationships

$$\beta_o = w' + \theta, \qquad \kappa_o = \theta'$$

Also recall from Eqs. 10.25 that the linearized balance of momentum equations can be written in the form

$$\rho A \ddot{w} - V' - q = 0, \qquad \rho I \ddot{\theta} - M' + V - m = 0 \qquad (12.42)$$

Finally, the linear, elastic constitutive equations for shear and moment can be expressed as

$$V = GA\beta_o, \qquad M = EI\kappa_o$$

These equations represent a complete set of equations for the motion of a Timoshenko beam. Note that we are not including the axial part of the problem, which is uncoupled from the transverse part in linear beam theory.

12.3.1   Virtual Work for the Timoshenko Beam

We can develop a virtual work functional for these equations by multiplying the residual of the first of Eqs. 12.42 by a virtual displacement $\bar{w}(x)$ and the residual of the second by a virtual rotation $\bar{\theta}(x)$, adding the two together, and integrating over the domain. To wit,

$$\mathcal{G} = \int_0^L \left[(\rho A \ddot{w} - V' - q)\bar{w} + (\rho I \ddot{\theta} - M' + V - m)\bar{\theta} \right] dx$$

If $\mathcal{G} = 0$ for all $\bar{w}$ and $\bar{\theta}$, then the original equations of motion are satisfied, as guaranteed by the fundamental theorem of the calculus of variations. We can integrate the two terms that involve derivatives of V and M by parts to get the equivalent expression

$$\mathcal{G} = \int_0^L \left(V\bar{w}' + M\bar{\theta}' + V\bar{\theta} \right) dx - V\bar{w} \Big|_0^L - M\bar{\theta} \Big|_0^L$$

$$- \int_0^L \left(q\bar{w} + m\bar{\theta} \right) dx + \int_0^L \left(\rho A \ddot{w}\bar{w} + \rho I \ddot{\theta}\bar{\theta} \right) dx$$

If we insist that the virtual displacements satisfy the homogeneous essential boundary conditions, then the last two terms on the first line vanish for all boundary conditions. Substituting the constitutive equations, i.e., $V = GA\left(w' + \theta \right)$

and $M = EI\theta'$, and rearranging terms, we arrive at the final virtual work functional for the Timoshenko beam

$$\mathcal{G} = \int_0^L \left(GA(w'\bar{w}' + \theta\bar{w}' + w'\bar{\theta} + \theta\bar{\theta}) + EI\theta'\bar{\theta}' \right) dx$$

$$- \int_0^L (q\bar{w} + m\bar{\theta}) \, dx + \int_0^L (\rho A\ddot{w}\bar{w} + \rho I\ddot{\theta}\bar{\theta}) \, dx \tag{12.43}$$

Again, if $\mathcal{G} = 0$ for all $\bar{w}$ and $\bar{\theta}$ then the virtual work equation is equivalent to the classical equations of motion. The main novelty of the Timoshenko beam is that there are two independent motion variables. This fact will have an impact on the selection of Ritz functions and the performance of the numerical approximation that emanates from the finite element method.

12.3.2   Finite Element Discretization

The virtual work functional, again, allows us to discretize the system with Ritz functions. We will go directly to the finite element version of the discretization, breaking the beam into M elements. Because the highest derivative that appears in Eq. 12.43 is a first derivative, we can use C^0 finite element functions to build our Ritz basis (as we did for the axial bar problem in Chap. 9). We will set up the formulation to allow for a higher-order interpolation within the element to get a good representation of shear and moment throughout the beam. Appendix D gives the details on how to go about constructing C^0 bubble functions and setting up the hp finite elements, including the implementation in the code.

We have two fields requiring interpolation: $w(x, t)$ and $\theta(x, t)$. The element degrees of freedom segregate into degrees of freedom associated with the interpolation of w (the transverse displacements $W_{e(i)}$ for $i = 1, \ldots, n$) and degrees of freedom associated with interpolation of θ (the rotations $\Theta_{e(i)}$ for $i = 1, \ldots, m$), where n and m are the number of interpolation functions associated with w and θ, respectively. Note that it is not necessary to have $m = n$, and we will find that the elements perform much better if the order of interpolation of the rotations is one polynomial degree less than the interpolation of the displacements.

To assist the formulation, we define two $\mathbf{B}_e$ matrices. The function of these matrices is to pick out the global degrees of freedom associated with the element, respectively for transverse displacement and for rotation. To wit,

$$\mathbf{W}_e(t) = \mathbf{B}_e^w \mathbf{u}(t), \qquad \mathbf{\Theta}_e(t) = \mathbf{B}_e^\theta \mathbf{u}(t) \tag{12.44}$$

where the discrete element transverse displacements and rotations are, respectively,

$$\mathbf{W}_e = \begin{Bmatrix} W_{e(1)} \\ \vdots \\ W_{e(n)} \end{Bmatrix} \quad \text{and} \quad \mathbf{\Theta}_e = \begin{Bmatrix} \Theta_{e(1)} \\ \vdots \\ \Theta_{e(m)} \end{Bmatrix} \qquad (12.45)$$

As usual, $\mathbf{u}(t)$ is the array of global displacements and rotations stored in accord with the global DOF numbering convention. The matrix $\mathbf{B}_e^w$ implied by Eq. 12.44(a) has n rows and $\mathcal{N}$ columns (the total number of degrees of freedom in the system). Similarly, the matrix $\mathbf{B}_e^\theta$ has m rows and $\mathcal{N}$ columns. A strategy for recording the associations needed for assembly when including non-nodal degrees of freedom is described in detail in Appendix D. The element to global DOF mapping that we call eDOF accomplishes the task. Row e of eDOF gives the global DOF numbers for the displacement and rotation degrees of freedom associated with element e and a means of distinguishing them from each other. The ordering is established by convention, as shown in Appendix E in Fig. E.3.

With these definitions, interpolate the unknown fields in element e as

$$w_e(\xi, t) = \mathbf{\psi}^T(\xi)\mathbf{W}_e(t), \qquad \theta_e(\xi, t) = \mathbf{\varphi}^T(\xi)\mathbf{\Theta}_e(t) \qquad (12.46)$$

and their virtual counterparts as

$$\bar{w}_e(\xi) = \bar{\mathbf{W}}_e^T \mathbf{\psi}(\xi), \qquad \bar{\theta}_e(\xi) = \bar{\mathbf{\Theta}}_e^T \mathbf{\varphi}(\xi) \qquad (12.47)$$

We now have two shape function arrays

$$\mathbf{\psi}(\xi) = \begin{Bmatrix} \psi_1(\xi) \\ \vdots \\ \psi_n(\xi) \end{Bmatrix}, \qquad \mathbf{\varphi}(\xi) = \begin{Bmatrix} \varphi_1(\xi) \\ \vdots \\ \varphi_m(\xi) \end{Bmatrix} \qquad (12.48)$$

the first associated with displacements and the second associated with rotations. To define the shape functions, we employ the usual change of variable $\xi = (x - x_{i(e)})/L_e$, where $x_{i(e)}$ is the coordinate of the start node associated with element e and L_e is the length of the element. Note that

$$\frac{d\mathbf{\psi}}{dx} = \frac{1}{L_e}\frac{d\mathbf{\psi}}{d\xi}, \qquad \frac{d\mathbf{\varphi}}{dx} = \frac{1}{L_e}\frac{d\mathbf{\varphi}}{d\xi}$$

The derivatives that appear in Eq. 12.43 are with respect to x.

Returning to our virtual work functional and substituting the approximation, we can write

$$\mathcal{G} = \sum_{e=1}^{M} \left\{ \bar{\mathbf{W}}_e^T \left[\mathbf{k}_e^{11} \mathbf{W}_e + \mathbf{k}_e^{12} \boldsymbol{\Theta}_e - \mathbf{f}_e^1 + \mathbf{m}_e^1 \ddot{\mathbf{W}}_e \right] \right.$$

$$\left. + \bar{\boldsymbol{\Theta}}_e^T \left[\mathbf{k}_e^{21} \mathbf{W}_e + \mathbf{k}_e^{22} \boldsymbol{\Theta}_e - \mathbf{f}_e^2 + \mathbf{m}_e^2 \ddot{\boldsymbol{\Theta}}_e \right] \right\} \tag{12.49}$$

The element applied force vectors are

$$\mathbf{f}_e^1 = \int_0^1 q\, \boldsymbol{\psi} L_e\, d\xi, \qquad \mathbf{f}_e^2 = \int_0^1 m\, \boldsymbol{\varphi} L_e\, d\xi \tag{12.50}$$

where $q(x, t)$ and $m(x, t)$ represent the applied transverse load and applied moment, respectively. The element mass matrices are

$$\mathbf{m}_e^1 = \int_0^1 \rho A\, \boldsymbol{\psi} \boldsymbol{\psi}^T L_e\, d\xi, \qquad \mathbf{m}_e^2 = \int_0^1 \rho I\, \boldsymbol{\varphi} \boldsymbol{\varphi}^T L_e\, d\xi \tag{12.51}$$

And the element stiffness matrices are

$$\mathbf{k}_e^{11} = \int_0^1 \frac{GA}{L_e^2} \boldsymbol{\psi}' \boldsymbol{\psi}'^T L_e\, d\xi$$

$$\mathbf{k}_e^{12} = \int_0^1 \frac{GA}{L_e} \boldsymbol{\psi}' \boldsymbol{\varphi}^T L_e\, d\xi$$

$$\mathbf{k}_e^{21} = \int_0^1 \frac{GA}{L_e} \boldsymbol{\varphi} \boldsymbol{\psi}'^T L_e\, d\xi \tag{12.52}$$

$$\mathbf{k}_e^{22} = \int_0^1 \left(GA\, \boldsymbol{\varphi} \boldsymbol{\varphi}^T + \frac{EI}{L_e^2} \boldsymbol{\varphi}' \boldsymbol{\varphi}'^T \right) L_e\, d\xi$$

where the prime on a shape function denotes a derivative with respect to ξ. To consolidate the assembly operation, define a combined $\mathbf{B}_e$ matrix

$$\mathbf{B}_e = \begin{bmatrix} \mathbf{B}_e^w \\ \mathbf{B}_e^\theta \end{bmatrix}$$

which is $(n+m) \times \mathcal{N}$. Combine the components of the element stiffness, mass, and force as

$$\mathbf{k}_e = \begin{bmatrix} \mathbf{k}_e^{11} & \mathbf{k}_e^{12} \\ \mathbf{k}_e^{21} & \mathbf{k}_e^{22} \end{bmatrix}, \quad \mathbf{m}_e = \begin{bmatrix} \mathbf{m}_e^1 & \mathbf{0} \\ \mathbf{0} & \mathbf{m}_e^2 \end{bmatrix}, \quad \mathbf{f}_e = \begin{Bmatrix} \mathbf{f}_e^1 \\ \mathbf{f}_e^2 \end{Bmatrix} \tag{12.53}$$

Now, the global stiffness, mass, and force are assembled as

$$\mathbf{K} = \sum_{e=1}^{M} \mathbf{B}_e^T \mathbf{k}_e \mathbf{B}_e, \quad \mathbf{M} = \sum_{e=1}^{M} \mathbf{B}_e^T \mathbf{m}_e \mathbf{B}_e, \quad \mathbf{f} = \sum_{e=1}^{M} \mathbf{B}_e^T \mathbf{f}_e \qquad (12.54)$$

As usual, these sums can be implemented with the standard assembly process. Once assembled, these matrices can be trimmed to include only the free degrees of freedom to account for the boundary conditions. With this notation, Eq. 12.49 reduces to

$$\mathcal{G} = \bar{\mathbf{u}}^T \left[\mathbf{M} \ddot{\mathbf{u}}(t) + \mathbf{K} \mathbf{u}(t) - \mathbf{f}(t) \right]$$

If $\mathcal{G} = 0$ for all $\bar{\mathbf{u}}$, we can conclude that the equations of motion are, as usual,

$$\mathbf{M} \ddot{\mathbf{u}}(t) + \mathbf{K} \mathbf{u}(t) = \mathbf{f}(t)$$

This equation should look pretty familiar, as we always seem to arrive at this same point in dynamics after discretizing our equations of motion in space but retaining the derivatives with respect to time. It is straightforward to add damping, as we have done previously. We can also use all of the techniques we have developed to solve this equation, both classical and numerical.

12.3.3   The Timoshenko Beam Code

The implementation of the Timoshenko beam is very similar to the finite element implementation of the axial bar. The routine that does element localization for the Timoshenko beam is located in Code E.6. It is very similar to the other codes that use bubble functions in the interpolation, relying on eDOF to execute the task that $\mathbf{B}_e$ does in the mathematical formulation. Also, the assembly function is nearly identical for the Timoshenko beam and the nonlinear beam in the next chapter. Therefore, we will defer the presentation of the complete code until then.

The main new function required for the Timoshenko beam is TimoElement, which does the computation of the element mass, stiffness, and force arrays. The listing of this function is provided in Code 12.6.

Code 12.6   MATLAB code to compute the element mass $\mathbf{m}_e$, stiffness $\mathbf{k}_e$, and force $\mathbf{f}_e$ arrays for the Timoshenko beam

```
function [me,ke,fe] = TimoElement(x,d,Load,S)
%   Form element mass, stiffness, and force matrices, Timoshenko beam.
%
%         x : Nodal coordinates
%         d : Constitutive properties [L,E,G,A,I,rho]
%      Load : Distributed load information [Loadtype,q,m]
%         S : Problem size parameters (nExt,nQpts,wt,st)

%. Extract load magnitudes and load type from Load
   LoadType = Load(1);              % spatial load form
```

```
    qo = Load(2:3);                          % [qo,mo] load magnitudes

%. Establish zero matrix for use in mass matrix
    Zed = zeros(2+S.nExt(1),2+S.nExt(2));

%. Compute location of left end of element e, initialize matrices
    L = d(1);   x0 = x(1,1);   Le = x(2,1)-x(1,1);

%. Initialize matrices
    nDOFe = 4+sum(S.nExt);                   % Total DOF per element
    me = zeros(nDOFe,nDOFe);                 % Element mass matrix
    ke = zeros(nDOFe,nDOFe);                 % Element stiffness matrix
    fe = zeros(nDOFe,1);                     % Element force vector

%. Compute element tangent and residual by numerical integration
    for i=1:S.nQpts

%... Set element coordinate for integration point, get load
        xi = S.st(i);   xx = (x0 + xi*Le)/L;
        q = qo*LoadFunction(xx,1,LoadType);

%... Compute shape functions and kernel matrices
        [h,dh] = BubbleFcns(xi,S.nExt(1));
        [g,dg] = BubbleFcns(xi,S.nExt(2));
        DhDh = (dh*dh')/Le;   Dhg = (dh*g');   hh = (h*h')*Le;
        DgDg = (dg*dg')/Le;   gDh = (g*dh');   gg = (g*g')*Le;

%... Compute cross sectional moduli and mass, d = [L,E,G,A,I,rho]
        GA = d(3)*d(4);   rhoA = d(4)*d(6);
        EI = d(2)*d(5);   rhoI = d(5)*d(6);

%... Compute the contributions to the element stiffness
        ke = ke + [GA*DhDh, GA*Dhg;...
                   GA*gDh,  GA*gg+EI*DgDg]*S.wt(i);

%... Compute the contributions to the element mass
        me = me + [rhoA*hh,     Zed;...
                   Zed',  rhoI*gg]*S.wt(i);

%... Compute the load vector from distributed loads
        fe = fe + [ q(1)*h*Le; ...
                    q(2)*g*Le]*S.wt(i);

    end
end
```

The contents of the array $S.nExt$ is the number of bubble functions specified for the element in the input function. For the Timoshenko beam there are two numbers; the first is the number of bubble functions associated with w and the second is the number of bubble functions associated with θ. These values are used to compute the total number of degrees of freedom in the element $nDOFe$, which defines the size of the element arrays. The specification of the distributed external loads comes in through the array Loads, which specifies the load type (e.g., constant, linear ramp up, sinusoidal) along with the magnitudes of q and m.

The element arrays are formed by numerical integration. The number of integration points is $S.nQpts$, the weights are $S.wt$, and the stations are $S.st$. Those values are specified in the problem input function through a call to NumInt, Code F.1, and are contained in the 'struct' S. The loop is over the integration stations. Within each loop, the value of the shape functions and their first derivatives are returned from the function `BubbleFcns`, which is contained in the appendix as Code D.2. The contribution of that integration point to each matrix follows Eqs. 12.50, 12.51, and 12.52 fairly closely.

To generate plots of displacement, rotation, moment, and shear, the discrete computed variables (i.e., the motion at the DOF) must be interpolated using the shape functions used to generate the element matrices. It is instructive to see how that is done in the code. The function that takes in the global $\mathbf{u}(t)$ and fills in the response fields for the beam, by interpolation with the shape functions, is given in Code 12.7.

Code 12.7   MATLAB code to compute the displacement, rotation, moment, and shear, from the global $\mathbf{u}(t)$ for the Timoshenko beam

```
function [z,w,th,mo,sh] = TimoFill(u,x,ix,eDOF,S,nPts)
%  Fill out element displacement, rotation, moment, and shear
%  from discrete values using shape functions.
%
%          u : Global displacement array
%          x : Nodal coordinates
%         ix : Element connections
%       eDOF : Global DOF for each element
%        S.  : Struct with problem size parameters (nElms, nExt)
%       nPts : Number of points between nodes to fill in

%. Initialize the displacement, and position arrays
    z  = zeros(nPts,1);         % Axial coordinate
    w  = zeros(nPts,1);         % Transverse displacement
    th = zeros(nPts,1);         % Rotation
    mo = zeros(nPts,1);         % Moment
    sh = zeros(nPts,1);         % Shear

%. Loop over elements, compute fields in each element
    n = 0;
    for e=1:S.nElms

%... Find element displacements, rotations, and coordinates
        [We,Te,xe,~] = TimoLocalize(e,u,x,ix,eDOF,S);
        x0 = xe(1,1);
        Le = xe(2,1) - xe(1,1);

%... Fill in the field by interpolation
        for i=1:nPts
            n = n + 1;
            xi = (i-1)/(nPts-1);
            [h,dh] = BubbleFcns(xi,S.nExt(1));
            [g,dg] = BubbleFcns(xi,S.nExt(2));
            w(n)  = dot(We,h);
            th(n) = dot(Te,g);
            mo(n) = dot(Te,dg)/Le;
            sh(n) = dot(We,dh)/Le + th(n);
            z(n)  = x0 + xi*Le;
        end
    end

end
```

This routine is used to generate the output graphics. The local element displacements are retrieved with the function `TimoLocalize`, which takes the global vector $\mathbf{u}(t)$ and picks out the values associated with element e for the transverse displacement w in the outputs `We`, and the rotations θ in the output `Te`. The values are recorded along with the axial coordinate values and are passed back in the arrays z, w, th, mo, and sh for x, $w(x, t)$, $\theta(x, t)$, $M(x, t)$, and $V(x, t)$, respectively. In commercial codes, the output is often filtered in some way (e.g., reporting stresses at the Gaussian quadrature points). Here we want to see exactly what our interpolation gives us.

12.3.4   Verification of Element Performance

Before we look at dynamic response, let us first consider the static problem as a means of verifying the performance of the finite element interpolation. We will consider various approximations. Because we have two fields, $w(x)$ and $\theta(x)$, we have the possibility of interpolating them differently. It turns out that equal order interpolation of the two fields results in very poor performance and gives rise to some fairly unpleasant numerical artifacts.

Since the main issue is the formation of the stiffness matrix, static analysis should be sufficient to see what is at stake. For statics, we solve $\mathbf{Ku} = \mathbf{f}$. To be specific, consider a fixed–fixed beam of length $L = 10$ with uniform transverse load $q(x) = 2$. Let's see if the Timoshenko beam can replicate the results of the Bernoulli–Euler beam, which has the constraint of zero shear deformation. Mathematically, the Bernoulli–Euler beam is the result of the physical assumption that $GA \to \infty$. We cannot go to the limit with the numerical model, but we can select a value of GA that is very large relative to the bending modulus EI. We will use $EI = 500$ and $GA = 2 \times 10^6$. Figure 12.5 shows the results for an interpolation on a mesh with 12 elements with linear interpolation of w and θ (figures on top) and six elements with quadratic interpolation of w and θ (figures on bottom). The linear elements were integrated with a 2-point Gauss–Legendre rule and the quadratic with a 3-point Gauss–Legendre rule (both sufficient for exact integration).

There are a couple of interesting features to note here. First, notice the magnitudes of the response fields for the linear elements. They are orders of magnitude smaller than the exact response ($w_{max} = 0.104$). This phenomenon is called *shear locking*, which is a well-known pathology of low-order finite elements for problems attempting to model a kinematic constraint (no shear deformation in this case).

It is easy to explain what is happening here. Because the system is attempting to give zero shear strain (due to the high GA value), it is trying to achieve the condition $\theta = -w'$. Since w is linear within each element, w' is constant. At the boundaries, the fixed–end conditions enforce $\theta(0) = 0$ and $\theta(L) = 0$. Therefore, w' has to be close to zero at those points and, hence, w' is approximately zero over the entire element. The displacement w is linear in those elements, but with nearly zero slope. Hence, the displacement at the other ends of those two elements is very small. You can continue this argument based upon the results of the elements adjacent to the two boundaries and conclude a similar small displacement at the ends of the next element, etc. The constraint, coupled with the limitation of the interpolation, essentially locks the element. Note also that while the moment diagram is essentially the right shape, albeit with significantly reduced magnitudes, the shear diagram is

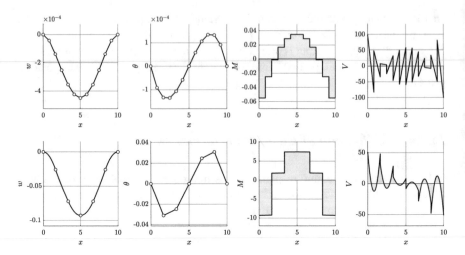

Fig. 12.5 *Static response—equal order interpolation of w and* θ. The fixed–fixed beam of length $L = 10$ subjected to uniform load $q = 2$, $EI = 500$, and $GA = 2 \times 10^6$. The top series of plots has 12 linear elements. The bottom series of plots has 6 quadratic elements. Both approximations have 22 global degrees of freedom. The nodal displacements and rotations are shown as dots

nonsense, oscillating wildly about the actual solution, which should be a straight line from 10 to -10).

The equal-order quadratic interpolation resolves the shear locking problem, as evidenced by the midspan displacement being closer to the exact value. With a quadratic interpolation of the rotations and displacements, the moments should be linear within each element. However, we see that the moment is essentially constant over each element. Despite the fact that the rotations *can* be quadratic, they are not. The reason they are not is because the element is trying to make $\theta = -w'$ and w' is linear. The shear diagram is still nonsensical, but better than it was for the linear elements. Clearly, this is still not a good approximation for this problem.

Figure 12.6 shows the results for elements with unequal interpolation of w and θ. The top case of Fig. 12.6 is identical to the bottom case of Fig. 12.5, except that the rotation is interpolated *with one order less*. The unequal interpolation solves the shear locking problem. However, it is evident that the quadratic/linear element still produces a rough solution to the problem, with the moment coming out piecewise constant. The solution could be improved simply by taking more elements with the same interpolation, but it can also be improved by using a higher-order interpolation with fewer elements (to keep the number of DOF roughly the same). Observe that the problem with the shear force diagram is resolved.

The bottom case of Fig. 12.6 uses elements with a quartic interpolation of w and a cubic interpolation of θ. It is important to keep the order of interpolation of θ one degree less than the order for w, mainly for the same reasons mentioned in the shear locking discussion. The quartic/cubic interpolation gives the exact response for this problem and does it with fewer degrees of freedom than the other cases. In fact, a single quartic/cubic element achieves the exact result for this problem. The Gauss–

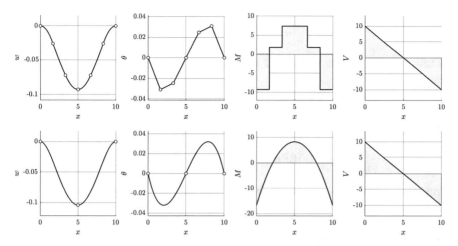

Fig. 12.6 *Static response—unequal interpolation of w and θ.* The fixed–fixed beam of length $L = 10$ subjected to uniform load $q = 2$, $EI = 500$, and $GA = 2 \times 10^6$. The top series of plots has 6 elements with quadratic interpolation of w and linear interpolation of θ. The bottom series of plots has 2 elements with quartic interpolation of w and cubic interpolation of θ. The models have 16 and 12 global degrees of freedom, respectively. The nodal displacements and rotations are shown as dots

Legendre quadrature rule is set to be two more than the interpolation of the w field (e.g., six quadrature points for the quartic interpolation of w).

This study suggests that when we move on to dynamics problems, we should use higher order interpolation because the fields within the elements are likely to be complex, especially in wave propagation problems. Also, we will generally not push to the limit of zero shear deformation but will consider values of GA that are in line with physical reality.

12.3.5   Wave Propagation in the Timoshenko Beam

The Timoshenko beam has been implemented in a modal dynamics code similar to the one presented in Sect. 9.6 for the axial bar. The initial conditions for the Timoshenko beam must include the specification of both $w_o(x)$ and $\theta_o(x)$. The discrete initial global displacement can be computed from the functions $w_o(x)$ and $\theta_o(x)$ as

$$\mathbf{u}_o = \mathbf{M}^{-1} \sum_{e=1}^{M} \left[\mathbf{B}_e^{wT} \mathbf{w}_e^o + \mathbf{B}_e^{\theta T} \boldsymbol{\theta}_e^o \right] \qquad (12.55)$$

where $\mathbf{M}$ is the mass matrix, defined in Eq. 12.54, and the element discrete displacement and rotation arrays are

$$\mathbf{w}_e^o = \int_0^1 \rho A \boldsymbol{\psi} w_e^o L_e \, d\xi, \qquad \boldsymbol{\theta}_e^o = \int_0^1 \rho I \boldsymbol{\varphi} \theta_e^o L_e \, d\xi \qquad (12.56)$$

where $w_e^o(\xi)$ is the part of $w_o(x)$ in element e and $\theta_e^o(\xi)$ is the part of $\theta_o(x)$ in element e.

To see how this result comes about, recall that Eqs. 12.44 and 12.46 at time zero imply that the interpolated values $\tilde{w}_e^o(\xi)$ and $\tilde{\theta}_e^o(\xi)$ can be computed from the discrete initial values $\mathbf{u}_o = \mathbf{u}(0)$ as

$$\tilde{w}_e^o(\xi) = \boldsymbol{\psi}^T(\xi)\mathbf{B}_e^w \mathbf{u}_o, \qquad \tilde{\theta}_e^o(\xi) = \boldsymbol{\varphi}^T(\xi)\mathbf{B}_e^\theta \mathbf{u}_o \qquad (12.57)$$

In essence, what these equations are saying is that if we knew $\mathbf{u}_o$ then we could compute $\tilde{w}_e^o(\xi)$ and $\tilde{\theta}_e^o(\xi)$ in element e. The problem we are trying to solve is actually the reverse of that. We know $w_e^o(\xi)$ and $\theta_e^o(\xi)$ and we want to compute $\mathbf{u}_o$, which is an overdetermined problem. How do we find the single array $\mathbf{u}_o$ that gives the best fit of the prescribed wave function over all of the elements?

One way to solve this problem is to project the given initial functions $w_e^o(\xi)$ and $\theta_e^o(\xi)$ onto the interpolation basis using a least-squares minimization of the difference between the actual and computed values. Consider the function

$$J(\mathbf{u}_o) = \sum_{e=1}^M \int_0^1 \tfrac{1}{2}\Big(\rho A\big(\tilde{w}_e^o(\mathbf{u}_o) - w_e^o\big)^2 + \rho I\big(\tilde{\theta}_e^o(\mathbf{u}_o) - \theta_e^o\big)^2\Big) L_e \, d\xi$$

where, for the purposes of this computation, we can view $\tilde{w}_e^o(\xi)$ and $\tilde{\theta}_e^o(\xi)$ as being functions of the unknown global initial displacement $\mathbf{u}_o$. The function J is minimized when its directional derivative in some arbitrary direction $\bar{\mathbf{u}}$ is zero. Thus, setting the derivative $DJ(\mathbf{u}_o) \cdot \bar{\mathbf{u}} = 0$ we get

$$\sum_{e=1}^M \int_0^1 \Big(\rho A(\tilde{w}_e^o - w_e^o)D\tilde{w}_e^o + \rho I(\tilde{\theta}_e^o - \theta_e^o)D\tilde{\theta}_e^o\Big)L_e \, d\xi = 0$$

where $D(\bullet)$ is short for $D(\bullet) \cdot \bar{\mathbf{u}}$. From Eqs. 12.57, it is straightforward to compute the directional derivatives of $\tilde{w}_e^o(\xi)$ and $\tilde{\theta}_e^o(\xi)$ in the direction $\bar{\mathbf{u}}$ as (see Appendix B)

$$D\tilde{w}_e^o(\mathbf{u}) \cdot \bar{\mathbf{u}} = \boldsymbol{\psi}^T(\xi)\mathbf{B}_e^w \bar{\mathbf{u}} \qquad D\tilde{\theta}_e^o(\mathbf{u}) \cdot \bar{\mathbf{u}} = \boldsymbol{\varphi}^T(\xi)\mathbf{B}_e^\theta \bar{\mathbf{u}}$$

Substituting these results and the expressions for $\tilde{w}_e^o(\xi)$ and $\tilde{\theta}_e^o(\xi)$ from Eq. 12.57, and carrying out the integrals gives

$$\bar{\mathbf{u}}^T \sum_{e=1}^M \Big[\mathbf{B}_e^{wT}\big(\mathbf{m}_e^1 \mathbf{B}_e^w \mathbf{u}_o - \mathbf{w}_o^e\big) + \mathbf{B}_e^{\theta T}\big(\mathbf{m}_e^2 \mathbf{B}_e^\theta \mathbf{u}_o - \boldsymbol{\theta}_o^e\big)\Big] = 0$$

where $\mathbf{m}_e^1$ and $\mathbf{m}_e^2$ are defined in Eq. 12.51 and $\mathbf{w}_e^o$ and $\boldsymbol{\theta}_e^o$ are defined in Eq. 12.56. Since $\bar{\mathbf{u}}$ is arbitrary, the sum of the terms in square brackets must be equal to zero.

Thus, we can write

$$\sum_{e=1}^{M} \Big[\mathbf{B}_e^{wT} \mathbf{m}_e^1 \mathbf{B}_e^w + \mathbf{B}_e^{\theta T} \mathbf{m}_e^2 \mathbf{B}_e^\theta \Big] \mathbf{u}_o = \sum_{e=1}^{M} \big[\mathbf{B}_e^{wT} \mathbf{w}_e^o + \mathbf{B}_e^{\theta T} \boldsymbol{\theta}_e^o \big]$$

The left side is the global mass matrix $\mathbf{M}$ times $\mathbf{u}_o$, thereby proving that Eq. 12.55 is the optimal value of the discrete initial displacement $\mathbf{u}_o$.

The code to carry out this computation is very similar to the one for the axial bar and the Rayleigh beam. Code 12.8 shows how the computation is organized. Note that the global mass matrix $\mathbf{M}$ is already computed and passed into the function (unlike the Rayleigh beam). The form of the initial wave is specified in WaveFunction, which is common to all of the codes that do wave propagation.

Code 12.8   MATLAB code to compute the initial conditions for the Timoshenko beam

```
function [uo,vo] = TimoICs(V,M,x,d,ix,eDOF,S,DF)
%  Establish the initial conditions for Timoshenko beam
%
%       V : Matrix of eigenvectors
%       M : Mass matrix
%       x : Nodal coordinates
%       d : Physical properties [L,E,G,A,I,rho]
%      ix : Element connections
%    eDOF : Element global DOF numbers
%       S : Struct with problem size parameters
%           S.Mode = 0 uo and vo set from function WaveForm
%                  > 0 uo and vo set as mode shape 'S.Mode' from V
%           S.Uo and S.Vo are magnitudes of uo and vo
%      DF : Struct for managing figures

    if S.Mode==0  % Set ICs in accord with initial wave

%... Form initial displacement and rotation array
     uo = zeros(S.nDOF,1);
     for e=1:S.nElms
        [~,~,xe,jj] = TimoLocalize(e,uo,x,ix,eDOF,S);
        [ue] = TimoElementIC(xe,d,S);
        uo(jj) = uo(jj) + ue;
     end

%... Trim to unconstrained DOF
     uo = uo(1:S.nFree);
     uo = M\uo;

    else  %. Set ICs to mode shape
     uo = V(:,S.Mode);
    end

%. Set initial velocities the same as uo
    vo = S.Vo*uo;
    uo = S.Uo*uo;

    TimoPlotWaveForm(S,DF);

end

%--------------------------------------------------------------
function [ue] = TimoElementIC(x,d,S)
%  Compute and assemble initial conditions for initial displacement.
%
```

```
%       x : Nodal coordinates fpr element
%       d : Physical properties [L,E,G,A,I,rho]
%       S : Problem size parameters (nExt,nQpts,wt,st)

%. Compute location of left end of element e
   L = d(1);    x0 = x(1,1);   Le = x(2,1)-x(1,1);

%. Initialize
   nDOFe = 4+sum(S.nExt);                 % Total DOF per element
   ue = zeros(nDOFe,1);                   % Element initial wave

%. Compute element tangent and residual by numerical integration
   for i=1:S.nQpts

%... Set element coordinate for integration point, get value of wave
      xi = S.st(i);    xx = x0 + xi*Le;
      [wo,to] = WaveFunction(xx,L,S);

%... Compute shape functions and kernel matrices
      [h,~] = BubbleFcns(xi,S.nExt(1));
      [g,~] = BubbleFcns(xi,S.nExt(2));
      rhoA = d(4)*d(6);
      rhoI = d(5)*d(6);

%... Compute the load vector from distributed loads
      ue = ue + [ rhoA*wo*h*Le; ...
                  rhoI*to*g*Le]*S.wt(i);

   end

end
```

The initial conditions of the beam are set based upon the value of S.Mode.
If S.Mode is zero, then the initial conditions come from a specified waveform via
WaveFunction (Code 9.4). If it is greater than zero, then the initial conditions are
set to mode shape number S.Mode. The values S.Uo and S.Vo are the magnitude
of the initial displacement and velocity, respectively. This feature allows exploration
of vibration in natural modes.

Timoshenko Beam Main Program The main program for carrying out the
dynamic analysis of the Timoshenko beam is presented in Code 12.9. The purpose of
this code is to do wave propagation with the linear Timoshenko beam. Hence, modal
analysis is the most efficient approach. One can also write a Newmark version of the
code to allow more general analysis scenarios, including applied forces. We defer
that discussion until the next chapter.

Code 12.9 MATLAB code to compute the dynamic response of the Timoshenko beam by modal
analysis

```
%   Timoshenko Dynamic Beam (modal analysis)
    clear; clc;

%. Get physical properties and set up mesh
   [x,ix,id,eDOF,d,Load,F,damp,S,DF] = TimoInput;

%. Create nodal force vector, initialize u and uPre
   [f,u,uPre] = TimoForce(F,id,S,DF);

%. Form mass, stiffness and load matrices
```

```
    [M,K,~] = TimoAssemble(u,x,d,eDOF,ix,Load,S);

%. Compute natural frequencies and mode shapes
    [Evec,Eval] = eig(K,M);
    freq = diag(sqrt(Eval));

%. Set up arrays to save results
    tf = 0.3;                           % Final time
    dt = 0.01;                          % Output interval
    nSteps = ceil(tf/dt)+1;             % Number of time steps
    Time = zeros(nSteps,1);             % History of time
    Disp = zeros(nSteps,S.nDOF);        % History of displacements

%. Establish initial conditions
    [uo,vo] = TimoICs(Evec,M,x,d,ix,eDOF,S,DF);

%. Compute coefficients for classical solution
    A = Evec'*M*uo;
    B = (Evec'*M*vo)./freq;

%. Compute motion by modal decomposition
    for n=1:nSteps
        t = tf*(n-1)/(nSteps-1);
        u = Evec*(A.*cos(freq.*t) + B.*sin(freq.*t));
        Time(n,:) = t;
        Disp(n,:) = [u; uPre]';
    end

%. Plot response
    TimoDynamicsGraphs(DF,x,ix,Time,Disp,id,eDOF,S,freq);
```

The input from function `TimoInput` consists of the nodal coordinates x, the element nodal connections ix, the global degrees of freedom associated with the nodes id, the global degrees of freedom associated with the elements eDOF, the material properties d, the array Load, the nodal forces F, and the 'struct' S, which contains the problem size parameters. The arrays Load and F are not used in this code, but are included to be consistent with input files for the nonlinear beam codes in the next chapter. The 'struct' DF manages the figures.

While there are no forces included, `TimoForce` initializes the displacements, which are needed in the assembly of the global matrices. The mass and stiffness matrices are created by calling the function `TimoAssemble`, which calls `TimoElement` (Code 12.6). The computation of the modal coefficients and the execution of the analysis is fairly self-explanatory. The results are generated by the function `TimoGraphs`, which takes care of all of the output chores.

Example An example of wave propagation is shown in Fig. 12.7. We seek evidence that waves propagate at two different wave speeds, as suggested in Fig. 11.2. The beam is fixed at both ends and has physical properties $L = 10$, $E = 80$, $G = 10$, $\rho = 0.6$, $A = 0.05$, and $I = 0.2$. The shear modulus is selected to be a very low value to accentuate the difference between flexure and shear. An initial wave is specified as Eq. 11.35 with $a = 4$, and $b = 6$. Because there is no initial rotation, the moment is initially zero, but the presence of shearing causes moment to develop quickly. As expected, waves propagate in both directions. It is evident that the wave associated with moment M propagates faster than the wave associated with shear V, which tracks with the propagation of the displacement. The rotation wave spreads out between the shear and moment waves but is roughly constant between them.

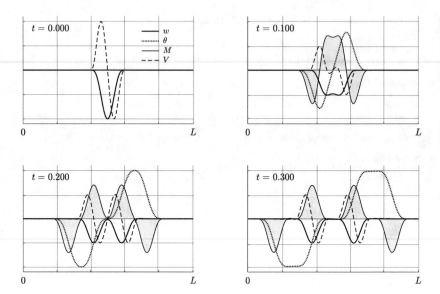

Fig. 12.7 *Wave propagation in the Timoshenko beam.* The fixed–fixed beam of length $L = 10$ with $E = 80$, $G = 10$, $\rho = 0.6$, $A = 0.05$, and $I = 0.2$. The beam has 20 elements with five bubble functions for w and four for θ. Four snapshots at different times with w, θ, M, and V shown

These results can be compared with those obtained with separation of variables in Fig. 11.15. The two approaches give identical results.

There are *many* different combinations that emanate from a theory with so many parameters. The purpose of developing the computational tool is to create a means of exploring those possibilities. This simple example gives a hint at the sorts of response one can investigate. The finite element approach makes the inclusion of different boundary conditions, loading forms, initial conditions, and beam properties straightforward. We will explore those generalizations in more detail in the subsequent chapters.

Chapter 13
Nonlinear Dynamic Analysis of Planar Beams

In this chapter we develop an approach to solve the fully nonlinear equations governing the dynamic response of beams derived in Chap. 10. Of course, the only route forward is numerical analysis. We will use Newmark's method for the temporal integration and Newton's method to solve the resulting nonlinear algebraic equations. The main additional task in setting up and solving the nonlinear problem is to compute the residual and tangent for Newton's method. As was the case for the truss, the most complicated part of the derivation is in the internal resistance, which is present in both a static and dynamic analysis. This chapter shows how the techniques of the previous chapter can be generalized to deal with nonlinearity, opening up the possibility of modeling scenarios that involve large motions and instability.

The derivation of the equations used in this chapter are done in Chap. 10. The complete equations are summarized in Sect. 10.5.

13.1 Equations of Motion

To get up to speed for this chapter, let us recall the equations of motion for the planar beam derived in Chap. 10. The strain–displacement relationships for the planar beam are

$$\mathbf{w}' = \left(1 + \varepsilon_o\right)\mathbf{n} + \beta_o\mathbf{m} - \mathbf{e}_1, \qquad \kappa_o = \theta' \tag{13.1}$$

Electronic Supplementary Material   The online version of this chapter (https://doi.org/10.1007/978-3-030-89944-8_13) contains supplementary material, which is available to authorized users.

where $\mathbf{w}(x,t) = u(x,t)\mathbf{e}_1 + w(x,t)\mathbf{e}_3$ is the displacement of the centroidal axis of the beam at location x and time t, and $\theta(x,t)$ is the rotation of the cross section relative to the initial orientation. The axial strain is ε_o, the shear strain is β_o, and the curvature is κ_o. The unit vector $\mathbf{n}$ is normal to the cross section, and $\mathbf{m}$ lies in the plane of the cross section. These quantities are described in Fig. 10.2 in Chap. 10.

Balance of linear momentum holds if

$$\mathbf{R}' + \mathbf{q} = \rho A\, \ddot{\mathbf{w}} \tag{13.2}$$

where $\mathbf{R}(x,t)$ is the net internal force at location x at time t. The applied loads are $\mathbf{q}(x,t) = p(x,t)\mathbf{e}_1 + q(x,t)\mathbf{e}_3$. The cross-sectional area is A and ρ is the density of the material.

Balance of angular momentum holds if

$$M' - \mathbf{\Gamma} \cdot \mathbf{R} + m = \rho I \ddot{\theta} \tag{13.3}$$

where $M(x,t)$ is the net internal bending moment and the applied distributed moments are $m(x,t)$. The second moment of the area about the centroid is denoted I. For convenience, we have defined the vector

$$\mathbf{\Gamma} = (1 + \varepsilon_o)\,\mathbf{m} - \beta_o \mathbf{n} \tag{13.4}$$

The internal resultant can be expressed in components as $\mathbf{R} = N\mathbf{n} + V\mathbf{m}$. Therefore,

$$\mathbf{\Gamma} \cdot \mathbf{R} = (1 + \varepsilon_o)\,V - \beta_o N$$

When the strains and rotations are small, $\mathbf{\Gamma} \to \mathbf{e}_3$, and $\mathbf{\Gamma} \cdot \mathbf{R} \to V$, and Eq. 13.3 reverts to the equation of balance of angular momentum for (linear) Timoshenko beam theory.

Finally, in Chap. 10, we wrote the constitutive equations in the form

$$N = \hat{N}(\varepsilon_o), \qquad V = \hat{V}(\beta_o), \qquad M = \hat{M}(\kappa_o) \tag{13.5}$$

We will extend the constitutive equations to include inelasticity in the next chapter, but these equations can model nonlinear elastic response. These three sets of equations, taken together, comprise the equations of motion of the beam. The equations are *geometrically exact* within the context of the kinematic hypothesis.

13.2   The Principle of Virtual Work

To set up the computation using the finite element method, we need to put the equations of motion into variational form. The method of weighted residuals is a simple way to do that. Define the functional

$$\mathcal{G} = \int_0^L \left[\left(\rho A \ddot{\mathbf{w}} - \mathbf{R}' - \mathbf{q} \right) \cdot \bar{\mathbf{w}} + \left(\rho I \ddot{\theta} - M' + \mathbf{R} \cdot \boldsymbol{\Gamma} - m \right) \bar{\theta} \right] dx \qquad (13.6)$$

where $\bar{\mathbf{w}}(x) = \bar{u}(x)\mathbf{e}_1 + \bar{w}(x)\mathbf{e}_3$ is the *virtual* displacement field and $\bar{\theta}(x)$ is the *virtual* rotation field. The *dynamic principle of virtual work* is the *weak form* of the equations of balance of momentum. If $\mathcal{G} = 0$ for all $\bar{\mathbf{w}}$ and $\bar{\theta}$, then Eqs. 13.2 and 13.3 hold. The proof is through the fundamental theorem of the calculus of variations, which is simply a statement that, for example, if $\bar{\mathbf{w}}$ can be anything, then the term in parentheses that multiplies it must be zero because one could simply choose $\bar{\mathbf{w}}$ to be equal to the term in parentheses. The only function whose square integrates to zero is the function that is exactly zero itself. Making the argument for both terms in $\mathcal{G}$ proves the theorem.

We will think of $\mathcal{G}$ as having arguments $(\mathbf{w}, \theta, \bar{\mathbf{w}}, \bar{\theta})$. In doing so, we are implicitly assuming that the strain–displacement and constitutive equations will be enforced strongly (i.e., by substituting them directly into the equations when needed). As such, the forces and moments should be viewed as being functions of the motion variables, e.g., $M(\kappa_o(\theta))$. To keep the notation simple, we will suppress this dependence in most of the derivations.

We can put the functional $\mathcal{G}$ into a form better suited to our purposes by integrating the second term in each residual by parts once. To wit,

$$\mathcal{G} = \int_0^L \left(M\bar{\theta}' + \mathbf{R} \cdot \bar{\mathbf{w}}' + \mathbf{R} \cdot \boldsymbol{\Gamma} \bar{\theta} \right) dx - M\bar{\theta} \Big|_0^L - \mathbf{R} \cdot \bar{\mathbf{w}} \Big|_0^L$$
$$- \int_0^L \left(\mathbf{q} \cdot \bar{\mathbf{w}} + m\bar{\theta} \right) dx + \int_0^L \left(\rho A \ddot{\mathbf{w}} \cdot \bar{\mathbf{w}} + \rho I \ddot{\theta}\bar{\theta} \right) dx$$

If we insist that the virtual displacements $(\bar{\mathbf{w}}, \bar{\theta})$ satisfy the homogeneous essential boundary conditions, then the last two terms in the first line vanish for all boundary conditions. Hence, the final form of the virtual work functional is

$$\mathcal{G} = \mathcal{G}_{int} - \mathcal{G}_{ext} + \mathcal{G}_{dyn} \qquad (13.7)$$

where

$$\mathcal{G}_{int} = \int_0^L \left(M\bar{\theta}' + \mathbf{R} \cdot \bar{\mathbf{w}}' + \mathbf{R} \cdot \boldsymbol{\Gamma} \bar{\theta} \right) dx$$
$$\mathcal{G}_{ext} = \int_0^L \left(\mathbf{q} \cdot \bar{\mathbf{w}} + m\bar{\theta} \right) dx \qquad (13.8)$$
$$\mathcal{G}_{dyn} = \int_0^L \left(\rho A \ddot{\mathbf{w}} \cdot \bar{\mathbf{w}} + \rho I \ddot{\theta}\bar{\theta} \right) dx$$

Some interpretation is in order. The first integral, $\mathcal{G}_{int}$, is the internal resistance and is often called the *internal virtual work*. The second integral, $\mathcal{G}_{ext}$, is the work done by the applied loads and is often called the *external virtual work*. The last term, $\mathcal{G}_{dyn}$, is associated with the inertial resistance and represents the *dynamic* part of the functional. Observe that the last term is the only one that involves time derivatives. The statement $\mathcal{G} = 0$ for all virtual displacements is generally called the *dynamic principle of virtual work* to distinguish it from an energy principle like Hamilton's principle of least action.

This virtual work functional is similar to the one we used for Timoshenko beam theory in the sense that it does not eliminate the shear deformation. For the geometrically exact kinematics, it does not make sense to introduce the constraint of zero shear deformation (i.e., $\beta_o = 0$) without also introducing the constraint of zero axial deformation (i.e., $\varepsilon_o = 0$) because they are coupled in the equations. For a static analysis, it is possible to make both assumptions, resulting in what is usually called *Euler's elastica*. The result is a differential equation for θ alone because the displacements w and u can be eliminated by the two constraint equations. Even for the static case, it is difficult to incorporate boundary conditions like the simple support because they involve the variables that were eliminated. For dynamics, the inertial terms provide an additional complication. The interesting conclusion is that it is simpler to deal with the full geometric nonlinearity than the constrained problem or some sort of second-order approximation. To make progress, we need to approach the problem numerically, following an approach similar to what we did for linear beam theory in Chap. 12.

It should be evident that if we know the functions $\mathbf{w}$ and θ, then we can compute the associated strains ε_o, β_o, and κ_o. With the strains we can compute the internal forces N, V, and M (note that $\mathbf{R}$ is built from N and V). With the internal forces, we can check to see if balance of momentum is satisfied. Such a formulation is called *motion driven*. In the dynamic setting, we generally iterate on the accelerations at the next time step, computing displacements and velocities using Newmark's method, until balance of momentum is satisfied.

13.3   Tangent Functional

In chapters that dealt with discrete systems like the shear building and the truss, we did not have an interpolation phase of the derivation. For the linear systems like the axial bar in Chap. 9 and the beams in Chap. 12, we linearized the equations before we implemented the interpolation. Now we have both tasks to do. The organizational strategy we adopt here is to linearize the functional and subsequently interpolate the fields with finite element functions. The main task in this chain of events is the derivation of the tangent functional.

We can compute the *tangent functional* as the directional derivative of the virtual work functional $\mathcal{G}$ in the direction of an arbitrary displacement $(\hat{\mathbf{w}}, \hat{\theta})$. Formally, the directional derivative is defined as (see Appendix B)

$$A = \frac{d}{d\epsilon}\left(\mathcal{G}(\mathbf{w} + \epsilon\hat{\mathbf{w}}, \theta + \epsilon\hat{\theta}, \bar{\mathbf{w}}, \bar{\theta})\right)\Big|_{\epsilon=0} \tag{13.9}$$

Informally, we denote this derivative as $A = \mathcal{D}(\mathcal{G})$. We will use the shorthand $\mathcal{D}(\bullet)$ to mean the directional derivative, as defined above, for any of the various quantities ($\bullet$) that appear in our equations. The important thing to note is that we augment the *real* displacement $\mathbf{w}$ and rotation θ in the directional derivative, but not the virtual displacements $\bar{\mathbf{w}}$ and rotation $\bar{\theta}$. Thus, the tangent functional is a function of three different motion quantities: the real motion $(\mathbf{w}, \theta)$, the virtual motion $(\bar{\mathbf{w}}, \bar{\theta})$, and the arbitrary motion $(\hat{\mathbf{w}}, \hat{\theta})$ used to support the computation of the tangent.

The directional derivative of the first term in Eq. 13.7 (the internal virtual work) is $A_{int} = \mathcal{D}(\mathcal{G}_{int})$. Explicitly,

$$A_{int} = \int_0^L \left(\mathcal{D}(M)\,\bar{\theta}' + \mathcal{D}(\mathbf{R})\cdot\bar{\mathbf{w}}' + \mathcal{D}(\mathbf{R}\cdot\mathbf{\Gamma})\bar{\theta}\right) dx \tag{13.10}$$

The directional derivative of the moment $M(\kappa_o)$ is straightforward because the moment is a function of the curvature and the curvature is simply the derivative of the rotation (i.e., $\kappa_o = \theta'$). Therefore, by the chain rule,

$$\mathcal{D}(M) = \frac{\partial M}{\partial \kappa_o}\hat{\theta}' \tag{13.11}$$

Noting that $\mathbf{R} = N\mathbf{n} + V\mathbf{m}$, we can compute the directional derivative of $\mathbf{R}$ using the product rule as

$$\mathcal{D}(\mathbf{R}) = \mathcal{D}(N)\mathbf{n} + N\mathcal{D}(\mathbf{n}) + \mathcal{D}(V)\mathbf{m} + V\mathcal{D}(\mathbf{m}) \tag{13.12}$$

It is fairly easy to prove that $\mathcal{D}(\mathbf{n}) = -\hat{\theta}\mathbf{m}$ and $\mathcal{D}(\mathbf{m}) = \hat{\theta}\mathbf{n}$, because $\mathbf{n}$ and $\mathbf{m}$ are purely functions of θ. Furthermore, by the chain rule,

$$\mathcal{D}(N) = \frac{\partial N}{\partial \varepsilon_o}\mathcal{D}(\varepsilon_o), \qquad \mathcal{D}(V) = \frac{\partial V}{\partial \beta_o}\mathcal{D}(\beta_o) \tag{13.13}$$

which leaves us in need of computing the directional derivatives of the strains ε_o and β_o. To find these derivatives, first rearrange Eq. 13.1 to the form

$$1 + \varepsilon_o = (\mathbf{w}' + \mathbf{e}_1)\cdot\mathbf{n}, \qquad \beta_o = (\mathbf{w}' + \mathbf{e}_1)\cdot\mathbf{m} \tag{13.14}$$

Now take the directional derivatives of both sides of these two equations, using the product rule and noting the expression for the directional derivatives of $\mathbf{n}$ and $\mathbf{m}$. Finally, after simplifying the resulting equations, we get

$$\mathcal{D}(\varepsilon_o) = \hat{\mathbf{w}}'\cdot\mathbf{n} - \hat{\theta}\,\beta_o, \qquad \mathcal{D}(\beta_o) = \hat{\mathbf{w}}'\cdot\mathbf{m} + \hat{\theta}\,(1 + \varepsilon_o) \tag{13.15}$$

Putting all of the pieces together, and simplifying terms, we can write the directional derivative of $\mathbf{R}$ as

$$\mathcal{D}(\mathbf{R}) = \mathbf{A}\hat{\mathbf{w}}' + \mathbf{b}\hat{\theta} \tag{13.16}$$

where we have defined the matrix

$$\mathbf{A} = \mathbf{n}\frac{\partial N}{\partial \varepsilon_o}\mathbf{n}^T + \mathbf{m}\frac{\partial V}{\partial \beta_o}\mathbf{m}^T \tag{13.17}$$

and the vector

$$\mathbf{b} = \left(V - \beta_o\frac{\partial N}{\partial \varepsilon_o} \right)\mathbf{n} - \left(N - \left(1 + \varepsilon_o\right)\frac{\partial V}{\partial \beta_o} \right)\mathbf{m} \tag{13.18}$$

Two main effects are evident in these equations. First, the material tangent is present through the derivative of each internal force with respect to its associated deformation (i.e., $\partial N/\partial \varepsilon_o$ and $\partial V/\partial \beta_o$). In this form, any nonlinear constitutive model can be used to represent the relationship between the strains and the internal forces. Second, the so-called *geometric stiffness* is present in the remaining terms. The geometric effects give rise to instability effects if the axial forces are compressive.

Both of these effects represent important aspects of the inherent nonlinearity in the beam. While some authors draw categorical lines between these two sources of nonlinearity, it is evident that they are coupled together in the equations. If any nonlinearity is present, then we need Newton's method to solve the problem. Hence, there is no additional computational difficulty created by including the geometric nonlinearities along with constitutive nonlinearities.

Now let us compute the directional derivative of the last component of the internal resistance in Eq. 13.10. While we have written the internal resistance term as $\mathbf{R} \cdot \mathbf{\Gamma}$, we will not use the product rule to compute this derivative. Rather, since $\mathbf{R} \cdot \mathbf{\Gamma} = (1 + \varepsilon_o)V - \beta_o N$, we can write its directional derivative in the form

$$\mathcal{D}(\mathbf{R} \cdot \mathbf{\Gamma}) = \left[V\,\mathcal{D}(\varepsilon_o) + \left(1 + \varepsilon_o\right)\mathcal{D}(V) \right] - \left[N\,\mathcal{D}(\beta_o) + \beta_o\,\mathcal{D}(N) \right]$$

We have already computed all of the derivatives that appear in this equation. First, substitute Eqs. 13.13 to get

$$\mathcal{D}(\mathbf{R} \cdot \mathbf{\Gamma}) = \left(V - \beta_o\frac{\partial N}{\partial \varepsilon_o} \right)\mathcal{D}(\varepsilon_o) - \left(N - \left(1 + \varepsilon_o\right)\frac{\partial V}{\partial \beta_o} \right)\mathcal{D}(\beta_o)$$

Then, substitute the expressions for the directional derivative of the strains from Eq. 13.15 to get

$$\mathcal{D}(\mathbf{R} \cdot \mathbf{\Gamma}) = \mathbf{b}^T \hat{\mathbf{w}}' + c_1 \hat{\theta} \tag{13.19}$$

where $\mathbf{b}$ is the same vector as defined in Eq. 13.18 and c_1 is defined as

$$c_1 = -\left(V - \beta_o \frac{\partial N}{\partial \varepsilon_o}\right)\beta_o - \left(N - (1+\varepsilon_o)\frac{\partial V}{\partial \beta_o}\right)(1+\varepsilon_o) \tag{13.20}$$

Now, the internal part of the tangent can be put into a compact form

$$\mathcal{A}_{int} = \int_0^L \bar{\mathbf{r}}^T \mathbf{T}\hat{\mathbf{r}}\, dx \tag{13.21}$$

where

$$\mathbf{T} = \begin{bmatrix} \mathbf{A} & \mathbf{b} & \mathbf{0} \\ \mathbf{b}^T & c_1 & 0 \\ \mathbf{0}^T & 0 & c_2 \end{bmatrix}, \qquad \bar{\mathbf{r}} = \begin{Bmatrix} \bar{\mathbf{w}}' \\ \bar{\theta} \\ \bar{\theta}' \end{Bmatrix}, \qquad \hat{\mathbf{r}} = \begin{Bmatrix} \hat{\mathbf{w}}' \\ \hat{\theta} \\ \hat{\theta}' \end{Bmatrix} \tag{13.22}$$

where $\mathbf{0}$ is a 2×1 zero matrix and $c_2 = \partial M/\partial \kappa_o$. It is interesting to note that the matrix $\mathbf{T}$ is symmetric, which implies that the global tangent stiffness matrix will also be symmetric. Code 13.1 executes these calculations.

Code 13.1 MATLAB code called by NLBElement to compute $\mathbf{A}$, $\mathbf{b}$, c_1, c_2, and $\mathbf{R}$ from Eqs. 13.17, 13.18, and 13.20

```
function [A,b,c,R,S,E] = NLBTang(dw,th,dth,d)
%   Compute the element tangent, residual and stress at a point
%
%        dw : [u',w'] displacement derivatives
%        th : theta, dth = theta' rotation and derivative
%         d : Material properties[ConstType,E,G,A,I,rho,L]
%   A,b,c,R : Needed to compute element tangent

%. Establish unit vectors n and m
   n = [cos(th);-sin(th)];
   m = [sin(th); cos(th)];
   e1 = [1;0];

%. Compute strains from displacements
   beta = dot(dw+e1,m);                     % Shear strain
   lambda = dot(dw+e1,n);                   % Stretch
   epsilon = lambda-1;                      % Axial strain
   E = [epsilon; beta; dth];                % Strain vector

%. Extract the section stiffnesses from 'd', Compute stress.
   EA = d(2)*d(4);   GA = d(3)*d(4);   EI = d(2)*d(5);
   D = [EA; GA; EI];
   S = D.*E;                                % Stress vector

%. Tangent
   A(1:2,1:2) = n*D(1)*n' + m*D(2)*m';
   c1 = S(2) - beta*D(1);
   c2 = S(1) - lambda*D(2);
   b = c1*n - c2*m;
   c(1) = -c1*beta - c2*lambda;
   c(2) = D(3);

%. Residual
   R = S(1)*n + S(2)*m;
   RG = lambda*S(2) - beta*S(1);
   R = [R; RG; S(3)];

end
```

We will see in the next chapter that this function is key to converting the beam code to a frame code, where the initial orientation of the element varies from one element to the next. The function NLBTang is essentially where the constitutive relationships are implemented. This code only has the linear elastic constitutive model, but others could be easily implemented here.

13.4   Finite Element Discretization

As we have done before, we will break the beam up into M elements and interpolate the response fields $\{u, w, \theta\}$ with C^0 shape functions (see Appendix D). With this strategy, we can express the integral in Eq. 13.21 as

$$
\mathcal{A}_{int} = \sum_{e=1}^{M} \int_{\mathcal{L}_e} \bar{\mathbf{r}}_e^T \mathbf{T}_e \, \hat{\mathbf{r}}_e \, dx_e \tag{13.23}
$$

where $\mathcal{L}_e$ is the domain of element e. Let us denote the *global displacement vector* as $\mathbf{u}(t)$. This array contains all of the nodal and non-nodal displacement (and rotation) variables, listed in accord with the degree of freedom labeling scheme implied by the eDOF array, described in Appendix E.

We have three fields to interpolate: $u(x, t)$, $w(x, t)$ and $\theta(x, t)$. The degrees of freedom segregate into degrees of freedom associated with the interpolation of u, w, and θ. These element degrees of freedom are stored in arrays as

$$
\mathbf{U}_e = \begin{Bmatrix} U_{e(1)} \\ \vdots \\ U_{e(n)} \end{Bmatrix}, \quad \mathbf{W}_e = \begin{Bmatrix} W_{e(1)} \\ \vdots \\ W_{e(n)} \end{Bmatrix}, \quad \mathbf{\Theta}_e = \begin{Bmatrix} \Theta_{e(1)} \\ \vdots \\ \Theta_{e(m)} \end{Bmatrix} \tag{13.24}
$$

where n is the number of element DOF associated with both u and w, and m is the number of element DOF associated with θ. It is generally optimal to have $m = n - 1$, i.e., the interpolation of rotation is one polynomial order less than the interpolation of displacement. That implies that the displacements u and w must have at least one bubble function.

To assist the formulation, we define three $\mathbf{B}_e$ matrices, whose function is to pick out the degrees of freedom associated with the element, respectively for axial displacement, transverse displacement, and rotation, from the global array $\mathbf{u}(t)$, which is stored in accord with the global DOF numbering convention. To wit,

$$
\mathbf{U}_e(t) = \mathbf{B}_e^u \mathbf{u}(t), \qquad \mathbf{W}_e(t) = \mathbf{B}_e^w \mathbf{u}(t), \qquad \mathbf{\Theta}_e(t) = \mathbf{B}_e^\theta \mathbf{u}(t) \tag{13.25}
$$

The matrix $\mathbf{B}_e^u$ implied by Eq. 13.25(a) has n rows and $\mathcal{N}$ columns (the total number of degrees of freedom in the system). The ith row of $\mathbf{B}_e^u$ has a one in the column associated with the global DOF number that corresponds with the ith element DOF, and zeros for the remaining columns. The matrix $\mathbf{B}_e^w$ implied by Eq. 13.25(b) has n rows and $\mathcal{N}$ columns. Again, the ith row of $\mathbf{B}_e^w$ has a one in column associated with the global DOF number that corresponds with the ith element DOF, and zeros for the remaining columns. Finally, the matrix $\mathbf{B}_e^\theta$ has m rows and $\mathcal{N}$ columns, with similar construction.

The information required to form the $\mathbf{B}_e$ matrices is contained in the eDOF array in the code. As we pointed out previously, there will be no need to actually form the $\mathbf{B}_e$ matrices in the code. Rather, we will use the standard localization and assembly process, getting the appropriate indices from eDOF.

With these definitions we can interpolate the unknown fields in element e as

$$u_e(\xi, t) = \boldsymbol{\varphi}^T(\xi)\mathbf{U}_e(t)$$

$$w_e(\xi, t) = \boldsymbol{\varphi}^T(\xi)\mathbf{W}_e(t) \qquad (13.26)$$

$$\theta_e(\xi, t) = \boldsymbol{\psi}^T(\xi)\boldsymbol{\Theta}_e(t)$$

Similarly, their virtual counterparts can be interpolated in exactly the same way. To wit,

$$\bar{u}_e(\xi) = \bar{\mathbf{U}}_e^T \boldsymbol{\varphi}(\xi), \quad \bar{w}_e(\xi) = \bar{\mathbf{W}}_e^T \boldsymbol{\varphi}(\xi), \quad \bar{\theta}_e(\xi) = \bar{\boldsymbol{\Theta}}_e^T \boldsymbol{\psi}(\xi)$$

Note also that the quantities with a hat in the directional derivative are interpolated in the same way as the quantities with a bar. In all cases, the shape function arrays are

$$\boldsymbol{\varphi}(\xi) = \left\{ \begin{array}{c} \varphi_1(\xi) \\ \vdots \\ \varphi_n(\xi) \end{array} \right\}, \qquad \boldsymbol{\psi}(\xi) = \left\{ \begin{array}{c} \psi_1(\xi) \\ \vdots \\ \psi_m(\xi) \end{array} \right\} \qquad (13.27)$$

If we use bubble functions, then the individual functions that make up $\boldsymbol{\varphi}(\xi)$ and $\boldsymbol{\psi}(\xi)$ are the same. Only the number of functions (and therefore the dimension of the array) are potentially different. To define the shape functions, we employ the change of variable

$$\xi = \frac{x - x_{i(e)}}{L_e} \qquad (13.28)$$

where $x_{i(e)}$ is the coordinate of the *start node* associated with element e and L_e is the length of the element. The derivatives that show up in our formulation are with respect to x. By the chain rule,

$$\frac{d\varphi}{dx} = \frac{1}{L_e}\frac{d\varphi}{d\xi}, \qquad \frac{d\psi}{dx} = \frac{1}{L_e}\frac{d\psi}{d\xi}$$

It will be convenient to consolidate the nodal displacements $\mathbf{u}_e$, defining the global-to-local mapping $\mathbf{B}_e$ through the relationship

$$\mathbf{u}_e = \left\{ \begin{array}{c} \mathbf{U}_e \\ \mathbf{W}_e \\ \mathbf{\Theta}_e \end{array} \right\} = \left[\begin{array}{c} \mathbf{B}_e^u \\ \mathbf{B}_e^w \\ \mathbf{B}_e^\theta \end{array} \right] \mathbf{u} = \mathbf{B}_e \mathbf{u} \tag{13.29}$$

This notation will allow us to refer to all of the displacements in the element at once, which will come in handy for setting up the assembly process.

To complete the formulation let us write the element derivatives that appear in the virtual work and tangent functional as

$$\mathbf{r}_e = \left\{ \begin{array}{c} \mathbf{w}'_e \\ \theta_e \\ \theta'_e \end{array} \right\} = \mathbf{N}^T(\xi)\mathbf{B}_e \mathbf{u} \tag{13.30}$$

The same interpolation holds for $\bar{\mathbf{r}}_e$ and $\hat{\mathbf{r}}_e$. The shape function derivative matrix is defined as

$$\mathbf{N}(\xi) = \frac{1}{L_e}\left[\begin{array}{cccc} \varphi' & 0 & 0 & 0 \\ 0 & \varphi' & 0 & 0 \\ 0 & 0 & L_e\psi & \psi' \end{array} \right] \tag{13.31}$$

where φ' and ψ' are derivatives with respect to ξ and $\mathbf{0}$ is a matrix with the same dimensions as the nonzero elements in the row. The $\mathbf{N}$ matrix has $2n + m$ rows and 4 columns, where n is the number of shape functions in φ and m is the number of shape functions in ψ.

Now, the *residual* of the internal virtual work functional can be assembled from element contributions as

$$\mathcal{G}_{int} = \bar{\mathbf{u}}^T \sum_{e=1}^{M} \mathbf{B}_e^T \mathbf{g}_e(\mathbf{u}) \tag{13.32}$$

where the internal resistance for element e is

$$\mathbf{g}_e(\mathbf{u}) = \int_0^1 \mathbf{N}(\xi)\,\boldsymbol{\gamma}_e(\mathbf{u})\,L_e\,d\xi \tag{13.33}$$

where $\boldsymbol{\gamma}_e = [\,H_e, Q_e, \mathbf{R}_e \cdot \mathbf{\Gamma}_e, M_e\,]^T$ is a 4×1 array containing the internal force quantities that appear in Eq. 13.7. The structure of the matrix $\mathbf{N}(\xi)$ is very sparse.

Therefore, we can carry out the matrix multiplications associated with the integrand $\mathbf{N}(\xi)\boldsymbol{\gamma}_e(\mathbf{u})$ by hand. To wit,

$$
\mathbf{g}_e(\mathbf{u}) = \int_0^1 \left\{ \begin{array}{c} H_e\,\boldsymbol{\varphi}' \\ Q_e\,\boldsymbol{\varphi}' \\ M_e\,\boldsymbol{\psi}' + L_e(\boldsymbol{\Gamma}_e \cdot \mathbf{R}_e)\,\boldsymbol{\psi} \end{array} \right\} d\xi
\tag{13.34}
$$

where H_e and Q_e are components of $\mathbf{R}_e = H_e\mathbf{e}_1 + Q_e\mathbf{e}_3$, relative to the standard basis $\{\mathbf{e}_1, \mathbf{e}_3\}$, and $\boldsymbol{\Gamma}_e \cdot \mathbf{R}_e = (1 + \varepsilon_o)V - \beta_o N$ for element e.

The external virtual work has the form

$$
\mathcal{G}_{ext} = \bar{\mathbf{u}}^T \sum_{e=1}^{M} \mathbf{B}_e^T \mathbf{p}_e
\tag{13.35}
$$

The element applied load vector in this equation is computed as

$$
\mathbf{p}_e = \int_0^1 \left\{ \begin{array}{c} p_e\,\boldsymbol{\varphi} \\ q_e\,\boldsymbol{\varphi} \\ m_e\,\boldsymbol{\psi} \end{array} \right\} L_e\,d\xi
\tag{13.36}
$$

where p_e, q_e, and m_e are the applied distributed axial, transverse, and moment loads, respectively, that act along the axis of element e. Recall that we have assumed that p_e acts in the $\mathbf{e}_1$ direction and q_e acts in the $\mathbf{e}_3$ direction. Thus, when we say "axial" and "transverse" we mean axial and transverse relative to the original configuration.

The directional derivative of the internal virtual work can be assembled from element contributions as

$$
\mathcal{A}_{int} = \bar{\mathbf{u}}^T \sum_{e=1}^{M} \mathbf{B}_e^T \mathbf{k}_e \mathbf{B}_e\,\hat{\mathbf{u}}
\tag{13.37}
$$

where the element stiffness matrix has the form

$$
\mathbf{k}_e = \int_0^1 \mathbf{N}(\xi)\,\mathbf{T}_e\,\mathbf{N}^T(\xi)\,L_e\,d\xi
\tag{13.38}
$$

The matrix $\mathbf{T}_e$ is defined by Eq. 13.22, and is specialized to element e. Again, there are lots of zeros in the $\mathbf{N}(\xi)$ matrix, so it is more efficient to write out the matrix multiplications explicitly. Once done, we can write the element stiffness matrix in the form

$$
\mathbf{k}_e = \int_0^1 \frac{1}{L_e} \begin{bmatrix} \boldsymbol{\varphi}' A_{11} \boldsymbol{\varphi}'^T & \boldsymbol{\varphi}' A_{12} \boldsymbol{\varphi}'^T & L_e\boldsymbol{\varphi}' b_1 \boldsymbol{\psi}^T \\ \boldsymbol{\varphi}' A_{21} \boldsymbol{\varphi}'^T & \boldsymbol{\varphi}' A_{22} \boldsymbol{\varphi}'^T & L_e\boldsymbol{\varphi}' b_2 \boldsymbol{\psi}^T \\ L_e\boldsymbol{\psi}\, b_1 \boldsymbol{\varphi}'^T & L_e\boldsymbol{\psi}\, b_2 \boldsymbol{\varphi}'^T & L_e^2\boldsymbol{\psi}\, c_1 \boldsymbol{\psi}^T + \boldsymbol{\psi}' c_2 \boldsymbol{\psi}'^T \end{bmatrix} d\xi
$$

where A_{ij} is the ijth component of the matrix $\mathbf{A}$ (from Eq. 13.17), and b_j is the jth component of the vector $\mathbf{b}$ (from Eq. 13.18). Of course, the matrix $\mathbf{k}_e$ can be computed by numerical quadrature. The shape functions are generally defined using the change of variable in Eq. 13.28. Note that the L_e factors floating around in the equation come from the derivatives of the shape functions.

By identification of terms, the *global tangent stiffness* matrix can be defined as

$$\mathcal{T} = \sum_{e=1}^{M} \mathbf{B}_e^T \mathbf{k}_e \mathbf{B}_e \tag{13.39}$$

such that the derivative of the internal virtual work is $\mathcal{A}_{int} = \bar{\mathbf{u}}^T \mathcal{T} \hat{\mathbf{u}}$. We can also define the global internal resistance and global applied load as

$$\mathbf{g}(\mathbf{u}) = \sum_{e=1}^{M} \mathbf{B}_e^T \mathbf{g}_e(\mathbf{u}), \qquad \mathbf{p} = \sum_{e=1}^{M} \mathbf{B}_e^T \mathbf{p}_e \tag{13.40}$$

so that the virtual work functionals for the internal resistance and external loads are $\mathcal{G}_{int} = \bar{\mathbf{u}}^T \mathbf{g}(\mathbf{u})$ and $\mathcal{G}_{ext} = \bar{\mathbf{u}}^T \mathbf{p}$. It is important to note that the load vector has been computed from the interpolation of the *distributed loads*. It is also possible to add *nodal loads* to the formulation.

Incorporation of Nodal Loads We will do a complete derivation of how to include nodal loads in the formulation in the next chapter. For now, let us simply observe that we only need to account for the virtual work done by the nodal loads in $\mathcal{G}_{ext}$ as follows:

$$\mathcal{G}_{ext} = \sum_{e=1}^{M} \int_{\mathcal{L}_e} \left(\mathbf{q}_e \cdot \bar{\mathbf{w}}_e + m_e \, \bar{\theta}_e \right) dx + \sum_{i=1}^{N} \left(\hat{\mathbf{p}}_i \cdot \bar{\mathbf{W}}_i + \hat{m}_i \bar{\Theta}_i \right) \tag{13.41}$$

where $\hat{\mathbf{p}}_i$ is the applied load and $\hat{m}_i$ the applied moment at node i, as shown in Fig. 13.1. The vector $\bar{\mathbf{W}}_i$ is the virtual nodal displacement and $\bar{\Theta}_i$ is the virtual nodal rotation at node i.[1]

Theoretically, we could have a discrete load applied at every node of the system. However, it does not make sense to have nodal loads applied to the interior non-nodal variables in an element. To see why this is true, consider a case where there is a discrete transverse load applied to an interior point in the element. A point load in the interior of the beam would cause a jump in the shear field at that point in the amount of the load. However, the interpolation functions we are using to represent

---

[1] We will distinguish the virtual displacement associated with w in element e as $\bar{\mathbf{W}}_e$ and the virtual displacement associated with node i as $\bar{\mathbf{W}}_i$. The same goes for $\bar{\Theta}_e$ and $\bar{\Theta}_i$. While e and i are just indices, the context should make clear which quantity it is in use. The two will never appear in the same sum.

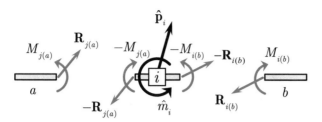

Fig. 13.1 *Nodal equilibrium.* Element a and element b contribute their end forces to node i, which has applied nodal loads $\hat{\mathbf{p}}_i$ and nodal moments $\hat{m}_i$. The forces and moments on the free body diagram of the node should sum to zero (or to nodal mass times nodal acceleration in the dynamic context)

the displacements (and therefore the internal force fields) are smooth within the domain of the element, precluding a jump in any internal field.

To set up how to include the nodal terms in the global equations, let us define a matrix $\mathbf{D}_i$ such that the nodal displacement and rotation can be computed from the global displacement vector. The nodal displacements and rotations are

$$\begin{Bmatrix} \mathbf{W}_i \\ \Theta_i \end{Bmatrix} = \mathbf{D}_i \mathbf{u} \tag{13.42}$$

The matrix $\mathbf{D}_i$ has dimensions $3 \times \mathcal{N}$, where $\mathcal{N}$ is the number of degrees of freedom in the system. It is similar to the $\mathbf{B}_e$ matrices in the sense that its role is simply to localize part of the global displacement vector. While the latter grabs motion information about elements, the former grabs motion information about nodal degrees of freedom. The nodal operator $\mathbf{D}_i$ is simpler than the element operators because it only requires information from the id array. The kth row of $\mathbf{D}_i$ has a one in column $\texttt{id}(i, k)$. The virtual nodal displacement and rotation obey the same relationship. Hence, we can compute

$$\sum_{i=1}^{N} \left(\hat{\mathbf{p}}_i \cdot \bar{\mathbf{W}}_i + \hat{m}_i \bar{\Theta}_i \right) = \bar{\mathbf{u}}^T \sum_{i=1}^{N} \mathbf{D}_i^T \check{\mathbf{F}}_i = \bar{\mathbf{u}}^T \mathbf{F}$$

where $\mathbf{F}$ is the global applied load array and

$$\check{\mathbf{F}}_i = \begin{Bmatrix} \hat{\mathbf{p}}_i \\ \hat{m}_i \end{Bmatrix}$$

concatenates the applied loads and moments into a single array. Therefore, if we construct a *nodal force vector* $\mathbf{F}$ that has the nodal forces entered in accord with the global degree of freedom numbering scheme implied by the id array, then the nodal forces are properly assembled into the equations of equilibrium. The virtual work functional for external loads, including nodal load, can be written as

$$\mathcal{G}_{ext} = \bar{\mathbf{u}}^T \left(\sum_{e=1}^{M} \mathbf{B}_e^T \mathbf{p}_e + \sum_{i=1}^{N} \mathbf{D}_i^T \check{\mathbf{F}}_i \right) = \bar{\mathbf{u}}^T \mathbf{f} \tag{13.43}$$

The global applied force vector can, by identification, be defined as

$$\mathbf{f} = \sum_{e=1}^{M} \mathbf{B}_e^T \mathbf{p}_e + \sum_{i=1}^{N} \mathbf{D}_i^T \check{\mathbf{F}}_i \tag{13.44}$$

Hence, all we need to do is add the contributions of the nodal loads to $\mathcal{G}_{ext}$.

13.5 Static Analysis of Nonlinear Planar Beams

Although our target is the dynamic analysis of the nonlinear beam, it is worthwhile to note that at this point we have the pieces of the formulation that support the *static* analysis of the nonlinear beam. One good reason to pause here for the static problem is that it provides a great way to verify the parts we have developed so far. That makes sense because these parts will not change when we add the inertial terms, so the verification will continue to hold for the dynamic problem.

To get a virtual work functional that applies to static equilibrium, all we need to do is neglect $\mathcal{G}_{dyn}$ in Eq. 13.7. With the finite element discretization, the static virtual work functional is

$$\mathcal{G}_{static} = \bar{\mathbf{u}}^T \left(\mathbf{g}(\mathbf{u}) - \mathbf{f} \right) \tag{13.45}$$

where $\mathbf{g}(\mathbf{u})$ is given in Eq. 13.40 and $\mathbf{f}$ is given in Eq. 13.44. If $\mathcal{G}_{static} = 0$ for all $\bar{\mathbf{u}}$, the fundamental theorem of the calculus of variations implies that static equilibrium is governed by the equation •

$$\mathbf{g}(\mathbf{u}) = \mathbf{f} \tag{13.46}$$

We can solve this nonlinear system of equations with Newton's method, as outlined in the next section.

13.5.1 Solution by Newton's Method

We can solve the global equilibrium equations $\mathbf{g}(\mathbf{u}) - \mathbf{f} = \mathbf{0}$ by Newton's method, with the iteration

$$\mathbf{u}^{i+1} = \mathbf{u}^i - \left[\mathcal{T}(\mathbf{u}^i) \right]^{-1} \left(\mathbf{g}(\mathbf{u}^i) - \mathbf{f} \right) \tag{13.47}$$

where $\mathbf{u}^i$ is the estimate of the discrete displacement vector $\mathbf{u}$ at iteration i. Note that $\mathcal{T}$ is the assembled tangent matrix and $\mathbf{g}$ is the assembled internal force, given in Eq. 13.40(a). Newton's method requires a starting guess $\mathbf{u}^0$ and continues until the norm of the residual is small enough, i.e., $\|\mathbf{g}(\mathbf{u}^i)\| < tol$, where tol is a suitably small number.

For a static problem, Newton's method might not converge with the full value of the loads applied to it. In that case the load can be applied gradually. If, for example, you want to apply uniform transverse distributed load q_o, the solution can be accomplished by applying $q_n = \lambda_n q_o$ at the nth load step, with $\lambda_n = n/N_s$, where N_s is the total number of load steps. With this strategy, the converged value of the displacement at load step n becomes the starting point for the Newton iteration at load step $n + 1$. That strategy helps convergence because Newton's method generally converges to a solution if the starting point is close enough. There are, of course, conditions for which Newton's method will not converge. The strategy of gradually incrementing the load works as long as the system does not exhibit a *limit load* (i.e., a maximum capacity after which the structure carries less load with increasing deformation). In such cases, one can implement *arc-length* methods (see, for example, K.D. Hjelmstad, *Fundamentals of Structural Mechanics*, Springer, 2005).

13.5.2   Static Implementation

Newton's method for the static nonlinear beam can be implemented with a few functions that will serve our needs when we get into dynamic analysis. The main program is given in Code 13.2.

Code 13.2   MATLAB code to execute static analysis of the nonlinear beam

```
%   Nonlinear Planar Beam (Static)
    clear; clc;

%.  Set parameters on Newton iteration and number of load steps
    Tol = 1.e-8; itMax = 15;   nSteps = 3;

%.  Get physical properties and set up mesh
    [x,ix,id,eDOF,d,Load,F,damp,S,DF] = NLBinput;

%.  Put nodal forces into DOF order
    [fn,u,uPre] = Force(F,id,S,DF);

%.  Compute forces associated with distributed loads
    uTot = [u; uPre];
    [~,~,~,fd] = NLBAssemble(x,uTot,d,eDOF,ix,Load,S);

%.  Total force is distributed plus nodal forces
    f = fd + fn;

%.  Plot the initial shape of the structure
    iFlag = 0;
    iFlag = NLBPlotShape(x,uTot,eDOF,id,ix,S,iFlag);
```

```
    for i=1:nSteps
      lambda = i/nSteps;
%... Newton iteration
        err = 1; its = 0;
        while (err > Tol) && (its < itMax)
%..... Form tangent and residual
          uTot = [u; lambda*uPre];
          [~,T,g,~] = NLBAssemble(x,uTot,d,eDOF,ix,Load,S);
          g = g - lambda*f;
%..... Compute newton update and error
          u = u - T\g;
          err = norm(g);
          its = its + 1;

        end
%... Plot shape for this load step
        uTot = [u; uPre];
        NLBPlotShape(x,uTot,eDOF,id,ix,S,iFlag);

    end

%. Plot response fields vs. x
    NLBPlotResults(x,uTot,eDOF,id,ix,S,d)
```

To accomplish the nonlinear static analysis, we must employ a Newton iteration that uses the residual and tangent computed by the NLBElement routine in Code 13.4 assembled into global equations in NLBAssemble in Code 13.3. We will describe these codes once the overall stage is set. For more information on Newton's method, see Appendix A).

The code starts by establishing the Newton tolerance Tol and the maximum number of iterations allowed itMax, which is just a fail-safe for cases where the solution diverges. We break the loading up into smaller load steps and iterate to convergence for each one. We set the number of load steps nSteps to establish how the loads will be applied.

The description of the structure comes from the function NLBinput (or some similar name). The nodal force vector is created from the inputs in the function Force. This is simply a recasting of values put in nodal order into values in global DOF order. The first call to NLBAssemble computes the force vector associated with the distributed loads. Once computed, the total load f is computed as the sum of nodal and distributed loads.

The graphical output is set up so that the deflected shape of the structure can be plotted at each load step without erasing the earlier shapes using the function NLBPlotShape. The beam is plotted again after each load step.

Next, we enter a loop over load steps. For each step, we increment the load by setting the value of lambda. From there we execute the Newton loop to iterate to convergence with that load level. The initial starting value for the Newton iteration is easy. We will simply use the undeformed configuration (i.e., $\mathbf{u} = \mathbf{0}$). Subsequent load steps simply use the converged value from the previous load step as the starting value for the next step.

Within the Newton loop, the call to NLBAssemble returns the tangent and residual for the current state. At that point the residual only includes the internal resistance. We subtract the applied external loads (scaled by the load level lambda) to complete the residual. Then we execute the Newton update and check the norm of the residual. Once the norm of the residual is small enough, we exit the loop, plot the deflected shape, and move on to the next load level. Once the analysis is complete, the axial, shear, and bending moment fields are plotted with the function NLBPlotResults.

The key to the analysis by Newton's method is the formation of the global tangent and residual. The MATLAB code that assembles the tangent and residual for the nonlinear beam is given in Code 13.3.

Code 13.3 MATLAB code to assemble the tangent and residual for the nonlinear beam

```
function [M,K,g,f] = NLBAssemble(x,u,d,eDOF,ix,Load,S)
%  Form mass, M, tangent K, and residual g for Nonlinear Beam
%
%        x : Nodal coordinates
%        u : Nodal displacements
%        d : Elements material properties
%     eDOF : Global DOF for each element
%       ix : Element nodal connections
%     Load : Distributed load parameters
%        S : Problem parameters (nElms,nExt,NFree,nDOF,etc)

%. Initialize global arrays
    M = zeros(S.nDOF,S.nDOF);     g = zeros(S.nDOF,1);
    K = zeros(S.nDOF,S.nDOF);     f = zeros(S.nDOF,1);

%. Loop over elements
    for e=1:S.nElms

%... Localize arrays and form element arrays
        [Ue,We,Te,xe,jj] = NLBLocalize(u,x,ix,e,eDOF,S);
        [me,ke,ge,fe] = NLBElement(xe,Ue,We,Te,d,Load,S);

%... Assemble element matrices into global arrays
        M(jj,jj) = M(jj,jj) + me;     g(jj) = g(jj) + ge;
        K(jj,jj) = K(jj,jj) + ke;     f(jj) = f(jj) + fe;

    end

%. Trim to unconstrained DOF
    M = M(1:S.nFree,1:S.nFree);     g = g(1:S.nFree);
    K = K(1:S.nFree,1:S.nFree);     f = f(1:S.nFree);

end
```

This is a standard assembly routine, with a loop over elements in which three things are done. First, the global arrays are localized in the function called NLBLocalize. This function isolates information relevant to the current element e and assembles the index array jj based on eDOF. Second, the element matrices are computed from NLBElement. Finally, the element contributions are assembled into the global matrices using the index array jj. Before returning to the main program, the global matrices are trimmed to eliminate the degrees of freedom restrained by boundary conditions. Code 13.4, the function NLBElement, shows how to organize the element computations for the nonlinear beam.

Code 13.4   MATLAB code to generate the element matrices required to form the global tangent and residual for the nonlinear beam

```
function [me,ke,ge,fe] = NLBElement(xe,Ue,We,Te,d,Load,S)
%  Compute element mass (me), stiffness (ke), internal residual (ge),
%  and distributed load (fe). Matrices are integrated numerically.
%
%      xe : Element nodal coordinates
%      Ue : Axial displacements at element DOF
%      We : Transverse displacements at element DOF
%      Te : Rotations at element DOF
%       d : Constitutive properties [ConstType,E,G,A,I,rho,L]
%    Load : Distributed load information
%       S : Problem parameters (nExt,nQpts,wt,st)

%. Compute location of left end of element e, element length
   L = d(7);   x0 = xe(1,1);   Le = xe(2,1)-xe(1,1);

%. Extract load magnitudes and load type from Load
   LoadType = Load(1);
   qo = Load(2:4);

%. Compute the element mass properties
   rhoA = d(4)*d(6);
   rhoI = d(5)*d(6);

%. Establish zero matrix for use in mass matrix
   Zed2 = zeros(2+S.nExt(1),2+S.nExt(1));
   Zed1 = zeros(2+S.nExt(1),2+S.nExt(3));

%. Initialize arrays
   nDOFe = 6+sum(S.nExt);                % Total DOF per element
   me = zeros(nDOFe,nDOFe);              % Element mass matrix
   ke = zeros(nDOFe,nDOFe);              % Element tangent stiffness
   ge = zeros(nDOFe,1);                  % Element residual
   fe = zeros(nDOFe,1);                  % Element distributed force

%. Compute element tangent and residual by numerical integration
   for i=1:S.nQpts

%... Set element coordinate for integration point, get load
       xi = S.st(i);      xx = (x0 + xi*Le)/L;
       q = qo*LoadFunction(xx,1,LoadType);

%... Compute shape functions and kernel matrices
       [h,dh] = BubbleFcns(xi,S.nExt(1));
       [g,dg] = BubbleFcns(xi,S.nExt(3));
       DhDh = (dh*dh')/Le;    Dhg = (dh*g');      hh = (h*h')*Le;
       DgDg = (dg*dg')/Le;    gDh = (g*dh');      gg = (g*g')*Le;

%... Compute the element displacements and derivatives
       dw = [dot(Ue,dh); dot(We,dh)]/Le;
       th = dot(Te,g);
       dth = dot(Te,dg)/Le;

%... Compute the core matrices for this integration point
       [A,b,c,R,~,~] = NLBTang(dw,th,dth,d);

%... Compute the contributions to the element tangent
       ke = ke + [A(1,1)*DhDh, A(1,2)*DhDh, b(1)*Dhg;...
                  A(2,1)*DhDh, A(2,2)*DhDh, b(2)*Dhg;...
                  b(1)*gDh,    b(2)*gDh,   c(1)*gg+c(2)*DgDg]*S.wt(i);

%... Compute the contributions to the element mass
       me = me + [rhoA*hh,     Zed2,     Zed1;...
                  Zed2',     rhoA*hh,    Zed1;...
                  Zed1',      Zed1',  rhoI*gg]*S.wt(i);
```

```
%... Compute the internal residual
      ge = ge + [R(1)*dh; R(2)*dh; R(3)*g*Le + R(4)*dg]*S.wt(i);

%... Compute the distributed load vector
      fe = fe + [q(1)*h; q(2)*h; q(3)*g]*Le*S.wt(i);

   end

end
```

Note that the weights and stations for the quadrature rule are specified in the input file and reside in the 'struct' S. These values are set in the function NumInt (see Appendix F), which creates the weights S.wt and stations S.st of the integration rule, depending on which type of rule is set in the input file. The integration is set up as a loop over the integration stations. At each trip through the loop, the contributions of that integration station are added to the total. At the end of the loop, the integration is complete.

Within the loop we compute the contributions to the element matrices by evaluating the shape functions at the current integration station and implementing them in the equations derived for these matrices. The notation should allow a fairly direct comparison with Eqs. 13.34, 13.36, and 13.38 (we have not yet discussed the element mass matrix).

The shape function routine BubbleFcn computes the values of all of the shape functions and their derivatives at the point ξ, as outlined in Appendix D. The element function calls Code 13.1, NLBTang, to form the components needed to compute tangent and residual (everything up to, but not including, the interpolation with the shape functions to compute the element matrices).

13.5.3   Verification of Static Code

To see that the static code is working properly, we put it to the task of computing the response for some problems where we know the answer. We consider two here plus an example to demonstrate how flexibility in shear looks for large deformations.

Beam Subjected to Pure Bending   Consider a cantilever beam of length $L = 10$ subjected to an end moment $M_o = 150\pi$. The beam has bending modulus $EI = 1000$ and shear and axial moduli of $GA = 500$ and $EA = 1000$. The beam was modeled using 4 quintic/quartic elements. The loading was applied in three steps and the deflected shape of the beam is shown at each load step in Fig. 13.2 (this is what the function NLBPlotShape produces). Note that the nodes are indicated by the open dots and the lines transverse to the axis of the beam show the rotation of the cross section at those points. Since the beam is in pure bending, the cross sections are perpendicular to the axis of the deformed beam, as expected.

The exact solution to this problem (at the final load step) is $M(x) = M_o$, with the angle subtended by the arc of the beam to be $\theta = M_o L / EI = 1.5\pi$. The shear and axial force should be exactly zero. The results for the displacements, rotations, and

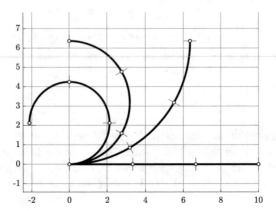

Fig. 13.2 *Beam subjected to pure bending* (Part 1). Cantilever beam subjected to end moment $M = 150\pi$, applied over three equal load steps. The displaced shape of the beam is shown at each step. Note that with the full load, the beam deforms into three quarters of a perfect circle. The physical properties are $L = 10$, $EI = 1000$, $GA = 500$, $EA = 1000$. The computation is done with 4 quintic/quartic elements

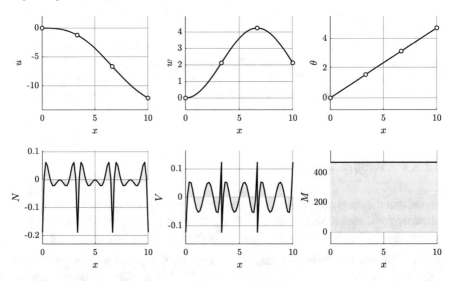

Fig. 13.3 *Beam subjected to pure bending* (Part 2). Response fields for the final state of the cantilever with end moment. The properties are the same as Fig. 13.2

internal forces at the final load level are shown in Fig. 13.3. Notice that the bending moment is constant and equal to the applied end moment, as it should be. The internal forces N and V are small, but not zero. These values can be made smaller with more elements or a higher-order interpolation. The rotation $\theta(x)$ varies linearly, which makes sense because the moment is proportional to the first derivative of the rotation and the moment is constant.

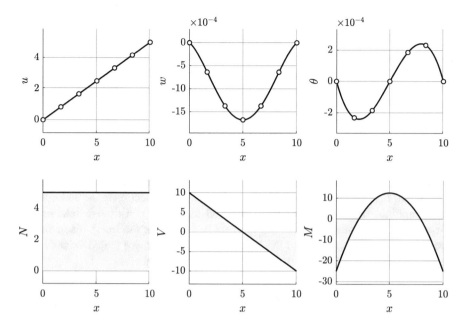

Fig. 13.4 *Bending of a stretched beam.* Response fields for the fixed–fixed beam of original length $L_o = 10$ subjected to transverse force $q_o = 2$ and stretched to a length of 15. The beam has properties $E = 1000$, $G = 5 \times 10^6$, $I = 100$, and $A = 0.01$

Bending of a Stretched Beam   Another way to verify the static code is to consider a beam subjected to transverse load and an axial force that stretches the beam to some significant fraction of its original length. The additional length should affect the moments and deflections, and we can compare the results with the linear results of the beam with the properties adjusted to the stretched lengths.

Consider a fixed–fixed beam of original length $L_o = 10$, subjected to a uniform transverse load $q_o = 2$ and stretched to a deformed length of $L = 15$ (by prescribing the motion of the node at the right end of the beam). The beam has properties $E = 1000$, $G = 5 \times 10^6$, $I = 100$, and $A = 0.01$. This set of values keeps the axial forces relatively small, the transverse deflections small (to avoid stiffening effects due to sagging in the presence of tensile axial force), and the shear stiffness large enough for the result not to be completely dominated by shear. The result of the analysis using 6 quartic/cubic elements is shown in Fig. 13.4. Note that the axial displacement is linear and reaches a value of five at the right end. The maximum deflection is at the middle of the beam and has the value $w_{max} = 1.67154 \times 10^{-3}$. The moments at the ends are equal to each other and equal to $M_{max} = 24.9956$.

We can compare this result with exact solution from linear Timoshenko beam theory if we account for the stretch. The exact solution to the fixed–fixed beam is

$$V_{max} = \frac{qL}{2}, \qquad M_{max} = \frac{qL^2}{12}, \qquad w_{max} = \frac{qL^4}{384EI}(1 + 4\beta)$$

where $\beta = 12EI/GAL^2$. Let λ be the stretch ratio, which is 1.5 in this example. To compare these results, we must note that the load $q = q_o/\lambda$ stretches out so that the total load remains the same and the length is stretched to $L = \lambda L_o$. Recall that the constitutive equations are affected by the large deformations. There is a discussion of this in the truss chapter, Sect. 8.5. Without going into the details, one can show that the moment–curvature relationship can be derived as

$$M = \int_A zP \, dA = \lambda \int_A z\Sigma \, dA = \lambda EI_o \kappa_o$$

where P is the first Piola–Kirchhoff stress, $\Sigma = P/\lambda$ is the second Piola–Kirchhoff stress, and I_o is the second moment of the area of the cross section in the undeformed configuration. In the code we have assumed the linear constitutive relationship $M = EI\kappa_o$. To compare results, the moment of inertia for the linear result should be scaled as $I = \lambda I_o$, as should the area $A = \lambda A_o$. Using these scaled values in the equations above, we find that the Timoshenko beam theory gives

$$V_{max} = 10, \qquad M_{max} = 25, \qquad w_{max} = 1.67188 \times 10^{-3}$$

These values are almost identical to the results computed by the program, providing a significant verification of the nonlinear response. This example also reminds us of the complications associated with writing constitutive equations for geometrically nonlinear problems.

Shear Flexibility in Large Deformations To show a case where shear deformations are important, consider a beam fixed at the left end and fixed at the right end (except that it can move freely in the horizontal direction). The beam is subjected to a uniform transverse load $q_o = 10$. The moduli are $EI = 1000$, $GA = 50$, and $EA = 1000$. The shear modulus is very small to accentuate the effects of shear deformation. The response was computed with 4 quintic/quartic elements in three load steps. The displaced structure under full load is shown in Fig. 13.5.

The displaced shape of the beam is plotted to scale, showing transverse displacements roughly one third of the initial length of the beam. Thus, the loading induces significant motion. The rotations of the beam are very small, indicating that the deflections are due primarily to shearing.

The response fields are shown in Fig. 13.6. The horizontal displacement $u(x)$ is restrained at the left end, but not at the right. You can see how this field develops over the length of the beam, and you can determine the motion of the right end to be about $u(L) = -0.27$. Note that the axial forces are not zero. The beam deflects so much that it engages axial force as a means of resistance (like a cable would). Overall, the length of the beam changes very little because the axial forces are fairly small, and the axial modulus is larger than the shear modulus. The beam is statically determinate and with a uniform load, the expected shear distribution is linear, and the internal moment distribution is quadratic.

Fig. 13.5 *Shear flexibility in large deformations* (Part 1). Displaced shape of a fixed–fixed beam of length $L = 10$ subjected to a uniformly distributed load of $q_o = 10$. Physical properties are $EI = 1000$, $GA = 50$, and $EA = 1000$. The right end is allowed to move horizontally. Displacements are to scale

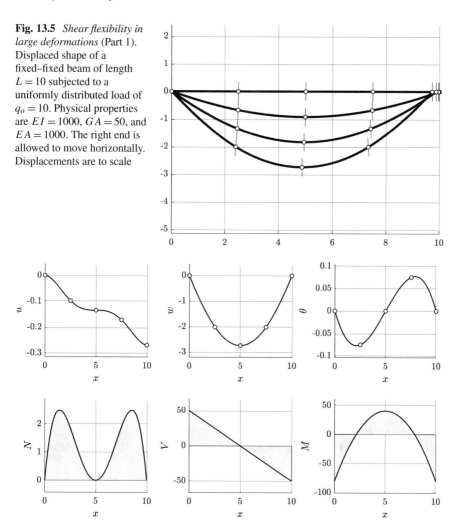

Fig. 13.6 *Shear flexibility in large deformations* (Part 2). Response fields for the fixed–fixed beam of Fig. 13.5

13.6 Dynamic Analysis of Nonlinear Planar Beams

Since we are ultimately interested in dynamic response, we will not pursue the additional machinery required to do robust nonlinear static analysis (e.g., arc-length control methods). In dynamics, we march through time and iterate at each time step to convergence of the equations of motion. At each time step, we have a very good starting guess for the Newton iteration (i.e., the acceleration at the previous time step). Therefore, we do not need to worry about load-stepping schemes in dynamics.

In this section we will derive the equations of motion of the nonlinear beam. Let us return to the virtual work functional of Eq. 13.7, which is

$$\mathcal{G} = \mathcal{G}_{int} - \mathcal{G}_{ext} + \mathcal{G}_{dyn}$$

Recall that the three component parts of the virtual work functional are defined in Eq. 13.8. We treated the internal and external virtual work terms in the previous section. In this section we will focus on the dynamic part

$$\mathcal{G}_{dyn} = \int_0^L \left(\rho A \, \ddot{\mathbf{w}} \cdot \bar{\mathbf{w}} + \rho I \ddot{\theta} \, \bar{\theta} \right) dx$$

which governs the inertial resistance. We can write the dynamic part of the functional as a sum over elements as

$$\mathcal{G}_{dyn} = \sum_{e=1}^M \int_{L_e} \left(\rho A \, \bar{u}_e \ddot{u}_e + \rho A \, \bar{w}_e \ddot{w}_e + \rho I \, \bar{\theta}_e \ddot{\theta}_e \right) dx_e$$

Now, using the interpolations defined in Eqs. 13.26, this functional can be expanded in terms of the element displacement and rotation parameters as

$$\mathcal{G}_{dyn} = \sum_{e=1}^M \int_{L_e} \left(\rho A \, \bar{\mathbf{U}}_e^T \boldsymbol{\varphi}\boldsymbol{\varphi}^T \ddot{\mathbf{U}}_e + \rho A \, \bar{\mathbf{W}}_e^T \boldsymbol{\varphi}\boldsymbol{\varphi}^T \ddot{\mathbf{W}}_e + \rho I \, \bar{\boldsymbol{\Theta}}_e^T \boldsymbol{\psi}\boldsymbol{\psi}^T \ddot{\boldsymbol{\Theta}}_e \right) dx$$

Recognizing that the element displacement and rotation parameters $\mathbf{U}_e$, $\mathbf{W}_e$, and $\boldsymbol{\Theta}_e$ and their virtual counterparts do not depend upon x, we can define mass matrices

$$\mathbf{m}_e^1 = \int_0^1 \rho A \, \boldsymbol{\varphi}\boldsymbol{\varphi}^T L_e \, d\xi, \qquad \mathbf{m}_e^2 = \int_0^1 \rho I \, \boldsymbol{\psi}\boldsymbol{\psi}^T L_e \, d\xi \qquad (13.48)$$

and put them into a complete element mass matrix as

$$\mathbf{m}_e = \begin{bmatrix} \mathbf{m}_e^1 & \mathbf{0} & \mathbf{0} \\ \mathbf{0} & \mathbf{m}_e^1 & \mathbf{0} \\ \mathbf{0} & \mathbf{0} & \mathbf{m}_e^2 \end{bmatrix} \qquad (13.49)$$

where the $\mathbf{0}$ matrices are of the appropriate size. Finally, noting the relationship between the element DOF parameters and the global displacement vector $\mathbf{u}$, the global mass matrix can be assembled from element contributions as

$$\mathbf{M} = \sum_{e=1}^M \mathbf{B}_e^T \mathbf{m}_e \mathbf{B}_e \qquad (13.50)$$

where $\mathbf{B}_e$ is defined in Eq. 13.29. Of course, the assembly process is carried out as usual using the eDOF array that contains the global node numbers for the elements, as discussed previously.

With this definition of mass, the total virtual work associated with the inertial terms is $\mathcal{G}_{dyn} = \bar{\mathbf{u}}^T \mathbf{M} \ddot{\mathbf{u}}$. Now, the complete virtual work functional takes the form

$$\mathcal{G} = \bar{\mathbf{u}}^T \left(\mathbf{M} \ddot{\mathbf{u}} + \mathbf{g}(\mathbf{u}) - \mathbf{f} \right) \tag{13.51}$$

where $\mathbf{f}$ is the global applied force vector defined by Eq. 13.44, and $\mathbf{g}(\mathbf{u})$ is the global internal resistance vector, given in Eq. 13.40. If $\mathcal{G} = 0$ for all $\bar{\mathbf{u}}$, then the fundamental theorem of the calculus of variations implies the discrete equations of motion

$$\mathbf{M} \ddot{\mathbf{u}}(t) + \mathbf{g}(\mathbf{u}(t)) = \mathbf{f}(t) \tag{13.52}$$

This is a nonlinear system of ordinary differential equations. We can solve this system by Newmark's method, with a Newton iteration at each time step to handle the nonlinear part. Note that the Newton iteration will be very similar to the static problem.

13.6.1 Solution of the Nonlinear Differential Equations

We can place the dynamic problem in the context of Newmark's method for time integration and Newton's method for solving the nonlinear algebraic problem that arises in each time step. In particular, let us assume that we have the state $\{\mathbf{u}_n, \mathbf{v}_n, \mathbf{a}_n\}$ at time step n and we want to compute the state at time step $n + 1$. First, we note the Newmark equations

$$\mathbf{u}_{n+1} = \mathbf{b}_n + \zeta \, \mathbf{a}_{n+1}$$
$$\mathbf{v}_{n+1} = \mathbf{c}_n + \eta \, \mathbf{a}_{n+1}$$

where, as usual, $\mathbf{b}_n = \mathbf{u}_n + h\mathbf{v}_n + h^2\beta \, \mathbf{a}_n$ and $\mathbf{c}_n = \mathbf{v}_n + h\gamma \, \mathbf{a}_n$ are the parts of the displacement and velocity at the new time step that can be computed from the state at time t_n. Note that β and γ are the Newmark parameters and that $\eta = (1 - \gamma)h$ and $\zeta = (0.5 - \beta)h^2$.

We can think of the primary unknown in this problem as $\mathbf{x} = \mathbf{a}_{n+1}$ because once we know the new acceleration, we can get the new displacement and velocity from Newmark. Define the dynamic residual as

$$\mathcal{R}(\mathbf{x}) = \mathbf{M}\mathbf{x} + \mathbf{g}(\mathbf{u}(\mathbf{x})) - \mathbf{f} \tag{13.53}$$

where $\mathbf{u}(\mathbf{x})$ implies that the displacement is a function of the acceleration through the Newmark relationship. The tangent associated with this residual is

$$\mathcal{T}_{dyn}(\mathbf{x}) = \frac{\partial \mathcal{R}}{\partial \mathbf{x}} = \mathbf{M} + \zeta \, \mathcal{T}(\mathbf{x}) \tag{13.54}$$

where $\mathcal{T}(\mathbf{x})$ is the static tangent, given by Eq. 13.39. The dynamic residual and tangent can be evaluated for any estimate of the acceleration $\mathbf{x}$.

The Newton iteration goes as follows. Set the initial guess of the new acceleration to be the converged acceleration of the previous times step, i.e., $\mathbf{x}^0 = \mathbf{a}_n$. Then iterate

$$\mathbf{x}^{i+1} = \mathbf{x}^i - \left[\mathcal{T}_{dyn}(\mathbf{x}^i) \right]^{-1} \mathcal{R}(\mathbf{x}^i) \tag{13.55}$$

Iteration continues until the norm of the residual $\|\mathcal{R}(\mathbf{x}^m)\| < tol$, where tol is a suitably small number and m is the iteration number at which convergence is achieved. The acceleration at the new time step is the converged value, i.e., $\mathbf{a}_{n+1} = \mathbf{x}^m$. We can summarize the dynamic analysis algorithm as follows:

Dynamic Analysis of Nonlinear Beams

1. Input structure geometry and loads.
2. Establish initial displacement $\mathbf{u}_o$ and velocity $\mathbf{v}_o$.
3. Compute initial acceleration from the equations of motion.
4. Loop over time steps

   - Compute new load $\mathbf{f}_{n+1}$.
   - Compute $\mathbf{b}_n$ and $\mathbf{c}_n$. Set $\mathbf{x}^0 \leftarrow \mathbf{a}_n$.
   - While $err > tol$

     – Compute $\mathbf{u}^i_{n+1}$ and $\mathbf{v}^i_{n+1}$ from Newmark.
     – Compute tangent $\mathcal{T}_{dyn}(\mathbf{x}^i)$ and residual $\mathcal{R}(\mathbf{x}^i)$.
     – Update acceleration $\mathbf{x}^{i+1} = \mathbf{x}^i - [\mathcal{T}_{dyn}(\mathbf{x}^i)]^{-1}\mathcal{R}(\mathbf{x}^i)$.
     – Compute norm of residual $err = \|\mathcal{R}(\mathbf{x}^i)\|$.
     – Increment counter $i \leftarrow i + 1$.

   - Converged value of $\mathbf{x}^i \rightarrow \mathbf{a}_{n+1}$ is the solution.
   - Compute final displacement and velocity $\mathbf{u}^i_{n+1}$ and $\mathbf{v}^i_{n+1}$.
   - Update internal variables (if applicable).
   - Update state $\mathbf{u}_n \leftarrow \mathbf{u}_{n+1}, \mathbf{v}_n \leftarrow \mathbf{v}_{n+1}, \mathbf{a}_n \leftarrow \mathbf{a}_{n+1}$.

5. Loop over time steps complete. Output graphics.

One thing to notice about the nonlinear algorithm for the dynamic problem is that the tangent is built around the mass matrix $\mathbf{M}$ with a small contribution from the static tangent $\mathcal{T}$ because ζ is a small number if the time step h is small. That stabilizes the calculation in a way that is not generally possible in the static setting. Also, we have a very good guess for the starting value for the iteration. In each time step we are iterating to find the new acceleration. A good starting value, then, is the converged acceleration from the previous time step. If that value is not close enough for Newton's method to converge, then the time step h should be reduced.

By now it should be evident that the approach to nonlinear dynamic analysis of structures follows a fairly standard path. The main differences from one theory to the next lies in the details of how to compute the mass, tangent stiffness, and residual at a given state of deformation. We can, of course, add damping to the model without significant difficulty (with the same caveats that have been relevant since the introduction of damping in the shear building). It is evident that the most complicated part of the formulation and implementation is associated with the static component of the residual because the inertia forces are always linearly proportional to the acceleration.

13.6.2 Dynamic Implementation

The implementation of the dynamic analysis will follow the framework laid out for the theories in the previous chapters very closely. In fact, one of the main messages at this point is how similar the implementation is for the different systems we have derived. The main program for dynamic analysis of nonlinear beams is given in Code 13.5.

Code 13.5 MATLAB code for dynamic analysis of nonlinear beams

```
%  Nonlinear Beam Dynamics (Newmark)
   clear; clc;

%. Get physical properties and set up mesh
   [x,ix,id,eDOF,d,Load,F,damp,S,DF] = NLBinput_SpinUp;

%. Put nodal forces into DOF order
   [fNodal,u,uPre] = Force(F,id,S,DF);

%. Final time, time step, retrieve Newmark parameters
   tf = 1.5;   h = 0.01;
   [beta,gamma,eta,zeta,nSteps,tol,itmax,tf] = NewmarkParams(tf,h);

%. Form mass, stiffness matrices and distributed load vector
   uTot = [u; uPre];
   [M,A,~,fDist] = NLBAssemble(x,uTot,d,eDOF,ix,Load,S);
   f = fNodal + fDist;

%. Compute damping matrix, natural frequencies and mode shapes
   [C,Evec,freq,damp] = DampingMatrix(M,A,damp,DF);

%. Set up arrays to save results
   Time = zeros(nSteps,1);        % History of time
   Disp = zeros(nSteps,S.nDOF);   % History of displacements

%. Establish initial conditions
   [uo,vo] = NLBICs(Evec,M,x,d,ix,eDOF,S,DF);

%. Initialize time, position, velocity, and acceleration
   t = 0;
   Ft = TimeFunction(t,S.TimeType,0);
```

```
    uold = uo;
    vold = vo;
    aold = M\(Ft*f - A*uo - C*vo);

%. Compute motion by Newmark integration
    for n=1:nSteps

%... Store results for plotting later
        Time(n,:) = t;
        Disp(n,:) = [uold; Ft*uPre]';

%... Compute the values for the next time step
        t = t + h; its = 0; err = 1;
        Ft = TimeFunction(t,S.TimeType,0);
        bn = uold + h*vold + beta*h^2*aold;
        cn = vold + gamma*h*aold;
        anew = aold;

%... Newton iteration to determine the new acceleration
        while (err > tol) && (its < itmax)
            its = its + 1;
            unew = bn + zeta*anew;
            vnew = cn + eta*anew;
            uTot = [unew; Ft*uPre];
            [~,A,g,~] = NLBAssemble(x,uTot,d,eDOF,ix,Load,S);
            g = M*anew + C*vnew + g - Ft*f;
            A = M + eta*C + zeta*A;
            anew = anew - A\g;
            err = norm(g);
        end

%... Compute velocity and displacement by numerical integration
        unew = bn + zeta*anew;
        vnew = cn + eta*anew;

%... Update state for next iteration
        aold = anew;
        vold = vnew;
        uold = unew;

    end

%. Generate graphical output of response
    NLBGraphs(DF,x,ix,Time,Disp,id,eDOF,S,freq);
```

13.6.3   Example

We will end this chapter with an example that involves large motions of a dynamic beam. We will consider the "spin-up" maneuver where a beam is rotated from horizontal to vertical in a specific amount of time. The scenario is illustrated in Fig. 13.7. The beam has $L = 10$, $E = 10000$, $G = 5000$, $A = 1$, $I = 0.1$, and $\rho = 0.1$.

A "spin-up" maneuver amounts to specifying the rotation at the base as a function of time. The beam responds dynamically to the forced motion of the base. The angle at the base is prescribed as the function $\theta_o(t) = \Theta_o f(t)$. In this case the

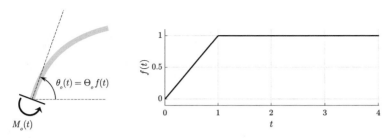

Fig. 13.7 *Example 3(a)*. Beam of length $L = 10$ with $E = 10000$, $G = 5000$, $A = 1$, $I = 0.1$, $\rho = 0.1$. The beam is spun up through an angle of $\Theta_o = \pi/2$ in a time in accord with the function $f(t)$ shown

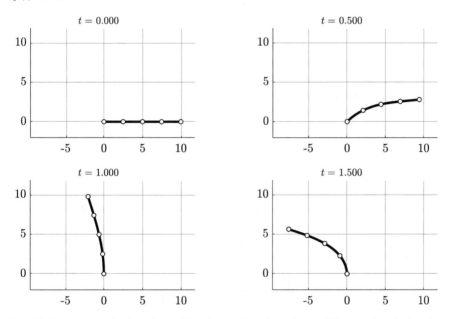

Fig. 13.8 *Example 3(b)*. Snapshots of the displaced configuration at different points in time for the spin-up problem of Fig. 13.7

angle is $\Theta_o = \pi/2$ and the time function is as shown in Fig. 13.7. Note that the moment required to execute the motion is an unknown reaction (i.e., if you specify the motion, the associated force must be a reaction).

The configuration of the beam is shown at four equally spaced time points between $t = 0$ and $t = 1.5$ in Fig. 13.8. The upper left graph shows the initial position of the beam. It starts from rest. At time $t = 0.5$ it is evident that the beam lags the prescribed rotation due to its inertia. At time $t = 1.0$ the beam has straightened and passed by the vertical location, indicating that the beam is vibrating and at that particular time the internal flexural mechanism has helped the beam to catch up

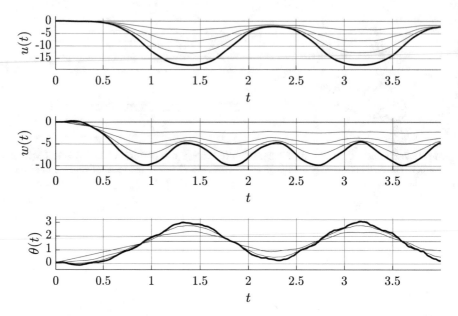

Fig. 13.9 *Example 3(c)*. Response history of the nodal degrees of freedom (the open circles in Fig. 13.8) for the spin-up problem of Fig. 13.7. The node at the end of the bar is the heavier line

with the driving rotation. The rotation at time $t = 1.5$ is well past vertical, indicating a significant inertial effect. From that point on, the beam oscillates about the vertical position.

Figure 13.9 shows the time history of response of the beam, by components. Note that u and w are defined with respect to the global coordinates. Hence, they do not remain "axial" and "transverse" throughout the motion. It appears that w oscillates with nearly twice the frequency of u and θ. But that is due to the fact that the vertical motion reaches a minimum at both the left and right and a maximum as it passes through the vertical position. The rotation angle θ shows a higher frequency component superimposed on the fundamental frequency. This example illustrates how the response quickly devolves into vibration, even though the excitation propagates as a wave, originating from the base.

13.7 Summary

The focus of this chapter was the implementation of the dynamic response of the fully nonlinear beam. The conceptualization of the problem as a beam of length L is similar to the previous two chapters. Of course, the real strength of structural analysis is the ability to describe a structure as an assembly of elements without the requirement that all elements lie along the horizontal axis initially. The next chapter will generalize the formulation of the nonlinear beam into planar frame structures. Most of the functions in the frame code are similar to those used in the nonlinear beam code. Thus, we have shown only a few of the functions in this chapter to illuminate the theory. A more complete accounting for the code will be done in the next chapter.

Chapter 14
Dynamic Analysis of Planar Frames

This chapter extends the formulation and solution methods of beam theory to two-dimensional (planar) framed structures. As was the case with the truss, frame structures can have multiple elements oriented in different directions sharing load at the places where they connect. Like the truss, we will establish balance of linear momentum at the nodes. Unlike the truss, angular momentum must also balance at the nodes to recognize that transmission of bending moment at the intersection points of elements is a fundamental feature of continuous frames.[1] This final chapter of the book gets us to a practical endpoint in our study of structural dynamics, providing a context to tie together all of the main themes of the book and yielding a computational framework in which many practical problems can be explored.

The main reason for restricting our attention to planar frames is to try to keep the bookkeeping tolerable for learning and the visualization of the results manageable. Unlike the truss, going from two dimensions to three for frames is not as simple as adding a third element to a few of the arrays. However, the conceptual framework for planar frames brings most of the key ideas for general frame analysis.

---

[1] The term *continuous frame* means that all elements share the same rotation at the joint where they meet. It is, of course, possible to have a structure where truss elements frame into a joint shared by frame elements. We will not pursue that more general modeling option in this book. Hence, the term *frame* will be used hereafter to imply a continuous frame.

Electronic Supplementary Material   The online version of this chapter (https://doi.org/10. 1007/978-3-030-89944-8_14) contains supplementary material, which is available to authorized users.

K. D. Hjelmstad, *Fundamentals of Structural Dynamics*,
https://doi.org/10.1007/978-3-030-89944-8_14

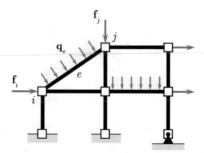

Fig. 14.1 *Typical frame structure.* A frame structure is a collection of nodes connected by elements. The connections between members at the nodes are rigid, and therefore transmit moment. The structure can be subjected to applied nodal loads, e.g., $\mathbf{f}_i$ and $\mathbf{f}_j$, or distributed element loads, e.g., $\mathbf{q}_e$. The motion of the structure is restrained at certain points through supports

14.1   What Is a Frame?

A frame is an assembly of beam elements connected together at joints (nodes). The main difference between a truss and a frame is that the joints connect the elements together rigidly, so that all elements that frame into a single joint must experience the same rotation at that point. As a result, the elements transmit bending moment to the joints and the balance of angular momentum at the joint determines, in part, how force flows through the structure.

A typical frame structure is shown in Fig. 14.1. The structure is a collection of *nodes* (with typical examples i and j labeled) connected by elements (with typical example e labeled in the figure). The nodes are represented by squares in the sketch as a reminder that the elements frame in rigidly. The nodes can be subjected to applied forces (with typical examples $\mathbf{f}_i$ and $\mathbf{f}_j$ labeled in the figure) and applied moments. The elements can be subjected to applied distributed loads (with typical example $\mathbf{q}_e$ labeled in the figure) and applied distributed moments. The members have properties E, G, A, I, and ρ, as considered in the previous chapters on beams. We will associate the nodes with having mass, and will designate the inertial properties for node i through the mass m_i and the rotational inertia J_i.[2] Attributing mass and rotational inertia to the nodes allows the incorporation of mass in the structure that is not associated directly with the elements.

The nodes are, from a mathematical perspective, rigid bodies, as described in Chap. 1. Because the motion is restricted to the plane of the structure, the motion of each node can be described with two components of translation and a rotation. Hence, each node has three degrees of freedom. We will consider the motion of the

---

[2] Note that the letter m gets a pretty hard workout in this chapter. We use it for the nodal mass, usually with a subscript i, as in m_i. We use it for the applied nodal moment, but always with a hat, as in $\hat{m}_i$. And we use it for the applied distributed moment with a subscript e, as in m_e. The context should usually make the usage clear.

Fig. 14.2 *Support conditions.* Pictorial representation of frame support conditions: (**a**) fixed, (**b**) pinned, (**c**) roller, (**d**) slide, and (**e**) free. The roller and slide conditions have two varieties, allowing translation in either the horizontal or vertical directions

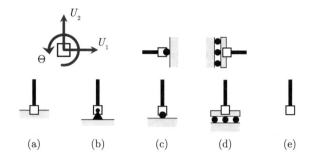

nodes, along with the motion parameters associated with non-nodal element bubble functions, to be the primary variables in our formulation.

The motion of the structure can be restrained at certain nodes, which provide the boundary conditions for the structure. At the supports we specify the motion at any of the three DOF, as illustrated in Fig. 14.2. A node can be either fixed (all three DOF restrained), pinned (translation restrained but rotation is free), roller (translation is restrained in one direction but rotation and translation in the other direction are free), slide (translation in one direction and rotation are restrained but translation in the other direction is free), or free (all three motion components are free). Note that two of the nodes in Fig. 14.1 are fixed and one is pinned. While it is possible to have an internal hinge (e.g., you could have a truss element in the structure with a pin at each end), we will not include that level of modeling complexity in the formulation or codes.

In dynamics, the loads and the response (i.e., the displacement, velocity, and acceleration of the elements and nodes and the internal axial force, shear force, and bending moment in the elements) depend upon time t. The physical properties of the structure will be considered constant.

As we saw for the truss, and for the finite element discretization for the axial bar and beam, we will satisfy the element-to-element continuity requirement by establishing global degrees of freedom at the nodes, and assert that the ends of the members framing into a certain node all share that specific motion. This approach is, perhaps, *the* big idea of structural analysis and is generally referred to as a kinematic, or motion-based, approach to formulating the equations of motion. From that perspective, we think of a configuration of the structure in terms of the motion of the nodes.[3] The elements must go along for the ride, and in doing so they develop forces associated with the motion they are asked to execute. The forces at the ends of the element contribute to the joint, which must satisfy nodal equations of motion.

---

[3] There are also global degrees of freedom associated with the bubble functions in each element, but these degrees of freedom do not play a role in establishing continuity of the structure because they are zero at the element ends.

Fig. 14.3 *Nodal equilibrium.*
Element *e* provides
contributions to the nodal
equilibrium of both nodes *i*
and *j*. Equilibrium is
established in the deformed
configuration. The
undeformed configuration of
element *e* has orientation
defined by unit vectors $\mathbf{e}_1^e$ and
$\mathbf{e}_3^e$

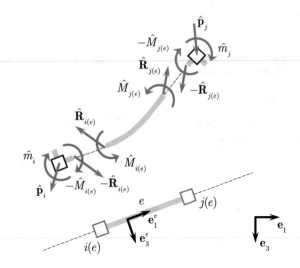

14.2   Equations of Motion

The basis for the analysis of structures is the *method of joints* which says that each
node must satisfy balance of linear and angular momentum. We will establish the
equations of motion in virtual work form. The virtual work functional that governs
the response of the entire structure is the sum of contributions from the elements
and the nodes. Consider an element in a frame and its associated nodes, as shown in
Fig. 14.3. Element *e* is associated with nodes $i(e)$ and $j(e)$, where node $i(e)$ is the
"start" node and $j(e)$ is the "end" node for element *e*, as determined by the element
base vectors $\{\mathbf{e}_1^e, \mathbf{e}_3^e\}$ that describe the orientation of the element in the undeformed
configuration.

As in the previous chapter, we will designate the resultant force on a cross section
as $\mathbf{R}(x, t)$, where x is measured from the i-end of the element and is defined in the
undeformed configuration, going from 0 to L_e. The resultant in element *e* can be
expressed as

$$\mathbf{R}_e(x, t) = N_e(x, t)\mathbf{n}_e(x, t) + V_e(x, t)\mathbf{m}_e(x, t)$$

where N_e and V_e are the axial force and shear force in the element and $\mathbf{n}_e$ and
$\mathbf{m}_e$ are unit vectors that point in the axial and transverse direction of the deformed
element for element *e*, i.e., $\mathbf{n}_e$ is normal to the deformed cross section and $\mathbf{m}_e$ is in
the plane of the deformed cross section (see Fig. 10.2). The resultant force can also
be expressed in terms of the base vectors associated with the original orientation of
the element. In particular,

$$\mathbf{R}_e(x, t) = H_e(x, t)\mathbf{e}_1^e + Q_e(x, t)\mathbf{e}_3^e$$

Fig. 14.4 *Traction sign convention.* The sign convention on the internal force $\mathbf{R}_e(x, t)$ derives from the sign convention on the internal forces and moments (e.g., N_e and V_e)

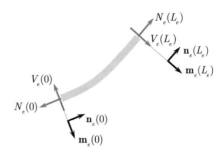

It is also possible to express this force (or any other vector) in the global basis $\{\mathbf{e}_1, \mathbf{e}_2, \mathbf{e}_3\}$.

We distinguish between the traction vector $\hat{\mathbf{R}}_{i(e)}$ and the internal force at the left end of the beam $\mathbf{R}_e(0)$. A *traction* is a force vector exposed by taking a cut, as in a free body diagram. It is the manifestation of the internal state as a force vector acting on the cut. There is a similar distinction for the right end. In fact, because of the nature of internal forces, and our sign convention on N, V, and M, we can write

$$\hat{\mathbf{R}}_{i(e)} = -\mathbf{R}_e(0), \qquad \hat{\mathbf{R}}_{j(e)} = \mathbf{R}_e(L_e)$$
$$\hat{M}_{i(e)} = -M_e(0), \qquad \hat{M}_{j(e)} = M_e(L_e)$$

(14.1)

To see why this is true, consider the axial force and shear force at the ends of the beam, shown in Fig. 14.4. The sign convention has axial force N_e positive in tension, which points in the $-\mathbf{n}_e$ direction at the left end of the beam and in the $\mathbf{n}_e$ direction at the right end. Similarly, the positive shear force convention has shear force V_e on the left end of the beam pointing in the $-\mathbf{m}_e$ direction and in the $\mathbf{m}_e$ direction at the right end. By designating the traction vectors as indicated, we can account for the difference between an internal stress resultant and a traction *vector*. Doing so enables us to use ordinary vector addition for balance of momentum, independent of which end of the beam is included. It is instructive to compare this with the similar strategy used in the derivation of equations of motion for the truss.

Following the above convention, balance of linear and angular momentum hold at node i if

$$m_i \ddot{\mathbf{W}}_i + \sum_{k \in \mathcal{K}(i)} \hat{\mathbf{R}}_k - \hat{\mathbf{p}}_i = \mathbf{0}, \qquad J_i \ddot{\Theta}_i + \sum_{k \in \mathcal{K}(i)} \hat{M}_k - \hat{m}_i = \mathbf{0} \qquad (14.2)$$

where $\mathcal{K}(i)$ is the set of elements that frame into node i. Each of the N nodes in the structure are governed by these equations. The vector $\mathbf{W}_i(t)$ is the displacement and $\Theta_i(t)$ is the rotation of node i. The applied nodal forces and moments are $\hat{\mathbf{p}}_i$ and $\hat{m}_i$, respectively.

We also know that element e satisfies balance of linear and angular momentum if (see Eqs. 13.2 and 13.3)

$$\rho A \ddot{\mathbf{w}}_e - \mathbf{R}'_e - \mathbf{q}_e = \mathbf{0}, \qquad \rho I \ddot{\theta}_e - M'_e + \mathbf{R}_e \cdot \mathbf{\Gamma}_e - m_e = 0 \qquad (14.3)$$

where $\mathbf{q}_e$ is the applied distributed load, m_e is the applied distributed moment, and $\mathbf{\Gamma}_e = (1 + \varepsilon_o)\mathbf{m}_e - \beta_o \mathbf{n}_e$. The functions $\mathbf{w}_e(x, t)$ and $\theta_e(x, t)$ are the displacement and rotation of element e. To be compatible, the nodal displacements and rotations must be equal to the element displacements and rotations at the ends of the element. To wit,

$$\mathbf{w}_e(0, t) = \mathbf{W}_{i(e)}(t) \quad \mathbf{w}_e(L_e, t) = \mathbf{W}_{j(e)}(t)$$
$$\theta_e(0, t) = \Theta_{i(e)}(t) \quad \theta_e(L_e, t) = \Theta_{j(e)}(t)$$

It is important to observe that we still have to assure this continuity, but that will be built into the numerical discretization by finite elements later.

Now, let us invoke the same virtual work arguments that we used for the truss problem (for the nodes) and the beam problem (for the elements). Let $\bar{\mathbf{W}}_i$ be a virtual displacement at node i and $\bar{\Theta}_i$ a virtual rotation. Similarly, let $\bar{\mathbf{w}}_e(x)$ be a virtual displacement field in element e and $\bar{\theta}_e(x)$ a virtual rotation field. The virtual displacements and rotations must satisfy the same compatibility conditions as the real ones, i.e.,

$$\bar{\mathbf{w}}_e(0) = \bar{\mathbf{W}}_{i(e)}, \quad \bar{\mathbf{w}}_e(L_e) = \bar{\mathbf{W}}_{j(e)}, \quad \bar{\theta}_e(0) = \bar{\Theta}_{i(e)}, \quad \bar{\theta}_e(L_e) = \bar{\Theta}_{j(e)}$$

Multiply the nodal balance of linear momentum equation by $\bar{\mathbf{W}}_i$, the nodal balance of angular momentum equation by $\bar{\Theta}_i$, the element balance of linear momentum equation by $\bar{\mathbf{w}}_e$, and the element balance of angular momentum equation by $\bar{\theta}_e$. Integrate the element equations over the domain of the element and sum over all elements. Similarly, sum the nodal equations over all N nodes. Add all of these together to arrive at the *virtual work functional*

$$\mathcal{G} = \sum_{i=1}^{N} \left[\left(m_i \ddot{\mathbf{W}}_i + \sum_{k \in \mathcal{K}(i)} \hat{\mathbf{R}}_k - \hat{\mathbf{p}}_i \right) \cdot \bar{\mathbf{W}}_i + \left(J_i \ddot{\Theta}_i + \sum_{k \in \mathcal{K}(i)} \hat{M}_k - \hat{m}_i \right) \bar{\Theta}_i \right]$$

$$+ \sum_{e=1}^{M} \int_0^{L_e} \left[\left(\rho A \ddot{\mathbf{w}}_e - \mathbf{R}'_e - \mathbf{q}_e \right) \cdot \bar{\mathbf{w}}_e + \left(\rho I \ddot{\theta}_e - M'_e + \mathbf{R}_e \cdot \mathbf{\Gamma}_e - m_e \right) \bar{\theta}_e \right] dx$$

If all four of the equations of motion hold, then this sum is obviously zero. By the fundamental theorem of the calculus of variations, if $\mathcal{G} = 0$ for all $\bar{\mathbf{W}}_i$, $\bar{\Theta}_i$, $\bar{\mathbf{w}}_e$, and $\bar{\theta}_e$, then all of the nodes and elements satisfy their individual equations of motion. This idea is generally called the method of weighted residuals or the *principle of virtual work*.

The virtual work functional can be put into a more useful form with a few mathematical simplifications. First note that sum over nodes of the sum over elements that frame into the nodes can be recast as the sum over elements of the sum of the two ends of each element. Hence, noting Eq. 14.1, we can write

$$\sum_{i=1}^{N} \sum_{k \in \mathcal{K}(i)} \hat{\mathbf{R}}_k \cdot \bar{\mathbf{W}}_i = \sum_{e=1}^{M} \left(\mathbf{R}_e(L_e) \cdot \bar{\mathbf{W}}_{j(e)} - \mathbf{R}_e(0) \cdot \bar{\mathbf{W}}_{i(e)} \right)$$

$$\sum_{i=1}^{N} \sum_{k \in \mathcal{K}(i)} \hat{M}_k \bar{\Theta}_i = \sum_{e=1}^{M} \left(M_e(L_e) \bar{\Theta}_{j(e)} - M_e(0) \bar{\Theta}_{i(e)} \right)$$

where the symbol M in the upper limit of the sums over elements e is the total number of elements in the structure (not internal moment).

We can integrate the element quantities by parts. The product rule for differentiation allows us to write

$$\int_0^{L_e} \left(\mathbf{R}_e \cdot \bar{\mathbf{w}}_e \right)' dx = \int_0^{L_e} \mathbf{R}_e' \cdot \bar{\mathbf{w}}_e \, dx + \int_0^{L_e} \mathbf{R}_e \cdot \bar{\mathbf{w}}_e' \, dx$$

$$\int_0^{L_e} \left(M_e \bar{\theta}_e \right)' dx = \int_0^{L_e} M_e' \cdot \bar{\theta}_e \, dx + \int_0^{L_e} M_e \cdot \bar{\theta}_e' \, dx$$

The integrals on the left side of these equations are integrals of exact differentials. Therefore, we can write

$$\int_0^{L_e} \left(\mathbf{R}_e \cdot \bar{\mathbf{w}}_e \right)' dx = \mathbf{R}_e(L_e) \cdot \bar{\mathbf{W}}_{j(e)} - \mathbf{R}_e(0) \cdot \bar{\mathbf{W}}_{i(e)}$$

$$\int_0^{L_e} \left(M_e \bar{\theta}_e \right)' dx = M_e(L_e) \bar{\Theta}_{j(e)} - M_e(0) \bar{\Theta}_{i(e)}$$

where we have noted the continuity of the element displacement and rotation quantities with the nodal displacement and rotations, respectively. From these relationships we can finally conclude that

$$\sum_{i=1}^{N} \sum_{k \in \mathcal{K}(i)} \hat{\mathbf{R}}_k \cdot \bar{\mathbf{W}}_i - \sum_{e=1}^{M} \int_0^{L_e} \mathbf{R}_e' \cdot \bar{\mathbf{w}}_e \, dx = \sum_{e=1}^{M} \int_0^{L_e} \mathbf{R}_e \cdot \bar{\mathbf{w}}_e' \, dx$$

$$\sum_{i=1}^{N} \sum_{k \in \mathcal{K}(i)} \hat{M}_k \bar{\Theta}_i - \sum_{e=1}^{M} \int_0^{L_e} M_e' \bar{\theta}_e \, dx = \sum_{e=1}^{M} \int_0^{L_e} M_e \bar{\theta}_e' \, dx$$

Substituting these relationships back into the original virtual work functional, we can write it in the form

$$\mathcal{G} = \sum_{e=1}^{M} \int_{0}^{L_e} \left(\mathbf{R}_e \cdot \bar{\mathbf{w}}'_e + M_e \bar{\theta}'_e + \mathbf{R}_e \cdot \mathbf{\Gamma}_e \bar{\theta}_e \right) dx$$

$$+ \sum_{i=1}^{N} \left(m_i \ddot{\mathbf{W}}_i \cdot \bar{\mathbf{W}}_i + J_i \ddot{\Theta}_i \bar{\Theta}_i \right) + \sum_{e=1}^{M} \int_{0}^{L_e} \left(\rho A \ddot{\mathbf{w}}_e \cdot \bar{\mathbf{w}}_e + \rho I \ddot{\theta}_e \bar{\theta}_e \right) dx$$

$$- \sum_{i=1}^{N} \left(\hat{\mathbf{p}}_i \cdot \bar{\mathbf{W}}_i + \hat{m}_i \bar{\Theta}_i \right) - \sum_{e=1}^{M} \int_{0}^{L_e} \left(\mathbf{q}_e \cdot \bar{\mathbf{w}}_e + m_e \bar{\theta}_e \right) dx$$

The first line represents the contribution of the *internal* virtual work. The second line represents the contributions of the *dynamic* or inertial terms. The third line represents the *external* virtual work done by the applied loads. It will be convenient to write the virtual work functional as

$$\mathcal{G} = \mathcal{G}_{int} + \mathcal{G}_{dyn} - \mathcal{G}_{ext} \tag{14.4}$$

where, by association, the three component parts are

$$\mathcal{G}_{int} = \sum_{e=1}^{M} \int_{0}^{L_e} \left(\mathbf{R}_e \cdot \bar{\mathbf{w}}'_e + M_e \bar{\theta}'_e + \mathbf{R}_e \cdot \mathbf{\Gamma}_e \bar{\theta}_e \right) dx$$

$$\mathcal{G}_{dyn} = \sum_{i=1}^{N} \left(m_i \ddot{\mathbf{W}}_i \cdot \bar{\mathbf{W}}_i + J_i \ddot{\Theta}_i \bar{\Theta}_i \right) + \sum_{e=1}^{M} \int_{0}^{L_e} \left(\rho A \ddot{\mathbf{w}}_e \cdot \bar{\mathbf{w}}_e + \rho I \ddot{\theta}_e \bar{\theta}_e \right) dx$$

$$\mathcal{G}_{ext} = \sum_{i=1}^{N} \left(\hat{\mathbf{p}}_i \cdot \bar{\mathbf{W}}_i + \hat{m}_i \bar{\Theta}_i \right) + \sum_{e=1}^{M} \int_{0}^{L_e} \left(\mathbf{q}_e \cdot \bar{\mathbf{w}}_e + m_e \bar{\theta}_e \right) dx$$

This expression for $\mathcal{G}$ is mathematically equivalent to the original one. Therefore, we can state the principle of virtual work as follows: If $\mathcal{G} = 0$ for all virtual displacements and rotations, then balance of linear and angular momentum of the nodes and elements are satisfied. Note that if we omit the term $\mathcal{G}_{dyn}$, then we have the appropriate virtual work functional to solve the static equilibrium problem.

It is worth reviewing our original task. We wanted to formulate the problem in a way that we could analyze frame structures. To do that we need to establish equations of motion for the nodes and the elements. By putting these into a weighted residual form and doing some mathematical manipulations of the expression— noting the compatibility between the nodal displacements and rotations and those same quantities at the ends of the elements—we arrived at the functional defined in Eq. 14.4, which can be used in a principle of virtual work. The only additions to what we had for the beam are the nodal quantities in $\mathcal{G}_{dyn}$ and $\mathcal{G}_{ext}$. It seems simple in retrospect, but it does point out the important role of compatibility between the nodal motion variables and the element motion variables.

It is interesting to note that the part of the functional associated with the internal quantities is identical to what we had for the beam alone in Chap. 13. The only additional consideration is the consequence of having elements of different orientations. But while we are in the neighborhood, let's throw another wrinkle into the problem. Up to this point we have only considered linear elastic constitutive models for the beam. In the next section we derive a model of inelasticity that extends our earlier model to beam theory. Once that model is in place, we will extend the analysis to include earthquake excitation.

14.3 Inelasticity

For the SDOF system, the shear building, and the truss, the constitutive response involved single-component systems where each element had a single stress component and a single associated measure of strain or deformation. In that context, the algorithm for inelasticity was relatively simple, and is outlined in Chap. 3 in Sect. 3.6.

The beam element is more complicated because there are three stress resultants (N, V, and M) and three associated strain resultants (ε_o, β_o, and κ_o). While it may be reasonable to consider the constitutive response of these resultants to be uncoupled for elasticity, it is not for inelasticity. A few simple thought experiments help to frame the issues.

The first reason to consider coupling of the response quantities is capacity. Imagine that a beam is in pure axial tension to the point where every fiber in the cross section is at the yield point. In this state, there would be no capacity left to resist any shearing or bending. Similarly, if a beam is at its plastic limit in bending, then it would have no capacity left for resisting shearing or axial force. In general, one can imagine that the capacity at a given cross section could be invested in a way that it could contribute resistance to all three of the stress resultants and that the capacity of each one in the presence of others must be less than the capacity of the cross section under the pure action of one in the absence of the others. This hypothesis gives rise to what we call a *yield surface*, illustrated in Fig. 14.5. The yield surface defines the capacity of the cross section.

Fig. 14.5 *Yield surface for planar beam.* The yield surface defines the capacity of the beam. The value where each axis pierces the yield surface is the capacity in that component in absence of the others

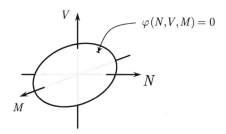

Most materials have the property that there exist a range of stress and strain states for which the response is *elastic* and a condition that initiates *plastic* response. The yield surface gives a way to mathematically distinguish those two regimes. Let us abbreviate the stress state as $\mathbf{s} = [N, V, M]^T$. Define a *yield function*, $\varphi(\mathbf{s})$, to have the property that if $\varphi(\mathbf{s}) < 0$, then the state $\mathbf{s}$ is elastic and plastic deformation does not occur. Plastic deformation can occur only if the state of stress is on the yield surface, i.e., if $\varphi(\mathbf{s}) = 0$. States outside of the yield surface (i.e., $\varphi(\mathbf{s}) > 0$) are not admissible. The yield function gives us a means of detecting yielding similar to what we did for the single-component model.

The yield surface provides a constraint on the stress state. Hence, we must deal with the constitutive equations as an inequality constraint. We will adopt the hypothesis that the strain is the sum of an elastic part and an inelastic part. Let the state of strain be abbreviated as $\mathbf{e} = [\varepsilon_o, \beta_o, \kappa_o]^T$. The additive decomposition of strain is

$$\mathbf{e} = \mathbf{e}^e + \mathbf{e}^p \tag{14.5}$$

where $\mathbf{e}^e$ is the elastic strain and $\mathbf{e}^p$ is the plastic strain. We will assume that the plastic strain evolves according to the *flow rule*

$$\dot{\mathbf{e}}^p = \gamma \, \mathbf{q}(\mathbf{s}) \tag{14.6}$$

where the *consistency parameter* γ governs how fast inelastic strains accrue, and the vector $\mathbf{q}(\mathbf{s})$ determines the direction of the plastic flow (i.e., how much relative plastic straining occurs in each of the three strain components). The consistency parameter and stress $\mathbf{s}$ must satisfy the *loading/unloading conditions*

$$\gamma \geq 0, \quad \varphi(\mathbf{s}) \leq 0, \quad \gamma \, \varphi(\mathbf{s}) = 0$$

and the *persistency condition*

$$\gamma \, \dot{\varphi}(\mathbf{s}) = 0$$

These conditions assure that states of stress will lie on or inside the yield surface. An *associative flow rule* takes $\mathbf{q}(\mathbf{s})$ to be normal to the yield surface, which can be achieved by taking

$$\mathbf{q}(\mathbf{s}) = \frac{\partial \varphi}{\partial \mathbf{s}} \tag{14.7}$$

Physically, what this assumption implies is that if any of the three stress resultants is active, then plastic strains associated with that action should develop in accord with the relative importance of that action. For example, if a cross section reaches yield with axial force and moment present, then some axial plastic strain and some plastic curvature should result. Conversely, if only one of the actions is present, then the plastic straining should only be associated with that one action.

The constitutive model for inelasticity can be written as a *rate equation* in the form

$$\dot{\mathbf{s}} = \mathbf{C}\big(\dot{\mathbf{e}} - \gamma\, \mathbf{q}(\mathbf{s})\big) \tag{14.8}$$

which essentially says that the rate of change of stress is proportional to the rate of change of the elastic strain (total strain minus plastic strain). The constant of proportionality is the matrix of elastic moduli. We will take

$$\mathbf{C} = \begin{bmatrix} EA & 0 & 0 \\ 0 & GA & 0 \\ 0 & 0 & EI \end{bmatrix}$$

which is consistent with the elastic constitutive model introduced in the previous chapters. The reason that the constitutive equation must be written in rate form is because of the distinct difference between the loading and unloading regimes. When the system is loading, inelastic strains develop. When the system is unloading, the response is elastic with inelastic straining ceasing immediately. Mathematically, this difference is captured by the rate equations.

14.3.1 Numerical Integration of the Rate Equations

Equation 14.8 is a rate equation and must be integrated numerically to get a relationship between stress and strain. For simplicity we will use the backward Euler method to do the integration.[4] Integrating from step n to step $n + 1$ gives

$$\mathbf{s}_{n+1} = \mathbf{s}_n + \mathbf{C}\Big(\mathbf{e}_{n+1} - \mathbf{e}_n - h\gamma_{n+1}\, \mathbf{q}(\mathbf{s}_{n+1})\Big) \tag{14.9}$$

where h is the time step. We will use this equation in a strain-driven framework where everything about state n is known, the new strain $\mathbf{e}_{n+1}$ is given, and the new stress state $\mathbf{s}_{n+1}$ is sought. This is a nonlinear equation with unknowns $\{\mathbf{s}_{n+1}, \gamma_{n+1}\}$. In the event of loading (i.e., $\gamma_{n+1} > 0$), we need an additional equation, which is provided by the loading condition

$$\gamma_{n+1}\, \varphi(\mathbf{s}_{n+1}) = 0 \tag{14.10}$$

---

[4] Any method can be used here, including the generalized trapezoidal rule.

Since $\gamma_{n+1} > 0$ in this case, it can be divided out of the equation. We can solve these equations by Newton's method to get the new stress. Define the residuals

$$\mathbf{g}_1(\bar{\gamma}, \mathbf{s}) = \mathbf{C}^{-1}(\mathbf{s} - \mathbf{s}^{tr}) + \bar{\gamma}\,\mathbf{q}(\mathbf{s})$$
$$g_2(\mathbf{s}) = \varphi(\mathbf{s}) \tag{14.11}$$

where

$$\mathbf{s}^{tr} = \mathbf{s}_n + \mathbf{C}(\mathbf{e}_{n+1} - \mathbf{e}_n)$$

is the *trial stress* and, for notational convenience, we have denoted $\bar{\gamma} = h\gamma_{n+1}$.[5] We aim to find values of $\{\bar{\gamma}, \mathbf{s}\}$ that make these residuals equal to zero. We can set up a Newton iteration by writing linearized versions of the above equations. To wit,

$$\mathbf{g}_1(\bar{\gamma}^i, \mathbf{s}^i) + \mathbf{q}^i\,\Delta\bar{\gamma} + \left[\mathbf{C}^{-1} + \bar{\gamma}^i\mathbf{Q}^i\right]\Delta\mathbf{s} = \mathbf{0}$$
$$g_2(\mathbf{s}^i) + \mathbf{q}^i \cdot \Delta\mathbf{s} = 0 \tag{14.12}$$

where i is the current iteration number, $\Delta\bar{\gamma} = \bar{\gamma}^{i+1} - \bar{\gamma}^i$, $\Delta\mathbf{s} = \mathbf{s}^{i+1} - \mathbf{s}^i$, and $\mathbf{Q} = \partial\mathbf{q}/\partial\mathbf{s}$. A superscript i on $\mathbf{q}$ or $\mathbf{Q}$ simply means that those functions are evaluated at the state $\mathbf{s}^i$. To consolidate the notation, let us define an *effective modulus* as

$$\bar{\mathbf{C}} = \left[\mathbf{C}^{-1} + \bar{\gamma}\,\mathbf{Q}\right]^{-1} \tag{14.13}$$

Now, we can solve the first linearized equation, Eq. 14.12(a), for $\Delta\mathbf{s}$ to get

$$\Delta\mathbf{s} = -\bar{\mathbf{C}}(\bar{\gamma}^i, \mathbf{s}^i)\left(\mathbf{g}_1(\bar{\gamma}^i, \mathbf{s}^i) + \mathbf{q}^i\,\Delta\bar{\gamma}\right) \tag{14.14}$$

Substitute this result into the second equation, Eq. 14.12(b), and solve for $\Delta\bar{\gamma}$ as

$$\Delta\bar{\gamma} = \frac{g_2(\mathbf{s}^i) - \mathbf{q}^i \cdot \bar{\mathbf{C}}(\bar{\gamma}^i, \mathbf{s}^i)\,\mathbf{g}_1(\bar{\gamma}^i, \mathbf{s}^i)}{\mathbf{q}^i \cdot \bar{\mathbf{C}}(\bar{\gamma}^i, \mathbf{s}^i)\,\mathbf{q}^i} \tag{14.15}$$

The result for $\Delta\bar{\gamma}$ can be put back into Eq. 14.14 to complete the solution for $\Delta\mathbf{s}$. Now, the new iterates can be determined as

$$\mathbf{s}^{i+1} = \mathbf{s}^i + \Delta\mathbf{s}, \qquad \bar{\gamma}^{i+1} = \bar{\gamma}^i + \Delta\bar{\gamma}$$

---

[5] There is a little bit more than notational convenience operating here. This model is what is usually called *rate independent*, which means that while the constitutive equations are rate equations, the outcome of the model is not sensitive to rate. One manifestation of this is that the time step size h is always associated with the consistency parameter and we solve for the product of the two. Hence the size of h does not play a role in the result.

The iteration continues until both $\| \mathbf{g}_1 \| < tol$ and $| g_2 | < tol$, where tol is a small number that governs how closely the solution is satisfied. In this way we compute the new state of stress and the plastic flow parameter associated with the prescribed strain state $\mathbf{e}_{n+1}$. Of course, this strain state was only a best guess coming from the primary Newton iteration on the global equations of motion. It is a Newton iteration within a Newton iteration.

14.3.2 Material Tangent

Not only do we need the stress values to move the primary Newton iteration forward, we also need the *material tangent* matrix

$$\mathbf{C}_t = \frac{\partial \mathbf{s}_{n+1}}{\partial \mathbf{e}_{n+1}} \tag{14.16}$$

This matrix expresses the instantaneous compliance of the cross section at the current state of stress (i.e., the converged state from the secondary Newton iteration to find the stress). Computing an exact tangent is crucial to the performance of the primary Newton iteration.

To find the tangent matrix, first take the derivative of Eq. 14.9 with respect to $\mathbf{e}_{n+1}$. With a slight rearrangement of terms, we get

$$\mathbf{C}^{-1}\mathbf{C}_t = \mathbf{I} - \mathbf{q}\,\mathbf{r}^T - \bar{\gamma}\,\mathbf{Q}\mathbf{C}_t$$

where $\mathbf{r}^T = \partial \bar{\gamma}/\partial \mathbf{e}_{n+1}$ and the chain rule has been invoked for $\mathbf{q}(\mathbf{s})$. This equation can be solved for the material tangent. To wit,

$$\mathbf{C}_t = \bar{\mathbf{C}}\left(\mathbf{I} - \mathbf{q}\,\mathbf{r}^T\right) \tag{14.17}$$

where we have invoked the definition of the effective modulus from Eq. 14.13. Next, take the derivative of the loading equation $\varphi(\mathbf{s}_{n+1}) = 0$ with respect to $\mathbf{e}_{n+1}$ to get

$$\mathbf{q}^T\mathbf{C}_t = \mathbf{0}$$

Now, premultiply both sides of Eq. 14.17 by $\mathbf{q}^T$ and note that the left side is zero, allowing us to solve for $\mathbf{r}^T$ as

$$\mathbf{r}^T = \frac{\mathbf{q}^T\bar{\mathbf{C}}}{\mathbf{q}^T\bar{\mathbf{C}}\mathbf{q}}$$

Substituting this result back into Eq. 14.17 gives the final expression for the tangent matrix as

$$\mathbf{C}_t = \bar{\mathbf{C}} - \frac{\bar{\mathbf{C}}\mathbf{q}\mathbf{q}^T\bar{\mathbf{C}}}{\mathbf{q}^T\bar{\mathbf{C}}\mathbf{q}} \tag{14.18}$$

Since the matrix $\bar{\mathbf{C}}$ depends upon the consistency parameter $\bar{\gamma}$, the tangent is influenced by the amount of inelasticity. Note also that while $\bar{\mathbf{C}}$ is often diagonal, $\mathbf{C}_t$ usually is not. Unlike the single-component model, the material tangent is not zero when the cross section is yielding. While this might seem counterintuitive, the stiffness comes from the possibility of altering the composition of stress components to adjust to whatever the loading on the system allows.

14.3.3   Internal Variables

The final issue in the computation of the stress state is detecting whether or not the state is plastic or elastic. The trial stress can be used to detect yielding, just as it was in the single-component case. If $\varphi(\mathbf{s}^{tr}) < 0$, then the new state is elastic. That implies that $\bar{\gamma} = 0$ and, thus, the new stress is equal to the trial stress and the tangent is equal to $\mathbf{C}$. On the other hand, if $\varphi(\mathbf{s}^{tr}) > 0$, then the new state is plastic and the computations outlined above apply.

In order to do the computations, we must know the state $\{\mathbf{s}_n, \mathbf{e}_n\}$ to compute the trial stress. However, we do not need to store both the stress and the strain to track the progress of plastic deformation. Assume that we have converged at the end of a loading step with state $\{\mathbf{s}_n, \mathbf{e}_n\}$. If we unload from there to a stress-free state, i.e., to the state $\{\mathbf{0}, \mathbf{e}^p\}$ then the elastic constitutive model would have (note that $\bar{\gamma} = 0$ because we are "unloading")

$$\mathbf{0} - \mathbf{s}_n = \mathbf{C}(\mathbf{e}^p - \mathbf{e}_n)$$

where $\mathbf{e}^p$ is the residual or unrecoverable strain associated with the stress-free state. Solve for the residual plastic strain as

$$\mathbf{e}^p = \mathbf{e}_n - \mathbf{C}^{-1}\mathbf{s}_n$$

If we keep track of $\mathbf{e}^p$, then we have enough information to compute the trial stress in the next step. In particular,

$$\mathbf{s}^{tr} = \mathbf{s}_n + \mathbf{C}(\mathbf{e}_{n+1} - \mathbf{e}_n) = \mathbf{C}(\mathbf{e}_{n+1} - \mathbf{e}^p)$$

Thus, the internal variable that we need to track in the inelastic computation is the plastic strain $\mathbf{e}^p$. That quantity remains constant during the entire two-level iterative process for a given time step. Only upon convergence of the equations of motion for the time step should the internal variable be updated.

This approach is identical to the one we used for the single-component inelasticity algorithm, where we had a single internal variable that allowed us to track the progress of inelasticity. In the present context, we must track three quantities because we have three strain components.

14.3.4 Specific Model for Implementation

For the purposes of implementation in the current context, we will assume a yield function of the form

$$\varphi(\mathbf{s}) = n^2 + v^2 + m^2 - 1 \tag{14.19}$$

where $n = N/N_o$, $v = V/V_o$, and $m = M/M_o$. The values N_o, V_o, and M_o are the full capacity of the cross section in axial, shear, and bending in absence of the others. Now, $\mathbf{q} = \mathbf{Qs}$ with the matrix $\mathbf{Q} = \partial\mathbf{q}/\partial\mathbf{s}$ defined implicitly as

$$\mathbf{Q}^{-1} = \tfrac{1}{2} \begin{bmatrix} N_o^2 & 0 & 0 \\ 0 & V_o^2 & 0 \\ 0 & 0 & M_o^2 \end{bmatrix} \tag{14.20}$$

There are many other models that have been proposed to capture the yield characteristics of particular cross sections. For example, a model that gives good results for a rectangular cross section is

$$\varphi(\mathbf{s}) = |m| + n^2\left(1 + v^2\right) + v^4 - 1$$

In essence, each of these yield functions represents a shape somewhere between an octahedron with vertices of 1 and -1 on the stress resultant axes and a cube with sides located at values of 1 and -1 on the stress axes. The former represents the linear relationship associated with first yield while the latter represents no interaction among the stress components. There are, of course, materials that exhibit other interesting properties. For example, concrete has greater bending capacity in the presence of some compressive axial force.

There are also many additional features that can enhance the fidelity of the material model. For example, introducing additional internal variables to allow for modeling features like Bauchinger's effect and strain hardening can be introduced. Such extensions are beyond the scope of this book.

The code to execute the computation of the stress and material tangent from the strain is given in Code 14.1.

Code 14.1   MATLAB code for the element tangent and residual computations for the planar beam element with yield function given by Eq. 14.19

```matlab
function [Ct,S,Ep] = NLFConstModels(E,Ep,d)
%  Compute constitutive tangent Ct, stress S, and plastic strain EP
%  from the given total strain E at current integration point.
%  .
%     E : Total strain
%     Ep : Plastic strain (internal variable)
%      d : Material properties [Type,E,G,A,I,rho,p,q,m,gravity,No,Vo,Mo]

%. Extract model type
   Type = d(1);

%. Set Newton parameters
   itmax = 20; tol = 1.e-8;

%. Extract the section stiffness C from 'd'.
   EA = d(2)*d(4);
   GA = d(3)*d(4);
   EI = d(2)*d(5);
   C = [EA; GA; EI];
   So = d(11:13)';               % Yield levels [No, Vo, Mo]

   switch Type
     case 1   % Linear Elastic
       S = C.*E;
       Ct = diag(C);

     case 2   % Inelastic
       St = C.*(E-Ep);            % Trial stress
       s = St./So;                % Normalized trial stress
       phi = dot(s,s) - 1;        % Yield function for trial stress

%..... If inside yield surface, then response is elastic
       if phi<0
         Ct = diag(C);
         S = St;

       else % Inelastic

%....... Establish fixed quantities. Initialize Newton iteration.
         it = 0; err = 1;
         Cinv = 1./C;
         Q = 2./So.^2;
         S = St; gamma = 0;

%....... Newton iteration
         while (err>tol) && (it<itmax)
           Cbar = 1./(Cinv + gamma*Q);
           q = Q.*S;   s = S./So;
           g1 = Cinv.*(S-St) + gamma*q;
           g2 = dot(s,s) - 1;
           dgamma = (g2 - dot(q,Cbar.*g1))/dot(q,Cbar.*q);
           ds = -Cbar.*(g1 + dgamma*q);
           S = S + ds;
           gamma = gamma + dgamma;
           err = norm(g1) + norm(g2);
           it = it + 1;
         end % Newton
         if(it==itmax)
             fprintf('\n Iteration limit reached ConstModels');
         end

%....... Compute the material tangent Ct and plastic strain Ep
         Cbar = 1./(Cinv + gamma*Q);
         q = Q.*S;   p = Cbar.*q;
```

```
        Ct = diag(Cbar) - p*p'/dot(q,p);
        Ep = E - Cinv.*S;

      end % phi

   end % switch

end
```

This code has two models: (1) linear elasticity and (2) inelasticity. The material properties needed to execute the computation are passed in through the array d. Note that the matrix $\mathbf{C}$ is diagonal, as is the matrix $\mathbf{Q}$ from Eq. 14.20. Thus, the inversion of these matrices is very simple. The notation in the code follows the notation from this section fairly closely. The secondary Newton loop to solve the nonlinear equations for the inelasticity model in this code follows the structure of the primary Newton iteration in the main code.

14.4 Element Matrices

One area that needs attention in the context of inelasticity is the computation of the element tangent matrix. When we formulated the tangent for beams, we assumed that each force component depended only on its associated strain component. For example, N was a function only of ε_o and not β_o or κ_o. Because the plastic strain accrues with a direction normal to the yield surface, each of the stress components depends on *all* of the strain components. Now, the directional derivatives of the forces must be written as

$$
\mathcal{D}(N) = \frac{\partial N}{\partial \varepsilon_o}\mathcal{D}(\varepsilon_o) + \frac{\partial N}{\partial \beta_o}\mathcal{D}(\beta_o) + \frac{\partial N}{\partial \kappa_o}\mathcal{D}(\kappa_o)
$$

$$
\mathcal{D}(V) = \frac{\partial V}{\partial \varepsilon_o}\mathcal{D}(\varepsilon_o) + \frac{\partial V}{\partial \beta_o}\mathcal{D}(\beta_o) + \frac{\partial V}{\partial \kappa_o}\mathcal{D}(\kappa_o) \tag{14.21}
$$

$$
\mathcal{D}(M) = \frac{\partial M}{\partial \varepsilon_o}\mathcal{D}(\varepsilon_o) + \frac{\partial M}{\partial \beta_o}\mathcal{D}(\beta_o) + \frac{\partial M}{\partial \kappa_o}\mathcal{D}(\kappa_o)
$$

Let us first note that the tangent matrix that we derived in the previous section is what we mean by

$$
\mathbf{C}_t = \begin{bmatrix} \partial N/\partial \varepsilon_o & \partial N/\partial \beta_o & \partial N/\partial \kappa_o \\ \partial V/\partial \varepsilon_o & \partial V/\partial \beta_o & \partial V/\partial \kappa_o \\ \partial M/\partial \varepsilon_o & \partial M/\partial \beta_o & \partial M/\partial \kappa_o \end{bmatrix} \tag{14.22}
$$

Following through the calculation that resulted in Eq. 13.21, but accounting for the above coupling, we can write Eq. 13.21 as

$$\mathcal{A}_{int} = \int_{\mathcal{B}} \bar{\mathbf{r}}^T \mathbf{T} \hat{\mathbf{r}} \, dx \tag{14.23}$$

where $\mathcal{B}$ is the entire domain of the structure. Recall that $\bar{\mathbf{r}}$ and $\hat{\mathbf{r}}$ are defined as follows

$$\bar{\mathbf{r}}^T = \begin{bmatrix} \bar{\mathbf{w}}'^T & \bar{\theta} & \bar{\theta}' \end{bmatrix}, \qquad \hat{\mathbf{r}}^T = \begin{bmatrix} \hat{\mathbf{w}}'^T & \hat{\theta} & \hat{\theta}' \end{bmatrix} \tag{14.24}$$

We will look at building the **T** matrix from various components, as we did before. To prepare for that, let us set up the following notation

$$\boldsymbol{\Omega} = \begin{bmatrix} n_1 & m_1 \\ n_2 & m_2 \end{bmatrix}, \qquad \boldsymbol{\lambda} = \left\{ \begin{matrix} -\beta_o \\ 1+\varepsilon_o \end{matrix} \right\}, \qquad \mathbf{f} = \left\{ \begin{matrix} V \\ -N \end{matrix} \right\} \tag{14.25}$$

where n_1 and n_2 are the components of $\mathbf{n}$ and m_1 and m_2 are the components of $\mathbf{m}$, relative to the element basis $\{\mathbf{e}_1^e, \mathbf{e}_3^e\}$. Thus, $\boldsymbol{\Omega}$ is a matrix whose columns are $\mathbf{n}$ and $\mathbf{m}$, respectively. Further, let us subdivide the material tangent matrix into parts as

$$\mathbf{C}_t = \begin{bmatrix} \mathbf{H} & \mathbf{h} \\ \mathbf{h}^T & h \end{bmatrix} \tag{14.26}$$

where $\mathbf{H}$ is the upper left 2×2 submatrix, $\mathbf{h}$ is the third column and first two rows of $\mathbf{C}_t$, and h is the (3,3) component. With this notation, we can define

$$\mathbf{A} = \boldsymbol{\Omega}\mathbf{H}\boldsymbol{\Omega}^T, \quad \mathbf{b} = \boldsymbol{\Omega}(\mathbf{H}\boldsymbol{\lambda} + \mathbf{f}), \quad \mathbf{c} = \boldsymbol{\Omega}\mathbf{h}$$

$$d = \boldsymbol{\lambda}^T\mathbf{h}, \quad e = \boldsymbol{\lambda}^T(\mathbf{H}\boldsymbol{\lambda} + \mathbf{f})$$

The reason for making these definitions will be clear soon. With these quantities defined as they are, we can construct the **T** matrix as

$$\mathbf{T} = \begin{bmatrix} \mathbf{A} & \mathbf{b} & \mathbf{c} \\ \mathbf{b}^T & e & d \\ \mathbf{c}^T & d & h \end{bmatrix} \tag{14.27}$$

This matrix forms the core of the tangent stiffness of the element. The remaining task is interpolation to take care of $\bar{\mathbf{r}}$ and $\hat{\mathbf{r}}$. The matrix **T** has 4 rows and 4 columns. The code that executes these computations is given in Code 14.2.

Code 14.2 MATLAB code for computing the core pieces of the element tangent and residual for the frame

```
function [T,g,S,E,Ep] = NLFTang(dw,th,dth,d,Rot,Ep)
%   Compute the tangent T, internal residual g, stress S, strain E,
%   and plastic strain Ep at current quadrature point
%
```

```
%    dw = Displacement derivatives [u',w']
%    th = Rotation, theta
%   dth = Derivative of rotation, theta'
%     d = Material properties [Type,E,G,A,I,rho,p,q,m,gravity,No,Vo,Mo]
%   Rot = Rotation from local to global coordinates
%    Ep = Plastic strain (internal variable)
%
%     T = Element tangent
%     g = Element internal residual [H, Q, Gamma.R, M]
%   S,E = Stress (N,V,M), Strain(epsilon, beta, curvature)

%. Convert displacement derivatives to local frame
   dw = Rot'*dw;

%. Establish unit vectors n and m
   n = [cos(th);-sin(th)];
   m = [sin(th); cos(th)];
   e1 = [1;0];

%. Compute strains from displacements
   beta = dot(dw+e1,m);
   epsilon = dot(dw+e1,n) - 1;
   lambda = 1 + epsilon;
   E = [epsilon; beta; dth];

%. Compute stress associated with E and material tangent Ct
   [Ct,S,Ep] = NLFConstModels(E,Ep,d);

%. Tangent
   Omg = [n, m]; Lambda = [-beta; lambda];  f = [S(2); -S(1)];
   H = Ct(1:2,1:2); h = Ct(1:2,3);
   A = Omg*H*Omg';
   b = Omg*(H*Lambda + f);
   c = Omg*h;
   d = dot(Lambda,h);
   e = dot(Lambda,H*Lambda + f);

%. Net force R and R.Gamma
   R = S(1)*n + S(2)*m;
   RGamma = lambda*S(2) - beta*S(1);

%. Transform from local to global frame
   R = Rot*R;
   A = Rot*A*Rot';
   b = Rot*b;
   c = Rot*c;

%. Element tangent and residual
   T = [A, b, c; b', e, d; c', d, Ct(3,3)];
   g = [R; RGamma; S(3)];

end
```

Observe that we transform dw, which contains the components of the displacement derivatives, from global to local coordinates as the very first step. These are needed to compute ε_o and β_o. Once the strains are computed, the stresses and material tangent are obtained from NLFConstModels, which is given in Code 14.1. You can see that **R** is computed from $\mathbf{R} = N\mathbf{n} + V\mathbf{m}$. The values of **A**, **b**, **c**, d, and e are also conveniently computed from the material tangent, stresses, and strains. The last step done in the function is to convert **R** (basically H and Q), **A**, **b**, and **c** from local to global coordinates. These transformations are the only ones needed to take care of member orientation and are discussed in the next section.

14.4.1   Finite Element Discretization

To complete the task of computing the element tangent matrix to assemble into the global tangent stiffness matrix, we need to interpolate. Similar to what we did for the nonlinear beam, let us define the interpolation matrix

$$\mathbf{N}(\xi) = \frac{1}{L_e} \begin{bmatrix} \varphi' & 0 & 0 & 0 \\ 0 & \varphi' & 0 & 0 \\ 0 & 0 & L_e\psi & \psi' \end{bmatrix} \tag{14.28}$$

where φ is the shape function array associated with the translational degrees of freedom and ψ is the array associated with the rotational degrees of freedom, as defined in Eq. 13.27. The matrix $\mathbf{N}$ has 4 columns and the number of rows is equal to twice the length of φ plus the length of ψ. Now, the element tangent stiffness can be computed as the integral

$$\mathbf{k}_e = \int_0^1 \mathbf{N}(\xi)\,\mathbf{T}_e\,\mathbf{N}^T(\xi)\,L_e\,d\xi \tag{14.29}$$

where $\mathbf{T}_e$ is from Eq. 14.27, specialized to element e. The integral will, as usual, be done by numerical quadrature. It is not necessary to form the matrix $\mathbf{N}$ when implementing Eq. 14.29. It is advantageous to recognize the structure of the matrix and to avoid the zero multiplications, as we did in Eq. 13.38 in Chap. 13.

As before, the global tangent stiffness matrix can be assembled from element contributions as

$$\mathbf{K}_t = \sum_{e=1}^M \mathbf{B}_e^T \mathbf{k}_e \mathbf{B}_e \tag{14.30}$$

where $\mathbf{B}_e$ is defined in Eq. 13.29. The matrix $\mathbf{K}_t$ is a representation of the second derivative functional A_{int} in the sense that

$$A_{int} = \hat{\mathbf{u}}^T \mathbf{K}_t \bar{\mathbf{u}}$$

All of the quantities defined so far are relative to an element-bound coordinate system, as was the case for the nonlinear beam of Chap. 13. For the frame we also need to reconcile the fact that different elements in the structure have different initial orientations. To account for that we will transform certain quantities from the global frame to the local frame, and vice versa, in the execution of the assembly process.

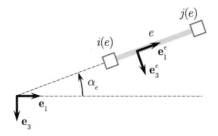

Fig. 14.6 *Local and global coordinate systems.* The undeformed configuration of element e has orientation defined by unit base vectors $\mathbf{e}_1^e$ and $\mathbf{e}_3^e$. The global coordinate system is defined by the base vectors $\mathbf{e}_1$ and $\mathbf{e}_3$. The angle that the element makes relative to horizontal is α_e

14.4.2 Local to Global Transformation

We have dealt with all of the terms in the global virtual work functional before. If we use the same conceptual framework, then most of the calculations should be similar. The main question centers on how to deal with the fact that the elements have different orientations.

We only need to consider the vectors $\mathbf{R}_e$, $\mathbf{w}_e$, and $\bar{\mathbf{w}}_e$. If these vectors were expressed in terms of components relative to the global coordinate system, then we could add the components together in the functional. Hence, we need to find the appropriate coordinates. Let's look at $\mathbf{w}_e$ as an example. The component form is

$$\mathbf{w}_e = u_e \mathbf{e}_1^e + w_e \mathbf{e}_3^e = u_e^g \mathbf{e}_1 + w_e^g \mathbf{e}_3 \tag{14.31}$$

where u_e and w_e are the components in the local element coordinate system, and u_e^g and w_e^g are the components in the global coordinate system. Since the nodal coordinates are expressed relative to the global reference frame, it is simple to compute $\Delta\mathbf{x}_e = \mathbf{x}_{j(e)} - \mathbf{x}_{i(e)}$ and then $\mathbf{e}_1^e = \Delta\mathbf{x}_e / \|\Delta\mathbf{x}_e\|$. Then, the vector $\mathbf{e}_3^e$ can be constructed as a vector perpendicular to $\mathbf{e}_1^e$ (see Fig. 14.6).

Taking the dot product of Eq. 14.31 with respect to $\mathbf{e}_1$ and $\mathbf{e}_3$, successively, we find the relationship

$$\begin{Bmatrix} u_e^g \\ w_e^g \end{Bmatrix} = \begin{bmatrix} \mathbf{e}_1^e \cdot \mathbf{e}_1 & \mathbf{e}_3^e \cdot \mathbf{e}_1 \\ \mathbf{e}_1^e \cdot \mathbf{e}_3 & \mathbf{e}_3^e \cdot \mathbf{e}_3 \end{bmatrix} \begin{Bmatrix} u_e \\ w_e \end{Bmatrix} \tag{14.32}$$

Hence, we can define the rotation matrix

$$\mathbf{\Lambda}_e = \begin{bmatrix} \mathbf{e}_1^e \cdot \mathbf{e}_1 & \mathbf{e}_3^e \cdot \mathbf{e}_1 \\ \mathbf{e}_1^e \cdot \mathbf{e}_3 & \mathbf{e}_3^e \cdot \mathbf{e}_3 \end{bmatrix} = \begin{bmatrix} \cos\alpha_e & -\sin\alpha_e \\ \sin\alpha_e & \cos\alpha_e \end{bmatrix} \tag{14.33}$$

This matrix allows the conversion from local to global as $(\bullet)^g = \Lambda_e (\bullet)^e$. The conversion from global to local is $(\bullet)^e = \Lambda_e^T (\bullet)^g$, where $(\bullet)$ represents the matrix of components of any vector.

There are only a few places where we need to make changes to the NLBeam codes. First, the element residual is defined as (see Eq. 13.32)

$$\mathbf{g}_e = \int_0^1 \left\{ \begin{array}{c} H_e \varphi' \\ Q_e \varphi' \\ M_e \psi' + L_e (\boldsymbol{\Gamma}_e \cdot \mathbf{R}_e)\psi \end{array} \right\} d\xi \tag{14.34}$$

where the finite element shape functions φ and ψ are defined in Eq. 13.27. Note that since $\boldsymbol{\Gamma} \cdot \mathbf{R} = (1 + \varepsilon_o)V - \beta_o N$ is a scalar, there is no conversion needed. Also, moment and rotation always point in the vector direction $\mathbf{e}_2^e = \mathbf{e}_2$, which does not change. The only thing we need to do, then, is convert H_e and Q_e to global coordinates. Similarly, the part of the force vector associated with the applied axial and transverse loads, $p(x)$ and $q(x)$ is, from Eq. 13.36,

$$\mathbf{p}_e = \int_0^1 \left\{ \begin{array}{c} p_e \varphi \\ q_e \varphi \\ m_e \psi \end{array} \right\} L_e d\xi \tag{14.35}$$

Again, the moment is not affected by the local to global transformation, but the loads are. If we think of p_e and q_e as being along and perpendicular to the initial configuration of the beam, then we need to transform them to global coordinates before forming and assembling the force vector. Therefore, we compute

$$\left\{ \begin{array}{c} H_e^g \\ Q_e^g \end{array} \right\} = \Lambda_e \left\{ \begin{array}{c} H_e \\ Q_e \end{array} \right\}, \qquad \left\{ \begin{array}{c} p_e^g \\ q_e^g \end{array} \right\} = \Lambda_e \left\{ \begin{array}{c} p_e \\ q_e \end{array} \right\} \tag{14.36}$$

Now the global residual and force can be computed as

$$\mathbf{g}_e = \int_0^1 \left\{ \begin{array}{c} H_e^g \varphi' \\ Q_e^g \varphi' \\ M_e \psi' + L_e (\boldsymbol{\Gamma}_e \cdot \mathbf{R}_e)\psi \end{array} \right\} d\xi, \quad \mathbf{p}_e = \int_0^1 \left\{ \begin{array}{c} p_e^g \varphi \\ q_e^g \varphi \\ m_e \psi \end{array} \right\} L_e d\xi \tag{14.37}$$

Note that in forming the residual (and tangent), we need to compute the element strains and resultant forces (i.e., N, V, and M). The quantity that we are tracking in the global displacement vector is the displacement in the global coordinate system. When we localize those for the element, they are still in the global system. Hence, we need to convert the element displacements from global to local

$$\left\{ \begin{array}{c} u_e \\ w_e \end{array} \right\} = \Lambda_e^T \left\{ \begin{array}{c} u_e^g \\ w_e^g \end{array} \right\} \tag{14.38}$$

Those transformed displacements can then be used to compute element strains. The tangent matrix, Eq. 14.27, can be transformed to global coordinates as

$$\mathbf{T}^g = \begin{bmatrix} \mathbf{A}_g & \mathbf{b}_g & \mathbf{c}_g \\ \mathbf{b}_g^T & e & d \\ \mathbf{c}_g^T & d & w \end{bmatrix} \tag{14.39}$$

where the matrices $\mathbf{A}_g$, $\mathbf{b}_g$, and $\mathbf{c}_g$ are the global versions of the matrices $\mathbf{A}$, $\mathbf{b}$, and $\mathbf{c}$ defined in Eq. 14.27. These matrices can be converted from local to global simply as

$$\mathbf{A}_g = \boldsymbol{\Lambda}_e \mathbf{A} \boldsymbol{\Lambda}_e^T, \qquad \mathbf{b}_g = \boldsymbol{\Lambda}_e \mathbf{b}, \qquad \mathbf{c}_g = \boldsymbol{\Lambda}_e \mathbf{c} \tag{14.40}$$

The values of d, e, and w in Eq. 14.27 are scalars and need no conversion. All of the conversions can be completed prior to interpolation. Once these terms have been converted to global coordinates, the element tangent stiffness can be computed as

$$\mathbf{k}_e^g = \int_0^1 \mathbf{N}(\xi)\, \mathbf{T}_e^g\, \mathbf{N}^T(\xi) L_e \, d\xi \tag{14.41}$$

The changes to the function NLBElement are minimal. The routine NLFElement (frame rather than beam) is shown as Code 14.3.

Code 14.3   MATLAB code for the element tangent and residual computations for the frame

```
function [me,ke,ge,fe,se,Qe,SE,Ep] = NLFElement(xe,Ue,We,Te,de,S,Ep)
%    Form element mass me, stiffness ke, internal residual ge, applied
%    distributed load fe, self-weight load se, EQ vector Qe. Compute
%    stress and strain SE and plastic strain EP.
%
%       xe : Element nodal coordinates
%       Ue : Axial displacements at element DOF
%       We : Transverse displacements at element DOF
%       Te : Rotations at element DOF
%       de : Properties [ConstType,E,G,A,I,rho,p,q,m,gravity,No,Vo,Mo]
%        S : Struct with problem size parameters
%       Ep : Plastic strain at integration points for element e

%. Establish zero matrices for use in mass matrix
   Zed1 = zeros(2+S.nExt(1),2+S.nExt(3));
   Zed2 = zeros(2+S.nExt(1),2+S.nExt(1));
   Zed3 = zeros(2+S.nExt(1),1);
   Zed4 = zeros(2+S.nExt(3),1);

%. Compute Local/Global rotation matrix
   nvec = xe(2,:) - xe(1,:);  Le = norm(nvec);  nvec = nvec/Le;
   mvec = [nvec(2), -nvec(1)];
   Rot  = [nvec; mvec];

%. Compute element properties
   qe = de(7:9)';                    % [po, qo, mo] element loads
   qe(1:2) = Rot*qe(1:2);            % Transform loads to local coordinates
   self = de(6)*de(4)*de(10);        % rho g A, weight per unit length
   rhoA = de(6)*de(4);               % rho A, mass per unit length
   rhoI = de(6)*de(5);               % rho I, inertia per unit length
```

```
%. Initialize element arrays
   nDOFe = 6+sum(S.nExt);          % Total DOF per element
   me = zeros(nDOFe,nDOFe);        % element mass matrix
   ke = zeros(nDOFe,nDOFe);        % element tangent stiffness matrix
   ge = zeros(nDOFe,1);            % element internal residual
   fe = zeros(nDOFe,1);            % element distributed force
   se = zeros(nDOFe,1);            % element self-weight
   Qe = zeros(nDOFe,2);            % element EQ effective load
   SE = zeros(1,S.nQpts*6);        % stress and strain

%. Compute element tangent and residual by numerical integration
   for i=1:S.nQpts

%... Set element coordinate for integration point, get load
     xi = S.st(i);

%... Compute shape functions and kernel matrices
     [h,dh] = BubbleFcns(xi,S.nExt(1));
     [g,dg] = BubbleFcns(xi,S.nExt(3));

%... Compute the element displacements and derivatives
     dw = [dot(Ue,dh); dot(We,dh)]/Le;
     th = dot(Te,g);
     dth = dot(Te,dg)/Le;

%... Establish pointers for SE and Ep
     jj = (i-1)*6+1:(i-1)*6+6;                % pointers for SE
     kk = (i-1)*3+1:(i-1)*3+3;                % pointers for Ep

%... Extract internal variables associated with integration point
     Epi = Ep(kk)';

%... Compute the core matrices for this integration point
     [T,G,So,Eo,Epi] = NLFTang(dw,th,dth,de,Rot,Epi);

%... Store stress and strain in SE and plastic strain in Ep
     SE(jj) = [So',Eo'];
     Ep(kk) = Epi';

%... Form submatrix components for shape functions
     DhDh = (dh*dh')/Le;   Dhg = dh*g';
     DgDg = (dg*dg')/Le;   gDg = g*dg';
     DhDg = (dh*dg')/Le;   Dgg = gDg';
     gg = (g*g')*Le;       hh = (h*h')*Le;

%... Compute the contributions to the element tangent
     ke11 = T(1,1)*DhDh;
     ke12 = T(1,2)*DhDh;
     ke22 = T(2,2)*DhDh;
     ke13 = T(1,3)*Dhg + T(1,4)*DhDg;
     ke23 = T(2,3)*Dhg + T(2,4)*DhDg;
     ke33 = T(3,3)*gg  + T(3,4)*gDg + T(4,3)*Dgg + T(4,4)*DgDg;

     ke = ke + [ ke11,   ke12,   ke13;...
                 ke12',  ke22,   ke23;...
                 ke13',  ke23',  ke33 ]*S.wt(i);

%... Compute the contributions to the element mass
     me = me + [ rhoA*hh,    Zed2,     Zed1;...
                 Zed2',    rhoA*hh,    Zed1;...
                 Zed1',    Zed1',    rhoI*gg ]*S.wt(i);

%... Compute the internal resistance part of the load vector
     ge = ge + [G(1)*dh; G(2)*dh; G(3)*g*Le + G(4)*dg]*S.wt(i);

%... Compute the distributed load vector
```

```
        fe = fe + [qe(1)*h; qe(2)*h; qe(3)*g]*Le*S.wt(i);
%... Compute the self-weight
        se = se + [0*h; -self*h; 0*g]*Le*S.wt(i);

%... Compute the earthquake vector
        Qe = Qe + [ rhoA*h,    Zed3;...
                    Zed3,    rhoA*h;...
                    Zed4,     Zed4]*Le*S.wt(i);

    end

end
```

One of the changes to the input data concerns the element distributed loads. In a general setting, an element likely represents only part of a complete structural element. Hence, it is complicated to implement a global feature like a load that varies in accord with global coordinates rather than the local ones. We did not encounter this problem in the previous chapters, because the system was a single beam. In that setting we specified only a single load form (also, we only specified a single set of physical properties, E, G, A, I, and ρ, for the entire system). Establishing initial conditions in the form of a specified wave function is also cumbersome in structural analysis. Hence, we do not provide that option in this code.

In NLFElement we form the element load vector fe separately from the self-weight of the beam, which is assembled into the vector se. That allows loads to be specified relative to the local element axes while keeping a fixed direction for the gravity loads.

For the frame code, we specify physical properties for each element, and we include the distributed element loads as one of the element physical properties (and we only include a uniform load). We still use the array d for the physical properties, but now the number of rows in d is equal to the number of material sets. You can see in the code that we extract qe from d (that is p, q, and m). We then convert those quantities to global coordinates, in accord with Eq. 14.36. Note that we have also added d to the list of items we localize in the routine NLBlocalize. Only the values associated with element e are passed into NLFelement in the array de.

The other change to this routine is that we compute the local to global transformation matrix $\boldsymbol{\Lambda}_e$, which we call Rot in the code. Observe that we compute this matrix from the element coordinates $\mathbf{x}_{i(e)}$ and $\mathbf{x}_{j(e)}$, which are stored in the array xe. All of the transformations can be done in the routine NLFtang, given in Code 14.2.

14.5 Static Verification

As the only changes to the code involved the computation of the tangent and residual, we can verify the code with static problems. We need to verify that the code works properly for elements oriented some way other than exactly in the $\mathbf{e}_1$ direction. For that purpose, we can redo the two verification problems from the

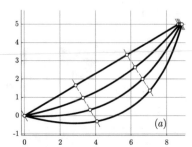 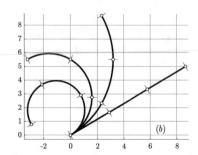

Fig. 14.7 *Example 1*. The two beams studied in Figs. 13.2 and 13.5 done again, but oriented at an angle of $\pi/6$. The properties are exactly the same. Number of elements and quintic/quadratic interpolation is the same

previous chapter with a beam oriented at an angle. The second feature we need to verify is that the code works for joints where more than two elements meet.

Example 1 Consider the static examples done in the previous chapter (see, for example, Figs. 13.2 and 13.5). These computations were done with the nonlinear beam code. The aim here is to demonstrate that we get the same results for a beam oriented at $\pi/6$ from the horizontal. The results are shown in Fig. 14.7. The beam has length $L = 10$, $EI = 1000$, and $EA = 1000$. Case (*a*) has $GA = 50$, $q_o = 10$, and no nodal loads. Case (*b*) has $GA = 500$, no distributed loads, and a counterclockwise applied moment at the right end of magnitude $M_o = 150\pi$.

Case (*b*) gives identical results, i.e., the beam bends into an exact three-quarter circle. Case (*a*) is not exactly the same. Notice that despite making the right boundary condition a "slide-*x*" case, the *x* direction is the global *x*, while the *x* in Fig. 13.5 was along the undeformed axis of the beam. The global-*x* slide condition did not allow the right end to move as much because the motion had to be horizontal. Otherwise, the response is very similar. These results suggest that the code works for beams at different orientations.

Example 2 Consider a case where more than two elements meet at a single node and axes of the elements do not align. This feature truly distinguishes a frame from a beam. Consider the two-story moment-resisting frame shown in Fig. 14.8. The bay width is 30 and the story height is 20, with a uniform load of $q_o = 1$ on the bottom beam, and a nodal load of 5 applied at the top right corner. The physical properties are $EA = 2 \times 10^4$, $GA = 10^4$, and $EI = 2 \times 10^3$. This is a relatively flexible frame. The displacements are shown to scale.

This structure was modeled using two quartic/cubic elements per structural member. The response is quite nonlinear. The loads were applied in a single step. It took eight Newton iterations to converge to within an error tolerance of 10^{-8}. The two examples appear to confirm the integrity of the frame code.

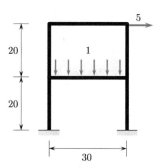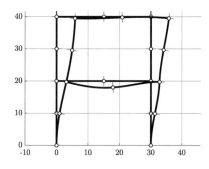

Fig. 14.8 *Example 2.* Structure that has more than two elements meeting at a single node. The loading is as shown in the sketch at left. The response is shown at right. The structure performs as expected

14.6 Dynamics of Frames

The setup for doing dynamics for the nonlinear beam in the previous chapter provides a great template for extending the dynamics to the more general frame context. In fact, there are no adjustments needed for the element inertial terms to deal with member orientation. The main reason for that is the inertia for both components of translation is exactly the same. Hence, orientation of the element does not matter. The only difference is the inertia associated with the nodes. To see what is at stake, we need only look at the dynamic part of the virtual work functional

$$\mathcal{G}_{dyn} = \sum_{i=1}^{N} \left(m_i \ddot{\mathbf{W}}_i \cdot \bar{\mathbf{W}}_i + J_i \ddot{\Theta}_i \bar{\Theta}_i \right) + \sum_{e=1}^{M} \int_{\mathcal{L}_e} \left(\rho A \ddot{\mathbf{w}}_e \cdot \bar{\mathbf{w}}_e + \rho I \ddot{\theta}_e \bar{\theta}_e \right) dx$$

The approach for including the nodal mass contributions in the formulation of the mass matrix is identical to what was done for the truss structure. We will develop the mass matrix in the next section where we formulate the analysis of frames subjected to earthquake ground motions.

14.6.1 Earthquake Ground Motion

Consider the scenario illustrated in Fig. 14.9, which shows a structure subjected to a ground motion at the supports. As we have done before, we assume that all supports experience the same ground motion. We will further assume that the ground motion is a scalar earthquake time function multiplying a constant vector direction (we made the same assumption for the truss formulation).

Fig. 14.9 *Frame with earthquake ground motion.* The structure is subjected to a ground motion at the supports. We assume that all supports experience the same ground motion

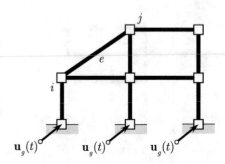

As we saw for the truss, it is straightforward to modify our formulation to account for the ground motion. The absolute displacement $\mathbf{w}_e^a$ of a point in element e is

$$\mathbf{w}_e^a(x, t) = \mathbf{u}_g(t) + \mathbf{w}_e(x, t) \tag{14.42}$$

where $\mathbf{w}_e(x, t)$ is the relative displacement, as before, and $\mathbf{u}_g(t)$ is the motion of the relative frame with respect to an inertial frame that is not moving. The absolute acceleration is, then,

$$\ddot{\mathbf{w}}_e^a(x, t) = \ddot{\mathbf{u}}_g(t) + \ddot{\mathbf{w}}_e(x, t)$$

The nodal displacements follow a similar pattern. Specifically, node i has absolute displacement and acceleration of

$$\mathbf{W}_i^a(t) = \mathbf{u}_g(t) + \mathbf{W}_i(t), \qquad \ddot{\mathbf{W}}_i^a(t) = \ddot{\mathbf{u}}_g(t) + \ddot{\mathbf{W}}_i(t)$$

Observe that the rotations are not affected by the ground motion. It is important to realize that both $\ddot{\mathbf{u}}_g$ and $\ddot{\mathbf{w}}_e$ are vectors, and those vectors can be expressed either in local element components or global components.

Tracing the derivation of the virtual work functional, one can note that the mass times acceleration terms that appear in the equations require the absolute acceleration. Thus, if we substitute the absolute accelerations just derived in each place where the acceleration appears, we will get a virtual work functional that accounts for the ground motion. To wit,

$$\mathcal{G}_{dyn} = \sum_{i=1}^{N} \left(m_i \left(\ddot{\mathbf{u}}_g + \ddot{\mathbf{W}}_i \right) \cdot \bar{\mathbf{W}}_i + J_i \ddot{\Theta}_i \bar{\Theta}_i \right)$$

$$+ \sum_{e=1}^{M} \int_{L_e} \left(\rho A \left(\ddot{\mathbf{u}}_g + \ddot{\mathbf{w}}_e \right) \cdot \bar{\mathbf{w}}_e + \rho I \ddot{\theta}_e \bar{\theta}_e \right) dx \tag{14.43}$$

Let us first consider the parts of $\mathcal{G}_{dyn}$ that do not include the earthquake ground motion. The part of $\mathcal{G}_{dyn}$ that involves the distributed element masses gives

$$\sum_{e=1}^{M} \int_{\mathcal{L}_e} \left(\rho A \ddot{\mathbf{w}}_e \cdot \bar{\mathbf{w}}_e + \rho I \ddot{\theta}_e \bar{\theta}_e \right) dx = \bar{\mathbf{u}}^T \sum_{e=1}^{M} \mathbf{B}_e^T \mathbf{m}_e \mathbf{B}_e \ddot{\mathbf{u}}$$

where the distributed element mass matrix $\mathbf{m}_e$ is given in Eq. 13.49. Recall that the matrix $\mathbf{D}_i$ is defined through the relationship

$$\left\{ \begin{matrix} \mathbf{W}_i \\ \Theta_i \end{matrix} \right\} = \mathbf{D}_i \mathbf{u}$$

which picks out the displacement and rotation for node i from the global parameter array $\mathbf{u}(t)$. This transformation is similar to the one used for the truss in Eq. 8.34, and can be constructed with the id array. The part of $\mathcal{G}_{dyn}$ that involves the nodal masses gives

$$\sum_{i=1}^{N} \left(m_i \ddot{\mathbf{W}}_i \cdot \bar{\mathbf{W}}_i + J_i \ddot{\Theta}_i \bar{\Theta}_i \right) = \bar{\mathbf{u}}^T \sum_{i=1}^{N} \mathbf{D}_i^T \check{\mathbf{m}}_i \mathbf{D}_i \ddot{\mathbf{u}}$$

where the nodal mass matrix is defined as

$$\check{\mathbf{m}}_i = \begin{bmatrix} m_i & 0 & 0 \\ 0 & m_i & 0 \\ 0 & 0 & J_i \end{bmatrix} \tag{14.44}$$

where m_i is the nodal mass and J_i is the nodal mass moment of inertia of node i. Combining the nodal contributions to the mass with the element contributions, we arrive at the global mass matrix through the process of assembly as

$$\mathbf{M} = \sum_{e=1}^{M} \mathbf{B}_e^T \mathbf{m}_e \mathbf{B}_e + \sum_{i=1}^{N} \mathbf{D}_i^T \check{\mathbf{m}}_i \mathbf{D}_i \tag{14.45}$$

where $\mathbf{m}_e$ is the mass matrix associated with element e, given by Eq. 13.49. The local-to-global transformation matrix $\mathbf{\Lambda}_e$ does not appear in the mass matrix $\mathbf{M}$ because each term involves the dot product of vectors, which can be expressed in either the element or global basis.

Let us now turn our attention to the terms involving the earthquake acceleration $\ddot{\mathbf{u}}_g$. Distributing the terms in Eq. 14.43, it is evident that the only change to the original functional is the addition of the earthquake ground motion terms

$$\mathcal{G}_{eq} = \sum_{e=1}^{M} \int_{\mathcal{L}_e} \rho A \, \ddot{\mathbf{u}}_g \cdot \bar{\mathbf{w}}_e \, dx + \sum_{i=1}^{N} m_i \, \ddot{\mathbf{u}}_g \cdot \bar{\mathbf{W}}_i$$

Recall from Eq. 13.29 that the element displacements can be extracted from the global displacement vector $\mathbf{u}(t)$ as

$$\left\{ \begin{array}{c} \mathbf{U}_e \\ \mathbf{W}_e \\ \mathbf{\Theta}_e \end{array} \right\} = \mathbf{B}_e \mathbf{u}$$

where $\mathbf{U}_e$, $\mathbf{W}_e$, and $\mathbf{\Theta}_e$ are the discrete parameters associated with element e.[6] Because the array $\mathbf{u}(t)$ contains the global components of the motion parameters, so do $\mathbf{U}_e$ and $\mathbf{W}_e$. The element displacement components, in global coordinates, can be interpolated as

$$\left\{ \begin{array}{c} u_e^g(\xi, t) \\ w_e^g(\xi, t) \end{array} \right\} = \mathbf{H}^T(\xi)\, \mathbf{B}_e \mathbf{u}(t)$$

where $\xi = (x - x_{i(e)})/L_e$ and the interpolation matrix $\mathbf{H}$ is defined in terms of the shape functions $\boldsymbol{\varphi}(\xi)$, first defined in Eq. 13.27, as

$$\mathbf{H}(\xi) = \begin{bmatrix} \boldsymbol{\varphi} & \mathbf{0} \\ \mathbf{0} & \boldsymbol{\varphi} \\ \mathbf{0} & \mathbf{0} \end{bmatrix} \tag{14.46}$$

This matrix is $\mathcal{N} \times 2$. Notice that this matrix does not involve derivatives of the shape functions and has a block of zeros in the column associated with the rotation degrees of freedom. The virtual displacements can be interpolated in the same way. Substituting the global displacements into this portion of the virtual work functional gives

$$\mathcal{G}_{eq} = \bar{\mathbf{u}}^T \left(\sum_{e=1}^{M} \mathbf{B}_e^T \int_0^1 \rho A\, \mathbf{H}(\xi)\, L_e\, d\xi + \sum_{i=1}^{N} m_i\, \bar{\mathbf{D}}_i^T \right) \ddot{\mathbf{u}}_g$$

where $\bar{\mathbf{D}}_i$ is the first two columns of $\mathbf{D}_i$ and the notation $\ddot{\mathbf{u}}_g$ means the components of the ground acceleration vector $\ddot{\mathbf{u}}_g$ relative to the global basis. We need this distinction because the interpolation of $\bar{\mathbf{w}}_e$ produced the global components of that vector (same for $\bar{\mathbf{W}}_i$) and the computation of the dot product must be done in one coordinate system or the other. Thus, the element contribution $\mathbf{Q}_e$ can be computed as

$$\mathbf{Q}_e = \int_0^1 \rho A\, \mathbf{H}(\xi)\, L_e\, d\xi \tag{14.47}$$

---

[6] Note that we are using $\mathbf{W}_i$ for nodal displacements and $\mathbf{W}_e$ for element transverse displacement parameters. The context will generally distinguish the usage.

which allows the global $\mathbf{Q}$ matrix to be assembled from element and nodal contributions as

$$\mathbf{Q} = \sum_{e=1}^{M} \mathbf{B}_e^T \mathbf{Q}_e + \sum_{i=1}^{N} m_i \bar{\mathbf{D}}_i^T \tag{14.48}$$

Now, the dynamic portion of the virtual work functional is

$$\mathcal{G}_{dyn} = \bar{\mathbf{u}}^T \left(\mathbf{M}\ddot{\mathbf{u}} + \mathbf{Q}\ddot{\mathbf{u}}_g \right)$$

where $\mathbf{M}$ is the global mass matrix, defined in Eq. 14.45. The contribution of $\mathcal{G}_{dyn}$ can be added to $\mathcal{G}_{int}$ and $\mathcal{G}_{ext}$ to obtain the complete virtual work functional, which can be written as

$$\mathcal{G} = \bar{\mathbf{u}}^T \left(\mathbf{M}\ddot{\mathbf{u}} + \mathbf{g}(\mathbf{u}) - \mathbf{f} + \mathbf{Q}\ddot{\mathbf{u}}_g \right) \tag{14.49}$$

where $\mathbf{f}$ is the global applied force vector, defined by Eq. 13.44 and $\mathbf{g}(\mathbf{u})$ is the global internal resistance vector, defined in Eq. 13.40. If $\mathcal{G} = 0$ for all $\bar{\mathbf{u}}$, then the fundamental theorem of the calculus of variations yields the discrete equations of motion

$$\mathbf{M}\ddot{\mathbf{u}}(t) + \mathbf{g}(\mathbf{u}(t)) = \mathbf{f}(t) - \mathbf{Q}\ddot{\mathbf{u}}_g \tag{14.50}$$

This is a nonlinear system of ordinary differential equations. We can solve this system by Newmark's method, with a Newton iteration at each time step to handle the nonlinear part. Note that the Newton iteration is very similar to the static problem.

While is it not necessary, we will generally make the simplification that the earthquake ground motion acts in a constant direction. Hence, we can write $\mathbf{u}_g = u_g(t)\mathbf{n}_g$, where $\mathbf{n}_g$ is an array of components of the fixed direction of the motion relative to the global frame. In general, earthquakes have independent ground acceleration components, but for the code we will stick with the simplified assumption.

The code for forming the mass matrix $\mathbf{M}$ and the $\mathbf{Q}$ matrix is given in Code 14.4.

Code 14.4 MATLAB code for forming the mass matrix $\mathbf{M}$ and matrix $\mathbf{Q}$ to support earthquake analysis

```
function [M,Q] = NLFMassMatrix(mass,Ep,x,u,d,ix,id,eDOF,S)
%    Assemble global mass matrix M and matrix Q for EQ analysis
%
%       mass : Nodal masses
%         Ep : Internal variables
%          x : Nodal coordinates
%          u : Global displacement
%          d : Material properties
%         ix : Element connectivities
%         id : Global DOF numbers for nodes
```

```
%       eDOF : Global DOF numbers for elements
%          S : Struct with problem size parameters

%. Initialize global arrays
   M = zeros(S.nDOF,S.nDOF);            % Mass matrix
   Q = zeros(S.nDOF,2);                 % Effective EQ force matrix
   I = eye(2);                          % Identity matrix

%. Assemble M and Q
   for n = 1:S.nNodes
      ii = id(n,1:2); kk = id(n,3);
      Q(ii,1:2) = Q(ii,1:2) + mass(n,1)*I;
      M(ii,ii) = M(ii,ii) + mass(n,1)*I;
      M(kk,kk) = M(kk,kk) + mass(n,2);
   end

%. Loop over elements
   for e=1:S.nElms

%... Form element mass matrix and effective EQ vector
     [Ue,We,Te,xe,de,Epe,jj] = NLFLocalize(e,u,Ep,x,d,ix,eDOF,S);
     [me,~,~,~,~,qe,~,~] = NLFElement(xe,Ue,We,Te,de,S,Epe);

%... Assemble element matrices into global arrays
     M(jj,jj) = M(jj,jj) + me;
     Q(jj,:) = Q(jj,:) + qe;

   end

%. Trim to unconstrained DOF
   M = M(1:S.nFree,1:S.nFree);
   Q = Q(1:S.nFree,1:2)*S.EQdir;

end
```

This code first puts the nodal mass contributions into M and Q. This part of the code is identical to the code for the truss that forms these matrices. The element contributions are a little more complex because they require integration, as outlined in Eq. 14.47. These operations are very similar to those associated with the formation of the tangent matrix and residual. Hence, they use the routines NLFLocalize and NLFElement, which are described in the next section. As indicated, the only element matrices needed from NLFElement are me and qe. Before returning to the main program, the global matrices are trimmed to include only the free degrees of freedom and **Q** is multiplied by the earthquake direction vector.

14.6.2   *Implementation*

The implementation of the formulation described in this chapter follows the same strategy of the codes developed for the shear building, truss, and beam codes presented in earlier chapters. One of the reasons for showing the frame codes is to see those similarities, but to also appreciate the key differences that make it possible to analyze complex frame structures. Code 14.5 is the main program for dynamic analysis of planar frames.

Code 14.5   MATLAB code for dynamic analysis of nonlinear planar frames

```
%   Nonlinear 2D Frame Dynamics (Newmark)
    clear; clc;

%. Get physical properties and set up mesh
    [x,ix,id,eDOF,d,F,Ep,mass,damp,S,DF] = FrameInput;

%. Put nodal forces into DOF order
    [fn,u,uPre] = Force(F,id,S);

%. Specify final time, time step, output frequency
    tf = 6.0;   h = 0.01;   nOutput = 2;

%. Retrieve parameters for Newmark integration
    [beta,gamma,eta,zeta,nSteps,tol,itmax,tf] = NewmarkParams(tf,h);

%. Initialize ouput storage scheme
    [nS,Hist] = NLFCreateHist(nOutput,nSteps,S);

%. Form stiffness matrix and element and self-weight load vectors
    uTot = [u; 0*uPre];
    [T,~,fEL,fSW,SE,Ep] = NLFAssemble(x,uTot,d,eDOF,ix,Ep,S);
    f = fn + fEL;

%. Compute mass, damping matrices and Q, D Earthquake matrices
    [M,Q] = NLFMassMatrix(mass,Ep,x,uTot,d,ix,id,eDOF,S);
    [C,Evec,freq,damp] = DampingMatrix(M,T,damp,DF);

%. Get earthquake ground motion
    [EQaccel,~,EQdispl] = EQGroundMotion(h,nSteps,S.EQon,DF);

%. Establish initial conditions
    [uo,vo] = NLFICs(Evec,S);

%. Initialize position, velocity, and acceleration
    t = 0; Ft = TimeFunction(t,S.TimeType,1);
    uold = uo;   vold = vo;
    aold = M\(Ft*f + fSW - T*uo - C*vo);

%. Compute motion by modal decomposition
    for i=1:nSteps

%... Take care of the storage of the time history
        Eq = EQdispl(i);
        u = [uold; Ft*uPre];
        [nS,Hist] = NLFStoreHist(nS,Hist,t,u,Eq,Ft*f,SE,Ep);

%... Compute the values for the next time step
        t = t + h; its = 0; err = 1;
        bn = uold + h*vold + beta*h^2*aold;
        cn = vold + gamma*h*aold;
        Ft = TimeFunction(t,S.TimeType,0);
        fEQ = EQaccel(i)*Q;
        force = Ft*f + fSW - fEQ;
        anew = aold;

%... Newton iteration to determine the new acceleration
        while (err > tol) && (its < itmax)
          unew = bn + zeta*anew;
          vnew = cn + eta*anew;
          uTot = [unew; Ft*uPre];
          [T,g,~,~,~,~] = NLFAssemble(x,uTot,d,eDOF,ix,Ep,S);
          g = M*anew + C*vnew + g - force;
          T = M + eta*C + zeta*T;
          anew = anew - T\g;
          err = norm(g); its = its + 1;
```

```
        end

%... Compute velocity and displacement by numerical integration
        unew = bn + zeta*anew;
        vnew = cn + eta*anew;

%... Update internal variables, record stress and strain
        uTot = [unew; Ft*uPre];
        [~,~,~,~,SE,Ep] = NLFAssemble(x,uTot,d,eDOF,ix,Ep,S);

%... Update state for next iteration
        aold = anew;
        vold = vnew;
        uold = unew;

     end

%. Generate graphical output of excitation and response
     NLFGraphs(DF,x,d,ix,id,eDOF,S,Hist,freq);
```

Of course, the first thing needed is the description of the physical properties of the structure. The function FrameInput is similar to previous codes. We have added the array Ep, which contains the plastic strains. The input function simply initializes the plastic strains to zero. Note that we need to keep track of three plastic strains at each one of the numerical quadrature points for each element. As usual, we need to specify the duration of the analysis, the load form (including the type of time function that drives the applied nodal loads and the earthquake acceleration). The time integration (Newmark) parameters are the same as all other codes that use numerical time integration. The formation of the mass matrix is described in Code 14.4. The formation of the damping matrix is the same code introduced for the shear building (and also used for the truss). The specification of the earthquake ground motions is also the same one used throughout the book. Code 14.6 creates the initial conditions for the frame code. It only provides the option of giving an initial displacement or velocity in the shape of an eigenvector or, more commonly, specifying *at rest* initial conditions. The initial waves used in the beam codes are not included in the frame code.

Code 14.6 MATLAB code for specifying initial conditions for the nonlinear planar frame code

```
function [uo,vo] = NLFICs(V,S)
%   Establish the initial displacement and velocity
%
%          V : Matrix of eigenvectors
%          S : Struct with problem size parameters (nFree)
%     S.Mode : = 0 uo and vo set from function WaveForm
%              > 0 uo and vo set as mode shape 'S.Mode' from V

    if (S.Mode==0)
      uo = zeros(S.nFree,1);     % Set uo to zero
    else
      uo = V(:,S.Mode);          % Set uo to mode S.Mode
    end

%. Scale displacements and velocities
    vo = uo*S.Vo;
    uo = uo*S.Uo;

end
```

The dynamic analysis by Newmark integration follows the same outline as previous codes. Observe that we call the function NLFAssemble to form the global tangent stiffness matrix and global residual force vector. The additional consideration for this code is the need to call NLFAssemble, given in Code 14.9, upon convergence of the primary Newton iteration to update the internal variables. The update is computed every time the function NLFElement is invoked, but the global array is only permanently changed when Ep is passed up to the main program. Thus, if inelasticity has occurred during the time step, the plastic strains have changed.

All of the graphical output of results is relegated to the function NLFGraphs, which takes all of the response quantities stored in the 'struct' Hist and plots them. The formation of Hist is managed by the functions NLFCreateHist and NLFStoreHist. There are two reasons for setting up a strategy for storing output results for later plotting. First, the 'struct' provides a simple way for passing the stored information to the graphic processing functions. Second, it is generally not necessary to store every time point computed to obtain a good graphical representation of the output. The function NLFStoreHist takes care of storing the response every nOutput time steps. Codes 14.7 and 14.8 provide the storage scheme.

Code 14.7 MATLAB code for establishing the 'struct' Hist

```
function [nS,Hist] = NLFCreateHist(nOutput,nSteps,S)
%. Handle the storage of the response.
%
%    nOutput : Output frequency
%     nSteps : Number of time steps
%          S : Struct with problem size parameters
%
%         nS : [out,iout,nOutput,nReport,nSteps]
%       Hist : Struct containing response history

%. Set up a history array to save results ever once in a while
   nReport = floor(nSteps/nOutput);            % Number output steps
   Hist.t  = zeros(nReport,1);                 % Time
   Hist.u  = zeros(nReport,S.nDOF);            % Displacement
   Hist.E  = zeros(nReport,1);                 % Ground displacement
   Hist.F  = zeros(nReport,S.nFree);           % Applied forces
   Hist.SE = zeros(nReport,S.nElms,6*S.nQpts); % Stress/Strain
   Hist.Ep = zeros(nReport,S.nElms,3*S.nQpts); % Plastic strain

%. Storage control array
   nS = [0,0,nOutput,nReport,nSteps];

end
```

Code 14.8 MATLAB code for storing results in the 'struct' Hist

```
function [nS,Hist] = NLFStoreHist(nS,Hist,t,u,Eq,F,SE,Ep)
%. Handle the storage of the response
%
%         nS : [out,iout,nOutput,nReport,nSteps]
%       Hist : Struct containing response history
%          t : time
%          u : Displacement (full DOF, absolute)
```

```
%         Eq : Earthquake ground displacement
%          F : Applied forces (at free DOF)
%         SE : Stress and strain
%         Ep : Plastic strains (internal variables)


%. Write the values out to history, if appropriate
   if (nS(1)==0)
      nS(2) = nS(2) + 1;                    % Increment output counter
      iout = nS(2);
      Hist.t(iout,:) = t;                   % Time
      Hist.u(iout,:) = u';                  % Displacement
      Hist.E(iout,:) = Eq;                  % Ground displacement
      Hist.F(iout,:) = F';                  % Dynamic force
      Hist.SE(iout,:,:) = SE;               % Stress/Strain
      Hist.Ep(iout,:,:) = Ep;               % Plastic strain
      nS(1) = nS(3);
   end
   nS(1) = nS(1) - 1;

end
```

The core of the program is the assembly of the global stiffness matrix, which is given in Code 14.9.

Code 14.9   MATLAB code for assembly of the tangent and residual

```
function [K,g,f,s,SE,Ep] = NLFAssemble(x,u,d,eDOF,ix,Ep,S)
%    Form tangent K, internal residual g, applied distributed force f,
%    stress and strain SE, and plastic strain Ep
%
%         x : Nodal coordinates
%         u : Nodal displacements
%         d : Material properties
%      eDOF : Global DOF for each element
%        ix : Element nodal connections
%        Ep : Plastic strain (internal variables)
%         S : Struct with problem parameters (nElms,nFree,nDOF,nQpts)

%. Initialize global arrays
   K = zeros(S.nDOF,S.nDOF);         % Tangent stiffness
   g = zeros(S.nDOF,1);             % Residual
   f = zeros(S.nDOF,1);             % Nodal forces from element loads
   s = zeros(S.nDOF,1);             % Nodal forces from self weight
   SE = zeros(S.nElms,6*S.nQpts);   % Stresses and strains (for plotting)

%. Loop over elements
   for e=1:S.nElms

%... Find displacements, coordinates, and assembly pointers
      [Ue,We,Te,xe,de,Epe,jj] = NLFLocalize(e,u,Ep,x,d,ix,eDOF,S);

%... Form element matrices
      [~,ke,ge,fe,se,~,SEe,Epe] = NLFElement(xe,Ue,We,Te,de,S,Epe);

%... Store stress, strain, and plastic strain
      SE(e,:) = SEe;
      Ep(e,:) = Epe;

%... Assemble element matrices into global arrays
      K(jj,jj) = K(jj,jj) + ke;
      g(jj) = g(jj) + ge;
      f(jj) = f(jj) + fe;
      s(jj) = s(jj) + se;
```

```
    end
%. Trim to unconstrained DOF
    K = K(1:S.nFree,1:S.nFree);
    g = g(1:S.nFree);
    f = f(1:S.nFree);
    s = s(1:S.nFree);
end
```

The structure of the assembly code is similar to the assembly codes described previously. The key to assembly is knowing where to assemble the element matrices, which are computed in NLFElement (see Code 14.3). The rows and columns of the element matrices are placed according to their associated global degree-of-freedom number. These global indices are extracted from the array eDOF in the function NLFLocalize as shown in Code E.5 in Appendix E. The global displacements associated with element e are extracted from the global displacement vector $\mathbf{u}$. They are also segregated into $\mathbf{U}_e$, $\mathbf{W}_e$, and $\boldsymbol{\Theta}_e$. The nodal coordinates $\mathbf{x}_{i(e)}$ and $\mathbf{x}_{j(e)}$ are extracted from the global array $\mathbf{x}$. Finally, the material properties and the plastic strains of the element are extracted. These localized values are passed to the function NLFElement for processing.

14.6.3 Examples

The possible structures one can investigate with the 2D frame code is endless. We have the full range of dynamic excitation possibilities available, including modal vibration with initial displacements or velocities in a pure mode, resonant vibration under sinusoidal loading, response to various loading types (e.g., constant and blast), and earthquake excitation with any direction for the earthquake acceleration. Within those, we can examine how nonlinear geometry affects phenomena like resonance and how inelasticity affects response to any type of loading. Here we will consider a small set of examples to demonstrate a portion of this range of possibility.

Pendulum Revisited To verify the overall dynamics with large motions, let us reconsider the pendulum problem that we did in Fig. 8.11 in Chap. 8. Consider a beam of length $L = 10$ oriented at an angle of $\alpha = 0.3$ radians from vertical. The density of the beam is $\rho = 0.1$. The material properties $E = 10000$ and $G = 5000$ are selected so that the beam is fairly stiff, limiting vibration of the beam. The area and moment of inertia are $A = 1$ and $I = 0.1$, respectively. The response of the system is shown in Fig. 14.10. Observe that the response is identical to that shown in Fig. 8.11.

This verification example is significant because the motions are very large. We did not need to add anything to the formulation, like we did for the truss, because the self-weight is, essentially, a distributed load like q_o and p_o. The distributed loads in the frame code are oriented relative to the initial orientation of the element, whereas the self-weight always acts vertically (and is constant).

Fig. 14.10 *Pendulum.* Beam of length $L = 10$, hinged at one end and subjected to self weight, gives the response of a pendulum. The properties are $\rho = 0.1$, $E = 10000$, $G = 5000$, $I = 0.1$, and $A = 1$

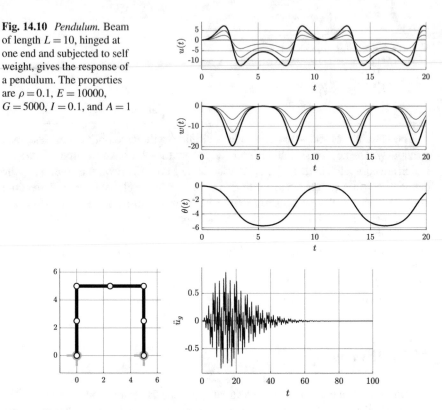

Fig. 14.11 *Frame subjected to earthquake.* Square frame with dimensions 5 m by 5 m, fixed at both base supports and subjected to a horizontal earthquake ground motion. The earthquake acceleration is shown

Frame Subjected to Earthquake The second example illustrates most of the features of the nonlinear frame model. Consider the frame shown in Fig. 14.11 subjected to the same earthquake defined in 3.15, scaled to twice the peak acceleration. The direction of the earthquake is *horizontal*. The duration of the earthquake is 100 s. The analysis only considers the first 40 s because the intensity of the motion is not significant after that. The beams in the frame are each five meters long. All elements have the same properties, which are given in Table 14.1.

The structure has Rayleigh damping with damping ratios of $\xi = 0.05$ specified in modes 1 and 10. The modes between 1 and 10 have less damping and the modes above 10 have more damping, with modes greater than 25 having over 20% damping. Newmark integration is used with $\beta = 0.25$ and $\gamma = 0.5$. The time step is $h = 0.01$. There are no applied loads and the self-weight of the elements is neglected. The finite element discretization uses six elements, each with two bubble modes for the displacements and one bubble mode for the rotation. The element matrices are integrated with a three-point Gauss–Legendre rule.

Table 14.1 *Physical properties of example frame.* The physical properties of the frame subjected to earthquake ground motions

Property	Symbol	Value
Density	ρ	7800 kg/m^3
Young's modulus	E	200 GPa
Shear modulus	G	100 GPa
Cross-section area	A	0.0100 m^2
Moment of inertia	I	0.0002 m^4
Axial yield	N_o	2000 kN
Shear yield	V_o	400 kN
Moment yield	M_o	300 kN-m

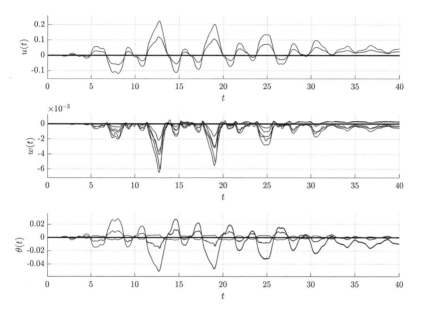

Fig. 14.12 *Frame example. Displacements.* The horizontal and vertical displacements and rotations of the nodes of the frame subjected to earthquake ground motion

The displacements vs. time for the seven nodes of the structure are shown in Fig. 14.12. Note that $u(t)$ is the horizontal displacement and $w(t)$ is the vertical displacement. There is one line for each node. It appears that there are fewer lines in the horizontal and rotation plots, but that is due to some symmetry in the system (in which case lines are plotting on top of each other). Clearly, the horizontal motion is the most significant response. The vertical motions are relatively small, generally because the columns are fairly stiff. There is evidence of higher frequency response in the vertical motion than the horizontal motion. The fundamental period of the system is $T = 2.2$ s (which is quite flexible).

Figure 14.13 shows the inelastic strains as a function of time. There are more lines on this plot than there were in the displacement plot because the inelastic strains are recorded at each integration point within each element. Obviously, the

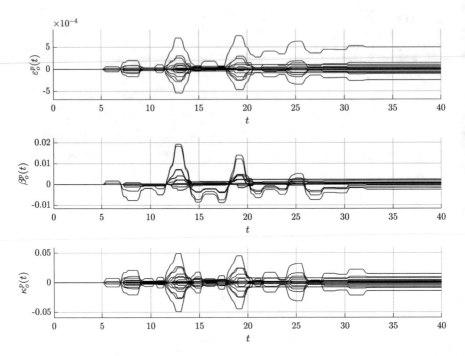

Fig. 14.13 *Frame example. Inelastic strains.* The inelastic axial, shear, and curvature strains at the Gauss points of the elements of the frame subjected to earthquake ground motion

system exhibits inelastic action due to the ground shaking. While the yielding is primarily due to the bending moments reaching capacity, there is plastic straining in all components, in accord with the normal flow rule associated with the model. It is evident that the inelastic shear strains are much larger than the inelastic axial strains. That is explained by the fact that N_o is much larger than V_o. Hence, the normal flow is less sensitive to the axial components. The graphs show numerous flat spots, suggesting that inelastic action is followed by unloading (during which the response is elastic). Once the earthquake dies down, the plastic strains become constant. Those are the non-recoverable strains.

Figure 14.14 shows the evolution in time of the internal stress resultants. Again, the stresses are recorded at the Gauss points, so there are three lines per element in these plots. The most important feature to notice in these results is that the moments are limited by the yield moment (which is 300). The shear forces get to about half of their yield value and the axial forces to only a tenth. As noted previously, there is still inelastic straining in all components, but it is clear that inelastic flexure is the most prevalent response mode. It is also evident that only a few of the locations get to the yield level. The rest are shielded by the yielding elements.

This example is simple but shows some of the features of the code. Even with this simple geometry, the analysis and response are very complex.

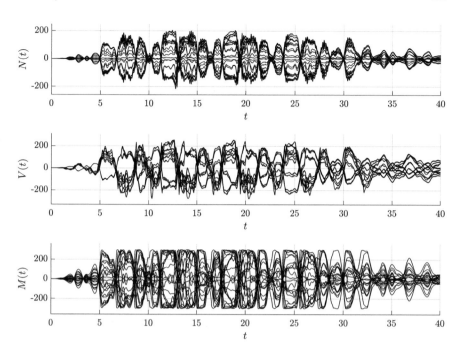

Fig. 14.14 *Frame example. Internal stress resultants.* The axial force, shear force, and bending moment at the Gauss points of the elements of the frame subjected to earthquake ground motion

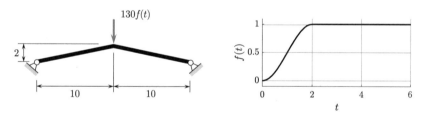

Fig. 14.15 *Shallow arch.* The basic geometry of the shallow arch example structure. The arch is subjected to a load of $130 f(t)$ at the apex

Snap-Through Buckling As the formulation of the nonlinear beam is geometrically exact, it is capable of capturing instability effects. In many ways, structural instability is much more naturally investigated as a dynamic phenomenon because instability often suggests a precipitous loss of internal load resistance capacity due to geometric configuration. One case that illustrates the loss of stability well is the snap-through buckling of the shallow arch shown in Fig. 14.15.

The arch consists of two straight segments rigidly connected at the apex and pin-supported at the base. The arch has a horizontal width of 20 and a height at the apex of 2. The system is subjected to a concentrated load of magnitude $130 f(t)$, where the time variation is as shown in the figure. In essence, the load ramps up from zero

Table 14.2 *Physical properties of the shallow arch.* The physical properties of the shallow arch subjected to vertical forces

Property	Symbol	Value
Density	ρ	0.2
Young's modulus	E	10,000
Shear modulus	G	5000
Cross-section area	A	4.0
Moment of inertia	I	0.5

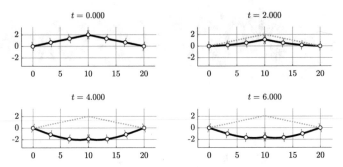

Fig. 14.16 *Deformed shape at different times.* The deformed configurations of the shallow arch at four different times are shown in the figure. The outline of the undeformed configuration is shown as a gray dotted line in each plot

to the full value of 130 within 2 s and then holds steady after that. The ramp up avoids most of the dynamic effects of suddenly applied loads.

The properties of the elements of the arch are given in Table 14.2. These values were chosen to make the arch stiff enough to carry the load through axial compression until the applied loads reach their full values. The constitutive model is elastic so there is no need to specify the yield values. The domain is represented with six elements, each with three bubble functions for the translations and two for the rotations. The element matrices were integrated with 4-point Gauss–Legendre integration. The analysis time step was $h = 0.01$.

Because the two ends of the arch are pinned, the arch acts like a truss initially, carrying most of the load through axial compression. As the members shorten, the arch deflects, and a bending mode of resistance engages. Near the point where the arch reaches a roughly horizontal position, the compressive axial forces start to drive the vertical motion downward. This phenomenon is often called *snap-through buckling*. Without the inertial resistance, snap-through is instantaneous (i.e., if analyzed statically). However, in a dynamic setting the motion accelerates and engages the inertial resistance.

Figure 14.16 shows the position of the arch at four times during the motion. Observe that at $t = 2$ the load achieves its full value. At that time, the arch is still

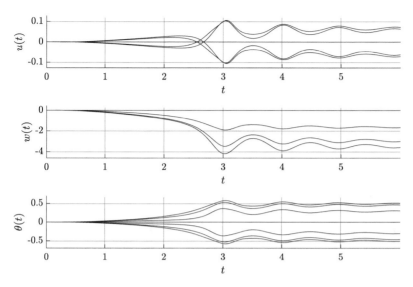

Fig. 14.17 *Displacement vs. time.* The three components of displacement of the nodes of the arch are plotted vs. time. Note that $u(t)$ is horizontal motion, $w(t)$ is vertical motion, and $\theta(t)$ is rotation at the nodes

above the horizontal position, but has momentum in the downward direction. That momentum is enough to drive the arch into snap-through. At times $t = 4$ and $t = 6$ the arch is below the horizontal position.

Figure 14.17 shows the nodal displacements of the structure as a function of time. Observe that the motions are not yet very large at $t = 2$ when the load reaches its full value of 130. The members have downward momentum, though, and that momentum pushes the structure past the horizontal position, thereby activating the snap-through mechanism. The vertical motion reaches its maximum value at around $t = 3$. The horizontal displacements are symmetric, as expected, and represent the displacement required to shorten the bars enough to snap through the horizontal position.

Figure 14.18 gives the variation of the internal stress resultants as a function of time. Observe that there is very little shear and bending in the elements prior to 2 s. The axial forces in all members are roughly the same and ramp up in accord with the function $f(t)$, as one would expect. The maximum compression force in the elements is about $N = -500$, but by $t = 3$ the bars have snapped through into tension of nearly $N = 700$.

The system is damped with a Rayleigh damping model with damping ratios of $\xi = 0.1$ specified in modes 1 and 10. The damping causes the oscillations to die out. By $t = 20$ the motion is static. In the static position the axial forces vary from about $N = 78$ to $N = 110$. The axial forces are in tension because the structure is carrying its load primarily through cable action. The bending in the elements in the static

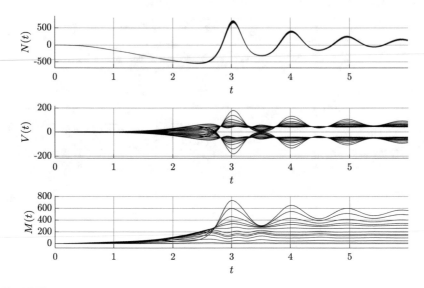

Fig. 14.18 *Internal stress resultants vs. time.* The three components of internal stress resultants at the integration points are plotted vs. time

position is relatively high, indicating that flexure is also an important load-carrying mechanism in the snapped-through position.

The snap-through buckling example shows how the formulation can capture the dynamic response associated with buckling. Dynamic results often defy intuition based upon static thinking. The dynamic analysis tells us about the rate at which instabilities transition from the point of buckling to oscillation about a stable static configuration.

14.6.4   Sample Input Function

The last detail to attend to in this chapter is the specification of the input for the frame structures (which are very similar to the beam of the previous chapters, too). The input function for the snap-through buckling example is given in Code 14.10.

Code 14.10   MATLAB code to specify the input of the structures for the snap-through buckling example

```
function [x,ix,idG,eDOF,d,F,Ep,mass,Damp,S,DF] = FrameInput
%  Shallow arch for snap through example
%
%  Problem input for Nonlinear Beam codes
%          x : Nodal coordinates [x, y]
%         ix : Element connections [I-node, J-node, Matl. Set]
%        idG : Global DOF numbers for each node [U,W,Theta]
%       eDOF : Global DOF numbers for each element
%          d : [type,E,G,A,I,rho,po,qo,mo,gravity,No,Vo,Mo]
```

```
%              F : Nodal applied forces in node order
%             Ep : Plastic strain (internal variables)
%           mass : Nodal massses
%          Damp. (struct for damping model)
%             S. (struct with problem dimensions)
%            DF. (struct to manage output)

%. Numerical discretization (input)
   S.nNodes = 7;                    % Number of nodes
   S.nElms = 6;                     % Number of elements
   S.nMat = 1;                      % Number of material sets
   nE = 3;                          % Number of extra DOF for U,W

%. Numerical discretization (computed)
   S.nExt = [nE,nE,nE-1];           % Extra DOF for [U,W,T]
   exDOF = sum(S.nExt);             % Total extra DOF per element
   S.nDOFpn = [1,1,1];              % DOF per node fro [U,W,T]
   S.nDOF = 3*S.nNodes + S.nElms*exDOF;  % Total number of DOF

%. Physical parameters
   Model = 1;                       % =1 Elastic, =2 Inelastic
   rho = 0.2;                       % Mass Density
   A = 4.0;                         % Cross-sectional area
   I = 0.5;                         % Moment of inertia
   E = 10000;                       % Young's modulus
   G = 5000;                        % Shear modulus
   q0 = [0,0,0];                    % Load magnitude [p,q,m]
   So = [100,100,100];              % Yield levels [No,Vo,Mo]
   gravity = 0;                     % =0 to neglect self-weight

%. Store physical parameters
   d(1,:) = [Model, E, G, A, I, rho, q0, gravity, So];

%. Establish the nodal coordinates
   L = 10; H = 2;
   x = [ 0,0; 1*L/3, 1*H/3; 2*L/3, 2*H/3; L, H;...
              4*L/3, 2*H/3; 5*L/3, 1*H/3; 2*L, 0];

%. Compute element connection array
   ix = zeros(S.nElms,3);
   for e=1:S.nElms
     ix(e,:) = [e:e+1,1];
   end

%. Establish BCs =0 DOF free, =1 DOF restrained [U,W,Theta]
   id = zeros(S.nNodes,3);
   id(1,:) = [1,1,0];  id(S.nNodes,:) = [1,1,0];

%. Compute global DOF numbers and store back in id array
   [eDOF,idG,S] = GlobalDOF(id,ix,S);

%. Nodal force, each row is (x-force, y-force, moment) in node order
   F = zeros(S.nNodes,3);
   F(4,:) = [0,-130,0];

%. Nodal masses
   Mo = 0.0;                    % Translational nodal mass
   Jo = 0.0;                    % Rotational nodal mass
   mass = [Mo*ones(S.nNodes,1), Jo*ones(S.nNodes,1)];

%. Damping
   Damp.type = 1;               % =1 Rayleigh, =2 Caughey, =3 Modal
   Xi = 0.10;                   % Reference damping ratio
   Damp.modes = [1,10];         % Modes for specified damping
   Damp.xi = Xi*ones(2,1);      % damping ratio in each specified mode

%. Initial condition, Time function, and EQ data
```

```
    S.Mode = 0;                    % Mode for initial condition
    S.Uo = 0;                      % Magnitude of initial displacement
    S.Vo = 0;                      % Magnitude of initial velocity
    S.TimeType = 4;                % =0 none =1 Constant, =2 Sinusoid, etc
    S.EQon = 0;                    % =0 No EQ, =1 Yes EQ
    S.EQdir = [1.0; 0.0];          % EQ direction

%. Numerical integration parameters
    S.QType = 1;                   % = 1 G-Leg, =2 G-Lob
    S.nQpts = nE+1;                % Number of integration points
    [S.wt,S.st,S.nQpts] = NumInt(S.QType,S.nQpts);

%. Initialize internal variables
    Ep = zeros(S.nElms,3*S.nQpts);

%. Echo structure properties to command window
    NLFEchoStructure(S,x,ix,id,d,F)

%. Manage output, render structure geometry
    [DF] = NLFManageFigures;
    PlotInitialStructure(x,F,ix,id,S,DF)

end
```

The function starts by specifying the size of the problem with the number of nodes (nNodes), the number of elements (nElms), the number of material sets (nMat), and the number of bubble functions (nE) to use for the translational degrees of freedom. From those values, the number of bubble functions (nExt) is set for all three variables. Notice that the two translations are the same and the rotation uses one fewer bubble functions. This also implies that it is not possible to specify $nE = 0$ because the rotation cannot have a negative number of bubble functions. The total number of extra (bubble) degrees of freedom (exDOF) and the total number of degrees of freedom in the structure (nDOF) are stored in the 'struct' S, which is used in almost all of the functions to carry out the analysis.

The next section of the code specifies the physical parameters. All of this information eventually finds its way into the array d, but it is nice to have the information with comments to keep track of the values. In this case there is only one row in the d array because all elements have the same properties.

The nodal coordinates (x) are established using the length and height parameters L and H. This is just a convenience. The element connection array ix is very simple in this case and can be generated in a loop. The boundary conditions are specified through the id array, which sets the two translational components to zero at the first and last node. Once the id array is specified, we call the function GlobalDOF, which is discussed in Appendix D. This function converts the id array into global DOF numbers for each node and it creates the eDOF array that contains the global node numbers associated with each element. The id and eDOF arrays govern the localization and assembly process for creating the residual, tangent, and mass matrices.

The nodal forces are specified in the array F, which is organized in node order with the three columns being associated with horizontal force, vertical force, and moment at the node. Note that if a force in the F array is associated with a fixed degree of freedom in the id array, then the value specified in F is considered to be a prescribed displacement.

Next, the nodal masses are specified. The translational masses must have the same value, but the rotation mass is independent of that. Therefore, only two values are specified for the array mass, Mo and Jo. In this case, the values are set to be the same for all nodes, but the values are set to zero for the snap-through example.

The damping information is conveyed in a 'struct' called Damp. The type of damping (either Rayleigh, Caughey, or modal) is specified in Damp.type. Then, depending on which type of damping is selected, the damping values and the modes in which those values are specified are put into Damp.xi and Damp.modes, respectively. This information is expected in the function DampingMatrix, which is used in all codes from NDOF, to truss, to the beam codes.

Next, the values needed for the initial conditions are set, including the mode number and magnitudes for initial displacement and velocity, the type of time variation, a flag to determine if the earthquake motion is active or not, and the earthquake direction vector.

In the next code segment, the numerical integration rule is established, and the stations and weights are stored in S. In this case we specify Gauss–Legendre integration (see Appendix F), and we set the number of integration points to be one more than the number of bubble functions specified for the translational degrees of freedom, with the assumption that the highest order of integral that will need to be evaluated will involve the product of the shape functions with themselves (e.g., in the mass matrix). The highest order shape function has polynomial order one greater than the number of bubble functions. Hence, the greatest order of polynomial that needs to be integrated accurately is $2(nE+1)$, which can be accomplished with a rule with $nE+1$ quadrature points.

Next, the internal variables (plastic strains) are initialized to zero. Notice that each element requires the tracking of the internal variables at all of the quadrature points and that we have three strains (ε_o, β_o, and κ_o) that can accrue plastic strain.

The last two actions of this function are to echo all of the input values to the command window with the function NLFEchoStructure. There is a similar function for all of the codes in the book. It is important to echo the input to make sure it is properly specified and to provide a record of it. Also, if the flag is set in DF to plot the initial structure, it is rendered with the function PlotInitialStructure. This function plots the structure with loads and boundary conditions shown and also with node and element numbers, which helps to verify that the structure geometry is input correctly.

Correction to: Fundamentals of Structural Dynamics

Correction to:
K. D. Hjelmstad, *Fundamentals of Structural Dynamics*,
https://doi.org/10.1007/978-3-030-89944-8

This book was inadvertently published without updating the following corrections:

Chapter 9
On page 252–253, the entire section "Distributed load" has been revised.

Chapter 11
On page 346, the corrections have been made in the "Code 11.5 MATLAB code to find the eigenvalues and the coefficients of the eigenfunction $\varphi_n(x)$ for a Rayleigh beam with different boundary conditions."

The updated original version for this book can be found at
https://doi.org/10.1007/978-3-030-89944-8_9
https://doi.org/10.1007/978-3-030-89944-8_11

Appendix A
Newton's Method for Solving Nonlinear Algebraic Equations

Newton developed a method for solving nonlinear algebraic equations. Consider a single equation in a single unknown, written in the form

$$g(x) = 0 \qquad\qquad (A.1)$$

An example function is $g(x) = x + 3 \sin \pi x$, shown in Fig. A.1. There is no algebraic manipulation that allows you to solve for x, and hence there is no possibility of getting a closed-form solution, i.e., some sort of equivalent to the quadratic formula for quadratic equations. So, what do we do?

A good way to start the quest of finding solutions to Eq. A.1 is to plot the function $g(x)$ to see where it is equal to zero, as shown in Fig. A.1 for the example function. To do so gives you a clear picture of the task. Unlike linear equations, there may be multiple solutions to the nonlinear equations, so uniqueness of solution is something we cannot count on. A graph will show you how many solutions there are and roughly where they are located. This approximate location might prove very useful in starting Newton's method, as we shall soon see. Unfortunately, the graphical method works for $g(x)$ but does not extend well to vector functions of vector variables.

Without a direct solution approach, we must use some sort of iterative approach wherein we guess a value of x and evaluate $g(x)$ to see if it is equal to zero. If it isn't, then we need to find a better guess and repeat the process. The success of the method relies on the quality of the guessing strategy. That is exactly what Newton's method does well.

© The Author(s), under exclusive license to Springer Nature Switzerland AG 2022
K. D. Hjelmstad, *Fundamentals of Structural Dynamics*,
https://doi.org/10.1007/978-3-030-89944-8

Fig. A.1 *Example nonlinear function.* To see what is at stake in solving nonlinear algebraic problems, consider the function $g(x) = x + 3 \sin \pi x$. The function has five roots, shown by the open circles, so the solution to $g(x) = 0$ is not unique

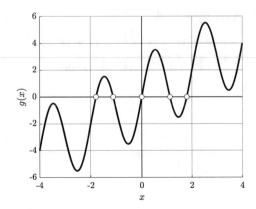

A.1   Linearization

Newton had a great idea. Since we know how to solve linear problems, let's create a linear problem that is close to the nonlinear one. Then solve the linear problem. How do we do that? We know from calculus that every function can be expanded in a Taylor series. Let us expand $g(x)$ around some known value x_o. To wit,

$$g(x) = g(x_o) + (x - x_o)\, g'(x_o) + \tfrac{1}{2}(x - x_o)^2 g''(x_o) + \cdots$$

where a prime denotes differentiation of the function with respect to x. The first term is constant, the second term is linear in x, the third term is quadratic, etc. We can create a linear function "close" to the original nonlinear one by just taking the first two terms of the Taylor series expansion. Let us call that linear function

$$\tilde{g}(x) = g(x_o) + (x - x_o)\, g'(x_o)$$

It is a geometric fact that a curve remains close to a line tangent to it over some distance. The curvature determines how large the region is where the two functions are approximately the same. In this neighborhood, the linear function does a good job of representing its nonlinear progenitor.

You can get an idea of the relationship between the original function $g(x)$ and the associated linear function $\tilde{g}(x)$ from Fig. A.2. The original function is the heavy line, and the linear function is the straight line that is tangent to it. The arbitrarily selected point x_o is shown as a gray dot on the abscissa, and the value of the original function $g(x_o)$ is shown as a gray dot on the ordinate. The gray circle where the curve and line intersect is the place where the original function and its linear counterpart have the same value, i.e., $\tilde{g}(x_o) = g(x_o)$. Note that the linear function is tangent to the nonlinear one at that point, so $\tilde{g}'(x_o) = g'(x_o)$, too. The open circle on the x-axis is the solution we seek. Since $g(x_o) \neq 0$, we know that x_o is *not* the solution we seek.

Fig. A.2 *Newton's method.*
We approximate the original
function $g(x)$ with a linear
function that matches the
value and tangent of the
original function at the point
x_o. We then solve the linear
equation to get a better
estimate of the solution

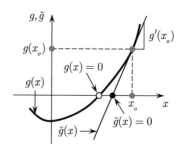

Newton's great idea was to recognize that we could actually solve the equation that results from setting the linear function equal to zero, i.e.,

$$\tilde{g}(x) = g(x_o) + (x - x_o)\, g'(x_o) = 0 \qquad (A.2)$$

The value of x found by solving this equation is the solid black dot on the abscissa in the figure. This value depends upon x_o. Taking the result of one iteration as the x_o for the next iteration establishes an iteration

$$x_{i+1} = x_i - \frac{g(x_i)}{g'(x_i)} \qquad (A.3)$$

that generates a sequence of values $\{x_0, x_1, x_2, \ldots, x_n\}$. The sequence converges to a solution to the original nonlinear equation, if all goes well. While Newton's method is not guaranteed to converge, when it does the convergence is quadratic in the neighborhood of the solution. For problems where convergence is an issue, other strategies can be used.

Example 1 *Scalar equation with scalar unknown.* To see how the process works, let us apply Newton's method to a simple nonlinear equation. Consider the equation

$$g(x) = x^2 - 3 = 0$$

The beauty of this nonlinear equation is that we can actually solve it (spoiler alert, the answer is $\pm\sqrt{3}$ or, approximately, ± 1.732050808). The Newton iteration equation takes the form

$$x_{i+1} = x_i - \frac{g(x_i)}{g'(x_i)} = x_i - \frac{x_i^2 - 3}{2x_i}$$

From any starting value (except for $x_0 = 0$) we can compute a sequence of estimates. Table A.1 shows how the iteration goes starting from the value $x_0 = 10$. We could probably guess that 10 is not anywhere near the solution we seek, but this choice shows how quickly Newton's method converges to the true solution.

Table A.1 *Results of Newton iteration on simple example. With a starting point of $x_0 = 10$ Newton's method solves $g(x) = x^2 - 3 = 0$ to within an error tolerance of $tol = 10^{-8}$ in 6 iterations*

n	x_n	$g(x_n)$
0	10.000000000	97.000000000
1	5.150000000	23.522500000
2	2.866262136	5.215458632
3	1.956460732	0.827738595
4	1.744920939	0.044749084
5	1.732098271	0.000164421
6	1.732050808	0.000000002

To run Newton's method, you need to specify the starting value and an error tolerance that tells the algorithm when to stop. For the example in Table A.1, the tolerance was $tol = 10^{-8}$. It is also a good idea to limit the number of allowable iterations in case the sequence diverges.

For the simple function in this example, the convergence is monotonic, which means that the error gets smaller in each successive iteration. The values of $g(x_n)$ are the error because the aim is for the function to be zero. Notice that the errors in the last three iterations are on the order 10^{-2}, 10^{-4}, and 10^{-9}, which is the hallmark of quadratic convergence. Why didn't the method find the other solution? Newton's method can only go to one solution at a time. To find the other root, a different starting point must be specified (any negative value will work in this case).

A.2   Systems of Equations

We can derive a similar method for systems of nonlinear equations, which is where Newton's method really has power and utility. In such situations, we will have N equations with N unknowns, which we write as

$$\mathbf{g}(\mathbf{x}) = \mathbf{0} \tag{A.4}$$

where $\mathbf{x} = [x_1, x_2, \ldots, x_N]^T$ is the vector variable of the vector-valued function $\mathbf{g}$. The specific details of each of the N equations vary from one application to the next. Newton's method is constructed in a similar manner by recognizing that we can create a linear function related to the original function $\mathbf{g}(\mathbf{x})$ using the Taylor series expansion of the function. To wit,

$$\tilde{\mathbf{g}}(\mathbf{x}) = \mathbf{g}(\mathbf{x}_o) + \nabla\mathbf{g}(\mathbf{x}_o)(\mathbf{x} - \mathbf{x}_o) \tag{A.5}$$

where $\mathbf{x}_o$ is simply a specific value of the vector $\mathbf{x}$ and the gradient of the function $\mathbf{g}$ is simply the matrix of partial derivatives with the rows of the matrix $[\nabla\mathbf{g}]_{ij}$ associated with the function g_i and the columns associated with the independent variable x_j. The matrix representation of the function, which we call the *residual*, is

$$\mathbf{g}(\mathbf{x}) = \left\{ \begin{array}{c} g_1\,(x_1, x_2, \ldots, x_N) \\ g_2\,(x_1, x_2, \ldots, x_N) \\ \vdots \\ g_N\,(x_1, x_2, \ldots, x_N) \end{array} \right\} \tag{A.6}$$

and the matrix of partial derivatives, which we call the *tangent matrix*, is

$$\nabla \mathbf{g} = \begin{bmatrix} \dfrac{\partial g_1}{\partial x_1} & \cdots & \dfrac{\partial g_1}{\partial x_N} \\ \vdots & \ddots & \vdots \\ \dfrac{\partial g_N}{\partial x_1} & \cdots & \dfrac{\partial g_N}{\partial x_N} \end{bmatrix} = \mathbf{A} \tag{A.7}$$

The order of terms in the equation matters and the rules of matrix multiplication apply. The term "residual" comes from the fact that we are trying to find solutions to $\mathbf{g}(\mathbf{x}) = 0$. For any value $\mathbf{x}_o$, then, $\mathbf{g}(\mathbf{x}_o)$ measures how far away from that goal we are. The residual measures the deviation from zero.

Solve the Linear Equations We have found a linear function that approximates the original nonlinear function in the neighborhood of the point $\mathbf{x}_o$. If we set this linear function equal to zero, then we can solve a linear system of equations to find $\mathbf{x}$. To wit,

$$\tilde{\mathbf{g}}(\mathbf{x}) = \mathbf{g}(\mathbf{x}_o) + \mathbf{A}(\mathbf{x}_o)(\mathbf{x} - \mathbf{x}_o) = 0$$

where $\mathbf{A} = \nabla \mathbf{g}$. Solving those linear equations, we can find $\mathbf{x}$, which we can take as the starting point for the next iteration. If we repeat the process, we get Newton's method

$$\mathbf{x}_{i+1} = \mathbf{x}_i - \left[\mathbf{A}(\mathbf{x}_i) \right]^{-1} \mathbf{g}(\mathbf{x}_i) \tag{A.8}$$

As was true for the scalar version, this new value $\mathbf{x}_{i+1}$ is taken as a "better" estimate of the solution than $\mathbf{x}_i$ was. This calculation can be done repeatedly, until we arrive at an estimate $\mathbf{x}_n$ after n repetitions of the process. The measure of error in this case is $\| \mathbf{g}(\mathbf{x}_i) \|$.

When should we stop? We will never get the exact answer, but we will get close. Remember that the goal was to solve the equation $\mathbf{g}(\mathbf{x}) = 0$. So, if the norm of the residual vector $\mathbf{g}(\mathbf{x}_n)$ is less than some preset tolerance `tol`, then we are probably close enough. Since Newton's method is not guaranteed to converge, we will probably want to specify a maximum number of iterations we are willing to try (call it `itMax`) in our programs. We stop if the iteration has converged or if we have exhausted the maximum number of iterations we have specified. Setting a maximum number of iterations is a fail-safe if the iteration diverges. The algorithm fails if the matrix $\mathbf{A}$ is singular.

Example 2 *Two equations in two unknowns.* To get an idea of how Newton's method works for systems of equations, consider the equations

$$x^3 - 3xy + 2y^2 + x - 6 = 0$$

$$y^3 + 2xy - 4x^2 + y + 10 = 0$$

The residual is

$$\mathbf{g}(x, y) = \left\{ \begin{array}{c} x^3 - 3xy + 2y^2 + x - 6 \\ y^3 + 2xy - 4x^2 + y + 10 \end{array} \right\}$$

The tangent matrix is simple to compute because the component functions are polynomials. To wit,

$$\mathbf{A} = \nabla \mathbf{g}(\mathbf{x}) = \begin{bmatrix} 3x^2 - 3y + 1 & -3x + 4y \\ 2y - 8x & 3y^2 + 2x + 1 \end{bmatrix} \tag{A.9}$$

The calculation can be implemented as shown in Code A.1.

Code A.1 MATLAB code for Newton's method applied to Example 2

```
%. Set parameters for Newton's method
   itMax = 20; tol = 1.e-8;

%. Initialize error and iteration counter
   err = 1.0; it = 0;

%. Set initial value of z=[x,y]
   z = [1; 2];

%. Carry out Newton iteration
   while (err > tol) && (it < itMax)
       it = it + 1;
       x = z(1); y = z(2);
       g = [ x^3 - 3*x*y + 2*y^2 + x - 6 ;...
             y^3 + 2*x*y - 4*x^2 + y + 10 ];
       A = [ 3*x^2 - 3*y + 1 , -3*x + 4*y ;...
                2*y - 8*x    ,  3*y^2 + 2*x + 1 ];
       z = z - A\g;
       err = norm(g);
   end
```

The array z is used for the vector **x** because there are no bold and italic fonts in MATLAB and we need to keep **x** distinct from the scalar variable x in the program. The program uses x and y to make the correspondence with the equation clearer. The starting values are $x_0 = 1$ and $y_0 = 2$. The iteration converges in 14 steps to a tolerance of 10^{-8}. The MATLAB program produces the results shown in Table A.2. Notice that things got worse before they got better, but Newton's method finally converged once it got on the right track. This phenomenon is not uncommon but depends upon the nature of the equations. Picking a starting value closer to the solution will generally reduce the number of iterations required and it will increase the likelihood of convergence.

Table A.2 *Results of Newton iteration for Example 2.* With a starting point of $x_0 = 1$ and $y_0 = 2$, the solution converges in 14 iterations

n	x_n	y_n	$\|\mathbf{g}(\mathbf{z}_n)\|$
0	1.000000000	2.000000000	2.01e+01
1	−12.000000000	−2.800000000	1.90e+03
2	−6.021091716	−35.379440189	4.41e+04
3	−6.868172912	−23.624905344	1.31e+04
4	−5.519256647	−15.769369207	3.88e+03
5	−4.178715180	−10.527310729	1.15e+03
6	−3.040506723	−7.036276958	3.40e+02
7	−2.115792429	−4.735015921	9.88e+01
8	−1.377426962	−3.278499799	2.72e+01
9	−0.816000324	−2.472691351	6.33e+00
10	−0.473272815	−2.152159597	1.05e+00
11	−0.361829741	−2.077810191	7.47e–02
12	−0.352113414	−2.071714268	5.32e–04
13	−0.352044274	−2.071670060	2.72e–08
14	−0.352044270	−2.071670058	3.55e–15

Fig. A.3 *Example 2.* Plot of the two functions $g_1(x, y) = 0$ and $g_2(x, y) = 0$, showing the points of intersection (the circles), which are the solution to the equations

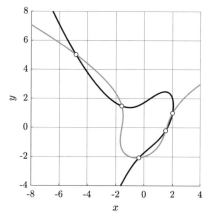

This example problem is more interesting than it might first seem. Figure A.3 shows a plot of the two curves defined by $g_1(x, y) = 0$ and $g_2(x, y) = 0$. The circles at the points of intersection are the solutions we seek—i.e., places where both functions are zero. It is clear that there are five solutions to this problem (at least in the window shown). We have found the one indicated at the bottom of the plot. The other solutions can be found by starting the Newton iteration at different starting points, as shown in Table A.3. With the advantage of seeing the graph, it is possible to select starting points closer to the solutions. Therefore, the number of iterations was fewer for those cases than for the case shown in Table A.2.

Newton's method exhibits *quadratic convergence* in the neighborhood of a solution, wherein the exponent on the error cuts in half with each successive iteration. Quadratic convergence has the advantage that setting a tight tolerance does not increase the number of iterations much. The tolerance can be set too tight,

Table A.3 *Other solutions to Example 2.* The remaining four solutions to Example 2 can be found by using different starting points

x_0	y_0	x_n	y_n	Iterations, n
2.0	0.0	1.513412584	−0.206014829	5
2.0	2.0	2.000000000	1.000000000	6
−2.0	2.0	−1.588099656	1.495065721	4
−5.0	5.0	−4.822770645	5.018970365	4

though. When it is, the roundoff error in the computer (double precision arithmetic is good to about 15 digits of accuracy) dominates the norm of the error. In such a situation, the error will appear to fluctuate randomly, and the iteration will not converge.

When we apply Newton's method in dynamics, we will have a great starting point. The dynamics algorithm steps through time at fairly closely spaced time steps. In each step we will be iterating to find the acceleration of the system at the new time point. To start the Newton iteration, we can take the initial guess to be the final converged acceleration from the previous time step. Because this guess is so good, we will seldom need many iterations to converge for these problems, which is particularly important for large problems because we solve a system of equations at each iteration.

Appendix B
The Directional Derivative

When we use Newton's method to solve systems of nonlinear algebraic equations, we need the residual $\mathbf{g}(\mathbf{x})$ and the tangent matrix $\mathbf{A}(\mathbf{x})$. The tangent matrix is the gradient of the function $\mathbf{g}(\mathbf{x})$, which is the matrix of partial derivatives, as shown in Eq. A.7 in Appendix A. In many circumstances, directly computing partial derivatives is not the most efficient way to find the tangent matrix. The directional derivative provides a good alternative approach to organizing the computation. It is also possible to extend the notion of the directional derivative to functionals. This appendix reviews the directional derivative and provides examples of how to apply it.

B.1 Ordinary Functions

The *directional derivative* of a function $\mathbf{g}(\mathbf{x})$ is defined as

$$\mathcal{D}\mathbf{g}(\mathbf{x}) \cdot \mathbf{h} = \frac{d}{d\varepsilon}\Big[\mathbf{g}(\mathbf{x} + \varepsilon\mathbf{h})\Big]_{\varepsilon=0} \tag{B.1}$$

The notation reads "the derivative of the function $\mathbf{g}$ at location $\mathbf{x}$ in the direction of the vector $\mathbf{h}$." The vector $\mathbf{h}$ is arbitrary and constant but has the same character as $\mathbf{x}$. In fact, the value $\varepsilon\mathbf{h}$ is a *perturbation* of $\mathbf{x}$—a little nudge to test how $\mathbf{g}$ changes in the neighborhood of $\mathbf{x}$. This approach reduces the rate of change of $\mathbf{g}$ to be an ordinary scalar derivative with respect to ε. Once the differentiation step is complete, we set $\varepsilon = 0$, giving the rate of change as we just start moving in the direction $\mathbf{h}$.

The directional derivative gives the rate of change in a specific direction, but since we did not restrict $\mathbf{h}$ in any way, it can represent *any* direction. In that sense, the directional derivative provides general information about the rate of change of the function. The reason the directional derivative is of interest to us is because

© The Author(s), under exclusive license to Springer Nature Switzerland AG 2022
K. D. Hjelmstad, *Fundamentals of Structural Dynamics*,
https://doi.org/10.1007/978-3-030-89944-8

$$\mathcal{D}g(\mathbf{x}) \cdot \mathbf{h} = \mathbf{A}(\mathbf{x})\mathbf{h} \tag{B.2}$$

In other words, if we can factor the $\mathbf{h}$ out of the final expression for the directional derivative, the matrix that multiplies it is the tangent matrix.

Example 1 *Scalar function of scalar variable.* Consider the scalar function

$$g(x) = 3x^2 + x$$

The derivative of this function is $g'(x) = 6x + 1$ by ordinary rules of differentiation. For comparison, we can take the directional derivative as

$$\mathcal{D}g(x) \cdot h = \frac{d}{d\varepsilon}\left[3(x + \varepsilon h)^2 + (x + \varepsilon h)\right]_{\varepsilon=0}$$

$$= \left[6(x + \varepsilon h)h + h\right]_{\varepsilon=0}$$

$$= \left[6xh + h\right] = (6x + 1)h$$

Thus, $6x + 1$ is the tangent of the function $3x^2 + x$. In this case the tangent is the same as the ordinary derivative, as we have verified.

Example 2 *Scalar function of vector variable.* Consider the function

$$r(\mathbf{x}) = \sqrt{\mathbf{x} \cdot \mathbf{x}}$$

This function simply computes the length of a vector as being the square root of the vector dotted with itself. Compute the directional derivative of $r(\mathbf{x})$:

$$\mathcal{D}r(\mathbf{x}) \cdot \mathbf{h} = \frac{d}{d\varepsilon}\left[r(\mathbf{x} + \varepsilon\mathbf{h})\right]_{\varepsilon=0}$$

$$= \frac{d}{d\varepsilon}\left[\left((\mathbf{x} + \varepsilon\mathbf{h}) \cdot (\mathbf{x} + \varepsilon\mathbf{h})\right)^{\frac{1}{2}}\right]_{\varepsilon=0}$$

$$= \left[\frac{1}{2}\left((\mathbf{x} + \varepsilon\mathbf{h}) \cdot (\mathbf{x} + \varepsilon\mathbf{h})\right)^{-\frac{1}{2}}\left(2(\mathbf{x} + \varepsilon\mathbf{h}) \cdot \mathbf{h}\right)\right]_{\varepsilon=0}$$

$$= \frac{\mathbf{x} \cdot \mathbf{h}}{\sqrt{\mathbf{x} \cdot \mathbf{x}}} = \frac{\mathbf{x}}{r(\mathbf{x})} \cdot \mathbf{h}$$

We have invoked the product rule and the chain rule in taking these derivatives with respect to ε. Note that the vector $\mathbf{x}/r$ is a unit vector pointing in the direction $\mathbf{x}$. We can identify $\mathbf{x}/r$ as the gradient of the function $r(\mathbf{x})$. In this case, the natural factoring out of the vector $\mathbf{h}$ is through the dot product. The directional derivative produces a scalar rate of change. The gradient of a scalar function of a vector variable is a vector. The dot product of vectors is a scalar.

Example 3 *Vector function of vector variable.* Consider the function

$$\mathbf{g}(\mathbf{x}) = r(\mathbf{x})\mathbf{x}$$

where $r(\mathbf{x}) = \sqrt{\mathbf{x} \cdot \mathbf{x}}$ is the same function as the previous example. Before we get started, first note that the directional derivative of the function $\mathbf{x}$ is simply $\mathbf{h}$. The proof is

$$\mathcal{D}\mathbf{x} \cdot \mathbf{h} = \frac{d}{d\varepsilon} \Big[\mathbf{x} + \varepsilon\mathbf{h}\Big]_{\varepsilon=0} = \mathbf{h} \tag{B.3}$$

The directional derivative of $r(\mathbf{x})\mathbf{x}$ can be computed with the product rule of differentiation, which works the same for directional derivatives as it does for ordinary derivatives. Since we have already derived the directional derivatives of each component, we can compute

$$\mathcal{D}\mathbf{g}(\mathbf{x}) \cdot \mathbf{h} = \Big(\mathcal{D}r(\mathbf{x}) \cdot \mathbf{h}\Big)\mathbf{x} + r(\mathbf{x})\Big(\mathcal{D}\mathbf{x} \cdot \mathbf{h}\Big) = \Big(\frac{\mathbf{x}}{r} \cdot \mathbf{h}\Big)\mathbf{x} + r(\mathbf{x})\mathbf{h}$$

How do we factor out the vector $\mathbf{h}$ so that we can identify the tangent matrix? First, note that in matrix notation the dot product is $\mathbf{a} \cdot \mathbf{b} = \mathbf{a}^T\mathbf{b}$. Also, note that the scalar that results from the dot product can be moved to the right side of $\mathbf{x}$ in the first term without violating the rules of matrix algebra. Thus, we can write

$$\mathbf{x}\Big(\frac{1}{r}\mathbf{x}^T\mathbf{h}\Big) + r\mathbf{h} = \Big[\frac{1}{r}\mathbf{x}\mathbf{x}^T + r\mathbf{I}\Big]\mathbf{h}$$

where $\mathbf{I}$ is the identity matrix of dimension equal to the dimension of $\mathbf{x}$ (which we will call n here). We put the identity matrix in because the identity times any vector returns the same vector. In order to group the two terms together, they must have the same matrix character. The term $\mathbf{x}\mathbf{x}^T$ is an $n \times n$ matrix, which can be added to an $n \times n$ identity matrix. Finally, we can identify the tangent matrix

$$\mathbf{A}(\mathbf{x}) = r(\mathbf{x})\Big[\mathbf{I} + \mathbf{n}(\mathbf{x})\mathbf{n}^T(\mathbf{x})\Big]$$

where $\mathbf{n}(\mathbf{x}) = \mathbf{x}/r$ is a unit vector.

B.2 Functionals

The directional derivative can also be applied to a functional (i.e., a function of a function). Let us assume that we have a functional $J(u)$, where the argument $u(x)$ is itself a function. A functional takes in a function as an argument and produces a number as output. For example, consider the functional

$$J(u) = \int_0^1 \left(u^2(x) + 4\sin(u(x)) \right) dx \tag{B.4}$$

For a given function $u(x)$, $J(u)$ produces a number. As a specific example for the functional $J(u)$, take $u(x) = x + 1$. Compute the value of the functional

$$J(x+1) = \int_0^1 \left((x+1)^2 + 4\sin(x+1) \right) dx = 6.15913$$

Each different function $u(x)$ will give a different value of J. The *directional derivative* asks how quickly does the value of J change as we move in the direction of a different function $v(x)$, and is defined as

$$DJ(u) \cdot v = \frac{d}{d\varepsilon} \left[J\big(u(x) + \varepsilon v(x)\big) \right]_{\varepsilon=0} \tag{B.5}$$

This derivative computes the rate of change of J at the function $u(x)$ just as we start to move in the direction of the function $v(x)$. The directional derivative allows the computation of this rate using the ordinary derivative with respect to ε. For our example functional, we can compute the directional derivative as

$$DJ(u) \cdot v = \frac{d}{d\varepsilon} \left[\int_0^1 \left((u + \varepsilon v)^2 + 4\sin(u + \varepsilon v) \right) dx \right]_{\varepsilon=0}$$

$$= \left[\int_0^1 \frac{d}{d\varepsilon} \left((u + \varepsilon v)^2 + 4\sin(u + \varepsilon v) \right) dx \right]_{\varepsilon=0}$$

$$= \left[\int_0^1 \left(2(u + \varepsilon v)v + 4\cos(u + \varepsilon v)v \right) dx \right]_{\varepsilon=0}$$

$$= \int_0^1 \left(2uv + 4\cos(u)v \right) dx$$

Observe that the functions $u(x)$ and $v(x)$ are not affected by differentiation with respect to ε. This derivative is useful for continuous problems in the context of the principle of virtual work.

Appendix C
The Eigenvalue Problem

In structural dynamics, the *eigenvalue problem* shows up in many contexts, both discrete and continuous. As such, it is a fundamental building block for understanding the subject. This appendix provides a review of the mathematics of the *matrix eigenvalue problem* as well as a description of some approaches to solving eigenvalue problems, including the QR algorithm and *subspace iteration* for large eigenvalue problems where we do not need all eigenvalues and eigenvectors.

C.1 The Algebraic Eigenvalue Problem

There are two statements of the algebraic eigenvalue problem that come up in structural mechanics. The *standard eigenvalue problem* takes the form

$$\left[\mathbf{A} - \lambda \mathbf{I}\right]\boldsymbol{\varphi} = \mathbf{0} \tag{C.1}$$

or $\mathbf{A}\boldsymbol{\varphi} = \lambda\boldsymbol{\varphi}$, where $\mathbf{A}$ is an $N \times N$ matrix and $\mathbf{I}$ is the $N \times N$ *identity* matrix. We assume that the matrix $\mathbf{A}$ is known and that we seek λ and $\boldsymbol{\varphi}$ as the solution to the eigenvalue problem, Eq. C.1. We restrict our attention to *symmetric* eigenvalue problems (which means that the matrix $\mathbf{A}$ is a symmetric matrix).

The *generalized eigenvalue problem* has the form

$$\left[\mathbf{K} - \lambda \mathbf{M}\right]\boldsymbol{\varphi} = \mathbf{0} \tag{C.2}$$

or $\mathbf{K}\boldsymbol{\varphi} = \lambda\mathbf{M}\boldsymbol{\varphi}$, where $\mathbf{K}$ and $\mathbf{M}$ are symmetric $N \times N$ matrices. It is possible to convert the generalized eigenvalue problem to standard form by premultiplying the equation by $\mathbf{M}^{-1}$, but the matrix $\mathbf{M}^{-1}\mathbf{K}$ is not necessarily symmetric and there is a substantial difference between symmetric and unsymmetric eigenvalue problems. In structural dynamics, the mass matrix $\mathbf{M}$ is generally guaranteed to be symmetric

© The Author(s), under exclusive license to Springer Nature Switzerland AG 2022
K. D. Hjelmstad, *Fundamentals of Structural Dynamics*,
https://doi.org/10.1007/978-3-030-89944-8

and positive definite.[1] Therefore, it is possible to factorize it with the *Cholesky decomposition* as $\mathbf{M} = \mathbf{L}\mathbf{L}^T$, where $\mathbf{L}$ is a lower triangular matrix. Now, we can convert the problem to standard form by premultiplying Eq. C.2 by $\mathbf{L}^{-1}$ and letting $\hat{\boldsymbol{\varphi}} = \mathbf{L}^T\boldsymbol{\varphi}$. The final form is the standard eigenvalue problem

$$\left[\hat{\mathbf{K}} - \lambda\mathbf{I}\right]\hat{\boldsymbol{\varphi}} = \mathbf{0}$$

where $\hat{\mathbf{K}} = \mathbf{L}^{-1}\mathbf{K}\mathbf{L}^{-T}$ is guaranteed to be symmetric because it is constructed with a similarity transformation. We have noted that $\mathbf{L}^{-1}\mathbf{M}\mathbf{L}^{-T} = \mathbf{I}$. Thus, the generalized symmetric eigenvalue problem can always be converted to a symmetric standard eigenvalue problem. The following discussion focuses on the standard eigenvalue problem.

Characteristic Equation The condition for a solution to the eigenvalue problem other than the trivial solution $\boldsymbol{\varphi} = \mathbf{0}$ is that the determinant of the coefficient matrix must be zero, i.e.,

$$\det\left[\mathbf{A} - \lambda\mathbf{I}\right] = 0 \tag{C.3}$$

This equation is called the *characteristic equation* associated with the eigenvalue problem. It is an Nth order polynomial equation in λ. Notice that this equation does not involve $\boldsymbol{\varphi}$. The zeros of the polynomial are called the *eigenvalues* of $\mathbf{A}$. For small systems, solving the characteristic equation is an effective approach to finding the eigenvalues.

As an example, consider the matrix

$$\mathbf{A} = \begin{bmatrix} 5 & 2 & 1 & 0 & 0 \\ 2 & 6 & 3 & 1 & 0 \\ 1 & 3 & 7 & 2 & 1 \\ 0 & 1 & 2 & 8 & 2 \\ 0 & 0 & 1 & 2 & 9 \end{bmatrix} \tag{C.4}$$

The plot of $\det\left[\mathbf{A} - \lambda\mathbf{I}\right]$ versus λ is shown in Fig. C.1. The roots of the characteristic equation are shown with open circles in the plot and have the values

$$\lambda = 2.891, \quad 4.388, \quad 6.233, \quad 9.060, \quad 12.428$$

These roots can be found by any root-finding algorithm for a scalar equation of a scalar unknown (e.g., interval halving or *regula falsi*). The challenge of this approach is that these algorithms require the specification of bounds on the search. Without the plot, it is difficult to establish these bounds.

---

[1] A matrix $\mathbf{M}$ is *positive definite* if it has the property that $\mathbf{v}^T\mathbf{M}\mathbf{v} > 0$ for all $\mathbf{v}$.

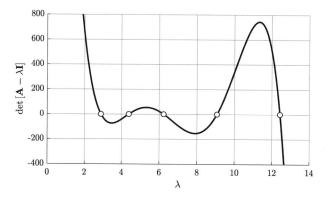

Fig. C.1 Plot of $\det\begin{bmatrix} \mathbf{A} - \lambda\mathbf{I} \end{bmatrix}$ versus λ. The eigenvalues are the roots of the characteristic equation

Once the eigenvalues λ_i for $i = 1, \ldots, N$ are known, the *eigenvectors* can be found by solving the homogeneous linear system of equations

$$\begin{bmatrix} \mathbf{A} - \lambda_i\mathbf{I} \end{bmatrix}\boldsymbol{\varphi}_i = \mathbf{0}, \qquad i = 1, \ldots, N \tag{C.5}$$

The coefficient matrix for each of these equations is singular (that is how we got the λ_i values). Therefore, solving these equations must assume some knowledge of at least one of the components of $\boldsymbol{\varphi}_i$ (more than one if the eigenvalue is repeated). Notice that we could multiply $\boldsymbol{\varphi}_i$ by any scalar (including negative values) and the resulting vector would still satisfy the eigenvalue problem. For that reason, the eigenvectors are often called *mode shapes* because they represent a shape whose magnitude is indeterminate. It is common to normalize the eigenvectors to unit length, i.e., $\boldsymbol{\varphi}_i^T \boldsymbol{\varphi}_i = 1$, in which case they are called *orthonormal*. The following example shows how the computation of the eigenvectors can be done.

Example Consider the matrix $\mathbf{A}$ in Eq. C.4, again. Find the eigenvalues by solving the characteristic equation. With each λ_i, find the associated eigenvector by solving a linear system of equations as follows. First, partition the problem as

$$\begin{bmatrix} \mathbf{A} - \lambda_i\mathbf{I} \end{bmatrix}\hat{\boldsymbol{\varphi}}_i = \begin{bmatrix} b_i & \mathbf{b}_i^T \\ \mathbf{b}_i & \mathbf{B}_i \end{bmatrix} \begin{Bmatrix} 1 \\ \mathbf{x}_i \end{Bmatrix} = \begin{Bmatrix} 0 \\ \mathbf{0} \end{Bmatrix}$$

where b_i and $\mathbf{b}_i$ make up the first column, b_i and $\mathbf{b}_i^T$ the first row, and $\mathbf{B}_i$ the lower right 4×4 submatrix of $[\mathbf{A} - \lambda_i\mathbf{I}]$. Solve for $\mathbf{x}_i$ using the last four equations, $\mathbf{b}_i + \mathbf{B}_i\mathbf{x}_i = \mathbf{0}$, and construct the eigenvector $\boldsymbol{\varphi}_i$ as

$$\mathbf{x}_i = -\mathbf{B}_i^{-1}\mathbf{b}_i, \quad \rightarrow \quad \hat{\boldsymbol{\varphi}}_i = \begin{Bmatrix} 1 \\ \mathbf{x}_i \end{Bmatrix} \quad \rightarrow \quad \boldsymbol{\varphi}_i = \frac{\hat{\boldsymbol{\varphi}}_i}{\|\hat{\boldsymbol{\varphi}}_i\|}$$

Repeat this computation for $i = 1, \ldots, 5$, as shown in Code C.1.

Code C.1 MATLAB code for computing the eigenvalues and eigenvectors of a matrix by solving the characteristic equation to find the eigenvalues and then solving a linear system of equations to find the eigenvectors

```
%   Solve eigenvalue problem with characteristic equation approach
    clear; clc;

%.  Input matrix
    A = [5, 2, 1, 0, 0;
         2, 6, 3, 1, 0;
         1, 3, 7, 2, 1;
         0, 1, 2, 8, 2;
         0, 0, 1, 2, 9];
    N = size(A,1);
    I = eye(N);

%.  Find eigenvalues by solving characteristic equation
    syms z
    f=@(z) det(A-z*I);
    lambda = double(vpa(solve(f(z)==0)));

%.  Find eigenvectors by solving linear system of equations
    PHI = zeros(N,N);
    for i=1:N
      B = A-lambda(i)*I;
      b = B(2:N,1);
      B = B(2:N,2:N);
      x = -B\b;
      phi = [1;x];
      phi = phi/norm(phi);
      PHI(:,i) = phi;
    end
```

This process assumes that the first component of the eigenvector is not zero. We assume a value of one for that element, solve the remaining four equations for x_i, and then normalize the vector to unit length. For this matrix, all five of the eigenvectors are amenable to this approach. If any eigenvector has a zero for that first component, then a different element of the eigenvector should be assumed as one with the appropriate change to the partitioning scheme. At least one of the components must be nonzero. If the element assumed as one is actually supposed to be zero, then the matrix $\mathbf{B}_i$ will be singular. If an eigenvalue is repeated, then additional values in the eigenvector must be prescribed, again with corresponding changes to the partitioning scheme. The eigenvalues are stored in the columns of the matrix PHI. The results for this example are

$$
\Phi = \begin{bmatrix}
0.4978 & 0.7010 & 0.3477 & 0.3289 & 0.1779 \\
-0.7445 & 0.0997 & 0.1980 & 0.4932 & 0.3916 \\
0.4389 & -0.6283 & 0.0329 & 0.3489 & 0.5383 \\
0.0023 & 0.3223 & -0.7478 & -0.1954 & 0.5466 \\
-0.0726 & -0.0035 & 0.5287 & -0.6990 & 0.4759
\end{bmatrix}
$$

To test the eigenvectors found, compute $\boldsymbol{\Phi}^T\boldsymbol{\Phi}$ and $\boldsymbol{\Phi}^T\mathbf{A}\boldsymbol{\Phi}$. Both are diagonal matrices, the first being the identity matrix and the second having the eigenvalues as the diagonal elements in order from λ_1 to λ_5.

Orthogonality of Eigenvectors The eigenvectors are orthogonal if the eigenvalues are distinct. The proof is straightforward. Take two different eigenpairs $(\lambda_n, \boldsymbol{\varphi}_n)$ and $(\lambda_m, \boldsymbol{\varphi}_m)$. Each pair satisfies Eq. C.5. Premultiply the equation with index n by $\boldsymbol{\varphi}_m^T$ and the equation with index m by $\boldsymbol{\varphi}_n^T$ and subtract the first equation from the second one to get

$$\boldsymbol{\varphi}_n^T\big[\mathbf{A} - \lambda_m\mathbf{I}\big]\boldsymbol{\varphi}_m - \boldsymbol{\varphi}_m^T\big[\mathbf{A} - \lambda_n\mathbf{I}\big]\boldsymbol{\varphi}_n = 0 \qquad\qquad (C.6)$$

The result equals zero because of Eq. C.5. The matrix $\mathbf{A}$ is symmetric, so

$$\boldsymbol{\varphi}_n^T\mathbf{A}\boldsymbol{\varphi}_m = \boldsymbol{\varphi}_m^T\mathbf{A}\boldsymbol{\varphi}_n$$

Of course, $\boldsymbol{\varphi}_n^T\boldsymbol{\varphi}_m = \boldsymbol{\varphi}_m^T\boldsymbol{\varphi}_n$, too. Therefore, Eq. C.6 reduces to

$$\big(\lambda_n - \lambda_m\big)\boldsymbol{\varphi}_n^T\boldsymbol{\varphi}_m = 0$$

which proves orthogonality, i.e., $\boldsymbol{\varphi}_n^T\boldsymbol{\varphi}_m = 0$, as long as the eigenvalues are not equal, i.e., $\lambda_n \neq \lambda_m$. If two or more eigenvalues *are* equal, then the associated eigenvectors form a subspace in which any linear combination of vectors satisfies the eigenvalue problem. Say, for example, that $\lambda_n = \lambda_m = \hat{\lambda}$ are repeated eigenvalues associated with the distinct eigenvectors $\boldsymbol{\varphi}_n$ and $\boldsymbol{\varphi}_m$. In other words, $\mathbf{A}\boldsymbol{\varphi}_n = \hat{\lambda}\boldsymbol{\varphi}_n$ and $\mathbf{A}\boldsymbol{\varphi}_m = \hat{\lambda}\boldsymbol{\varphi}_m$. Let $\hat{\boldsymbol{\varphi}} = a\boldsymbol{\varphi}_n + b\boldsymbol{\varphi}_m$ and compute

$$\begin{aligned}
\mathbf{A}\hat{\boldsymbol{\varphi}} &= \mathbf{A}\big(a\boldsymbol{\varphi}_n + b\boldsymbol{\varphi}_m\big) \\
&= a\mathbf{A}\boldsymbol{\varphi}_n + b\mathbf{A}\boldsymbol{\varphi}_m \\
&= a\hat{\lambda}\boldsymbol{\varphi}_n + b\hat{\lambda}\boldsymbol{\varphi}_m \\
&= \hat{\lambda}\big(a\boldsymbol{\varphi}_n + b\boldsymbol{\varphi}_m\big) = \hat{\lambda}\hat{\boldsymbol{\varphi}}
\end{aligned}$$

Thus, the vector $\hat{\boldsymbol{\varphi}}$, which is a linear combination of $\boldsymbol{\varphi}_n$ and $\boldsymbol{\varphi}_m$, satisfies the eigenvalue problem for any values of a and b. The vectors $\boldsymbol{\varphi}_n$ and $\boldsymbol{\varphi}_m$ might not be orthogonal, but it is always possible to select vectors from the subspace that are orthogonal (or two select two vectors and orthogonalize them).

It is also straightforward to prove that the eigenvectors are $\mathbf{A}$-orthogonal. Premultiply Eq. C.5 (with index m) by $\boldsymbol{\varphi}_n$ to get

$$\boldsymbol{\varphi}_n^T\big[\mathbf{A} - \lambda_m\mathbf{I}\big]\boldsymbol{\varphi}_m = 0$$

Since $\boldsymbol{\varphi}_n^T\boldsymbol{\varphi}_m = 0$ if $\lambda_n \neq \lambda_m$, it must also be true that $\boldsymbol{\varphi}_n^T\mathbf{A}\boldsymbol{\varphi}_m = 0$. Also, note that for orthonormal eigenvectors $\boldsymbol{\varphi}_n^T\mathbf{A}\boldsymbol{\varphi}_n = \lambda_n$.

The Spectral Representation The matrix $\mathbf{A}$ can be put into *spectral* form using the eigenvalues and eigenvectors. In particular, for orthonormal eigenvectors, we can write

$$\mathbf{A} = \sum_{n=1}^{N} \lambda_n \boldsymbol{\varphi}_n \boldsymbol{\varphi}_n^T$$

The proof is simple. The eigenvector $\boldsymbol{\varphi}_m$ should satisfy $\mathbf{A}\boldsymbol{\varphi}_m = \lambda_m \boldsymbol{\varphi}_m$. Use the spectral representation to compute

$$\mathbf{A}\boldsymbol{\varphi}_m = \sum_{n=1}^{N} \lambda_n \boldsymbol{\varphi}_n \boldsymbol{\varphi}_n^T \boldsymbol{\varphi}_m = \lambda_m \boldsymbol{\varphi}_m$$

Because the eigenvectors are orthogonal, only the mth term of the sum survives and $\boldsymbol{\varphi}_m^T \boldsymbol{\varphi}_m = 1$. The inverse of $\mathbf{A}$ can also be represented in spectral form. For orthonormal eigenvectors, we can write

$$\mathbf{A}^{-1} = \sum_{n=1}^{N} \frac{\boldsymbol{\varphi}_n \boldsymbol{\varphi}_n^T}{\lambda_n}$$

The proof is similar to the previous one. First, note that $\mathbf{A}^{-1}\mathbf{A} = \mathbf{I}$ and that $\mathbf{I}\boldsymbol{\varphi}_m = \boldsymbol{\varphi}_m$. Now, compute

$$[\mathbf{A}^{-1}\mathbf{A}]\boldsymbol{\varphi}_m = \sum_{n=1}^{N} \frac{\boldsymbol{\varphi}_n \boldsymbol{\varphi}_n^T \mathbf{A}\boldsymbol{\varphi}_m}{\lambda_n} = \frac{\boldsymbol{\varphi}_m \lambda_m}{\lambda_m} = \boldsymbol{\varphi}_m$$

The second equality is true because the eigenvectors are orthogonal and, since the eigenvectors have unit length, $\boldsymbol{\varphi}_m^T \mathbf{A}\boldsymbol{\varphi}_m = \lambda_m$.

C.2   The QR Algorithm

Another class of methods to solve the eigenvalue problem is based upon sequential transformations that are aimed at creating a diagonal matrix. The motivation for these methods is that the eigenvectors diagonalize the original matrix. If we have an $N \times N$ matrix $\boldsymbol{\Phi}$ whose kth column is the eigenvector $\boldsymbol{\varphi}_k$, then

$$\boldsymbol{\Phi}^T \mathbf{A} \boldsymbol{\Phi} = \begin{bmatrix} \lambda_1 & & \\ & \ddots & \\ & & \lambda_N \end{bmatrix} \tag{C.7}$$

if the eigenvectors are normalized such that $\boldsymbol{\varphi}_k^T \boldsymbol{\varphi}_k = 1$ for all k. Thus, if we can find a sequence of transformations, each one of which tries to drive the matrix $\mathbf{A}$ and its subsequent iterates toward a diagonal matrix, then that sequence should converge to the matrix $\boldsymbol{\Phi}$. There are several algorithms in this class. We will look at one of those, QR, as exemplifying this approach.

The QR algorithm operates by generating a sequence of orthogonal similarity transformations. Any matrix $\mathbf{A}$ can be decomposed into the product of two matrices

$$\mathbf{A} = \mathbf{QR}$$

where $\mathbf{Q}$ is an orthogonal matrix (i.e., $\mathbf{Q}^T \mathbf{Q} = \mathbf{I}$) and $\mathbf{R}$ is an upper triangular matrix. It is simple to show that the matrix $\mathbf{RQ}$ is a similarity transformation of $\mathbf{A}$. Since $\mathbf{Q}$ is orthogonal, $\mathbf{Q}^T = \mathbf{Q}^{-1}$, i.e., its transpose is equal to its inverse. Therefore, we can compute $\mathbf{R} = \mathbf{Q}^T \mathbf{A}$. Now, we can compute

$$\mathbf{A}' = \mathbf{RQ} = \mathbf{Q}^T \mathbf{AQ}$$

which is a similarity transformation. A similarity transformation preserves the eigenvalues of a matrix. Therefore, $\mathbf{A}'$ has the same eigenvalues as $\mathbf{A}$. If we repeat the operation, we can form the sequence

$$\mathbf{A}^{(k+1)} = \mathbf{R}^{(k)} \mathbf{Q}^{(k)} \tag{C.8}$$

where $\mathbf{Q}^{(k)} \mathbf{R}^{(k)}$ is the QR decomposition of the matrix $\mathbf{A}^{(k)}$. The matrix sequence $\mathbf{A}^{(k)}$ converges to a diagonal matrix for sufficiently large k. Since the similarity transformation preserves the eigenvalues, the diagonal values of $\mathbf{A}^{(k)}$ are the eigenvalues of the original matrix $\mathbf{A}$. Furthermore, we can write

$$\mathbf{A}^{(k+1)} = \mathbf{Q}^{(k)T} \cdots \mathbf{Q}^{(1)T} \mathbf{AQ}^{(1)} \cdots \mathbf{Q}^{(k)} \tag{C.9}$$

The product of orthogonal matrices is an orthogonal matrix. The proof is simple. Let $\mathbf{P} = \mathbf{Q}^{(1)} \mathbf{Q}^{(2)}$

$$\begin{aligned}
\mathbf{P}^T \mathbf{P} &= \left(\mathbf{Q}^{(1)} \mathbf{Q}^{(2)} \right)^T \left(\mathbf{Q}^{(1)} \mathbf{Q}^{(2)} \right) \\
&= \mathbf{Q}^{(2)T} \mathbf{Q}^{(1)T} \mathbf{Q}^{(1)} \mathbf{Q}^{(2)} \\
&= \mathbf{Q}^{(2)T} \left(\mathbf{I} \right) \mathbf{Q}^{(2)} = \mathbf{I}
\end{aligned} \tag{C.10}$$

From this observation we can conclude that the product

$$\boldsymbol{\Phi} = \mathbf{Q}^{(1)} \mathbf{Q}^{(2)} \cdots \mathbf{Q}^{(k)} \tag{C.11}$$

is orthogonal and, therefore, is a matrix whose columns are the eigenvectors of $\mathbf{A}$.

This approach represents a way to solve the eigenvalue problem distinctly different from the method of first finding the roots of characteristic equation. The QR algorithm is an iterative approach where the iteration is essentially on the eigenvectors. Throughout the process we never explicitly compute the eigenvalues.

The success of the QR algorithm depends upon an ability to compute $\mathbf{Q}$. One way to do it is with Gram–Schmidt orthogonalization. Let $\mathbf{a}_k$ be the kth column of the matrix $\mathbf{A}$. Take $\mathbf{q}_1 = \mathbf{a}_1$. Now compute the second vector by subtracting the part of $\mathbf{a}_2$ that lies along $\mathbf{q}_1$. That can be accomplished as

$$\mathbf{q}_2 = \mathbf{a}_2 - \left(\frac{\mathbf{a}_2 \cdot \mathbf{q}_1}{\mathbf{q}_1 \cdot \mathbf{q}_1}\right)\mathbf{q}_1 \tag{C.12}$$

It is easy to verify that $\mathbf{q}_2 \cdot \mathbf{q}_1 = 0$ (just dot the above equation with $\mathbf{q}_1$). We can repeat the process for the third vector as

$$\mathbf{q}_3 = \mathbf{a}_3 - \left(\frac{\mathbf{a}_3 \cdot \mathbf{q}_1}{\mathbf{q}_1 \cdot \mathbf{q}_1}\right)\mathbf{q}_1 - \left(\frac{\mathbf{a}_3 \cdot \mathbf{q}_2}{\mathbf{q}_2 \cdot \mathbf{q}_2}\right)\mathbf{q}_2 \tag{C.13}$$

This vector is orthogonal to both $\mathbf{q}_1$ and $\mathbf{q}_2$. The general case is captured by the equation

$$\mathbf{q}_i = \mathbf{a}_i - \sum_{j=1}^{i-1}\left(\frac{\mathbf{a}_i \cdot \mathbf{q}_j}{\mathbf{q}_j \cdot \mathbf{q}_j}\right)\mathbf{q}_j, \qquad i = 2, \ldots, N \tag{C.14}$$

which we start by seeding $\mathbf{q}_1 = \mathbf{a}_1$. It is important to normalize the orthogonal vectors at each step to prevent them from growing in magnitude. If we normalize the vectors so that $\mathbf{q}_i \cdot \mathbf{q}_i = 1$, that also simplifies the denominator of each of the terms in the Gram–Schmidt process.

It is evident from Eq. C.14 why $\mathbf{R}$ is upper triangular. If you move the sum to the left side, you get a representation of the ith column of $\mathbf{A}$. Since it is equal to $\mathbf{QR}$, you can see that the ith column of $\mathbf{R}$ provides the coefficients that determine the linear combination of columns of $\mathbf{Q}$ that make up $\mathbf{a}_i$. From Gram–Schmidt, that only includes vectors $\mathbf{q}_1$ through $\mathbf{q}_i$. The vectors $\mathbf{q}_{i+1}$ through $\mathbf{q}_N$ are not included and hence correspond to zero elements in those columns of $\mathbf{R}$.

Implementation A MATLAB code that executes the QR algorithm is shown in Code C.2. The input matrix $\mathbf{A}$ is included as a specific example, but the remainder of the program is general. The purpose of presenting this code is to show how to compute with this algorithm.

Code C.2 MATLAB code for computing the eigenvalues and eigenvectors of a matrix by the QR algorithm. This code solves the example problem

```
%. QR algorithm
  clear; clc;
  N = 5; err = 1; tol = 1.e-8; nSteps = 100; k = 0;
```

```
%. Input matrix
   A = [5, 2, 1, 0, 0;
        2, 6, 3, 1, 0;
        1, 3, 7, 2, 1;
        0, 1, 2, 8, 2;
        0, 0, 1, 2, 9];

%. Initialize the eigenvector matrix
   PHI = eye(N);

   while (err>tol) && (k<nSteps)
     k = k + 1;

%... Compute Q by Gram-Schmidt
     Q = A;
     for i=1:N
       for j=1:i-1
         Q(:,i) = Q(:,i) - Q(:,j)*dot(Q(:,i),Q(:,j));
       end
       Q(:,i) = Q(:,i)/norm(Q(:,i));
     end

%... Compute R and update A
     R = Q'*A;
     A = R*Q;

%... Accumulate the product of orthogonal matrices
     PHI = PHI*Q;

%... Check for convergence
     E = A - diag(diag(A));
     D = norm(diag(A));
     err = norm(E,'fro')/D;
     fprintf(' It = %5i   err = %12.5e\n',k,err)
   end
```

There are a few things to notice about this code. We execute the iteration using a *while* loop and we compute the *Frobenius norm*, i.e., the square root of the sum of the squares, of the off-diagonal elements of $\mathbf{A}^{(i)}$. When that sum divided by the norm of the diagonals is less than `tol`, then the matrix is sufficiently diagonal. The tighter the tolerance, the closer the matrix is to diagonal, and the more iterations it takes to get there. For the example problem it takes 57 iterations to get that sum to be less than 10^{-8}.

We include a fail-safe on the loop by counting the iterations and terminating the process if that number exceeds the preset value `nSteps`. Running the code verifies that we get the same eigenvalues as we did by finding the roots of the characteristic equation.

The QR algorithm is an approach to finding the eigenvalues and eigenvectors of a full matrix. There are several variations on the original theme, aimed at speeding up the convergence of the QR algorithm. For example, QR works really well on tightly banded matrices so one enhancement is to couple QR with an algorithm that first reduces $\mathbf{A}$ to tridiagonal form. We present only the basic version of the algorithm here because MATLAB has a highly optimized built-in function $[V, D] = \text{eig}(A)$ that computes all eigenvalues D and eigenvectors V of a matrix $\mathbf{A}$. It is also the case that MATLAB can solve the generalized eigenvalue problem with $[V, D] = \text{eig}(K, M)$.

C.3   Eigenvalue Problems for Large Systems

Solving eigenvalue problems is computationally intense and the computational time required grows quickly with the size of the matrices involved. For large structural systems it may be prohibitive to compute the complete set of eigenvectors and eigenvalues. One can argue that most of the solution can be captured with a basis comprising a subset of the eigenvectors associated with the lowest eigenvalues. This is especially relevant for computing the response to earthquake excitation (see Sect. 7.1 for an application of this idea to the shear building). In this section we consider an algorithm known as *subspace iteration* that gives a route to computing some of the eigenvalues and eigenvectors.

Ritz Vectors We would like to find a subspace for carrying out the dynamics computations as

$$\mathbf{u}(t) = \sum_{i=1}^{n} \boldsymbol{\psi}_i a_i(t)$$

where n is much smaller than the dimension of $\mathbf{u}(t)$. The vectors $\boldsymbol{\psi}_i$ are called *Ritz vectors*. Let us first consider how we might create a set of Ritz vectors. Inspired by the earthquake problem for the shear building, we might select our first Ritz vector as

$$\hat{\boldsymbol{\psi}}_1 = \mathbf{K}^{-1}\mathbf{M1} \quad \rightarrow \quad \boldsymbol{\psi}_1 = \frac{\hat{\boldsymbol{\psi}}_1}{\|\hat{\boldsymbol{\psi}}_1\|} \tag{C.15}$$

where

$$\|\mathbf{v}\| = \sqrt{\mathbf{v}^T \mathbf{M} \mathbf{v}}$$

is the length of $\mathbf{v}$ relative to the metric $\mathbf{M}$. What we are really doing here is recognizing that $\mathbf{M1}$ is the pattern of loads associated with the earthquake acceleration and we are simply taking the first Ritz vector to be the static displacement under the load $\mathbf{M1}$. This static displacement usually imitates the first mode shape fairly well, which is why it is a good starting point for generating the subspace. For the truss and 2D frame we replaced $\mathbf{M1}$ with $\mathbf{Q}$ (not the same as the $\mathbf{Q}$ in the QR algorithm). We can generate a sequence of Ritz vectors from the first one with the recursion

$$\hat{\boldsymbol{\psi}}_{i+1} = \mathbf{K}^{-1}\mathbf{M}\boldsymbol{\psi}_i \quad \rightarrow \quad \boldsymbol{\psi}_{i+1} = \frac{\hat{\boldsymbol{\psi}}_{i+1}}{\|\hat{\boldsymbol{\psi}}_{i+1}\|} \tag{C.16}$$

One can prove that these vectors are linearly independent and, therefore, span a subspace of dimension n. The vectors are normalized to have unit modal mass. Otherwise, they will grow in magnitude with each iteration. The problem with these

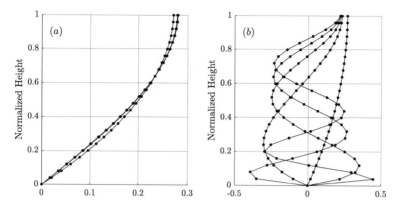

Fig. C.2 *Example of Ritz vectors.* Structure is 25-story shear building with $k = 10$ and $m = 1$ in all levels. Plot of 5 Ritz vectors: **(a)** plain Ritz vectors, **(b)** orthogonalized Ritz vectors

vectors is that they very quickly start to look like each other. So, while they are linearly independent, they are ill-conditioned as a basis. The case shown in Fig. C.2 is a 25 DOF shear building with 5 Ritz vectors computed. It is difficult to visually discern the difference between them.

The plain Ritz vectors are poorly conditioned, but we can produce a better version by making each new vector orthogonal to all of the previous ones. We can use the Gram–Schmidt orthogonalization process similar to the one described for the QR algorithm. Assume that we have generated $\psi_1, \psi_2, \ldots, \psi_{n-1}$. Compute

$$\tilde{\psi}_n = \mathbf{K}^{-1}\mathbf{M}\psi_{n-1} \tag{C.17}$$

Subtract off bits of our existing vectors to get

$$\hat{\psi}_n = \tilde{\psi}_n - \sum_{i=1}^{n-1} a_i \psi_i \tag{C.18}$$

To compute the constants a_i we will force the new vector to be orthogonal to all of the existing vectors

$$\hat{\psi}_n^T \mathbf{M} \psi_j = 0, \quad j = 1, \ldots, n - 1 \tag{C.19}$$

Now substituting Eq. C.18, we have

$$\left(\tilde{\psi}_n - \sum_{i=1}^{n-1} a_i \psi_i \right)^T \mathbf{M} \psi_j = 0 \tag{C.20}$$

Noting the orthogonality of the previously orthogonalized vectors, we can solve for the constants

$$a_j = \frac{\tilde{\psi}_n^T \mathbf{M} \psi_j}{\psi_j^T \mathbf{M} \psi_j} \tag{C.21}$$

With these coefficients determined, we can complete the construction of the new vector and, once constructed, we can normalize it to unit length

$$\psi_n = \frac{\hat{\psi}_n}{\|\hat{\psi}_n\|} \tag{C.22}$$

Note that if we normalize the vectors, the denominator of Eq. C.21 is one (and therefore does not need to be computed again). This process allows us to generate a set of independent Ritz vectors that are orthogonal with respect to the mass matrix.

Figure C.2b shows the orthogonalized Ritz vectors for the 25-story shear building. Notice how these vectors look very different from each other. In fact, they are as different as possible (i.e., they are orthogonal). Contrast these vectors with the plain Ritz vectors. The orthogonal Ritz vectors provide a well-conditioned basis for representing the solution to the structural dynamics problem.

C.4   Subspace Iteration

The orthogonal Ritz vectors are nice, but they are not eigenvectors. We have a small subspace of vectors. Can we use these vectors to create a small subspace of eigenvectors? The answer is yes. The process is called *subspace iteration*. We start the process with an $N \times n$ matrix $\mathbf{\Psi}^{(0)}$ whose kth column is ψ_k, which is an $N \times 1$ vector that could, for example, be one of the orthogonal Ritz vectors mentioned in the previous section.

Let us assume that we have run this algorithm n times and we now have the matrix $\mathbf{\Psi}^{(n)}$. First, we will improve the Ritz vectors by computing

$$\tilde{\mathbf{\Psi}} = \mathbf{K}^{-1} \mathbf{M} \mathbf{\Psi}^{(n)} \tag{C.23}$$

Now, project the mass and stiffness matrices as

$$\tilde{\mathbf{M}} = \tilde{\mathbf{\Psi}}^T \mathbf{M} \tilde{\mathbf{\Psi}}, \qquad \tilde{\mathbf{K}} = \tilde{\mathbf{\Psi}}^T \mathbf{K} \tilde{\mathbf{\Psi}} \tag{C.24}$$

The reduced mass and stiffness matrices are $n \times n$ in size. Next, find the eigenvectors of the reduced matrices

$$\tilde{\mathbf{K}} \mathbf{\Phi}^{(n)} = \tilde{\mathbf{M}} \mathbf{\Phi}^{(n)} \mathbf{\Lambda}^{(n)} \tag{C.25}$$

If $n \ll N$, this will be a small $n \times n$ eigenvalue problem, which could be solved by the QR algorithm. Finally, improve the subspace vectors as

$$\mathbf{\Psi}^{(n+1)} = \tilde{\mathbf{\Psi}} \mathbf{\Phi}^{(n)} \tag{C.26}$$

and normalize all new vectors to unit length with respect to $\mathbf{M}$.

After a sufficient number of iterations, the vectors $\mathbf{\Psi}^{(n)}$ converge to the lowest n eigenvectors associated with $\mathbf{K}$ and $\mathbf{M}$. Also, the values of $\mathbf{\Lambda}^{(n)}$ converge to the actual eigenvalues associated with those vectors. The number of refinement iterations required depends upon the quality of the initial vectors in $\mathbf{\Psi}^{(0)}$. For example, if we start with the orthogonalized Ritz vectors, it will take fewer iterations than if we start with the plain Ritz vectors. You can show that it is possible to start with random vectors and still converge to the actual eigenvalues and eigenvectors. Be aware that it is possible to miss vectors if the initial vectors are orthogonal to one of the real eigenvectors. However, executing additional iterations will usually resolve the problem because computer roundoff eventually breaks down that initial orthogonality.

Since the actual eigenvectors are so easily obtained from the Ritz vectors, it is reasonable to simply do the subspace iteration refinements to get to the actual eigenvectors. The computational cost is not that much.

Code C.3 contains the algorithm for: (1) generating plain Ritz vectors starting with $\psi_1 = \mathbf{K}^{-1}\mathbf{M1}$, (2) generating orthogonal Ritz vectors starting with the plain Ritz vectors and executing Gram–Schmidt orthogonalization, and (3) refining initial vectors by subspace iteration. For the subspace option we can start from: (a) the plain Ritz vectors, (b) the orthogonal Ritz vectors, or (c) random vectors. This code can be used to study the performance of subspace iteration.

Code C.3 MATLAB code to generate plain Ritz vectors, orthogonal Ritz vectors, and enhanced vectors by subspace iteration

```
%  Subspace iteration
   clear; clc;

%. Establish problem and subspace size and number of refinements
   nDOF = 25;              % Number of DOF in system
   nRitz = 5;              % Number of Ritz Vectors
   nRefine = 5;            % Number of subspace refinements

%. Establish which vectors to initialize subspace iteration
   SubStart = 2;           % =1 Orth. Ritz, 2 = Plain Ritz, 3 Random

%. Limit number of Ritz vectors to nDOF
   nRitz = min(nRitz,nDOF);

%. Initialize arrays
   One = ones(nDOF,1);        % Vector of ones
   psi = zeros(nDOF,nRitz);   % Plain Ritz
   phi = zeros(nDOF,nRitz);   % Orthogonal Ritz

%. Form mass and stiffness matrices
   [M,K] = FormMatrices(nDOF);

%. Compute random vectors
   Rvecs = ones(nDOF,nRitz)-2.*rand(nDOF,nRitz);
```

```
%. Compute Plain Ritz (no orthogonalization)
   psi(:,1) = K\M*One;
   length = sqrt(dot(psi(:,1), M*psi(:,1)));
   psi(:,1) = psi(:,1)/length;
   for i=2:nRitz
     psi(:,i) = K\M*psi(:,i-1);
     length = sqrt(dot(psi(:,i),M*psi(:,i)));
     psi(:,i) = psi(:,i)/length;
   end

%. Compute Orthogonal Ritz Vectors by Gram-Schmidt
   phi(:,1) = K\M*One;
   length = sqrt(dot(phi(:,1),M*phi(:,1)));
   phi(:,1) = phi(:,1)/length;
   for i=2:nRitz
     phi(:,i) = K\M*phi(:,i-1);
     for j=1:i-1
       phi(:,i) = phi(:,i) - phi(:,j)*dot(phi(:,i), M*phi(:,j));
     end
     length = sqrt(dot(phi(:,i),M*phi(:,i)));
     phi(:,i) = phi(:,i)/length;
   end

%. Sequentially improve vectors by subspace iteration
   if SubStart==1; Init = phi; end
   if SubStart==2; Init = psi; end
   if SubStart==3; Init = Rvecs; end

   qq = Init;
   for j=1:nRefine
     q = K\M*qq;
     KK = q'*K*q;
     MM = q'*M*q;
     [VV,DD] = eig(KK,MM);
     qq = q*VV;
     for i=1:nRitz
       length = sqrt(dot(qq(:,i),M*qq(:,i)));
       qq(:,i) = qq(:,i)/length;
     end
   end

%. Sort eigenvectors in order of increasing eigenvalues
   [d,ind] = sort(diag(DD));
   q = qq(:,ind);

%. Produce graphicl output
   SubspacePlots(K,M,phi,psi,q,Init,nDOF,nRitz)
```

The function FormMatrices creates the mass and stiffness matrices **M** and **K**. The model used in this function for the subsequent study is the shear building, so the mass matrix is diagonal, and the stiffness matrix is tridiagonal. Any code that generates these matrices can be substituted. The function SubspacePlots simply generates the graphics to compare the results.

Appendix D
Finite Element Interpolation

For the finite element approach to setting up Ritz functions, there are numerous possibilities. For theories that have only first derivatives in the virtual work functionals, Ritz functions having C^0 continuity are not only admissible, but also the best choice from a physical standpoint. Additional smoothness prevents modeling behavior like an abrupt jump in the interpolated force fields. For theories that have second derivatives in the virtual work functional, C^1 continuity is required. The lowest order polynomials that are C^0 continuous are the piecewise linear functions. The lowest order polynomials for C^1 continuity are the cubic Hermitian functions. The lowest order interpolation is often inadequate to capture the derivatives of the interpolated fields. Higher order interpolation within the framework of C^0 continuity helps.

This appendix gives a brief outline of two approaches, one called *Lagrangian interpolation* and the other we call *bubble functions*. The former is more common in the finite element literature, but we will adopt the latter in the codes in this book because they are simpler to implement. There is no theoretical difference between the two approaches.

D.1 Polynomial Interpolation

Interpolation is the process of finding a continuous approximation of a discrete set of $n+1$ data points $\{f_0, \ldots, f_n\}$ defined at locations $\{x_0, \ldots, x_n\}$ using a certain set of base functions such that the approximation exactly matches the given data at the discrete points. In essence, interpolation aims to model the behavior of the process underlying the data to give an approximation of what happens between the data points.

One common set of base functions are the monomials $\{1, x, x^2, x^3, \ldots\}$. A polynomial of degree n can be built from the first $n+1$ terms of this monomial

© The Author(s), under exclusive license to Springer Nature Switzerland AG 2022
K. D. Hjelmstad, *Fundamentals of Structural Dynamics*,
https://doi.org/10.1007/978-3-030-89944-8

basis, and has the form

$$p_n(x) = a_0 + a_1 x + a_2 x^2 + \cdots + a_n x^n \tag{D.1}$$

Note that there are $n+1$ coefficients $\{a_0, \ldots, a_n\}$ that define the polynomial. We can determine those coefficients if we know the value of the function at $n+1$ points $p(x_i) = f_i$ for $i = 0, \ldots, n$. To find the coefficients we solve the system of equations

$$\begin{bmatrix} 1 & x_0 & \cdots & x_0^n \\ 1 & x_1 & \cdots & x_1^n \\ \vdots & \vdots & & \vdots \\ 1 & x_n & \cdots & x_n^n \end{bmatrix} \begin{Bmatrix} a_0 \\ a_1 \\ \vdots \\ a_n \end{Bmatrix} = \begin{Bmatrix} f_0 \\ f_1 \\ \vdots \\ f_n \end{Bmatrix} \tag{D.2}$$

The points do not need to be equally spaced.

Example As a specific example consider finding a quadratic function that passes through the values $\mathbf{f} = \{6, 8, 4\}$ at locations $\mathbf{x} = \{0, 1, 2\}$. The general quadratic function can be expressed as

$$q(x) = a_0 + a_1 x + a_2 x^2$$

The order of the polynomial is $n = 2$ and there are 3 constants a_0, a_1, and a_2 that must be specified to define the function. The system of equations that determine the coefficients, from Eq. D.2, is

$$\begin{bmatrix} 1 & 0 & 0 \\ 1 & 1 & 1 \\ 1 & 2 & 4 \end{bmatrix} \begin{Bmatrix} a_0 \\ a_1 \\ a_2 \end{Bmatrix} = \begin{Bmatrix} 6 \\ 8 \\ 4 \end{Bmatrix}$$

The solution of this system of equations is $a_0 = 6$, $a_1 = 5$, and $a_2 = -3$. So,

$$q(x) = 6 + 5x - 3x^2$$

It is easy to verify that the function passes through the three given values.

One of the downsides to this approach is that the coefficient matrix gets increasingly ill-conditioned as the number of terms increases, as shown in Fig. D.1. The coefficient matrices $\mathbf{A}_n$ (what we are calling the coefficient matrix in Eq. D.2) are computed assuming that the locations where the data are specified are $\mathbf{x} = \{0, 1, 2, \ldots, n-1\}$. Notice that we are plotting the log of the condition number, i.e., the ratio of the largest and smallest eigenvalues of the matrix.

There are other well-known pathologies of equally spaced interpolation using high-order polynomials. So, while this basic formulation of the interpolation problem is instructive, we need a more robust and efficient approach.

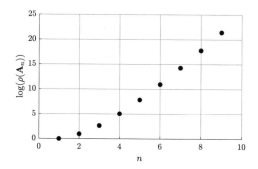

Fig. D.1 *Condition number for interpolation.* The coefficient matrix for interpolation in Eq. D.2 gets increasingly ill-conditioned as the order of interpolation n increases. The condition number $\rho(\mathbf{A})$ is the ratio of the maximum eigenvalue of $\mathbf{A}$ to the minimum eigenvalue, and is a measure of how well the solution of the system of equations will go numerically

D.2 Lagrangian Interpolation

Any nth order polynomial can be expressed in factored form as

$$p_n(x) = C(x - r_0)(x - r_1) \cdots (x - r_n) \tag{D.3}$$

where C is a constant and $\{r_0, r_1, \ldots, r_n\}$ are the roots (or zeros) of the polynomial (which can be repeated). This observation implies that a polynomial of degree n can be expressed as a product of $n+1$ linear functions $L_i(x) = x - r_i$. The factored form affords the opportunity to control where the polynomial is zero.

We can create a set of nth-order polynomials $\{\varphi_1(x), \ldots, \varphi_n(x)\}$ that have the property

$$\varphi_j(x_i) = \begin{Bmatrix} 1 & i = j \\ 0 & i \neq j \end{Bmatrix}$$

With these functions, the interpolation can be expressed as

$$p_n(x) = f_1\varphi_1(x) + f_2\varphi_2(x) + \cdots + f_n\varphi_n(x) \tag{D.4}$$

Each function $\varphi_i(x)$ has $n-1$ zeros. If you substitute $x = x_i$ into the equation, it is evident that $p_n(x_i) = f_i$ because $\varphi_i(x_i) = 1$ and $\varphi_j(x_i) = 0$ when $j \neq i$. This makes the problem of interpolation almost trivial because the data f_i that we want to interpolate shows up explicitly in the interpolation equation.

Lagrange realized that the functions

$$\varphi_i(x) = \prod_{j=1, j \neq i}^{n} \left(\frac{x - x_j}{x_i - x_j} \right), \quad i = 1, \ldots, n \tag{D.5}$$

are precisely the functions we seek. There are n such functions, where n is the number of nodes in the region. Note that the Π symbol means the product of the terms over the range (similar to the sum notation). The product excludes the term where $i=j$, so there are $n-1$ terms in the product. It is easy to see that when $x=x_i$, the value of the function is one. At the other nodes (i.e., where $x=x_j$) the value is zero.

Example (Revisited) Let us consider, again, the problem of finding a quadratic function that has values $\mathbf{f}=\{6, 8, 4\}$ at the locations $\mathbf{x}=\{0, 1, 2\}$. This time we will use Lagrangian interpolation to complete the task. The three Lagrangian functions are

$$\varphi_1(x) = \left(\frac{x-1}{0-1}\right)\left(\frac{x-2}{0-2}\right) = \tfrac{1}{2}(x-1)(x-2)$$

$$\varphi_2(x) = \left(\frac{x-0}{1-0}\right)\left(\frac{x-2}{1-2}\right) = -x(x-2)$$

$$\varphi_3(x) = \left(\frac{x-0}{2-0}\right)\left(\frac{x-1}{2-1}\right) = \tfrac{1}{2}x(x-1)$$

To interpolate the values $\mathbf{f}=\{6, 8, 4\}$ with these functions, we compute the interpolation as

$$q(x) = f_1\,\varphi_1(x) + f_2\,\varphi_2(x) + f_3\,\varphi_3(x)$$

$$= 6\left(\tfrac{1}{2}(x-1)(x-2)\right) + 8\left(-x(x-2)\right) + 4\left(\tfrac{1}{2}x(x-1)\right)$$

Distributing the products, collecting coefficients of like polynomial terms, and simplifying this expression gives $q(x)=6+5x-3x^2$, as before. Observe that this approach did not involve the solution of a system of equations. Hence, there is no possibility that ill-conditioning can creep into this process.

D.3   Ritz Functions with hp Interpolation

One of the merits of the Lagrangian representation of the interpolation problem is that it fits well with the framework for the Ritz approximation of the virtual work functionals in our various applications. In other words, the interpolation is represented as the product of constants and known functions. While we have framed the interpolation problem so far as finding the polynomial given the values f_i at certain points x_i, in the Ritz method the goal is to find the values of the constants that comprise the interpolation from a condition like the vanishing of the virtual work functional. In essence, we interpolate before we know the values of f_i and use the functions $\varphi_i(x)$ to build the machinery needed to compute the coefficients.

In a Ritz approximation, we assume that the unknown function, say $u(x)$ can be interpolated as

$$u(x) = \sum_{i=1}^{N} h_i(x)\, a_i$$

where the base functions $h_i(x)$ are the known Ritz functions and the constants a_i are to be determined from the application of the principle of virtual work. The key to the Ritz method is the selection of the base functions. In the case of a theory for which the highest derivative in the virtual work functional is a first derivative (e.g., the axial bar and the Timoshenko beam), only C^0 continuity is required. The "hat" functions are the lowest order C^0 functions. They ramp up from zero at the left end to one at the right end of one element and from one at the left end to zero at the right end of the next element and are zero everywhere outside of those two adjacent elements.

The problem with this lowest-order interpolation is that the first derivatives are piecewise constant. Generally, the element strains and stress resultants are proportional to the first derivative of the function. As such, the representation of those functions is poor. The approximation can be enhanced by including higher-order Ritz functions that do not change the C^0 nature of the approximation. This strategy is commonly known as the hp version of the finite element method. It allows improvement of accuracy both through mesh refinement (i.e., taking more elements) *and* through using a higher-order interpolation. Mesh refinement is the h-part and the higher-order interpolation is the p-part of the hp approach.

To set the context for the hp finite element strategy for constructing Ritz functions, consider the situation shown in Fig. D.2. We assume that the region $[\,0, L\,]$ is divided up into segments (called *elements*). The points where two elements join together are called *nodes*. While the region can be divided arbitrarily, we will assume that it is divided up into equal segments of length $L_e = L/M$, where M is the number of elements.

Figure D.2 shows Ritz functions built from cubic Lagrangian polynomials. The black dots are the two intermediate points required to define cubic Lagrangian functions. Observe that φ_1 and φ_2 have unit values at the ends of the element, but that φ_3 and φ_4 are zero at the ends. We will refer to the shape functions that are zero at both ends as *nodeless* shape functions. Because these nodeless functions are zero at the ends, they join with the zero function in adjacent elements. In the finite element approach, the Ritz functions are zero for all elements outside of those shown in the figure.

It is easy to see how the functions are only C^0 continuous at the nodes as each function has a kink at that point (meaning that the first derivative is discontinuous there). No matter how high the order of the functions is between the nodes, the continuity of the Ritz functions overall is limited by the kinks at the nodes. Because the Ritz functions must be continuous, they must be built from appropriate shape functions within neighboring elements. For example, the Ritz function $h_n(x)$ is

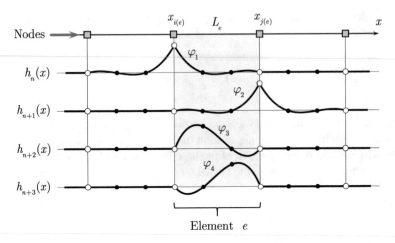

Fig. D.2 *Ritz functions and finite elements.* The region is divided into *elements* joined at *nodes.* Element *e* is shown along with the Ritz functions associated with it. In this case the Ritz functions are built from cubic Lagrangian polynomials. The black dots are the intermediate points that define the Lagrangian functions

made from φ_1 in element e, but joins up with φ_2 from element $e-1$. Similarly, Ritz function h_{n+1} is made from φ_2 in element e, but joins up with φ_1 in element $e+1$.

There is no limit to the order of interpolation in the domain of the element. We will look at two basic ways of specifying the nodeless functions and how those can be implemented in MATLAB codes. The first approach, Lagrangian functions, connects with our previous discussion on interpolation. The second approach, bubble functions, is actually simpler to implement. Both approaches give identical results when executed at the same order of interpolation.

D.4   Lagrangian Shape Functions

In applications it is convenient to scale the region of the element to $[\,0, 1\,]$ through a change of variable. To wit, let

$$\xi = \frac{x - x_{i(e)}}{L_e} \tag{D.6}$$

where $x_{i(e)}$ is the location of the leftmost node associated with element e and $L_e = x_{j(e)} - x_{i(e)}$ is the actual distance between the nodes. With this adjustment, the Lagrangian shape functions take the slightly simplified form

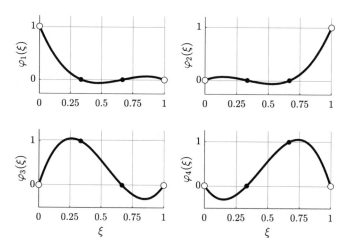

Fig. D.3 C^0 *finite element shape functions.* Lagrangian shape functions for a cubic element. Note that $\varphi_i(\xi)$ has unit value at node i and is zero at the other nodes. The end nodes are shown as open circles and the interior nodes are shown as black dots

$$\hat{\varphi}_i(\xi) = \prod_{\substack{j \neq i}}^{n} \frac{(n-1)\xi - j + 1}{i - j}, \quad i = 1, \dots, n \tag{D.7}$$

where we have altered the indices so that they start with 1 rather than 0 and are associated with points numbered from left to right. The actual Lagrangian shape functions $\varphi_i(\xi)$ are simply a reordered numbering of $\hat{\varphi}_i(\xi)$ so that the two functions that are nonzero at the ends are numbered first and the nodeless functions are numbered after that. Observe that $(n-1)\xi$ has integer values at the equally spaced Lagrangian points. For $n = 4$, the Lagrangian points would be at the one-third points and the ends. The Lagrangian shape functions for a cubic element are shown in Fig. D.3.

When applying the Ritz approximation in the principle of virtual work, the formulation will require the first derivatives of the shape functions. Using the product rule for differentiation, we have

$$\hat{\varphi}_i'(\xi) = \sum_{\substack{k \neq i}}^{n} \left(\prod_{\substack{j \neq i,k}}^{n} \frac{(n-1)\xi - j + 1}{i - j} \right) \frac{n-1}{i - k} \tag{D.8}$$

Of course, when you take the derivative of $\varphi_i(\xi)$ with respect to x, the chain rule applies. The result is simply multiplication of the result from Eq. D.8 by $1/L_e$. Now, when we integrate over an element, we can write it as

$$\int_{x_{i(e)}}^{x_{j(e)}} f(x)\, dx = \int_0^1 f(\xi) L_e\, d\xi$$

Equations D.7 and D.8 can be coded in general form to produce finite element shape functions of any order. Having this general function will allow us to explore the performance of our finite element formulations (e.g., of the Timoshenko beam), where it is not exactly clear what the best interpolation and mesh refinement strategy is for the dynamic problem. The MATLAB code that computes the Lagrangian shape functions and their first derivatives is given in Code D.1.

Code D.1 MATLAB code to evaluate C^0 Lagrangian shape functions and their derivatives for a Lagrangian function with N points at location x

```
function [shp,dshp] = LagrangianFcns(x,N)
% C0 Lagrangian shape function and derivative for N nodes per element.
%
%       x : location in range [0,1] to evaluate shape function
%       N : number of Lag. points per element (e.g., cubic has N=4)

%. Convert to range [0,N-1]
   eta = x*(N-1);

%. Initialize shape function and derivative arrays
   shp = zeros(N,1);                     % shape functions at xi
   dshp = zeros(N,1);                    % shape function derivative at xi

%. Set location of nodes and linear factors
   ii = 1:N;                             % eta+1 values at nodes
   lin = eta-ii+1;                       % All linear factors

%. Loop over nodes in the element, create shp and dshp for each one
   for i=1:N
     jj = ii; jj(i)=0;
     j = jj(jj>0);                       % Indices minus ith node
     mag = prod(ii(i)-ii(j));            % Denominator of shape function
     shp(i) = prod(lin(j))/mag;          % Shape function for node i

%... Compute the derivative of shape function by product rule
     dshpn = 0;
     for k=1:N-1
       kk = j; kk(k)=0;
       ik = kk(kk>0);                    % Indices minus ith & kth node
       dshpn = dshpn + prod(lin(ik));    % Derivative of shape function
     end
     dshp(i) = (N-1)*dshpn/mag;          % Dshape function for node i
   end

%. Reorder functions so that 1 and 2 are at ends
   a = shp(N);    shp(3:N) = shp(2:N-1);    shp(2)  = a;
   a = dshp(N);  dshp(3:N) = dshp(2:N-1);  dshp(2) = a;

end
```

The function LagrangianFcns takes as input the location x, where you want to evaluate the shape function and its derivative, and the number of Lagrangian points in the element N (which is one more than the order of interpolation). The result is the two $N \times 1$ arrays shp (the shape functions) and dshp (the derivatives of the shape functions), evaluated at the location x. We set up the function to evaluate only one point because we will typically be calling this function during a numerical integration loop where we only want the value of the functions at the location of the current integration station. Consequently, to produce the plots in Fig. D.3, we wrote a code that generated the values at numerous points along the ξ axis, evaluating the function values at those points one by one.

There are a couple of features that we introduce in this code to bridge between the theoretical formulation presented in this section and the needs of the code. The first step is to scale the variable x from the range $[0, 1]$ to the range $[0, N-1]$. That puts the Lagrange points at integer values. Second, at the end of the code we reorder the shape functions so that $\varphi_2(\xi)$ is the shape function with unit value at the right end. This will correspond with the specific bookkeeping scheme that we will use to order and access the global equation numbers when we localize the element information in the code.

D.5 C^0 **Bubble Functions**

We can formulate an alternative polynomial interpolation using what we call *bubble functions*—polynomial functions that are zero at their end points—in conjunction with the "hat" functions, which ramp linearly between zero and one at the ends. In particular, we will use a simple quadratic bubble in conjunction with a monomial enrichment to construct the shape functions. This approach avoids the identification of the Lagrangian points in the interior of the element and results in a slightly simpler formulation and code. The results are identical to the Lagrangian shape functions of the same order.

Consider interpolation functions to be associated with an element with its left end located at $x_{i(e)}$ and its right end located at $x_{j(e)}$. Again, define the change of variable

$$\xi(x) = \frac{x - x_{i(e)}}{L_e} \tag{D.9}$$

where L_e is the length of the element. Start with the first two base functions being the linear ramp down and linear ramp up. To wit,

$$\varphi_1(\xi) = 1 - \xi, \qquad \varphi_2(\xi) = \xi \tag{D.10}$$

To generate the remaining shape functions, let us define the basic *bubble function* as

$$b(\xi) = \xi \left(1 - \xi\right) \tag{D.11}$$

This function is quadratic and is equal to zero at both ends. Any function that includes $b(\xi)$ as a multiplier will also vanish at the ends. We will build additional functions as follows:

$$\varphi_{3+n}(\xi) = c_n \, b(\xi) \, \xi^n, \quad n = 0, \ldots, N - 1 \tag{D.12}$$

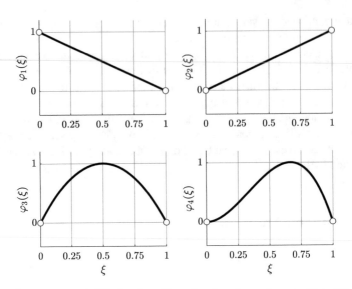

Fig. D.4 C^0 *finite element bubble functions.* Shape functions for an element with bubble function enrichment. The first four functions are shown here

where N is the number of bubble functions. Thus, there are $N+2$ shape functions. The scale factor c_n is selected to make the maximum value of the shape function equal to one. The value is

$$c_n = \frac{(2+n)^{2+n}}{(1+n)^{1+n}}, \quad n = 0, \ldots, N-1 \tag{D.13}$$

Note that the polynomial order of the highest shape function is $N+1$. For example, if $N=2$ the shape function $\varphi_4(\xi)$ is cubic. The first four bubble shape functions are shown in Fig. D.4.

These function skew to the right, but that does not affect the polynomial order and should not affect the conditioning of the functions too much. It is simple to create a more symmetric set of enhanced functions by alternating multiplication of $b(\xi)$ by ξ and $1-\xi$, but that complicates the code without increasing the accuracy or improving the conditioning.

One advantage that the bubble functions have over the Lagrangian functions is that increasing the order of the interpolation does not change any of the original functions, e.g., $\varphi_4(\xi)$ is the same no matter how many additional functions are included beyond four. The Lagrangian functions change entirely for each polynomial degree because they are forced to have the property of being equal to either zero or one at the Lagrange points, and the location of those points changes with the order of the polynomial degree. The MATLAB code that computes the shape functions and their first derivatives, evaluated at the input point x, is given in Code D.2.

Code D.2 MATLAB code to generate C^0 bubble shape functions of order $N+2$, where N is the number of bubble functions

```
function [shp,dshp] = BubbleFcns(x,N)
%   C0 Bubble functions and derivatives for N+2 functions.
%   defined on range [0,1]. Bubble function is b(x) = x*(1-x).
%   Shape functions as shp = [x, 1-x, b, b*x, b*x^2,..., b*x^(N-1)]
%
%
%       x : Location to evaluate function and derivative
%       N : Number of internal bubble functions

%. Establish number of shape functions
   Nshp = N + 2;

%. Initialize arrays for shp and dshp
   shp  = zeros(Nshp,1);              % Shape function
   dshp = zeros(Nshp,1);             % Derivative of shape function

%. Compute the linear shape functions
   shp(1) = 1-x;        dshp(1) = -1;
   shp(2) = x;          dshp(2) = 1;
   if N>0
      b = x*(1-x);     db = 1-2*x;
      for n=0:N-1
         c = ((2+n)^(2+n))/((1+n)^(1+n));
         shp(3+n) = c*b*x^n;
         dshp(3+n) = c*(db*x^n + n*b*x^max(0,n-1));
      end
   end

end
```

The degrees of freedom associated with the bubble functions do not interact with the shape functions from adjacent elements. Of course, the shape functions associated with the element end nodes do interact with the adjacent elements to provide the C^0 continuity. We can view the interior Lagrangian shape functions as being nodeless additional degrees of freedom, just as we do for the bubble functions. In that way, they are bubble functions themselves, and that explains why we reorder the Lagrangian functions as we do in the code LagrangianFcns.

D.6 C^1 Bubble Functions

In Chap. 12 we encounter the Bernoulli–Euler and Rayleigh beams. Both of these theories lead to virtual work functionals that require C^1 continuity in the interpolation. The cubic Hermitian shape functions, given in Sect. 12.1.4, Eqs. 12.16, provide the building blocks for enhancing the basis with bubble functions. These functions are

$$\varphi_1(\xi) = 1 - 3\xi^2 + 2\xi^3 \qquad \varphi_3(\xi) = 3\xi^2 - 2\xi^3$$
$$\varphi_2(\xi) = \left(\xi - 2\xi^2 + \xi^3\right)L_e \qquad \varphi_4(\xi) = \left(-\xi^2 + \xi^3\right)L_e \tag{D.14}$$

To create C^1 bubble functions, we must start with a function that is zero at both ends and has zero slope at both ends. The function that satisfies these conditions is

$$b(\xi) = \xi^2(1 - \xi)^2 \tag{D.15}$$

As we did for the C^0 bubble functions, we can define additional base functions to enhance the cubic Hermitian functions as

$$\varphi_{5+n}(\xi) = c_n \, b(\xi) \, \xi^n, \quad n = 0, \ldots, N-1 \tag{D.16}$$

where N is the number of bubble functions. To give the functions unit maximum value, the constants are

$$c_n = \frac{(4+n)^{4+n}}{4(2+n)^{2+n}}, \quad n = 0, \ldots, N-1 \tag{D.17}$$

The derivatives of these functions are straightforward to compute. Code D.3 shows how to compute the shape functions.

Code D.3 MATLAB code to generate C^1 bubble shape functions of order $N+3$, where N is the number of bubble functions

```
function [h0,h1,h2,h3]  = BubbleFcnsC1(x,M,Le)
%   Compute the C1 FE functions extended by bubbles
%         b = x^2*(1-x)^2   is the basic quartic bubble function
%      h(5+n) = cn*b*x^n,      n = 0,1,2,...,M
%
%   Incoming variables
%         x = Location to evaluate functions x = [0,1]
%         M = Total number bubble functions in basis
%         Le = Element length (for rotation shape functions)

%.  Initialize functions
    N = M + 4;                     % Total number of shape functions
    h0 = zeros(N,1);               % 0th derivative of h
    h1 = zeros(N,1);               % 1st derivative of h
    h2 = zeros(N,1);               % 2nd derivative of h
    h3 = zeros(N,1);               % 3rd derivative of h

%.  Cubic hermitian functions fcn, 1st der., 2nd der., 3rd der.
%   Local DOF 1: Translation at left end
    h10 = 1 - 3*x^2 + 2*x^3;    h11 = -6*x + 6*x^2;
    h12 = -6 + 12*x;            h13 = 12;

%   Local DOF 2: Rotation at left end
    h20 = x - 2*x^2 + x^3;      h21 = 1 - 4*x + 3*x^2;
    h22 = -4 + 6*x;             h23 = 6;

%   Local DOF 3: Translation at right end
    h30 = 3*x^2 - 2*x^3;        h31 = 6*x - 6*x^2;
    h32 = 6 - 12*x;             h33 = -12;

%   Local DOF 4: Rotation at right end
    h40 = -x^2 + x^3;           h41 = -2*x + 3*x^2;
    h42 = -2 + 6*x;             h43 = 6;

%.  Initialize the h arrays with cubic hermitian functions and derivatives
%   Note that h2 and h4 must be multiplied by the element length so that
%   the DOF is the actual rotation of the beam at the node.
    h0(1:4) = [h10; h20*Le; h30; h40*Le];
    h1(1:4) = [h11; h21*Le; h31; h41*Le];
    h2(1:4) = [h12; h22*Le; h32; h42*Le];
```

```
    h3(1:4) = [h13; h23*Le; h33; h43*Le];
%. Compute bubble functions only if M>0
    if M>0

%... Compute the bubble functions and derivatives
      for i=5:N
        n = i-5; nml = n-1;
        np1 = n+1; np2 = n+2; np3 = n+3; np4 = n+4;

%..... Constant to make magnitude of each function equal to one
        cn = (np4^np4)/(4*np2^np2);

%..... h0 is function, h1, h2, and h3 are 1st, 2nd, and 3rd derivatives
        h0(i) = cn*x^np2*(1 - 2*x + x^2);
        h1(i) = cn*x^np1*(np2 - 2*np3*x + np4*x^2);
        h2(i) = cn*x^n*(np1*np2 - 2*np2*np3*x + np3*np4*x^2);
        h3(i) = cn*x^n*(-2*np1*np2*np3 + np2*np3*np4*x);
        if n>0; h3(i) = h3(i) + cn*x^nml*n*np1*np2; end
      end
    end

end
```

The first four C^1 bubble functions are shown in Fig. D.5. The first four Hermitian shape functions are shown in Fig. 12.1 in Chap. 12. Observe that the functions and their first derivatives vanish at the end points.

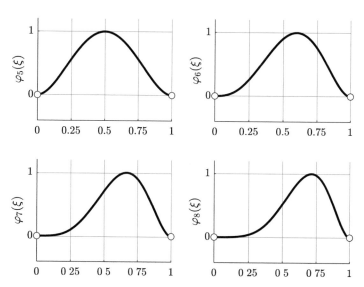

Fig. D.5 C^1 *finite element bubble functions.* Bubble functions for C^1 interpolation. The first four bubble functions are shown here. The four cubic Hermitian shape functions are not shown

Appendix E
Data Structures for Finite Element Codes

Modern methods of structural analysis are built on the idea that structures can be modeled as a collection of *nodes* connected together by *elements*. The problem of maintaining continuity of the fields from one element to the next is solved by establishing global degrees of freedom and then describing element response directly in terms of those global DOF. The formulation leads to the direct assembly of the equations of motion. All of the structural theories in this book are organized along those lines. This appendix provides a summary of how the computer codes organize the data that supports the assembly of the equations of motion.

The direct assembly process relies on the $\mathbf{B}_e$ arrays that pick out degrees of freedom associated with element e and the $\mathbf{D}_i$ arrays that pick out global degrees of freedom associated with node i from the global array of unknowns. These arrays are never formed in the code because their effect reduces to simple submatrix extraction of values in the localization process and submatrix insertion of values in the assembly process. However, we must still represent the information implicit in these arrays. In this appendix we give an overview of the organizational scheme to handle this bookkeeping issue. We start with the truss element, which is an example of a structural element absent nodeless degrees of freedom. We then extend the formulation to structures with elements (the axial bar and beams) that have nodeless DOF.

E.1   Structure Geometry and Topology

The geometry of the structure manifests first in the nodal coordinates, stored in the array x. The ith row of x contains the coordinates of node i. The number of columns of x defines the spatial dimension of the structure. The shear building, axial bar, and the beams have only one spatial dimension. The frame has two spatial dimensions,

© The Author(s), under exclusive license to Springer Nature Switzerland AG 2022
K. D. Hjelmstad, *Fundamentals of Structural Dynamics*,
https://doi.org/10.1007/978-3-030-89944-8

and the truss has either two or three spatial dimensions. The nodes are numbered from 1 to nNodes.

The array ix describes how the nodes are connected together by elements. Each row of ix gives the global node number of the two ends of the element. The ix array describes the *topology* of the structure, independent of the geometry. For the codes in the book, the first column of ix is the *i*-node, the second column is the *j*-node, and the third column is an index that refers to the material set number. The *i*-node and *j*-node establish the orientation of the element (i.e., the positive axial coordinate *x*). The elements are numbered from 1 to nElms.

Information on boundary conditions is transmitted through the array id, which starts out with zeros in the free DOF and ones in the restrained DOF (from user input) but is then recast to store the global DOF numbers in nodal order. Each row of id associates with a node. The number of columns is equal to the number of degrees of freedom per node. In the sequel we will put a finer point on how to number the global degrees of freedom and how to represent that information in the code so that enforcement of boundary conditions is simple.

E.2   Structures with Only Nodal DOF

Let us first consider structures that only have degrees of freedom associated with the nodes. Examples include the NDOF and truss structures. We will focus on the planar truss shown in Fig. E.1 as a case in point.

To establish the global DOF numbers we use the id array, which is populated initially with zeros in the free DOF and ones in the restrained DOF. Each row of id is associated with a node and the number of columns is equal to the number of DOF per node. For the planar truss in Fig. E.1, nodes 1 and 3 are free, node 2 is restrained in both the *x* and *y* directions, and node 4 is restrained in the *y* direction but free in the *x* direction. The id and ix arrays for the example structure are

Fig. E.1 *Example truss.*
Node numbers, element
numbers, boundary
conditions, and applied nodal
loads for an example truss

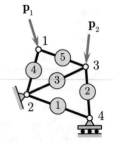

$$
id = \begin{bmatrix} 0 & 0 \\ 1 & 1 \\ 0 & 0 \\ 0 & 1 \end{bmatrix}, \qquad ix = \begin{bmatrix} 2 & 4 & 1 \\ 4 & 3 & 1 \\ 2 & 3 & 1 \\ 2 & 1 & 1 \\ 1 & 3 & 1 \end{bmatrix}
$$

assuming that all elements have material set 1. To establish the global DOF numbers, we traverse the nodes, numbering the nodal DOF sequentially in x,y order (or x,y,z order for three dimensions) starting at one. We keep track of the restrained DOF by numbering them sequentially with negative numbers, starting with -1. Once all nodes are accounted for, the total number of free DOF are known (we call it nFree in the codes). For the example nFree $= 5$. Now, we renumber the restrained DOF with sequential positive numbers following the free DOF. This sequence looks as follows:

$$
id = \begin{bmatrix} 0 & 0 \\ 1 & 1 \\ 0 & 0 \\ 0 & 1 \end{bmatrix} \quad \rightarrow \quad id = \begin{bmatrix} 1 & 2 \\ -1 & -2 \\ 3 & 4 \\ 5 & -3 \end{bmatrix} \quad \rightarrow \quad id = \begin{bmatrix} 1 & 2 \\ 6 & 7 \\ 3 & 4 \\ 5 & 8 \end{bmatrix}
$$

This strategy will allow us to partition our system matrices very simply. Code E.1 establishes the global DOF numbering for structures with only nodal degrees of freedom.

Code E.1 MATLAB code to generate the id array for truss structures

```
function [id,nFree] = Bound(id,nDim,nNodes)
%   Convert the boundary condition array to global DOF numbers
%
%         id : Coming in: Boundary conditions (1=fixed, 0=free)
%              Going out: Global DOF numbers
%       nDim : Dimension of geometric space (i.e., 2D or 3D)
%     nNodes : Number of nodes in structure

    nFree = 0;
    nRest = 0;

%. Loop over all nodes and dimensions in the structure
    for i=1:nNodes
      for j=1:nDim
        if(id(i,j)==0)
          nFree = nFree + 1;      % Augment positive DOF number
          id(i,j) = nFree;        % Record DOF number in id
        else
          nRest = nRest - 1;      % Diminish negative DOF numer
          id(i,j) = nRest;        % Record DOF number in id
        end
      end
    end

%. Reset negative equations numbers to be after the free DOF
    for i=1:nNodes
      for j=1:nDim
        if(id(i,j)<0); id(i,j) = nFree - id(i,j); end
      end
    end

end
```

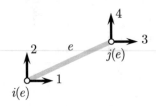

Fig. E.2 *Local DOF numbering scheme for 2D truss elements.* The element degrees of freedom for a planar truss are numbered as shown. Horizontal and vertical translation at the i-node are first, followed by the horizontal and vertical translation of the j-node

This code takes in the array id with boundary condition codes and returns the array id with global DOF numbers.

To make it possible that the function responsible for element calculations works for all elements, we must establish a local numbering scheme for the typical element. All degrees of freedom that affect the element must be included in the set of local degrees of freedom. The local DOF numbering scheme for the planar truss is shown in Fig. E.2. The three-dimensional truss numbers the three translations at the i-node followed by the three translations at the j-node. Finding the global DOF numbers associated with the local nodes identified in Fig. E.2 is simple. Since the ix array tells which nodes are attached to element e, and the id array tells which global DOF are associated with each node, we can find the global DOF for element e as id(ix(e,1),:) for the i-node and id(ix(e,2),:) for the j-node.

Because the lookup with ix is so simple, no additional data structures are needed. In the assembly process we must localize the information associated with each element (essentially carrying out the mission of $\mathbf{B}_e$). Code E.2 shows how this is accomplished for the truss.

Code E.2 MATLAB code to localize information for an element from the global arrays

```
function [xe,ue,de,ive,ii] = TrussLocalize(e,x,u,d,iv,ix,id)
%    Extract local element quantities from global arrays
%
%        e  : current element
%        x  : Nodal coordinates
%        u  : Nodal displacements
%        d  : Material properties
%       iv  : Internal variables
%       ix  : Element connectivities
%       id  : Global DOF numbers

%. Find the i-node, j-node, and material set for element e
    inode = ix(e,1);   jnode = ix(e,2);   mat = ix(e,3);

%. Localize quantities for element e
    xe = [x(inode,:); x(jnode,:)];        % Nodal coordinates
    ue(1,:) = u(id(inode,:));             % i-node displacement
    ue(2,:) = u(id(jnode,:));             % j-node displacement
    de(:,1) = d(mat,:);                   % Material properties
    ive(:,1) = iv(e,:);                   % Internal variables

%. Create indices for Global DOF for element e for assembly
    ii = [id(inode,:),id(jnode,:)];

end
```

This function is the heart of the data structure for the assembly process. We extract the nodal coordinates (xe), element displacements (ue), element material properties (de), internal variables (ive), and the index numbers needed for assembly (ii). The fact that we can manage the computations in this way is why we do not need to explicitly form the $\mathbf{B}_e$ matrices and the localization process represents a very efficient way to execute calculations represented as $\mathbf{u}_e = \mathbf{B}_e \mathbf{u}$ in the theory.

It is worth noting that the nodal force vector is also input to the code in nodal order in an array F. Each row of F is the applied load associated with that node. The number of columns of F is equal to the number of degrees of freedom at each node, in the order established for the theory at hand. For example, the 3D truss orders the node DOF in u_x, u_y, u_z order. For the 2D frame, the order is u, w, θ, where u is the horizontal displacement, w is the vertical displacement, and θ is the rotation. If the entry in F associates with an entry in id that has a one (i.e., a restrained DOF), then it is interpreted as a prescribed displacement. This strategy provides a very simple way to enforce the boundary conditions for a structure. The formulation of the equations requires that the nodal forces be in an array numbered in order of global DOF. Code E.3 puts the nodal forces in global DOF order.

Code E.3 MATLAB code to put nodal loads f into global DOF order as F

```
function [f,u,uPre] = Force(F,id,S)
%   Put force in global DOF order, initialize displacements.
%
%           F : applied force/prescribed displacement in node order
%          id : Global DOF in for each node
%           S : nFree,nNodes,nDOF

%. Put applied nodal forces in DOF order
    f = zeros(S.nDOF,1);
    N = size(id,2);
    for i=1:S.nNodes
      for j=1:N
        ii = id(i,j);
        f(ii) = F(i,j);
      end
    end

%. Initialize displacement and segregate free DOF from fixed
    u = zeros(S.nDOF,1);              % displacements at all DOF
    uPre = f(S.nFree+1:S.nDOF);       % prescribed displacements
    f = f(1:S.nFree);                 % applied nodal loads

end
```

This function applies to all of the codes in the book because the id array is always interpreted in the same way. Observe that the output is the nodal forces at the free degrees of freedom f, an initialized displacement array u (total DOF), and the prescribed displacements at the support points uPre. The main reason for creating u is that we generally want to create an initial *linear* stiffness matrix. The functions that assemble the stiffness matrix require a complete displacement array as input. In the course of solving the problem we keep track of uold and unew, which are the displacements at the free DOF. To construct the complete displacement, when needed, we form

$$u = [\,uold;\ uPre\,]$$

E.3   Structures with Non-nodal DOF

The formulations for the axial bar and the various beam theories include degrees of
freedom that are not associated with nodes, i.e., the degrees of freedom associated
with the bubble functions described in Appendix D. The task of associating
local degrees of freedom to global DOF is a bit more challenging in this case
because it must account for the internal nodeless DOF associated with each field
being interpolated and the different fields may have different numbers of nodeless
variables. The axial bar in Chap. 9 has only one field (u), the Timoshenko beam in
Chap. 12 has two (w and θ), and the nonlinear beam in Chap. 13 has three (u, w,
and θ).

The global DOF numbering scheme is similar to the one outlined for the truss
in the previous section. We first number the nodal DOF sequentially in node order
(with restrained DOF having negative numbers). Next, we number the nodeless DOF
in element order. Finally, the restrained nodes are given positive numbers after the
free degrees of freedom.

Each non-nodal degree of freedom is associated with a bubble shape function
and is associated with a global unknown. The local numbering scheme adopted here
is illustrated in Fig. E.3. To establish the global DOF numbers, the nodeless degrees
of freedom are all after the DOF associated with the nodes. While that is not the
best for the bandwidth, it is the simplest scheme for doing the bookkeeping. The
figure shows the examples of the Timoshenko beam and the nonlinear beam. The
examples show an element with three nodeless degrees of freedom associated with
u and w and two with θ.

In the code we define the array nExt, which stores the number of extra (non-
nodal) DOF for each interpolated variable. The dimension of nExt, then, indicates
how many fields are being interpolated. For example, the Timoshenko beam in
Fig. E.3 (on the left) has

$$\text{nExt} = [\,3,\ 2\,]$$

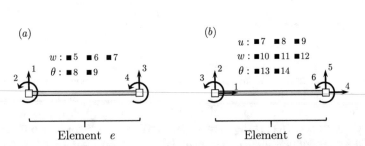

Fig. E.3 *Local DOF numbering scheme for elements with bubble functions.* The element degrees
of freedom are numbered as shown. Translation and rotation at the i-node are first, followed by
the translation and rotation of the j-node, followed by the degrees of freedom associated with the
bubble functions. Numbering for (**a**) a quintic Timoshenko beam element and (**b**) for a quintic
nonlinear beam or frame element

This mechanism gives the flexibility to specify a different degree of interpolation for each field, which is essential to get good results with the Timoshenko and nonlinear beam elements. We also need to know how many degrees of freedom there are at each node. In the code, we call this nDOFpn. For most of the elements that is the same as the number of interpolated variables, but for the C^1 beam elements, it is not. For example, the Rayleigh beam has two degress of freedom per node (rotation and transverse displacement) but only one field is interpolated (w).

In the implementation, it is important to identify which global degrees of freedom are associated with the local element degrees of freedom for assembly. To deal with this feature in the code, we introduce a new array eDOF that has one row for each element and the number of columns is equal to the number of degrees of freedom associated with the element (in the order indicated in Fig. E.3). We can create this array at the same time we create the id array.

We also add rows to eDOF after the last element to record the local DOF ordering associated with each variable—the number of additional rows being equal to the number of variables in the theory. For example, if we have five elements and two fields (e.g., transverse displacement and rotation in the Timoshenko beam), then the sixth row of eDOF would contain the local degrees of freedom associated with w, which are [1,3,5,6,7] in Fig. E.3a and the seventh row would contain the local degrees of freedom associated with θ, which are [2,4,8,9] in Fig. E.3a. Notice that the ordering must match the associated shape function. The first two shape functions, φ_1 and φ_2 are associated with the shape functions that are nonzero at the i-nodes and j-nodes of the elements, respectively. The ordering of the interior DOF is somewhat arbitrary (but you have to select it and stick with it throughout the code). Code E.4 creates the id and eDOF arrays for structures with non-nodal degrees of freedom.

Code E.4 MATLAB code to generate the id and eDOF arrays for use with the enhanced bubble shape functions

```
function [eDOF,id,S] = GlobalDOF(id,ix,S)
%   Establish the global DOF numbering system for equation assembly
%
%       id : Coming in, has 0 in free DOF and 1 in fixed DOF
%       ix : Element connections
%        S. struct with problem size parameters)
%           nExt   - Number of non-nodal DOF per element per variable
%           nDOFpn - Number of DOF per node per variable
%           nNodes - Number of nodes
%           nElms  - Number of elements
%
%     eDOF : Global DOF numbers in element order (plus local map)
%       id : Going out, has global DOF numbers in node order

%. Establish number of DOF per element
   nVars = length(S.nExt);             % Number of variables
   nDOFpnTot = sum(S.nDOFpn);          % Total DOF per node
   exDOF = sum(S.nExt);                % Total extra DOF per element
   neDOF = 2*nDOFpnTot+exDOF;          % Total DOF per element

%. Initialize eDOF array
   eDOF = zeros(S.nElms+nVars,neDOF);

%. Number free DOF positive, restrained negative in id array
   nFree = 0; nRest = 0;
```

```
    for n=1:S.nNodes
      for j=1:nDOFpnTot
        jj = id(n,j);
        if jj==0
          nFree = nFree + 1;
          id(n,j) = nFree;
        else
          nRest = nRest - 1;
          id(n,j) = nRest;
        end
      end
    end

%. Establish element DOF numbers for interior DOF
    for e=1:S.nElms
      for i=1:exDOF
        nFree = nFree + 1;
        eDOF(e,2*nDOFpnTot+i) = nFree;
      end
    end

%. Renumber the fixed nodes at end of DOF list
    for n=1:S.nNodes
      for j=1:nDOFpnTot
        if id(n,j)<0; id(n,j) = nFree - id(n,j); end
      end
    end

%. Establish element DOF numbers for left and right nodes
    for e=1:S.nElms
      for j=1:nDOFpnTot
        m = j+nDOFpnTot;
        eDOF(e,j) = id(ix(e,1),j);
        eDOF(e,m) = id(ix(e,2),j);
      end
    end

%. Record the local DOF numbering scheme in last nVars rows of eDOF
    k=2*nDOFpnTot;
    for n=1:nVars
      nn = zeros(1,neDOF);   kk = n;
      for m=1:S.nDOFpn(n)
        nn(m) = kk;
        nn(m+S.nDOFpn(n)) = kk+nDOFpnTot;
        kk = kk+1;
      end
      for i=1:S.nExt(n)
        k = k+1;
        nn(2*S.nDOFpn(n)+i) = k;
      end
      eDOF(S.nElms+n,:) = nn;
    end

%. Add nVars and nFree to struct S
    S.nVars = nVars;
    S.nFree = nFree;

end
```

The first segment of the code sweeps through the id array and relabels the zeros with sequentially increasing positive integers and the ones with sequentially decreasing negative integers. This part of the code is identical to Bound (see Code E.1). At the end of this sweep, nFree is equal to the number of free nodal DOF and nRest is equal to the negative of the number of restrained nodal DOF.

At this point, we number the nodeless DOF within each element. Notice that all nodeless degrees of freedom have global DOF numbers *after* the nodal DOF. Next, we number the restrained DOF after all of the nodeless DOF. Then, we complete the eDOF array by filling in the nodal DOF numbers. Finally, we record the local DOF numbering scheme in the rows of eDOF beyond row nElms. This last step makes the local-to-global DOF alignment completely self-contained in the array eDOF.

This routine allows us to handle the localization process in the same way for all of the finite element models in the book. The key point to recognize is that the only information we need for any of the theories is the set of indices for element *e* that tell us which rows and columns to assemble the element matrices into the global matrices and, for the nonlinear elements, the displacements (and rotations) associated with the element. The code that accomplishes this task for the nonlinear planar frame is given in Code E.5.

Code E.5 MATLAB code to localize information for the assembly process for the nonlinear planar frame

```
function [Ue,We,Te,xe,de,Epe,jj] = NLFLocalize(e,u,Ep,x,d,ix,eDOF,S)
%    Localize displacements, coordinates, material properties, plastic
%    strains, and establish assembly pointers for element e
%
%           e : Element number
%           u : Global displacement vector
%          Ep : Plastic strain (internal variables)
%           x : Nodal coordinates
%           d : Material properties
%          ix : Element connections
%        eDOF : Global DOF numbers for each element,
%                 Last three rows contain the local DOF numbers for element
%           S : Struct with problem size parameters (nElms)

%. Fetch the local DOF numbers for the element
    indU = eDOF(S.nElms+1,eDOF(S.nElms+1,:)>0);
    indW = eDOF(S.nElms+2,eDOF(S.nElms+2,:)>0);
    indT = eDOF(S.nElms+3,eDOF(S.nElms+3,:)>0);

%. Fetch the global DOF numbers
    jU = eDOF(e,indU);                   % U degrees of freedom
    jW = eDOF(e,indW);                   % W degrees of freedom
    jT = eDOF(e,indT);                   % Theta degrees of freedom
    jj = [jU, jW, jT];                   % All DOF

%. Fetch displacements associated with element e
    Ue = u(jU); We = u(jW); Te = u(jT);

%. Localize nodal coordinates
    inode = ix(e,1); jnode = ix(e,2);
    xe = [x(inode,:); x(jnode,:)];

%. Fetch material properties associated with element e
    mat = ix(e,3);
    de = d(mat,:);

%. Fetch internal variables associated with element e
    Epe = Ep(e,:);

end
```

The first step is to extract the local DOF numbers from the trailing rows of the eDOF array. In this case we have indices for *u* (indU), indices for *w* (indW), and indices for θ (indT). We then use these indices to extract the global DOF numbers for element e from the eDOF array, respectively, jU, jW, and jT for the three fields and pack the complete set of global DOF indices into the array jj. Finally, we extract local displacements and rotations for the DOF associated with the element, the node coordinates for the element, the material properties associated with the element, and the internal variables. The information that is returned is sufficient to carry out the element formation and global assembly process.

The localization functions for the axial bar and the Timoshenko beam are slightly different from Code E.5 because the number of variables is different. The modifications needed are illustrated by the Timoshenko beam case in Code E.6 and the axial bar case in Code E.7.

Code E.6 MATLAB code to localize information for the assembly process for the Timoshenko beam

```
function [We,Te,xe,jj] = TimoLocalize(e,u,x,ix,eDOF,S)
%   Fetch the global DOF numbers, element displacements, element
%   coordinates for the Timoshenko beam.
%
%
%          e : Element number
%          u : Global displacement array
%          x : Nodal coordinates
%         ix : Element connections array
%       eDOF : Global DOF numbers for each element,
%              Last 2 rows contain the local DOF numbers for element
%    S.nElms : Number of elements in structure

%. Fetch the local DOF numbers for the element
     localWDOF = eDOF(S.nElms+1,eDOF(S.nElms+1,:)>0);
     localTDOF = eDOF(S.nElms+2,eDOF(S.nElms+2,:)>0);

%. Fetch the global DOF numbers
     jW = eDOF(e,localWDOF);                  % W degrees of freedom
     jT = eDOF(e,localTDOF);                  % Theta degrees of freedom

%. Fetch local quantities associated with element e
     We = u(jW);                              % Local displacements
     Te = u(jT);                              % Local rotations
     xe = [x(ix(e,1),:);x(ix(e,2),:)];        % Local coordinates
     jj = [jW, jT];                           % Global DOF for element e

end
```

Code E.7 MATLAB code to localize information for the assembly process for the axial bar

```
function [Ue,xe,jj] = AxialLocalize(e,a,x,ix,eDOF,S)
%   Fetch the global DOF numbers, element displacements, element
%   coordinates for Axial Bar element
%
%
%          e : Element number
%          a : Global displacement array
%          x : Nodal coordinates
%         ix : Element connections array
%       eDOF : Global DOF numbers for each element,
%              Last row contains the local DOF numbers for element
%    S.nElms : Number of elements in structure

%. Fetch the local DOF numbers for the element
     localDOF = eDOF(S.nElms+1,eDOF(S.nElms+1,:)>0);
```

```
%. Fetch the global DOF numbers
   jU = eDOF(e,localDOF);                   % U degrees of freedom

%. Extract local quantities
   Ue = a(jU);                              % Local displacements
   xe = [x(ix(e,1),:);x(ix(e,2),:)];        % Local coordinates
   jj = jU;                                 % Global DOF for element e

end
```

Appendix F
Numerical Quadrature

In the study of continuous systems (e.g., the axial bar and beams), we encounter spatial integrals (i.e., integrals with respect to x, not t) that show up when we invoke the principle of virtual work and do the calculations associated with the Ritz method. These integrals are *definite integrals* for which the answer is just a number. The process of executing these integral if often called *quadrature*.

The definite integral of the function $g(x)$ between the limits a and b is shown in Fig. F.1. The geometric significance of this integral is that it is the shaded area under the curve between those two points. By the fundamental theorem of calculus, we can evaluate this integral as

$$\int_a^b g(x)\,dx = G(b) - G(a) \tag{F.1}$$

where $G(x)$ is the *antiderivative* of the function $g(x)$, i.e., $G'(x) = g(x)$, where the prime indicates differentiation with respect to x. When the antiderivative is simple to compute, that approach is ideal, because it is exact. However, if the function is complicated, or if you just want to write a general-purpose computer routine to do integrals, then numerical quadrature is a good alternative. The sketch on the right side of Fig. F.1 shows one way to look at numerical quadrature. Since the definite integral is an area, we can approximate that area by summing areas of simpler geometric figures, like a rectangle or trapezoid.

Numerical quadrature rules can be put into the general form

$$\int_a^b g(x)\,dx = \sum_{i=1}^N g(x_i)w_i \tag{F.2}$$

where x_i are called the *stations* and w_i are called the *weights*. Numerical quadrature, then, reduces the problem of integration to the evaluation of the function at certain points, multiplying those values by their respective weights, and adding them up.

© The Author(s), under exclusive license to Springer Nature Switzerland AG 2022
K. D. Hjelmstad, *Fundamentals of Structural Dynamics*,
https://doi.org/10.1007/978-3-030-89944-8

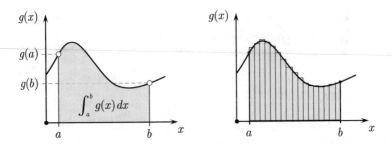

Fig. F.1 *Integral.* Geometrically, the integral of $g(x)$ from a to b is the area under the curve between those two points, as shown at left. The area can be approximated by summing areas of geometric figures that are easier to compute, as shown at right

The only distinction between one numerical quadrature rule and another is the specific values of the stations and weights.

In most of the places in this book where spatial integration is required, the integrand is not simply a scalar-valued function of a scalar variable. Most frequently, the integrand is a vector-valued or matrix-valued function of a scalar variable. For those cases, one can recognize that the integral is simply the area under the curve of each component of the vector or matrix. Hence, the concept of numerical quadrature generalizes to

$$\int_a^b \mathbf{g}(x)\,dx = \sum_{i=1}^N \mathbf{g}(x_i)w_i \tag{F.3}$$

where $\mathbf{g}(x)$ is either a vector or matrix function of x.

F.1   Trapezoidal Rule

The simplest version of numerical quadrature is the trapezoidal rule. The idea is based upon the formula for area of a trapezoid, which is illustrated in Fig. F.2, and has the explicit expression

$$A_{trap} = \tfrac{1}{2}(b-a)\big(g(b) + g(a)\big) \tag{F.4}$$

The vertices of the trapezoid lie on the curve $g(x)$. Hence, the area formula only involves the evaluation of the function at those two points. It is evident that in regions where the curvature of the function is negative (as shown in the figure), the area of the trapezoid underestimates the area under the curve. On the other hand, in regions where the curvature is positive, the opposite happens.

If we divide the region of integration into segments defined by *nodes* located at $\{x_0, \ldots, x_N\}$, then we can create a trapezoid between each pair of nodes. This gives

Fig. F.2 *Trapezoidal Rule.*
The trapezoidal rule is based
upon the area of the trapezoid
being $(b - a)(g(a) + g(b))/2$.
Notice that the shaded area of
the trapezoid underestimates
the area under the curve

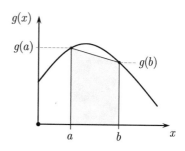

$N + 1$ nodes with N trapezoidal segments. Let us further assume that the nodes
are equally spaced, i.e., all $x_{i+1} - x_i = h$ are the same. If the goal is to integrate
from a to b, for example, then $h = (b - a)/N$. With this definition, we can write the
composite rule as

$$\int_a^b g(x)\,dx = \tfrac{1}{2}h \sum_{i=0}^{N} \big(g(x_{i+1}) + g(x_i)\big) \tag{F.5}$$

This equation can be put into the standard form of Eq. F.2 if we identify the weights
as

$$w_i = \tfrac{1}{2}h \begin{cases} 1 & i = 0 \text{ or } N \\ 2 & i \neq 0 \text{ or } N \end{cases} \tag{F.6}$$

with stations $x_i = a + ih$ for $i = 0, \ldots, N$. The reason that the interior points have
twice the weight is because they participate in two adjacent rectangles. This method
is called the composite *trapezoidal rule*.

F.2   Simpson's Rule

Polynomial functions are simple to antidifferentiate. If we pass a quadratic function
through three points, then we could approximate the area under the curve by integrating the quadratic function exactly. This approach is a way to derive *Simpson's
rule*. The setup is shown in Fig. F.3.

Imagine that we want to integrate $g(x)$ from a to c. The point b is halfway
between a and c, the distances between points being $h = b - a = c - b$. Rather than
integrate $g(x)$, we will find a quadratic function $q(x)$ (gray curve) and integrate that.
To simplify this calculation let us do a change of variable. Let

$$\xi = \frac{x - b}{h} \tag{F.7}$$

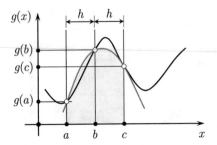

Fig. F.3 *Simpson's rule.* To derive the formula for Simpson's rule, we first find a quadratic function that is equal to the original function at a, b, and c. Then integrate the quadratic function (the shaded area) exactly. The dark line is the actual function. The gray line is the quadratic approximation

where $h = b - a$. Note that $dx = h\,d\xi$. Now the approximate quadratic function can be expressed as

$$q(\xi) = A + B\xi + C\xi^2 \tag{F.8}$$

The point a is at $\xi = -1$, b is at $\xi = 0$, and c is at $\xi = 1$. The values of A, B, and C are determined by making sure the quadratic function passes through the three points on the actual curve. To wit,

$$A - B + C = g(a)$$
$$A = g(b)$$
$$A + B + C = g(c)$$

We can solve this system of equations to give

$$A = g(b), \quad B = \tfrac{1}{2}g(c) - \tfrac{1}{2}g(a), \quad C = \tfrac{1}{2}g(a) - g(b) + \tfrac{1}{2}g(c)$$

Now, integrate the quadratic function to give

$$\int_{-1}^{1} q(\xi) h\,d\xi = h \int_{-1}^{1} \left(A + B\xi + C\xi^2 \right) d\xi = 2h\left(A + \tfrac{1}{3}C \right)$$

Finally, substitute the values of A and C to give the value of the integral

$$\int_{a}^{c} q(x)\,dx = \tfrac{1}{3}h\big[\, g(a) + 4g(b) + g(c) \,\big] \tag{F.9}$$

This is the Simpson approximation of the integral of $g(x)$. It should be evident that we can chop any integration region up into multiple pieces and then treat each piece just like this one. When you do that, the c point from the first region becomes the *a* point for the next one, etc. As with the trapezoidal rule, adjacent interior points contribute twice. Therefore, the weights for the composite Simpson rule is

$$w_i = \tfrac{1}{3}h \begin{cases} 1 & i = 0 \text{ or } N \\ 4 & i = \text{odd interior} \\ 2 & i = \text{even interior} \end{cases} \qquad (\text{F.}10)$$

with stations $x_i = a + ih$, where $h = (b - a)/N$, for $i = 0, \ldots, N$. Note that N must be an even number (because each Simpson segment involves three points). A simple way to generate the weights in MATLAB is to define M as the number of Simpson segments and then use the `repmat` function

$$w = [1, \texttt{repmat}([4, 2], 1, M), 4, 1]$$

to generate the values $\mathbf{w} = [\,1, 4, 2, 4, \ldots, 2, 4, 1\,]$. Notice that the submatrix $[\,4, 2\,]$ does not catch the first 1 and the last $[\,4, 1\,]$ in the sequence, which have to be added manually. Now compute $N = \text{length}(w)$ and the length of the Simpson segment as $2h = L/M$. Finally, complete the weights by multiplying $\mathbf{w}$ by $h/3$. That approach avoids the possibility of selecting a value of N that is even.

Any integral can be transformed with a change of variable $\xi = (x - a)/h$ so that

$$\int_a^b g(x)\,dx = (b - a) \int_0^1 g(\xi)\,d\xi$$

which makes the specification of the quadrature stations for trapezoidal rule and Simpson simply $\xi_i = i/N$ for $i = 1 \ldots, N$.

F.3   Gaussian Quadrature

It may seem obvious, but we want our numerical quadrature formulas to integrate as exactly as possible. One way to interpret that desire is to have the numerical quadrature formula integrate a certain set of simple functions exactly. For example, any numerical integrator should integrate the function $g(x) = 1$ exactly. If it cannot do that one, then what is the hope of integrating more complex functions? If you plug that function into Eq. F.2, you get

$$\sum_{i=1}^{N} w_i = b - a \qquad (\text{F.}11)$$

You can verify that the trapezoidal rule and Simpson's rule both satisfy that requirement.

Since any function can be represented as a polynomial through Taylor's series, we might also think of the accuracy of quadrature as being how many of the terms in the series it can integrate exactly. The trapezoidal rule can integrate a linear function exactly. Simpson's rule can integrate a quadratic function exactly (recall, that is how we derived the rule). The *order of accuracy* of an integrator is the exponent of the last term that it can integrate exactly. Thus, the trapezoidal rule is $O(h)$ and Simpson's rule is $O(h^2)$. Higher-order integration schemes are possible.

Gauss–Legendre Quadrature   When we derived Simpson's rule, we assumed that the integration stations were fixed, and we sought weights that would integrate the quadratic function exactly. Gauss had the great idea that you could optimize the integrator by adjusting both the stations x_i and the weights w_i to find the integrator that could exactly integrate a polynomial of the highest order possible. Let's consider a two-point quadrature rule

$$\int_{-1}^{1} g(\xi)\, d\xi = w_1 g(\xi_1) + w_2 g(\xi_2) \tag{F.12}$$

Note that the range of integration is not important, because we can always use a linear mapping for a change of variable

$$x = \tfrac{1}{2}(1-\xi)a + \tfrac{1}{2}(1+\xi)b$$

to map the range $[-1, 1]$ to the range $[a, b]$. We have four unknowns, so let's try to integrate the functions $\{1, \xi, \xi^2, \xi^3\}$ exactly. To wit,

$$\int_{-1}^{1} 1\, d\xi = 2 = w_1 + w_2 \qquad \int_{-1}^{1} \xi\, d\xi = 0 = w_1\xi_1 + w_2\xi_2$$

$$\int_{-1}^{1} \xi^2\, d\xi = \tfrac{2}{3} = w_1\xi_1^2 + w_2\xi_2^2 \qquad \int_{-1}^{1} \xi^3\, d\xi = 0 = w_1\xi_1^3 + w_2\xi_2^3$$

These equations are nonlinear, but easy to solve. The ξ equation shows that $w_1\xi_2 = -w_2\xi_2$. Substituting this into the ξ^3 equation gives $\xi_1 = -\xi_2$. Using this in the ξ equation gives $w_1 = w_2$. Now, the 1 equation gives $w_1 = 1 = w_2$. Finally, substituting this result into the ξ^2 equation gives $\xi_1 = 1/\sqrt{3}$ and $\xi_2 = -1/\sqrt{3}$. So, we have the weights and stations for two-point Gauss-Legendre quadrature

$$\mathbf{w} = \{1, 1\}, \qquad \xi = \left\{\tfrac{1}{\sqrt{3}}, -\tfrac{1}{\sqrt{3}}\right\} \tag{F.13}$$

Note that the weights sum to 2 because we defined the integral on the range $[-1, 1]$. This quadrature rule is $O(h^3)$, integrating cubic functions exactly with only two integration stations. Higher-order Gauss–Legendre quadrature rules can be derived from the Legendre polynomials. The integration stations x_i for an n-point Gauss–

Legendre rule lie at the zeros of the Legendre polynomial $P_n(x)$. The weights are then given by

$$w_i = \frac{2}{\left(1 - x_i^2\right)\left[P_n'(x_i)\right]^2}$$

The Legendre polynomials can be computed symbolically in MATLAB with the function legendreP(n,x), where x is declared symbolic.

Gauss–Lobatto Quadrature In some circumstances it is desirable to specify that two of the integration stations are at the ends of the integration region. Gaussian quadrature rules that include the end points are called *Gauss–Lobatto* quadrature. Since we are specifying the first and last station, that takes away two of our equations, and hence we can expect to integrate two orders less accurate than Gaussian quadrature (because the number of equations we generate must equal the number of unknowns, which dictates the highest-order polynomial we can integrate).

Consider, for example, a three-point Gauss–Lobatto rule

$$\int_{-1}^{1} g(\xi)\, d\xi = w_1 g(-1) + w_2 g(\xi_2) + w_3 g(1) \tag{F.14}$$

where $\xi_1 = -1$ and $\xi_3 = 1$ are already specified. Applying this rule to the same four functions as we did for the two-point Gauss–Legendre rule, we get

$$w_1 + w_2 + w_3 = 2$$
$$-w_1 + w_2\xi_2 + w_3 = 0$$
$$w_1 + w_2\xi_2^2 + w_3 = \tfrac{2}{3}$$
$$-w_1 + w_2\xi_2^3 + w_3 = 0$$

Adding the first two equations together and the last two equations together. Then substract the same two pairs of equation to get

$$w_2\left(1 + \xi_2\right) + 2w_3 = 2, \qquad w_2\xi_2^2\left(1 + \xi_2\right) + 2w_3 = \tfrac{2}{3}$$
$$w_2\left(1 - \xi_2\right) + 2w_1 = 2, \qquad w_2\xi_2^2\left(1 - \xi_2\right) + 2w_1 = \tfrac{2}{3}$$

Now subtract the second in each horizontal pair from the first to get

$$w_2\left(1 - \xi_2^2\right)\left(1 + \xi_2\right) = \tfrac{4}{3}$$
$$w_2\left(1 - \xi_2^2\right)\left(1 - \xi_2\right) = \tfrac{4}{3}$$

Finally, take the ratio of these two equation to get $1 + \xi_2 = 1 - \xi_2$, which shows that $\xi_2 = 0$. Substituting back gives $w_2 = 4/3$. Now you can go back to the other two equations involving w_1 and w_3 to find that $w_1 = w_2 = 2/3$. Thus, the three-point Gauss–Lobatto rule has

$$\mathbf{w} = \left\{ \tfrac{2}{3}, \tfrac{4}{3}, \tfrac{2}{3} \right\}, \qquad \boldsymbol{\xi} = \{-1, 0, 1\} \tag{F.15}$$

which, coincidentally, is the same as Simpson's rule. The integration stations x_i for an n-point Gauss–Lobatto rule lie at the zeros of the Legendre polynomial $P'_{n-1}(x)$ (plus the end points). The weights are then given by

$$w_i = \frac{2}{n(n-1)\left[P_{n-1}(x_i)\right]^2}$$

The weights of the end points are half of one minus the sum of the weights of the interior points. Again, the stations and weights can be computed symbolically in MATLAB.

F.4   Implementation

Numerical quadrature is used frequently in the codes in this book. We only need one function to get that job done because the main task is to specify the rule and then retrieve the weights and stations needed to integrate with that rule. The code that generates the weights and stations for various quadrature rules is given in the following function.

Code F.1 MATLAB code to retrieve the weights and stations for numerical quadrature by either Gauss–Legendre quadrature, Gauss–Lobatto quadrature, composite Simpson's rule, or composite trapezoidal rule

```
function [wt,st,nQpts] = NumInt(QType,n)
%. This function establishes the weights and stations for numerical
%   integration by various rules on segment x = [0,1]
%
%    QType = Specify quadrature rule
%        n = Number of integration points (nQpts going out)
%       wt = Weights
%       st = Stations
%- - - - - - - - - - - - - - - - - - - - - - - - - - - - - - - - - - -

    switch QType

        case 1    % Gauss-Legendre
            syms x
            P = legendreP(n,x);             % nth order Legendre polynomial
            dP = diff(P,x);                 % Compute first derivative of P
            st = vpasolve(P==0);            % Stations are roots of Pn
            a = vpa(subs(dP,x,st));         % Evaluate dP at stations
            wt = eval(1./((1-st.^2).*a.^2));% Weights for integration
            st = eval((st + 1)/2);          % Convert to region [0,1]
            clear x;                        % Release symbolic x
```

```
        nQpts = n;                        % Number of points in rule

    case 2    % Gauss-Lobatto
        syms x
        P = legendreP(n-1,x);             % n-1th order Legendre polynomial
        dP = diff(P,x);                   % Compute first derivative of P
        st = vpasolve(dP==0);             % Stations are roots of dP
        a = vpa(subs(P,x,st));            % Evaluate dP at stations
        wt = eval(1./(n*(n-1)*a.^2));     % Weights for integration
        st = eval((st + 1)/2);            % Convert to region [0,1]
        st = [0; st; 1];                  % Add end points
        R = (1-sum(wt))/2;                % Compute leftover weights
        wt = [R; wt; R];                  % Add weights for end points
        clear x;                          % Release symbolic x
        nQpts = n;                        % Number of points in rule

    case 3    % Composite Simpson
        m = ceil((n-3)/2);                % Number of simpson segments
        wt = repmat([4;2],m,1);           % [4,2,4,2,4,2...]
        wt = [1;wt;4;1];                  % Simpson weights
        nQpts = length(wt);              % No. Simpson Integration pts.
        wt = wt/(3*(nQpts-1));            % Simpson weights
        st = (0:1/(nQpts-1):1)';          % Simpson stations

    case 4    % Composite Trapezoidal
        wt = 2*ones(n-2,1);               % [2,2,2,...,2]
        wt = [1;wt;1];                    % Trapezoidal weights
        nQpts = length(wt);              % No. Trapezoidal Integration pts.
        wt = wt/(2*(nQpts-1));            % Trapezoidal weights
        st = (0:1/(nQpts-1):1)';          % Trapezoidal stations

    end

end
```

Note that this function assumes that the range of integration is [0, 1]. The routine works for any number of points. The number of integration points is specified in the input file for the problem.

F.5   Examples

To demonstrate the performance of the Gauss–Lobatto, Gauss–Legendre, Simpson, and trapezoidal quadrature rules, let us consider a couple of examples. The functions are shown in Fig. F.4. We will compute the integrals using quadrature rules of different order, N. Specifically,

$$
\begin{aligned}
\text{Gauss–Legendre} \quad & N = \{2, 3, 4, 5, 6\} \\
\text{Gauss–Lobatto} \quad & N = \{3, 4, 5, 6, 7\} \\
\text{Simpson's rule} \quad & N = \{3, 5, 7, 9, 11\} \\
\text{Trapezoidal rule} \quad & N = \{11, 21, 31, 41, 51\}
\end{aligned}
$$

The first case is a polynomial function $g(x) = -5x + 9x^3 + 10x^6$, shown in Fig. F.4 (the function on the left), integrated from $x = 0$ to $x = 1$. The results are given in Table F.1. It is evident that Gauss–Legendre gets the exact result with

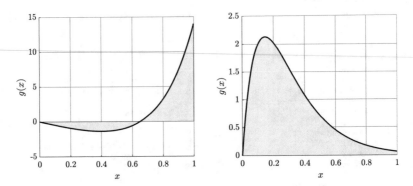

Fig. F.4 *Functions to compare numerical quadrature rules.* The function on the left, Example 1, is $g(x) = -5x + 9x^3 + 10x^6$. The results for this function are in Table F.1. The function on the right, Example 2, is $g(x) = 10\left(e^{-5x} - e^{-9x}\right)$. The results for this function are in Table F.2

Table F.1 *Results of numerical quadrature, Example 1.* Integration with Gauss–Legendre, Gauss–Lobatto, Simpson, and Trapezoidal quadrature rules integrating the function $g(x) = -5x + 9x^3 + 10x^6$ between the limits $x = 0$ and $x = 1$

N	G.–Legendre	N	G.–Lobatto	N	Simpson	N	Trapezoidal
2	0.9537037	3	1.5208333	3	1.5208333	11	1.2509050
3	1.9537037	4	1.1833333	5	1.2034505	21	1.1966860
4	1.1785714	5	1.1785714	7	1.1836134	31	1.1866249
5	1.1785714	6	1.1785714	9	1.1801809	41	1.1831020
6	1.1785714	7	1.1785714	11	1.1792333	51	1.1814712
Exact	1.1785714		1.1785714		1.1785714		1.1785714

$N = 4$ and Gauss–Lobatto does so with $N = 5$. This result is consistent with the observation that Gauss–Legendre will integrate a polynomial up to order $2N-1$ exactly, while Gauss–Lobatto will do one of order $2N-3$ exactly. Hence, the fourth order Gauss–Lobatto rule can get a fifth order polynomial, but not a sixth. Note that Simpson's rule is not exact, even up to 11 integration points. In fact, even with 59 integration points, Simpson's rule still has an error in the seventh digit. The trapezoidal rule is accurate only to three places with 51 integration points. Observe that Gauss–Lobatto and Simpson are the same for $N = 3$.

The second example is *not* a polynomial function, but rather it is an exponential rise followed by exponential decay, i.e., $g(x) = 10\left(e^{-5x} - e^{-9x}\right)$, as shown in Fig. F.4 (the function on the right). Again, the function is integrated from $x = 0$ to $x = 1$. The results are given in Table F.2. None of the quadrature rules get the integral exactly right. Gauss–Legendre and Gauss–Lobatto are right to five digits with 6 and 7 stations, respectively. Simpson and trapezoidal rule get 2 digits with 11 and 51 stations, respectively. The superiority of the Gaussian quadrature rules is evident.

The Gaussian integration rules are particularly well-suited for integrating polynomials and perform well on many functions with a small number of integration

Table F.2 *Results of numerical quadrature, Example 2.* Integration with Gauss–Legendre, Gauss–Lobatto, Simpson, and Trapezoidal quadrature rules integrating the function $g(x) = 10 \left(e^{-5x} - e^{-9x} \right)$ between the limits $x = 0$ and $x = 1$

N	G.–Legendre	N	G.–Lobatto	N	Simpson	N	Trapezoidal
2	1.0845107	3	0.4841976	3	0.4841976	11	0.8427671
3	0.9211570	4	0.8110151	5	0.8019802	21	0.8672011
4	0.8797249	5	0.8701934	7	0.8559785	31	0.8718266
5	0.8757772	6	0.8752738	9	0.8685867	41	0.8734531
6	0.8755585	7	0.8755403	11	0.8725302	51	0.8742073
Exact	0.8755501		0.8755501		0.8755501		0.8755501

stations. Most of the applications in this book fall in that category since the interpolation functions are polynomials.

Index

Printed in the United States
by Baker & Taylor Publisher Services